T0295497

CLIMATE IMPACTS ON EXTREME WEATHER

CLIMATE IMPACTS ON EXTREME WEATHER

Current to Future Changes on a Local to Global Scale

Edited by

VICTOR ONGOMA
Assistant Professor, Mohammed VI Polytechnic University, Morocco

HOSSEIN TABARI
Research Associate, KU Leuven, Belgium

Elsevier
Radarweg 29, PO Box 211, 1000 AE Amsterdam, Netherlands
The Boulevard, Langford Lane, Kidlington, Oxford OX5 1GB, United Kingdom
50 Hampshire Street, 5th Floor, Cambridge, MA 02139, United States

ISBN: 978-0-323-88456-3

For information on all Elsevier publications
visit our website at https://www.elsevier.com/books-and-journals

Publisher: Susan Dennis
Acquisitions Editor: Gabriela D. Capille
Editorial Project Manager: Michelle Fisher
Production Project Manager: Kumar Anbazhagan
Cover designer: Christian J. Bilbow

Typeset by STRAIVE, India

Contents

Contributors vii

1. Understanding weather and climate extremes

ERESANYA EMMANUEL OLAOLUWA, OLUFEMI SUNDAY DUROWOJU, ISRAEL R. ORIMOLOYE, MOJOLAOLUWA T. DARAMOLA, AKINYEMI AKINDAMOLA AYOBAMI, AND OLASUNKANMI OLORUNSAYE

1. Understanding weather and climate 1
2. Introduction to weather and climate extremes 3
3. Conclusion 13

References 13
Further reading 17

2. The role of climate datasets in understanding climate extremes

MALCOLM N. MISTRY

1. Introduction 19
2. CEIs: Definition, data quality control and methodology 21
3. Historical climate data for assembling the CEIs 28
4. Existing datasets of CEIs: Strengths and limitations 39
5. Conclusion and recommendations 41

Appendix 43
Acknowledgments 43
Conflict of interest 43
References 43

3. Advances in weather and climate extremes

VICTOR NNAMDI DIKE, ZHAO-HUI LIN, CHENGLAI WU, AND COLMAN CHIKWEM IBE

1. Introduction 49
2. Evolution of definitions of weather and climate extremes 50
3. Spatiotemporal scales of measurement of weather and climate extremes 52
4. Changes in frequency, intensity, duration of climate extremes 53
5. Innovations in weather and climate extremes prediction 57
6. Conclusion 57

Acknowledgments 58
References 59

4. Uncertainties in daily rainfall over West Africa: Assessment of gridded products and station gauges

IMOLEAYO E. GBODE, JOSEPH D. INTSIFUL, AKINTOMIDE AFOLAYAN AKINSANOLA, AKINTAYO T. ABOLUDE, AND KEHINDE O. OGUNJOBI

1. Introduction 65
2. Data and methods 66
3. Results and discussion 69
4. Summary and conclusion 75

Conflict of interest 81
References 81

5. Features of regional Indian monsoon rainfall extremes

HAMZA VARIKODEN AND M.J.K. REJI

1. Introduction 83
2. Data and methods 86
3. Results 87
4. Conclusions 94

Acknowledgment 95
References 95

6. Historical changes in hydroclimatic extreme events over Iran

VAHID NOURANI AND HESSAM NAJAFI

1. Introduction 101
2. Materials and methods 105
3. Results and discussion 109
4. Conclusions 112

References 113
Further reading 115

7. Intensification of precipitation extremes in the United States under global warming

AKINTOMIDE AFOLAYAN AKINSANOLA AND GABRIEL J. KOOPERMAN

1. Introduction 117
2. Data and methodology 118
3. Results and discussion 119
4. Summary and conclusion 124

Acknowledgments 126
Data availability statement 126
References 127

8. A review on observed historical changes in hydroclimatic extreme events over Europe

KRISTIAN FÖRSTER AND LARISSA NORA VAN DER LAAN

1. Climate variability in the past millennium 131
2. Precipitation extremes 133
3. Glacier mass balances 135
4. Floods 137
5. Droughts 139
6. Summary on past changes in hydroclimatic extremes 140

References 141

9. Meteorological droughts in semi-arid Eastern Kenya

CHARLES W. RECHA, GRACE W. KIBUE, AND A.P. DIMRI

1. Introduction 145
2. Materials and methods 146
3. Results and discussions 149
4. Limitations of the study 155
5. Conclusions 155

Acknowledgments 156
References 156

10. Drought across East Africa under climate variability

CHARLES ONYUTHA, BRIAN AYUGI, HOSSEIN TABARI, HAMIDA NGOMA, AND VICTOR ONGOMA

1. Introduction 159
2. Materials and methods 160
3. Results and discussion 164
4. Conclusions 171

Acknowledgment 171
Conflict of interest 171
References 171

11. Revisiting the impacts of tropical cyclone Idai in Southern Africa

COLLEN MUTASA

1. Introduction 175
2. Materials and methods 176
3. Results 177
4. Discussion and conclusion 184

References 186

12. Impacts of climate extremes over Arctic and Antarctic

MASOUD IRANNEZHAD, BEHZAD AHMADI, AND HANNU MARTTILA

1. Introduction 191
2. The new face of Arctic and Antarctic 194
3. Conclusion 207

Acknowledgments 209
References 209

13. The degradation of the Amazon rainforest: Regional and global climate implications

KERRY W. BOWMAN, SAMUEL A. DALE, SUMANA DHANANI, JEVITHEN NEHRU, AND BENJAMIN T. RABISHAW

1. Introduction 217
2. Methods 218
3. Discussion 219
4. Recommendations 227
5. Conclusion 230

References 230

14. Farmers' perceptions of climate hazards and coping mechanisms in Fiji

SAMROY LILIGETO AND NAOHIRO NAKAMURA

1. Introduction 235
2. Background 235
3. Methods 240
4. Results 241
5. Discussion and conclusion 247

Appendix 248
Acknowledgment 249
References 249

15. People's management of risks from extreme weather events in the Pacific Island region

EBERHARD WEBER

1. Introduction 253
2. Climate change, natural hazards and disasters in Pacific Island countries and territories 254
3. The 2020/2021 cyclone seasons and covid-19 in the Pacific Islands region 265
4. Conclusion 266

References 267

16. Management of extreme hydrological events

RUTH KATUI NGUMA AND VERONICA MWIKALI KILUVA

1. Introduction 271
2. Management measures 272
3. Challenges 280
4. Recommendations and a case study 282

References 284

Index 287

Contributors

Akintayo T. Abolude Nouveau Projects Limited, Lagos, Nigeria

Behzad Ahmadi WSP, Portland, OR, United States

Akintomide Afolayan Akinsanola Department of Geography, University of Georgia, Athens, GA; Environmental Science Division, Argonne National Laboratory, Lemont, IL, United States

Akinyemi Akindamola Ayobami Department of Meteorology and Climate Science, Federal University of Technology, Akure, Nigeria

Brian Ayugi Department of Civil Engineering, Seoul National University of Science and Technology, Republic of Korea; Organization of African Academic Doctors (OAAD), Nairobi, Kenya

Kerry W. Bowman Faculty of Medicine, School of the Environment, University of Toronto, Toronto, ON, Canada

Samuel A. Dale University of Toronto, Toronto, ON, Canada

Mojolaoluwa T. Daramola Department of Meteorology and Climate Science, Federal University of Technology, Akure, Nigeria; Institute of Geographic Sciences and Natural Resources Research, Chinese Academy of Sciences, Beijing, China

Sumana Dhanani University of Toronto, Toronto, ON, Canada

Victor Nnamdi Dike International Center for Climate and Environment Sciences, Institute of Atmospheric Physics, Chinese Academy of Sciences, Beijing, China; Energy, Climate, and Environment Sciences Group, Imo State Polytechnic Umuagwo, Ohaji, Imo State, Nigeria

A.P. Dimri School of Environmental Sciences, Jawaharlal Nehru University, New Delhi, India

Olufemi Sunday Durowoju Department of Geography, Faculty of Social Sciences, Osun State University, Osogbo, Nigeria

Kristian Förster Institute for Hydrology and Water Resources Management, Leibniz University Hannover, Hannover, Germany

Imoleayo E. Gbode West African Science Service Center on Climate Change and Adapted Land Use; Department of Meteorology and Climate Science, Federal University of Technology Akure, Akure, Nigeria

Colman Chikwem Ibe Energy, Climate, and Environment Sciences Group, Imo State Polytechnic Umuagwo, Ohaji, Imo State, Nigeria

Joseph D. Intsiful Division of Mitigation and Adaptation, Green Climate Fund, Songdo Incheon, Republic of Korea

Masoud Irannezhad Water, Energy and Environmental Engineering Research Unit, Faculty of Technology, University of Oulu, Oulu, Finland; School of Environmental Science and Engineering, Southern University of Science and Technology (SUSTech), Shenzhen, People's Republic of China

Grace W. Kibue Department of Natural Resources, Egerton University, Egerton, Kenya

Veronica Mwikali Kiluva Masinde Muliro University of Science and Technology, Kakamega, Kenya

Gabriel J. Kooperman Department of Geography, University of Georgia, Athens, GA, United States

SamRoy Liligeto School of Agriculture, Geography, Environment, Ocean and Natural Sciences, The University of the South Pacific, Laucala Campus, Suva, Fiji

Zhao-Hui Lin International Center for Climate and Environment Sciences, Institute of Atmospheric Physics, Chinese Academy of Sciences, Beijing, China; China-Pakistan Joint Research Center on Earth Sciences, Chinese Academy of Sciences-Higher Education Commission (CAS-HEC), Islamabad, Pakistan

Hannu Marttila Water, Energy and Environmental Engineering Research Unit, Faculty of Technology, University of Oulu, Oulu, Finland

Malcolm N. Mistry Department of Economics, Ca' Foscari University of Venice, Venice, Italy; Department of Public Health, Environments and Society, London School of Hygiene and Tropical Medicine (LSHTM), London, United Kingdom

Collen Mutasa UNESCO Regional Office for Southern Africa, Harare, Zimbabwe

Hessam Najafi Center of Excellence in Hydroinformatics and Faculty of Civil Engineering, University of Tabriz, Tabriz, Iran

Naohiro Nakamura School of Agriculture, Geography, Environment, Ocean and Natural Sciences, The University of the South Pacific, Laucala Campus, Suva, Fiji

Jevithen Nehru University of Toronto, Toronto, ON, Canada

Hamida Ngoma Department of Geosciences, University of Connecticut, Storrs, CT, United States

Ruth Katui Nguma Kenya Meteorological Department, Nairobi, Kenya

Vahid Nourani Center of Excellence in Hydroinformatics and Faculty of Civil Engineering, University of Tabriz, Tabriz, Iran

Kehinde O. Ogunjobi Department of Meteorology and Climate Science, Federal University of Technology Akure, Akure, Nigeria; West African Science Service Center on Climate Change and Adapted Land Use, Competence Center, Ouagadougou, Burkina Faso

Eresanya Emmanuel Olaoluwa South China Sea Institute of Oceanology, Chinese Academy of Sciences, Guangzhou, People's Republic of China; Department of Marine Science and Technology, Federal University of Technology, Akure, Nigeria; Organization of African Academic Doctors (OAAD), Off Kamiti Road, Nairobi, Kenya

Olasunkanmi Olorunsaye Department of Marine Science and Technology, Federal University of Technology, Akure, Nigeria

Victor Ongoma International Water Research Institute, Mohammed VI Polytechnic University, Benguerir, Morocco

Charles Onyutha Department of Civil and Environmental Engineering, Kyambogo University, Kyambogo, Kampala, Uganda

Israel R. Orimoloye Centre for Environmental Management, Faculty of Natural and Agricultural Sciences, University of the Free State, Bloemfontein; School of Social Science, The Independent Institute of Education, MSA, Johannesburg, South Africa

Benjamin T. Rabishaw University of Toronto, Toronto, ON, Canada

Charles W. Recha Department of Geography, Environment and Development Studies, Bomet University College, Bomet, Kenya

M.J.K. Reji Indian Institute of Tropical Meteorology, Ministry of Earth Sciences, Pune, India

Hossein Tabari Hydraulics Laboratory, KU Leuven, Leuven, Belgium

Larissa Nora van der Laan Institute for Hydrology and Water Resources Management, Leibniz University Hannover, Hannover, Germany

Hamza Varikoden Indian Institute of Tropical Meteorology, Ministry of Earth Sciences, Pune, India

Eberhard Weber School of Agriculture, Geography, Environment, Oceans and Natural Sciences, The University of the South Pacific, Suva, Fiji

Chenglai Wu International Center for Climate and Environment Sciences, Institute of Atmospheric Physics, Chinese Academy of Sciences, Beijing, China; China-Pakistan Joint Research Center on Earth Sciences, Chinese Academy of Sciences-Higher Education Commission (CAS-HEC), Islamabad, Pakistan

CHAPTER

1

Understanding weather and climate extremes

Eresanya Emmanuel Olaoluwa[a,b,c], *Olufemi Sunday Durowoju*[d], *Israel R. Orimoloye*[e,f], *Mojolaoluwa T. Daramola*[g,h], *Akinyemi Akindamola Ayobami*[g], *and Olasunkanmi Olorunsaye*[b]

[a]South China Sea Institute of Oceanology, Chinese Academy of Sciences, Guangzhou, People's Republic of China
[b]Department of Marine Science and Technology, Federal University of Technology, Akure, Nigeria
[c]Organization of African Academic Doctors (OAAD), Off Kamiti Road, Nairobi, Kenya
[d]Department of Geography, Faculty of Social Sciences, Osun State University, Osogbo, Nigeria
[e]Centre for Environmental Management, Faculty of Natural and Agricultural Sciences, University of the Free State, Bloemfontein, South Africa
[f]School of Social Science, The Independent Institute of Education, MSA, Johannesburg, South Africa
[g]Department of Meteorology and Climate Science, Federal University of Technology, Akure, Nigeria
[h]Institute of Geographic Sciences and Natural Resources Research, Chinese Academy of Sciences, Beijing, China

1 Understanding weather and climate

Weather and climate affect nearly all socioeconomic activities. Humans have been studying stars, moon, the sun, and other celestial bodies with their immediate environment from time immemorial. In fact, the importance of weather studies has made the assertion "what is weather worth" an underestimation. Nowadays, nobody asks such a question about the value of weather to human, his activities and life (Alexander, Zhang, Hegerl, & Seneviratne, 2016; Aremu, 2008). While environments have placed limitations on human activities, people have also changed their environment more productive for themselves by studying its atmospheric conditions. The atmosphere, the laboratory of studying weather covers mankind and all living things (Wallace & Hobbs, 1977). All atmospheric processes directly or indirectly influence humans and the entire environment. Weather influences humans, and living beings also influence weather through their actions, works, and lives (Ayoade, 1988; Daramola, Eresanya, & Erhabor, 2017). Thus, the knowledge of weather and climate to humans is a necessity to better live with it, by it, and in it (Mauder, 1970).

In any field of study, it is useful from time to time to reexamine the meanings attached to apparently simple words frequently used by both specialists and the public (Gibbs, 1987). A lot of blunders are usually committed by people in their level of understanding of words like weather and climate. The importance of weather and climate in human life is quite great. In fact, the study of weather occupies a central and important position within the broad field of environmental science (Aremu, 2008; Ayoade, 1988). The word "weather" (along with climate) has been misdefined by various authors. For example, the essential difference between the two terms is that weather only relates to the state of the atmosphere during one and only one specific period. For climate, it relates to the statistical likelihood of occurrence of various states of the atmosphere over a longer period. In the past, climate is defined as "average weather" and also the synthesis of weather of a given location over a period of at least 30 to 35 years. This definition may not necessarily be right in that a lot of ambiguity has developed with respect to the notion of weather and climate.

This problem of real definitions for weather and climate persist long until when World Climate Conference (1979) adopted the following definitions: *Weather* is associated with the complete state of the atmosphere at a particular instant in time and with the evolution of this state through the generation, growth, and decay of individual disturbances. On the other hand, *Climate* is the synthesis of weather events over the whole of a period statistically long

Climate Impacts on Extreme Weather
https://doi.org/10.1016/B978-0-323-88456-3.00008-3

enough to establish its statistical ensemble properties and largely independent of any instantaneous state (Huang, Zhang, Gao, & Sun, 2018; Masson, 1979; Luo, Tang, Zhong, Bian, & Heilman, 2013). Climate, therefore, refers to the characteristics condition of the atmosphere deduced from repeated observations over a long period. Climate includes more than the average weather conditions over a given area. It includes considerations of departures from average (i.e., variabilities), extreme conditions, and the probabilities of frequencies of occurrences of given weather conditions. Thus, climate represents a generalization, whereas weather deals with specific events.

1.1 The earth system

The term "Earth system" refers to the earth's interacting physical, chemical, and biological processes. The system consists of the land (lithosphere), water (hydrosphere), air (atmosphere) and living organisms (biosphere) (Kump et al., 2004). The first three of these spheres are abiotic while the last sphere is biotic. Abiotic describes substances that are made from nonliving materials. Biotic relates to living things like bacteria, birds, mammals, insects, and plants. The earth system includes the planet's natural cycles—the carbon, hydrological, nitrogen, sulfur, phosphorus and other cycles, and deep earth processes.

1.1.1 Atmosphere

The earth's atmosphere is the gaseous layer that envelopes the world. The commons term for the atmosphere is "air." This study is centered on the atmosphere as it houses all the elements of weather and climate. The earth's atmosphere has five main layers and a sixth layer, the ionosphere that overlaps the mesosphere, thermosphere, and exosphere. The bottom layer, which is the layer closest to the earth, is the densest of the five layers. This layer is known as the *troposphere*. This is the layer of the earth's atmosphere that humans live and breathe in. The troposphere starts at ground level and extends to 10 km in altitude. This layer mostly contains a mixture of mostly nitrogen (78%), oxygen (21%), and argon (0.9%) (Kump et al., 2004). At this level, water vapor, dust particles, contaminants, and pollen are also incorporated into the atmosphere. The higher the altitude, the thinner the atmosphere is. The next layer is the *stratosphere*. This layer is the layer that contains the earth's ozone layer. Unlike the troposphere, the stratosphere has no turbulence. Unlike the air in the troposphere, the air in the stratosphere gets warmer higher up in this layer. Above the stratosphere is the *mesosphere*. This layer in the Earth's atmosphere is the highest layer in which the gases are still mixed up rather than layered. The uppermost layer of the earth's atmosphere is the exosphere. The atmosphere is extremely thin in this layer with gases like hydrogen and helium.

1.1.2 Hydrosphere

All the water on earth is known collectively as the earth's hydrosphere. This includes surface water (such as rivers, lakes, and oceans), groundwater, ice and snow, and water in the atmosphere in the form of water vapor. Water is found in all three states on earth which are gas, liquid and solid. As gas, water is found as water vapor in the atmosphere. In liquid form, water is found in streams, rivers, lakes, ponds, and oceans along with mist in the air and as dew on the surface of the ground. Water is found in solid form as ice and snow.

1.1.3 Lithosphere

The lithosphere contains the elements of the earth's crust and part of the upper mantle that moves consistently over the weaker, convecting asthenosphere. This is the hard and rigid outer layer of the earth. The term is taken from the Greek word *lithos* meaning "rocky." This part of the earth includes soil. The lithosphere is divided into two main types: continental and oceanic lithosphere.

1.1.4 Biosphere

The biosphere covers all living organisms on earth. There are an estimated 20 million to 100 million different species in the world organized into the 100 phyla that make up the five kingdoms of life forms. These organisms can be found in almost all parts of the geosphere. The geosphere is the collective name for the earth's atmosphere, lithosphere, hydrosphere, and cryosphere. There are organisms in the air, soil, and water on earth.

1.2 Earth energy balance

The term "Earth Energy Budget" refers to the balance between the energy received from the sun and the energy that the earth radiates to space after passing through the components of the climate system. As a result, the earth's climate is highly dependent on the earth's energy budget, and any change in the earth's energy budget would alter the climate system. Energy from the sun reaches the earth's surface through a series of complex interactions with the atmosphere, ocean, and land surface, including scattering, absorption, transmission, and emission. According to the first law of thermodynamics, energy cannot be created or destroyed but must be converted from one form to another; in other words, energy is conserved. This means that the energy input into the earth is roughly balanced by the energy emitted on an annual basis. This pseudo-radiative equilibrium is in charge of keeping the earth's temperature relatively constant over time. The spherical shape of the Earth causes differential heating between the equator and the high latitudes. On the equator, more energy is incident, while at higher latitudes, less energy is incident. Several mechanisms are used to balance the amount of energy within the earth, including convection, wind motion, and ocean current/circulation, all of which are driven by differential heating and help to transport heat from the equator to the poles. Heat also drives evaporation of ocean water and the water cycle, and light energy is a major component in the process of photosynthesis, some of which is converted to electrical energy to power devices and machines. Other factors influencing energy balance within the earth include albedo (reflectivity), aerosols, greenhouse gases, cloud cover, vegetation, surface properties, and land use.

1.3 Hydrological cycle

There are numerous cycles in nature, including the carbon cycle, nitrogen cycle, and other biogeochemical cycles. Certain studies contend that the hydrological cycle is the most important of these cycles due to its impact on the earth's ecosystem, climate, energy budget, and socioeconomic system. The hydrological cycle is the continuous circulation of water through the earth's land, ocean, and atmosphere. Water's ability to exist in three states; liquid, solid (ice), and gas, facilitates this interaction (vapor). The cycle has no beginning nor the end because it is a continuous exchange of mass, matter, and energy within the earth system. Water makes up approximately 70% of the earth's surface, making it one of the most abundant elements on the planet. Furthermore, the quantity of water on the planet has remained constant as it moves from one storage to another, including lakes, streams, ocean, aquifers, glaciers, ice-caps, and the atmosphere. However, the quantity of water in these storages varies depending on seasonal variability. This movement between water storage renews the water supply to various parts of the earth for use by living organisms as well as to maintain balance; however, changes in this cycle may result in hydrological extremes mainly drought and flood, on a local or regional scale. The earth contains approximately 1,386,000,000 km^3 of water, with the ocean being the largest reservoir, containing approximately 97% (1,338,000,000 km^3) of the earth's water as saline water. Of the remaining 3% fresh water, 78% is stored in ice in Antarctica and Greenland, 21% is stored in aquifers or groundwater, less than 1% is stored in rivers, streams, and lakes combined, and 0.001% is stored in the atmosphere. The hydrological cycle is a simple complex recharge and discharge process that connects the atmosphere and the two major water storages, the ocean and the lithosphere.

2 Introduction to weather and climate extremes

As discussed in Section 1, weather and climate are part of the biophysical environment and can be exploited by living beings to satisfy their wants and improve their welfare. Weather and climate, to an extent, are resources. Climate can be a resource when and where its beneficial effects such as rain, sunshine, wind and radiation occur in the proper amount or intensity, while it can also be a resistance (hazard) when and where these same elements occur in the wrong amount or intensity giving rise to floods, droughts, heat and cold waves, hurricane winds, etc. In this regard, a climatic hazard is also referred to as climate extreme. According to Seneviratne et al. (2012), extreme weather or climate event is generally defined as the occurrence of a value of a weather or climate variable above or below a threshold value near the upper or lower ends (tails) of the range of observed values of the variable. Climate extremes may be the result of an accumulation of weather or climate events that are, individually, not extreme themselves (though their accumulation is extreme).

According to the Intergovernmental Panel on Climate Change (IPCC), climate and its extremes are changing (IPCC, 2013). A changing climate leads to changes in the frequency, intensity, duration, spatial extent and timing of weather and climate extremes, and can lead to unprecedented extremes (CRED, 2019). Likewise, weather or climate events, even if not extreme in a statistical sense, can still lead to extreme conditions or impacts, either by crossing a critical threshold in a social, physical, or ecological system or by occurring simultaneously with other events. However, not all extremes necessarily result to serious impacts but a weather system such as a tropical cyclone can have an extreme impact, depending on where and when it approaches landfall, even if the specific cyclone is not extreme relative to other tropical cyclones (Seneviratne et al., 2012). Hence, reliable predictions of extremes are needed on short and long time scales to reduce potential risks and damages that result from weather and climate extremes (Chen, Moufouma-okia, Zhai, & Pirani, 2018; IPCC, 2013; Seneviratne et al., 2012).

Understanding weather and climate extremes is recognized as a major area necessitating further studies in climate research and has thus been selected as one of the World Climate Research Program (WCRP) Grand Challenges, which is hereafter referred to as the Extremes Grand Challenge (Alexander, Zhang, Hegerl, & Seneviratne, 2016; Chen et al., 2018; Sillmann et al., 2017; Zhang et al., 2013). This will further provide academics, decision-makers, international development agencies, nongovernmental organizations, and civil society the necessary information for monitoring and giving early warning to prevent or minimize the risks associated with weather-related hazards. It is worth noting that many weather and climate extremes are the results of natural climate variability (including phenomena such as El Niño-Southern Oscillation (ENSO)), and natural decadal or multidecadal variations in the climate provide the background for anthropogenic climate changes. Studies show that even if there were no anthropogenic changes in climate, a wide variety of natural weather and climate extremes would still occur (Chatzopoulos, Pérez, Zampieri, & Toreti, 2019; Seneviratne et al., 2012).

The changes in the weather and climate extremes vary across regions and for different types of extremes depending on a weather variable of interest. Consequently, the demand for information is often at its greatest in an event's immediate aftermath, requiring a quick response from scholars. But apparently conflicting views can confuse the public, for example, that all weather events are affected by climate change (Daramola et al., 2017; Stott et al., 2016; Trenberth, 2011), or that it is not possible to ascribe an extreme weather event to climate change (Stott et al., 2016). The danger is that such potential confusion could subvert the credibility of the science of climate change. Consequently, there is a need for climate science to better inform decision-makers, keenly aware of the need to protect life and property from the impacts of extreme weather and climate. The purpose of this work, therefore, is to provide an overview for a wider audience of the society on the current state of weather and climate extremes and the potential ways forward based on the expert discussions.

2.1 Physical processes of climate extremes, timing, and types of extremes

The concept of climate change and its extremes has to be understood first in order to understand its physical processes. Climate change is a statistically significant variation in either the mean state of the climate or in its variability, persisting for an extended period (Huber & Gulledge, 2011). These changes influence the frequency and intensity of extreme conditions. However, all climate and weather conditions regardless of their severity can still have extreme impacts if a certain threshold is crossed or with persistent occurrence (IPCC, 2012; Richard, 2015).

On the other hand, climate extreme does not have a concise definition, extreme is ambiguous, it is relative in time and location (e.g., a hot day in the tropics would be different from a hot day in the mid-latitude). More so, it can be used to describe the property of a climate variable or its impact (Zwiers et al., 2013). However, climate extremes could refer to the values of a climate variable above or below a threshold near the tail of the variable distribution, which generally rarely occurs (Alexander et al., 2016; Chen et al., 2018; IPCC, 2012; Karoly, 2014; Sillmann et al., 2017; Zhang et al., 2013). The climate extremes are also regionally dependent and vary from one region to another. For instance, some areas may warm significantly more than others, some will receive more rainfall, while others will be subjected to more frequent climate hazards with different impacts on people and ecosystems.

Generally, climate extremes are climate events that rarely occur within a defined climate system with the inclusion of cyclones (Zwiers et al., 2013). More so, certain climate extremes such as droughts and floods, result from the accumulation of individual climate events that may not in themselves be extreme, although their accumulation is extreme (IPCC, 2012). Many other climate extremes are a result of natural climate variability (phenomenon such as ENSO and Monsoon) that occur on decadal times scales, this means that even without anthropogenically induced climate change, climate extremes would still occur naturally (IPCC, 2012). There are various kinds of climate extremes, having varying physical and environmental impacts and occurring at different space and times scales. These extremes can range from continental-scale drought to widespread heat waves lasting several days to weeks as well as short-term events such as flash floods and tornadoes due to short-lived storms (Zhang et al., 2013).

Furthermore, the relationship between the kinds of climate extremes is arbitrary, as not all climate extreme events lead to environmental impact if there is no exposure to vulnerability. More so, the impact of any event would depend on the season, duration, intensity, vulnerability and simultaneous occurrence of climate extremes such as drought and heat waves.

Understanding the systems of climate extremes has received a lot of public attention in recent decades because of its social-economic importance and impacts Huber & Gulledge, 2011; Müller & Kaspar, 2014). The adverse effects that stem from extreme climate events have affected many societies, this consequently has led to several studies. The impacts of climate extremes are more devastating in developing countries than in the developed countries (Chen et al., 2018; Huber & Gulledge, 2011; Richard, 2015). Low-income nations are the most vulnerable to climate variability and change because of multiple existing stresses and low adaptive capacity. Awareness of climate extremes are on the increase in recent years. The sustainability of the socioeconomic environment and its development depends on our understanding of climate extremes (Albert et al., 2009; Richard, 2015). In the 21st century, climate extreme is expected to be more frequent and intense (IPCC, 2007), some studies have attributed the observed increase in climate extreme events to global warming (Nicholls et al., 2012; Swain, Singh, Touma, & Diffenbaugh, 2020).

2.2 Measuring climate extremes

Generally, there is no unified definition for climate extremes or the extremeness of climate events. Müller and Kaspar (2014) reported that climate extremes are generally easy to recognize but difficult to define, due to the level of variability observed in climate extremes, which vary in duration, spatiotemporal coverage and socioeconomic impact. Climate extreme can be defined based on (a) rarity (b) intensity (c) severity (in terms of socioeconomic damage and number of casualties).

Some literatures, in their quest to define climate extreme, make use of extreme indices obtained from the probability of occurrence of a given factor or the extent to which it exceeds a given threshold (Chen et al., 2018; Zwiers et al., 2013). Zwiers et al. (2013) defined extreme indices based on the number of days with maximum or minimum temperature or precipitation, below the 1st, 5th, or 10th percentile or above the 90th, 95th, or 99th percentile, for a given time frame (days, month, year or season) relative to a reference period (Fig. 1.1). Several other definitions are based on duration above a given threshold or persistence of climate extreme. The major advantage of using climate extreme indices is the possibility of comparison across regions and across climate models (Chen et al., 2018).

2.3 Extreme weather climate variable

2.3.1 *Temperature extremes*

Temperature is associated with several kinds of climate extremes, ranging from heat waves to cold spells. These extremes have variety of impacts on the environment, human health and natural ecosystem. For accurate analysis of temperature extremes, they are usually estimated on daily time scales (daily or more) as they occur on weather time scales. Many studies have been conducted regarding temperature extremes, especially heat waves. Heat waves occur as a result of atmospheric blocking or quasi-stationary anticyclonic circulation anomalies (IPCC, 2012). There is a high probability that since 1950, there has been a global scale decrease in the number of cold days and nights and an increase in the number of warm days and nights, however with varying levels of confidence across continents (Chen et al., 2018).

Studies have shown that the current increase in warm spells is most likely a result of anthropogenic activities (Zwiers et al., 2013). Models projection has also shown that there would be an increase in the intensity, frequency and magnitude of hot extremes and a decrease in cold extremes over the course of the 21st century globally. Based on this projection, the frequency and magnitude of warm spell and heat waves are bound to increase with more warm days (and nights) and less cold nights (and days) (Eresanya, Ajayi, Daramola, & Balogun, 2018; IPCC, 2012).

Based on observation carried out, recent studies have categorized the hot days and night into three, daytime events (hot day–normal night), nighttime events (normal day–hot night), and complex events (hot day–hot night). More so, recent studies have discovered a new concept known as marine heat wave, which has had a devastating impact on the marine ecosystem. However, very few studies exist on the subject, making it difficult to assign a concrete definition (Chen et al., 2018). Aside from heat waves, certain parts of the world (eastern China, North America and Central Eurasia) have experienced increase cold waves during their winter season. Nonetheless, there are still opposing views on the cause of such cold waves. A consensus is yet to be established due to a lack of compelling evidence, thus making attribution and projections impossible (Chen et al., 2018).

2.3.2 *Precipitation extremes*

Precipitation refers to all forms of hydrometeor, in general, there are many kinds but the most significant is rain and snow. This section deals with the variation observed in daily precipitation or precipitation extremes. Precipitation

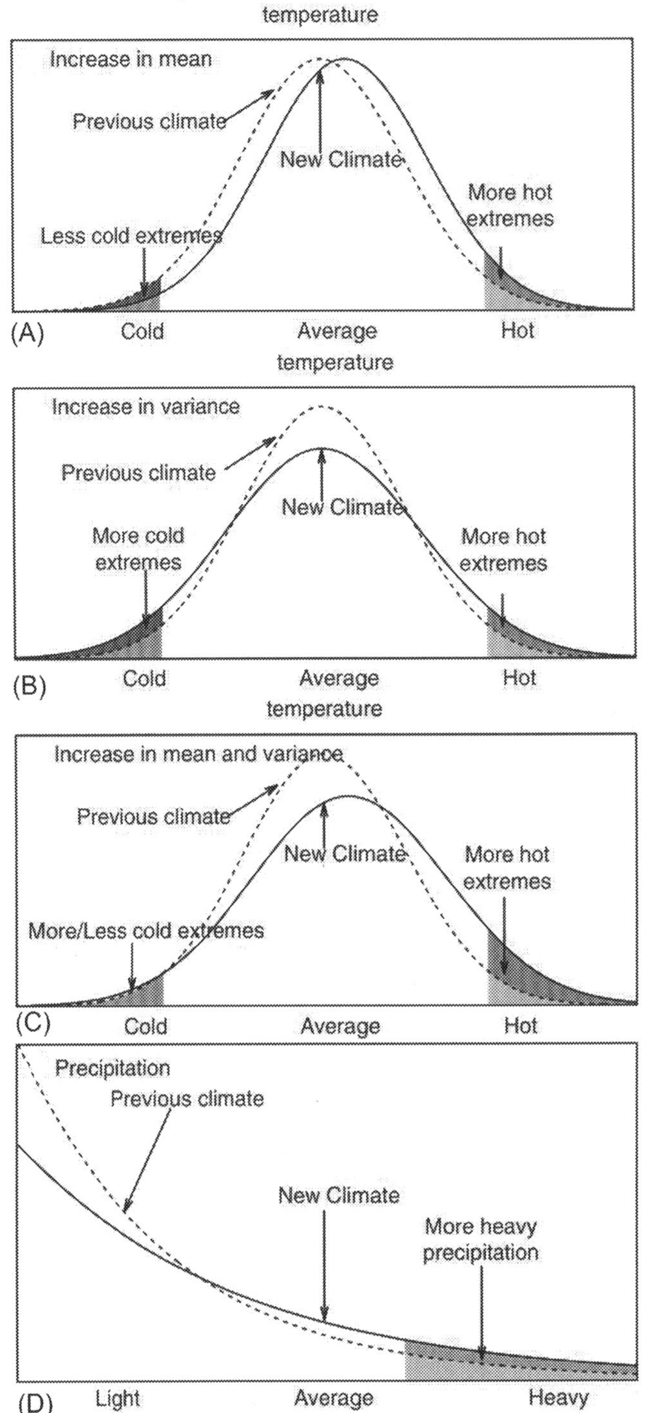

FIG. 1.1 Representation of the distribution of daily temperature, indicating regions of extreme temperature and precipitation. *(This figure was adopted from Zwiers, F.W., Lisa V.A., Gabriele C.H., Thomas R.K. , James P.K., Phillippe N., et al. (2013). Climate extremes: Challenges in estimating and understanding recent changes in the frequency and intensity of extreme climate and weather events. In G. Asrar, J. Hurrell (Eds.),* Climate science for serving society. *Springer Dordrecht. https://doi.org/10.1007/978-94-007-6692-1_13. - License No. 5150550526257.)*

extreme has the characteristic of being linked to different weather patterns and climate variations (El Nino and Monsoon), which usually defines their duration, intensity and trend (Chen et al., 2018). This variation in climate systems of different parts of the world makes it difficult to define precipitation extreme. Generally, studies make use of absolute threshold (50.8 mm/day in the United States or 100 mm/day in China) or relative threshold (the 95th percentile) (IPCC, 2012). Generally, results from precipitation extremes mostly have a medium confidence level due to the uncertainties arising from an inaccurate understanding of the underlining climate variations.

Statistically, current observation shows that it is more likely that precipitation extremes (heavy precipitation) has increased in more parts (regions) of the world than has decreased; however, there are certain regional, subregional, and seasonal variation in the trend (Karl & Easterling, 1999). Similarly, there is also a medium confidence level concerning attribution studies, claiming that the global increase in extreme precipitation is linked to anthropogenic activities (Durowoju, Olusola, & Anibaba, 2017; Zwiers et al., 2013). Projections from models indicate the frequency and intensity of precipitation extremes (heavy precipitation) is set to increase over the course of the 21st century in many parts of the world especially (tropics, northern mid-latitudes in winter season and high latitude), whereas with a decrease in total precipitation (Durowoju et al., 2017; IPCC, 2012; Zwiers et al., 2013).

2.3.3 *Wind extreme*

Winds, unlike temperature and precipitation, are most times considered in their extreme state, in the form of tropical and extratropical cyclones, tornadoes, and thunderstorms. Extreme winds speed can have a very severe impact on infrastructure, the maritime and aviation sector as wells as influence water availability by increasing evaporation rate which can lead to drought. Wind on water bodies can influence the coastal sea level, coastline stability and wave motion. Wind processes can also influence the formation and growth of arid and semiarid biomes, which defines the kinds of soil, and vegetation that emerges. The motion and position of forest fires are defined by the wind motion and in certain cases can lead to the formation of tornado genesis (IPCC, 2012; Osinowo, Okogbue, Eresanya, et al., 2017). Wind extreme may be defined based on parameters such as high percentile, maxima over a particular timescale (daily or yearly), wind gust (measure of highest wind in a short time interval, usually less than 20s), or storm-related highest. Wind extreme variability is mostly affected by changes in local convective activity, movement of large-scale circulation patterns or associated phenomena. Observation of wind trend is rather uncertain due to shortcomings associated with anemometer readings and reanalysis data. Typically, because there are very few studies on the relationship between wind speed trend and extreme wind trend there is low confidence on the projection of extreme wind except for changes associated with tropical cyclones. More so, low confidence in the projection of small-scale phenomena such as tornadoes arises from competing physical processes as well as the inability of models to simulate such phenomena (IPCC, 2012).

2.4 Climate phenomenon that influences the occurrence of extremes

2.4.1 *Modes of variability*

There exists much uncertainty as regards climate change in monsoon regions especially regarding circulation and precipitation, however with few exceptions to some monsoon regions. The conclusions are drawn from very few studies, as there is no consensus between the model representation of monsoon and its process creating a high level of uncertainty (low confidence) in the projection of monsoon change. However, there is a high possibility of precipitation increase in monsoon regions (not all monsoon regions) which may not be a result of changes in the monsoon characteristics (IPCC, 2012).

The El Nino-Southern Oscillation (ENSO) is a natural climate variability caused by equatorial ocean-atmosphere interaction in the tropical Pacific Ocean. The resulting oscillation is associated with variation in the sea surface temperature (SST) in the eastern equatorial pacific. The El-Nino phase is usually associated with warm SST in the eastern equatorial pacific while the La-Nina is associated with cool SST in the same region. The variation between the Ocean and atmosphere as a result of the El Niño and La-Nina is known as the ENSO and is usually associated with spatial patterns of weather extremes (heavy precipitation, extreme temperature, flood and drought). Invariably, monitoring and predicting ENSO using early warning system can lead to disaster risk reduction (IPCC, 2012).

Based on recent trends, there exists medium confidence toward more frequent El-Nino episodes in the central equatorial pacific, but too little evidence to make any deduction as regards ENSO. There is low confidence as regards projection of ENSO events because of inconsistencies in models that link its variability to increase in greenhouse gases. However, most global climate models project with medium confidence an increase in events within the central equatorial pacific which exhibits different patterns of climate variations than the classic eastern pacific (IPCC, 2012). There are several other climate variabilities that influence climate extremes aside from the monsoon and ENSO, they include; North Atlantic Oscillation (NAO), the Southern Annular Mode (SAM), and the Indian Ocean Dipole (IOD). There is low confidence in the ability of models to project changes in this circulation (NOA, SAM, and IOD). The uncertainty arises from the inability of models to attribute the variability to stratospheric ozone or greenhouse gases. However, it is likely that there has been anthropogenic influence on SAM (due to trends in stratospheric ozone rather than greenhouse gases), while there is uncertainty regarding the influence of anthropogenic activities on NAO (IPCC, 2012).

2.4.2 *Storms*

Storms are powerful cyclonic activities, usually driven by latent heat release that occurs in the atmosphere. Storms can occur on a range of scales from tornadoes to mesoscale convective complexes to tropical and extra-tropical cyclones. They cause severe damage that primarily results from high wind speed and heavy precipitation, additionally, damages could be compounded by flying debris, storm surges, high waves, drifting snow, wind-driven ice movements and many more (Zwiers et al., 2013).

2.4.3 *Tropical cyclones*

Tropical cyclones are the most common of extremes associated with wind speed, they occur over Tropical Ocean and pose major threat to population and infrastructure close to the coast, including offshore and shipping activities (IPCC, 2012). Since the inception of geostationary satellite, there remains, a constant number of 90 tropical cyclones observed annually. However, there is variability in the frequency and location of their tracks within individual ocean basin (IPCC, 2012; Zwiers et al., 2013).

Tropical cyclones cause more damage through storm surge and freshwater flooding than through heavy wind. However, the intensity of a tropical cyclone is measured based on its near surface wind speed usually on the Saffir–Simpsons scale, with the most dangerous wind in the category of (3, 4, and 5), although they do not occur often. Aside from the intensity, the impact of tropical cyclones is dependent on the structure and the areal extent of the wind field, especially in the case of storm surge. Other relevant measure includes the frequency, duration, precipitation and track of the cyclone. When tropical cyclone tracks poleward, they can become extra-tropical cyclones (IPCC, 2012; Zwiers et al., 2013).

There is low confidence as regards attributing changes in tropical cyclone activity to anthropogenic influences; this is because there are still lapses in understanding as regards the relationship between tropical cyclone and climate change, uncertainties in historical records of tropical cyclones, and the degree of tropical cyclone variability. Similarly, confidence remains low as regards the projection of the changes in tropical cyclone genesis, track, location, duration and area of impact. However, most models show that warming induced by greenhouse effect would likely increase the rainfall rate associated with tropical cyclones. It is also likely that the frequency of tropical cyclones would decrease or remain unchanged globally; similarly, it is likely that some tropical region would experience an increase in the maximum wind speed of tropical cyclones, more so it is likely that the intensity of severe storms would also increase in some ocean basin (IPCC, 2012).

2.4.4 *Extra-tropical cyclones*

Extratropical cyclones (synoptic-scale low-pressure systems) that occur within the mid-latitude in both hemisphere and are associated with extreme precipitation, storm surges, extreme winds, sea level and wave build up. They usually form over ocean basin within the proximity of upper tropospheric jets streams, through conversion from tropical to extra-tropical cyclones or as a result of flow over mountains. They majorly serve to convey heat and moisture from the tropics toward the poles and have a major impact on regional temperature and precipitation (Zwiers et al., 2013). Extra-tropical cyclones are formed and grow because of latent heat release from phase change of water and atmospheric instabilities (disturbance) along a zone of high-temperature contrast (baroclinic instability). Such zones of baroclinic instability are rich in potential energy that can be converted to kinetic energy associated with extra-tropical cyclones. Detecting changes in extratropical cyclones is, however, a major challenge because of inhomogeneity introduced through changes in the observing system. Observation indicates that over the last 50 years there exist the likelihood of a poleward shift in extratropical storms for both the north and Southern Hemisphere. Generally, there is medium confidence as regards the influence of anthropogenic activity on the changes to intensity, frequency and track of extra-tropical cyclones. Projection of extratropical cyclones is still quite uncertain because, differences in study techniques, different (threshold, physical quantities, storm track, and vertical level) of cyclone activity results in different projections (IPCC, 2012).

2.4.5 *Tornadoes*

Other kinds of small-scale severe weather associated with wind are tornadoes and small-scale storms. Tornadoes are extreme events that result from high local vorticity produced within a thunderstorm resulting from the convergence of angular momentum produced by rapid vertical motion. Although our understanding of the phenomena has greatly improved over time there is however very limited research as regards the frequency and intensity of tornadoes globally. The limitation stems from inhomogeneity in report method, time and observatory platforms. The projection of future changes in the frequency and intensity of tornadoes remains uncertain because the influence of climate forcing such as greenhouse warming on tornadoes occurrence is not properly understood.

2.4.6 Hydrological extremes

The primary hydrological extremes discussed here are flood and droughts, which affect numerous people every year. Drought is of major significance because it can occur over a continental scale with duration lasting years or longer (Orimoloye, Belle, Olusola, Busayo, & Ololade, 2021). However, some kinds of floods are localized and occur over a short duration while others can affect as large as a whole basin lasting for a month. These hydrological events although opposing in nature are not very mutually exclusive as they can occur simultaneously (Zwiers et al., 2013).

2.4.7 Flood

The overflow of water beyond the bounds of a water body or the gathering of water in locations where it is not normally present is referred to as a flood. Fluvial floods, flash floods, urban floods, pluvial floods, sewer floods, coastal floods, and glacial lake outburst floods are all examples of floods. Intense, frequent, and long-lasting precipitation, snowmelt, dam failure, and local storms are the most common causes of floods. Soil qualities, drainage basin conditions, the presence of snow and ice, urbanization, and the presence of dams, dike, and reservoirs are all elements that impact flood occurrence (Durowoju et al., 2017; IPCC, 2012).

Floods have gotten more regular and frequent in a number of countries (White, 2013), since the 2000s, the number of floods in the world has nearly doubled when compared to 1990s' figures (Guha-Sapir, Hargitt, & Hoyois, 2004). They are a serious threat with a large number of victims, floods harmed 50% of those affected by natural disasters in 2018 (CRED, 2019). Floods often have serious economic and material impacts (Kundzewicz et al., 2014). The increase in floods can be elucidated by various factors, including climate change (Hua et al., 2020), which causes changes in precipitation regimes and intensity (Bates, Kundzewicz, Wu, & Palutikof, 2008), and often manifests in heavy rains. In small river basins and rivers, heavy rains can create flooding (Kundzewicz et al., 2014). Extreme events in Africa (Kadomura, 2005), Europe (Marchi, Borga, Preciso, & Gaume, 2010), and Asia (Herring, Hoerling, Peterson, & Stott, 2014) are instances of climate change's role in increased floods.

2.4.8 Drought

Drought has no universal definition because drought definitions are region specific, reflecting differences in climatic characteristics as well as incorporating different variables such as physical, biological, and socioeconomic. Therefore, it is usually difficult to adopt definitions derived for one region. For example, come of the commonly used definitions are: (i) The World Meteorological Organization (World Meteorological Organization (WMO), 1975) defined drought as a sustained, extended deficiency in precipitation. (ii) Linsely., Kohler, and Paulhus (1959) defined drought as a sustained period of time without significant rainfall. (iii) The UN Convention to Combat Drought and Desertification (UN Secretariat General, 1994) defined drought as the naturally occurring phenomenon that exists when precipitation has been significantly below normal recorded levels, causing serious hydrological imbalances that adversely affect land resource production systems. (iv) Gumbel (1963) defined drought as the smallest annual value of daily streamflow. (v) Palmer (1965) described drought as a significant deviation from the normal hydrologic conditions of an area while the Food and Agriculture Organization (FAO, 1983) of the United Nations defines a drought hazard as the percentage of years when crops fail from the lack of moisture.

Wilhite and Glantz (1985) classified the definitions into four different categories; meteorological, agricultural, hydrological and socioeconomic drought. Meteorological drought is defined as a lack of precipitation over a region for a period of time. Precipitation has been commonly used for meteorological drought analysis (Chang, 1991; Eltahir, 1992; Estrela, Penarrocha, & Millan, 2000; Mishra & Singh, 2010). Hydrological drought is related to a period with inadequate surface and subsurface water resources for established water uses of a given water resources management system (Clausen & Pearson, 1995; Mishra & Singh, 2010). Agricultural drought is a period with declining soil moisture and consequent crop failure without any reference to surface water resources. A decline of soil moisture depends on several factors which affect meteorological and hydrological droughts along with differences between actual evapotranspiration and potential evapotranspiration while socioeconomic drought is associated with failure of water resources systems to meet water demands and thus associating droughts with supply of and demand for an economic good (water) (American Meteorological Society (AMS), 2004; Ayugi, Eresanya, & Onyango, 2022)

However, it is noteworthy to distinguish aridity from drought. Unlike drought or dryness, aridity is the characteristic of a preexisting climate condition (e.g., desert). Numerous studies have engaged the use of drought indices (proxies) to monitor and study changes in drought conditions as there are few direct observations of drought, such indices include but not limited to Palmer Drought Severity Index (PDSI, Palmer, 1965), Standardized Precipitation Index (SPI,

McKee, Doesken, & Kleist, 1993, 1995), National Rainfall Index (NRI, Gommes & Petrassi, 1994), Rainfall Anomaly Index (RAI, Van Rooy, 1965), Deciles (Gibbs & Maher, 1967), Crop Moisture Index (CMI, Palmer, 1968), and Bhalme and Mooley Drought Index (BMDI, Bhalme & Mooley, 1980), while Shafer and Dezman (1982) introduced Surface Water Supply Index (SWSI) and the Standardized Precipitation Evapotranspiration Index (SPEI) (IPCC, 2012; Zwiers et al., 2013). The WMO proposed the commonly used drought indicators/indices that are being used across drought-prone regions with the goal of advancing monitoring, early warning, and information delivery systems in support of risk-based drought management policies and preparedness plans (Svoboda, Fuchs, Poulsen, & Nothwehr, 2015). Feedback land-atmosphere interaction, drought has the potential to affect extremes of other weather and climate elements such as precipitation, temperature, and other variables. Based on the impact of anthropogenic activities on temperature and precipitation, there is medium confidence concerning the impact of anthropogenic activities on observed changes in drought conditions. Similarly, there is medium confidence that drought has become more severe in some parts of the world (South Europe and West Africa) and less severe in other parts of the world (North America and Northwest Australia).

Inconsistency in data and evidence makes it difficult to attribute observed changes in drought conditions on a regional scale. There is low to medium confidence as regards the projected increase in the duration and intensity of drought in some parts of the world. Observation challenge for drought arises because of definitional uncertainty and lack of data, while projection challenges arise from the same issues plus the inability of models to include all the factors responsible for drought occurrence, uncertainty as regards the influence of climate forcing (e.g., El Nino) on drought occurrence and the land-atmosphere feedback changes associated with drought.

2.4.9 *Extreme sea level*

Extreme sea levels are caused by severe weather events (tropical and extratropical cyclones) that can lead to storm surges and extreme wave height at the coast. Atmospheric storminess and means sea level rise may contribute to futuristic changes in extreme sea level, although nonuniform spatially across the globe. Other factors such as glacial isostatic adjustments, variation in wind change, changes in atmospheric pressure, water density, and rate of thermal expansion, rapid melt of ice sheets, ocean circulation, coastal engineering and changes in earth's gravitational field can influence sea level change along coastlines. There also exists eternal variability that has transient effects on extreme sea levels including; El-Nino Southern Oscillation, Pacific Oscillation, North Atlantic Oscillation and the position of the South Atlantic high (IPCC, 2012; Zwiers et al., 2013).

There are various methods used to characterize extreme sea levels, such as storm-related the highest values, annual maxima, or percentiles. Assessment based on limited data (low confidence) indicates a growing trend in coastal water that results from increase in mean sea level rather than change in storminess. It is also likely that anthropogenic influence on mean sea level is responsible for the observed increase in extreme coastal high waters. However, it is possible that change in storminess contribute to changes in sea level extreme; however, uncertainties arise because of limited geographical coverage of studies and limited understanding of storminess change. Projected studies suggest the likelihood of constant rise in coastal high waters resulting from increase in mean sea level based on observed trends (IPCC, 2012).

2.5 Compound and simultaneous extremes

While considering various kinds of climate extremes, it is of importance to consider the concepts of compound and simultaneous climate extremes. Although these climate events are less frequent than most extremes, their occurrences have however been spotted. Compound extremes can be defined as; (a) multiple extreme events occurring simultaneously or successively (b) combination of extreme events with underlying condition that intensifies the impact of the events. (c) Combination of events that are not in themselves extreme but their combination leads to extreme events or impact (Chen et al., 2018). For instance, the combination of heat wave and drought on wildfire increased risk of flooding from intense precipitation and sea-level rise. Various regions have been affected by climate-related disasters, for instance, drought in Africa (Orimoloye, Belle, & Ololade, 2021), pest invasion in China (Chu, Qu, & Guo, 2019), and flood in Southern Africa (Twumasi et al., 2017). Compound extremes can also result from the combination of nonextreme or moderately extreme events, or even contrasting events such as drought and flood (IPCC, 2012). However, simultaneous climate extremes are different from compound extremes in that compound extreme occurs in the same location while simultaneous extremes occur at adjacent locations coincidentally. Simultaneous extremes are notorious for their magnitude and spatial coverage, posing a major challenge for organizations coping with disasters (Chen et al., 2018).

2.6 Measurement, detection and attribution of extremes

Climate extremes are the potentially uncontrollable and undesired outcomes of climatic variations and happenings on earth. Climate extremes include unexpected, unusual, severe, or unseasonal weather; weather at the extremes of the historical distribution—the range that has been seen in the past. Usually, climate extremes are detected based on a location's recorded weather history and defined as lying in the rarest percentiles. There is evidence that human-induced global warming and subconscious heat activities are boosting the frequency and persistence of some extreme climate events (Rahmstorf & Coumou, 2011). This implies that the activities of humankind in regard to energy utilization and acclimation, directly and indirectly, affect the geosphere and concomitantly aggravates the extremity of climate changes.

A study by Dokken and Angelsen (2015), explains in more technical terms however that climate extremes are conventionally accepted as the occurrence of a value of a weather or climate variable above or below a limiting value near the higher or lesser end margins of the gamut of observed values of the variable. This is geometrically progressive and can only be appreciated on a statistical illustration. It explains that some forms of climate extremes (e.g., droughts, floods) may be the result of an accumulation of weather or climate events that are, distinctly, not extreme themselves. This further implies that climate extremes are additional results of morphing climate events that concatenate over time to become fill blown and seismic in statistical analysis and skews. As well, weather or climate events, even if not extreme in a statistical sense, can still lead to extreme conditions or impacts, either by crossing a critical threshold in a social, ecological, or physical system or by occurring simultaneously with other events. Another complementing issue in the monitoring and attribution of climatic extremes is the influence of tidal wave disparities along coastal lines.

Easterling, Kunkel, Wehner, and Sun (2016) in his research study on climate extremes opined that measuring, detecting and attributing climate extremes to globally and meteorologically accepted standards demands an insightful deposition that is evidenced by perpetual observation and meticulous team deliberation to create consensus. According to the state of the global climate report (WMO, 2020), assessing climate change is most commonly measured using the average surface temperature of the planet with alterations staying in the range of two standard deviations.

Further, study by Wu et al. (2021) shows that across the globe, land area as a whole there has been a measured overall decrease in the number of cold days and nights and overall increase in the number of warm days and nights. By effect, the diurnal balance has sunk shallow and this has made the outlook of climate stability erratic and technically unpredictable. Further, during the course of the research of Wu et al. (2021), more areas with increases than decreases in the frequency, intensity and/or amount of heavy rainfall were discovered with large parts of Europe, Asia, Eastern Africa and Australia and hinterland compartments of the globe as a whole have seen detectable increases in the frequency or length of warm periods. According to Easterling et al. (2016), extreme weather-related events, such as droughts, heat waves, wildfires, large magnitude and intensity storms, floods, and blizzards, also appear to be associated with global warming. Left unresolved, the impact on ecosystems and human quality of life may be devastating.

Sociocultural practices have contributed largely to the periodicity and intensity of climatic extremes. Whereas mortal civilization and activity has occurred during a period of what we now know has been a tolerable and relatively stable climate, the scale and efficiency with which we are extracting and burning carbon-rich fuel sources have created conditions outside the range of modern human experience. Risks to human health are among the most threatening of global warming-associated climate change, and are accelerating. These concerns in conjunction with impacts on ecosystems, urbanized areas, and community infrastructure contribute to heightened compulsion to act.

2.7 Environmental effects of the weather extremes

Global warming is responsible for climate change that influences weather and climate extremes. In turn, these impacts affect the severity and frequency of environmental threats, such as forest fires, hurricanes, heat waves, floods, droughts episodes, and land degradation, in some cases (Orimoloye et al., 2019; Orimoloye, Zhou, & Kalumba, 2021). Studies have shown that extreme weather events can decrease the carbon absorption potential of an ecosystem and create a dangerous cycle in which extreme weather fuels climate change by preventing carbon absorption by forests, causing more of it to stay in the atmosphere (Nunes, Meireles, Pinto Gomes, & Almeida Ribeiro, 2020). Records from IPCC show that since the 1970s, the global average temperature has risen by at least 0.4°C and that by 2100 it could rise above preindustrial temperatures to about 4°C (Horton et al., 2020; Koraim, Heikal, & Abozaid, 2011). While the global implications of climate change may seem too insignificant for people living around the world to notice, the consequences of climate change have already been witnessed through extreme weather events, including forest fires, hurricanes, droughts, heat waves, floods, and storms. Evaluation or modeling of real data has revealed that the intensity and frequency of these extreme events are influenced by climate change (Sarhadi & Soulis, 2017).

Environmental components and other human aspects have been identified to be modified or influenced by extreme events. For instance, risks for wildlife, a decline in ecosystem functioning, land degradation, biodiversity and wetland extinction among others (Table 1.1) have been influenced or affected by extreme events (Maxwell et al., 2019; Orimoloye, Kalumba, Mazinyo, & Nel, 2020). However, a lot still need to be done to confirm the actual relationship between climate change and extreme environmental events. Nevertheless, climate change or extreme events has not been proven to directly cause individual extreme environmental events, but has been identified to make these events more destructive, and likely happen more frequently, than they normally would be (Gasper, Blohm, & Ruth, 2011; Turco, Llasat, von Hardenberg, et al., 2014).

TABLE 1.1 Impacts of climate change and extreme events.

SN	Climate change and extreme events	References
1	**Increasing risk of drought**: High temperatures can lead to increased precipitation, evaporation and transpiration. Thus, as temperatures rise, the probability of hydrological and agricultural drought increases. Atmospheric rivers (narrow streams of moisture transported into the atmosphere) can also be influenced by a changing climate, which can particularly disrupt precipitation patterns in any region.	Vicente-Serrano, Beguería, and López-Moreno (2010), Vicente-Serrano, McVicar, Miralles, Yang, and Tomas-Burguera (2020), Stewart, Rogers, and Graham (2020)
2	**Increasing risk of biodiversity loss**: Plant and animal life would be more affected as climate change changes temperature and weather patterns continue to increase. As temperatures continue to rise, researchers predict that the number and variety of species that characterize biodiversity will decrease considerably. The loss of biodiversity due to climate change will alter the structures and functions of environmental or ecological systems.	Araújo and Rahbek (2006), Sintayehu (2018)
3	**Increasing risk of forest fires**: Climate change and the extreme event has been the key factors in increasing the risk and extent of wildfires in many regions of the world, for example, Western United States. Climate change causes forest fuels (the organic matter that burns and spreads wildfires) to become drier, and the number of major fires in the western United States between 1984 and 2015 has doubled. According to a recent NOAA-funded report, global change from rising greenhouse gas emissions is expected to significantly increase the likelihood of very big, destructive wildfires over the next few decades.	Barbero, Abatzoglou, Larkin, Kolden, and Stocks (2015), Jolly et al. (2015), Williams et al. (2015)
4	Decrease in crop yield: Temperature rises can cause yield decreases of between 2.5% and 10% across a variety of agronomic species (Hatfield et al., 2011). Other climate assessments on crop yield have provided differing results. Productivity can be dramatically impacted by temperatures that are deemed extreme and fall below or above particular thresholds at critical times during growth. Crop yield would be decreased by chronic exposure to high temperatures and other extreme events during the pollination stage of the potential grain or crop harvest.	Hatfield et al. (2011), Hatfield and Prueger (2015)
5	**Increase in mortality from heat waves**: Prolong periods of extreme heat, usually defined as heat waves, have been linked with a substantial increase in mortality, and specific events have been reported as public health disasters. Extreme heat can increase the risk of other types of disasters. Rising temperatures and other climatic components across the globe poses a threat to people, ecosystems and the economy.	Wolf, Adger, Lorenzoni, Abrahamson, and Raine (2010), Kovats and Kristie (2006), Ropo, Perez, Werner, and Enoch (2017)
6	**Increase in energy demand**: Climate change may alter the capacity for energy production and energy demands. Changes in the water cycle, for example, have an effect on hydropower, and high temperatures increase the energy demand for summer cooling, thus reducing the demand for winter heating.	Lipson, Thatcher, Hart, and Pitman (2019), Takane, Ohashi, Grimmond, Hara, and Kikegawa (2020)

TABLE 1.1 Impacts of climate change and extreme events—cont'd

SN	Climate change and extreme events	References
7	**Increase in multiple climatic hazards**: Studies have identified that climate change is already modifying the frequency and intensity of many weather-related hazards. Climate change magnifies disaster risk; it can increase the risk while reducing the resilience of households and societies at the same time.	IPCC (2014)
8	**Risk to the livelihood of indigenous people**: Extreme events due to climate change, which has a significant effect on the livelihoods of the rural population, declining agricultural yields in warmer environments can also contribute to migration to urban areas, increasing the population exposed to natural hazards in such locations. The risk of severe, widespread and irreversible impacts on humans and ecosystems is increased by increasing rates of warming. In communities with low development levels, risks are unevenly felt and are typically greater for vulnerable individuals and communities. It has been indicated that indigenous peoples are impacted by climate change in distinctive ways, as well as by policies or acts aimed at addressing it. Indigenous groups, on the other side, have a unique role to play in climate change through their traditional knowledge and professions, cutting across all climate mitigation and adaptation efforts, as well as just transition policies.	UNISDR, (2015), Alam (2017), Etchart (2017)
9	**High risk of species extinction**: There is a growing awareness of anthropogenic climate change as one of the major threats to biodiversity and to species. There are numerous and nuanced ways in which climate change is likely to impact biodiversity, including the loss or destruction of essential ecosystems and microhabitats, changes in environmental thresholds, such as temperature, water availability/quality, beyond those that can be tolerated by a species. The loss of significant interactions between two unrelated species, or the arrival of invasive species. Environmental disruption (e.g., reproduction or migration), direct loss of individual species or even populations as a result of extreme events.	Thomas et al. (2004); Dantas et al. (2020)

No permission required.

3 Conclusion

A growing population globally is likely to be driven into poverty without planning for climate change-induced environmental hazards. Since 2008, an average of about 22.5 million individuals have been displaced every year by climate or weather-related incidents. Climate extremes events such as droughts, floods, and heat waves have become more intense with increase in global temperature. These have led to loss of lives and destruction of properties. One way to plan for such significant environmental events is to utilize climatic evaluations that display the frequency and intensity of these events by using current and historical data records. It is also possible to use these assessments to predict when and where future incidents will take place and how disruptive they will be. With this detail, by warning individuals living in high-risk areas and providing disaster relief, we can plan for severe weather events. In combating climate-related impact and environmental extremes, the influence of climate change can also be observed by simulating the effects of various greenhouse gas concentrations on the environmental components. Consequently, it has been found that climate change has increased the risk of wildfires in some regions of the world, extreme heat in China, and drought in South Africa. Continuous study and development in the attribution of extreme events will help us find out more specifically how extreme environmental events are influenced by climate change and how we could change this trajectory.

References

Alam, G. M. (2017). Livelihood cycle and vulnerability of rural households to climate change and hazards in Bangladesh. *Environmental Management*, *59*(5), 777–791. https://doi.org/10.1007/s00267-017-0826-3.

Alexander, L. V., Zhang, X., Hegerl, G., & Seneviratne, S. I. (2016). Implementation plan for WCRP grand challenge on understanding and predicting weather and climate extremes. In *Extremes grand challenge*. https://www.wcrp-climate.org/images/documents/grand_challenges/WCRP_Grand_Challenge_Extremes_Implementation_Plan_v20160708.pdf. (Accessed 27 September 2021).

American Meteorological Society (AMS). (2004). Statement on meteorological drought. *Bulletin of the American Meteorological Society, 85*, 771–773.

Araújo, M. B., & Rahbek, C. (2006). How does climate change affect biodiversity? *Science, 313*(5792), 1396–1397.

Aremu, K. J. (2008). *Understanding weather and climate.* Kaduna, Nigeria: Cee Kay Bee Printers.

Ayoade, J. O. (1988). *Introduction to climatology for the tropics.* Ibadan: Spectrum. https://doi.org/10.1144/GSL.JGS.1907.063.01-04.19.

Ayugi, B., Eresanya, E., Onyango, A. O., et al. (2022). Review of meteorological drought in Africa: Historical trends, impacts, mitigation measures, and prospects. *Pure and Applied Geophysics.* https://doi.org/10.1007/s00024-022-02988-z.

Barbero, R., Abatzoglou, J. T., Larkin, N. K., Kolden, C. A., & Stocks, B. (2015). Climate change presents increased potential for very large fires in the contiguous United States. *International Journal of Wildland Fire, 24*(7), 892–899.

Bates, B. C., Kundzewicz, Z. W., Wu, S., & Palutikof, J. P. (2008). *Le Changement Climatique et l'eau. Doc. Tech. Publ. Par Groupe D'experts Intergouv. Sur L'évolution Clim. Secrétariat GIEC Genève Éd* (p. 236).

Bhalme, H. N., & Mooley, D. A. (1980). Large-scale droughts/floods and monsoon circulation. *Monthly Weather Review, 108*, 1197–1211.

Chang, T. J. (1991). Investigation of precipitation droughts by use of kriging method. *Journal of Irrigation and Drainage Engineering, 117*(6), 935–943.

Chatzopoulos, T., Pérez, I., Zampieri, M., & Toreti, A. (2019). Climate extremes and agricultural commodity markets: A global economic analysis of regionally simulated events. *Weather and Climate Extremes, 27.* https://doi.org/10.1016/j.wace.2019.100193, 100193.

Chen, Y., Moufouma-okia, W., Zhai, P., & Pirani, A. (2018). Recent progress and emerging topics on weather and climate extremes since the fifth assessment report of the intergovernmental panel on climate change. *Annual Review of Environment and Resources, 43*, 35–39.

Chu, D., Qu, W. M., & Guo, L. (2019). Invasion genetics of alien insect pests in China: Research progress and future prospects. *Journal of Integrative Agriculture, 18*, 748–757. https://doi.org/10.1016/S2095-3119(17)61858-6.

Clausen, B., & Pearson, C. P. (1995). Regional frequency analysis of annual maximum streamflow drought. *Journal of Hydrology, 173*, 111–130.

CRED. (2019). *Natural Disasters 2018.* Brussels, Belgium: CRED.

Dantas, B. F., Moura, M. S., Pelacani, C. R., Angelotti, F., Taura, T. A., Oliveira, G. M., & Seal, C. E. (2020). Rainfall, not soil temperature, will limit the seed germination of dry forest species with climate change. *Oecologia, 192*(2), 529–541.

Daramola, M. T., Eresanya, E. O., & Erhabor, S. C. (2017). Analysis of rainfall and temperature over climatic zones in Nigeria. *Journal of Geography, Environment and Earth Science International, 11*(2). https://doi.org/10.9734/JGEESI/2017/35304, JGEESI.35304.

Dokken, T., & Angelsen, A. (2015). Forest reliance across poverty groups in Tanzania. *Ecological Economics, 117*, 203–211.

Durowoju, O. S., Olusola, A. O., & Anibaba, B. W. (2017). Relationship between extreme daily rainfall and maximum daily river discharge within lagos metropolis. *Ethiopia Journal of Environmental Studies and Management*, 1998-0507. *10*(14), 492–504. https://doi.org/10.4314/ejesm.v10i4.7.

Easterling, D. R., Kunkel, K. E., Wehner, M. F., & Sun, L. (2016). Detection and attribution of climate extremes in the observed record. *Weather and Climate Extremes, 11*, 17–27. https://doi.org/10.1016/j.wace.2016.01.001.

Eltahir, E. A. B. (1992). Drought frequency analysis in central and Western Sudan. *Hydrological Sciences Journal, 37*(3), 185–199. https://doi.org/10.1080/02626669209492581.

Eresanya, E. O., Ajayi, V. O., Daramola, M. T., & Balogun, R. (2018). Temperature extremes over selected stations in Nigeria. *Physical Science International Journal, 20*(1), 1–10. https://doi.org/10.9734/PSIJ/2018/34637.

Estrela, M. J., Penarrocha, D., & Millan, M. (2000). Multi-annual drought episodes in the Mediterranean (Valencia region) from 1950-1996. A spatio-temporal analysis. *International Journal of Climatology, 20*, 1599–1618. https://doi.org/10.1002/1097-0088(20001115)20:13<1599::AID-JOC559>3.0.CO;2-Q.

Etchart, L. (2017). The role of indigenous peoples in combating climate change. *Palgrave Communications, 3*(1), 1–4.

Food and Agriculture Organization. (1983). Guidelines: Land evaluation for Rainfed agriculture. *FAO soils bulletin no., 52.* Rome: FAO.

Gasper, R., Blohm, A., & Ruth, M. (2011). Social and economic impacts of climate change on the urban environment. *Current Opinion in Environmental Sustainability, 3*(3), 150–157.

Gibbs, W. J. (1987). *A drought watch system. WMO/TD - no. 193; WCP - no. 134.* Geneva, Switzerland: World Meteorological Organization. https://library.wmo.int/index.php?lvl=notice_display&id=11690#.YVEU7ppByUk. (Accessed 26 September 2021).

Gibbs, W. J., & Maher, J. V. (1967). Rainfall deciles as drought indicators. *Bureau of Meteorology Bull. 48.* Melbourne, Australia: Commonwealth of Australia.

Gommes, R., & Petrassi, F. (1994). *Rainfall variability and drought in Sub-Saharan Africa since 1960. Agro-meteorology series working paper 9.* Rome, Italy: Food and Agriculture Organization.

Guha-Sapir, D., Hargitt, D., & Hoyois, P. (2004). *Thirty years of natural disasters 1974–2003: The numbers.* Ottignies-Louvain-la-Neuve, Belgium: Presses Univ. de Louvain. https://www.preventionweb.net/files/1078_8761.pdf. Accessed 10 October 2021.

Gumbel, E. J. (1963). Statistical forecast of droughts. *Bulletin - International Association of Scientific Hydrology, 8*(1), 5–23. https://doi.org/10.1080/02626666309493293.

Hatfield, J. L., Boote, K. J., Kimball, B. A., Ziska, L. H., Izaurralde, R. C., Ort, D., … Wolfe, D. (2011). Climate impacts on agriculture: Implications for crop production. *Agronomy Journal, 103*(2), 351–370. https://doi.org/10.2134/agronj2010.0303.

Hatfield, J. L., & Prueger, J. H. (2015). Temperature extremes: Effect on plant growth and development. *Weather and Climate Extremes, 10*, 4–10. https://doi.org/10.1016/j.wace.2015.08.001.

Herring, S. C., Hoerling, M. P., Peterson, T. C., & Stott, P. A. (2014). Explaining extreme events of 2013 from a climate perspective. *Bulletin of the American Meteorological Society, 95*, S1–S104.

Horton, B. P., Khan, N. S., Cahill, N., Lee, J. S., Shaw, T. A., Garner, A. J., … Rahmstorf, S. (2020). Estimating global mean sea-level rise and its uncertainties by 2100 and 2300 from an expert survey. *npj Climate and Atmospheric Science, 3*(1), 1–8. https://doi.org/10.1038/s41612-020-0121-5.

Hua, P., Yang, W., Qi, X., Jiang, S., Xie, J., Gu, X., … Krebs, P. (2020). Evaluating the effect of urban flooding reduction strategies in response to design rainfall and low impact development. *Journal of Cleaner Production, 242*, 118515.

Huang, D., Zhang, L., Gao, G., & Sun, S. (2018). Projected changes in population exposure to extreme heat in China under a RCP8. 5 scenario. *Journal of Geographical Sciences, 28*(10), 1371–1384.

Huber, D. G., & Gulledge, J. (2011). *Extreme weather and climate change: Understanding the link and managing the risk. Science and impacts program.* Arlington, VA: Center for Climate and Energy Solutions. http://www.c2es.org/publications/extreme-weather-and-climate-change.

IPCC. (2007). Climate change 2007: Synthesis report. In Core Writing Team, R. K. Pachauri, & A. Reisinger (Eds.), *Contribution of working groups I, II and III to the fourth assessment report of the intergovernmental panel on climate change* (p. 104). Geneva, Switzerland: IPCC.

IPCC. (2012). Managing the risks of extreme events and disasters to advance climate change adaptation. In C. B. Field, V. Barros, T. F. Stocker, D. Qin, D. J. Dokken, K. L. Ebi, et al. (Eds.), *A special report of working groups I and II of the intergovernmental panel on climate change* (p. 582). Cambridge, UK/ New York, NY: Cambridge University Press.

IPCC. (2013). Climate change 2013: The physical science basis. In T. F. Stocker, D. Qin, G.-K. Plattner, M. Tignor, S. K. Allen, J. Boschung, … P. M. Midgley (Eds.), *Contribution of Working Group I to the fifth assessment report of the intergovernmental panel on climate change* (p. 1535). Cambridge, United Kingdom and New York, NY, USA: Cambridge University Press.

IPCC. (2014). Climate change 2014: Mitigation of climate change. In O. Edenhofer, R. Pichs-Madruga, Y. Sokona, E. Farahani, S. Kadner, A. Seyboth, et al. (Eds.), *Contribution of working group III to the fifth assessment report of the intergovernmental panel on climate change*. Cambridge, UK/New York, NY: Cambridge University Press.

Jolly, W. M., Cochrane, M. A., Freeborn, P. H., Holden, Z. A., Brown, T. J., Williamson, G. J., & Bowman, D. M. (2015). Climate-induced variations in global wildfire danger from 1979 to 2013. *Nature Communications*, *6*(1), 1–11. https://doi.org/10.1038/ncomms8537.

Kadomura, H. (2005). Climate anomalies and extreme events in Africa in 2003, including heavy rains and floods that occurred during northern hemisphere summer. *African Study Monographs*. http://citeseerx.ist.psu.edu/viewdoc/download?doi=10.1.1.566.8442&rep=rep1&type=pdf. Accessed 10 October 2021.

Karl, T. R., & Easterling, D. R. (1999). Climate extremes: Selected review and future research directions. *Climatic Change*, *42*, 209–325.

Karoly, D. (2014). *WCRP Grand Challenges: Science underpinning the prediction and attribution of extreme events*. https://www.wcrp-climate.org/documents/GC_Extremes.pdf. (Accessed 26 September 2021).

Koraim, A. S., Heikal, E. M., & Abozaid, A. A. (2011). Different methods used for protecting coasts from sea level rise caused by climate change. *Current Development in Oceanography*, *3*(1), 33–66.

Kovats, R. S., & Kristie, L. E. (2006). Heatwaves and public health in Europe. *European Journal of Public Health*, *16*(6), 592–599.

Kump, L., et al. (2004). *The earth system* (2nd ed.). New Jersey: Prentice Hall, ISBN:978-0-13-142059-5.

Kundzewicz, Z. W., Kanae, S., Seneviratne, S. I., Handmer, J., Nicholls, N., Peduzzi, P., … Mach, K. (2014). Flood risk and climate change: Global and regional perspectives. *Hydrological Sciences Journal*, *59*, 1–28.

Linsely, R. K., Jr., Kohler, M. A., & Paulhus, J. L. H. (1959). *Applied hydrology*. New York: McGraw Hill.

Lipson, M. J., Thatcher, M., Hart, M. A., & Pitman, A. (2019). Climate change impact on energy demand in building-urban-atmosphere simulations through the 21st century. *Environmental Research Letters*, *14*(12). https://doi.org/10.1088/1748-9326/ab5aa5, 125014.

Luo, L., Tang, Y., Zhong, S., Bian, X., & Heilman, W. E. (2013). Will future climate favor more erratic wildfires in the Western United States? *Journal of Applied Meteorology and Climatology*, *52*(11), 2410–2417.

Marchi, L., Borga, M., Preciso, E., & Gaume, E. (2010). Characterisation of selected extreme flash floods in Europe and implications for flood risk management. *Journal of Hydrology*, *394*, 118–133.

Masson, B. J. (1979). The distinction between weather and climate. *The Meteorological Magazine*, *108*(1284), 211–212. HMSO London.

Mauder, W. J. (1970). *The value of weather Methuen*. London: Methuen.

Maxwell, S. L., Butt, N., Maron, M., McAlpine, C. A., Chapman, S., Ullmann, A., & Watson, J. E. (2019). Conservation implications of ecological responses to extreme weather and climate events. *Diversity and Distributions*, *25*(4), 613–625.

McKee, T. B., Doesken, N. J., & Kleist, J. (1993). The relationship of drought frequency and duration to time scales. In *Paper presented at 8th conference on applied climatology, American Meteorological Society, 17–22 January 1993*. California: Anaheim.

McKee, T. B., Doesken, N. J., & Kleist, J. (1995). *Drought monitoring with multiple time scales, paper presented at 9th conference on applied climatology*. Dallas, Texas: American Meteorological Society.

Mishra, A. K., & Singh, V. P. (2010). A review of drought concepts. *Journal of Hydrology*, *391*(1–2), 202–216. https://doi.org/10.1016/j.jhydrol.2010.07.012.

Müller, M., & Kaspar, M. (2014). Event-adjusted evaluation of weather and climate extreme. *Natural Hazards and Earth System Sciences*, *14*, 473–483. https://doi.org/10.5194/nhess-14-473-2014.

Nicholls, S. I. N., Easterling, D., Goodess, C. M., Kanae, S., Kossin, J., Luo, Y., et al. (2012). Changes in climate extremes and their impacts on the natural physical environment. In C. B. Field, V. Barros, T. F. Stocker, D. Qin, D. J. Dokken, K. L. Ebi, et al. (Eds.), *Managing the risks of extreme events and disasters to advance climate change adaptation. A Special report of working groups I and II of the intergovernmental panel on climate change (IPCC)* (pp. 109–230). Cambridge, UK/New York, NY: Cambridge University Press. https://www.ipcc.ch/site/assets/uploads/2018/03/SREX-Chap3_FINAL-1.pdf.

Nunes, L. J., Meireles, C. I., Pinto Gomes, C. J., & Almeida Ribeiro, N. (2020). Forest contribution to climate change mitigation: Management oriented to carbon capture and storage. *Climate*, *8*(2), 21. https://doi.org/10.3390/cli8020021.

Orimoloye, I. R., Belle, J. A., & Ololade, O. O. (2021). Drought disaster monitoring using MODIS derived index for drought years: A space-based information for ecosystems and environmental conservation. *Journal of Environmental Management*, *284*. https://doi.org/10.1016/j.jenvman.2021.112028, 112028.

Orimoloye, I. R., Belle, J. A., Olusola, A. O., Busayo, E. T., & Ololade, O. O. (2021). Spatial assessment of drought disasters, vulnerability, severity and water shortages: A potential drought disaster mitigation strategy. *Natural Hazards*, *105*, 2735–2754. https://doi.org/10.1007/s11069-020-04421-x.

Orimoloye, I. R., Kalumba, A. M., Mazinyo, S. P., & Nel, W. (2020). Geospatial analysis of wetland dynamics: Wetland depletion and biodiversity conservation of Isimangaliso wetland, South Africa. *Journal of King Saud University-Science*, *32*(1), 90–96.

Orimoloye, I. R., Ololade, O. O., Mazinyo, S. P., Kalumba, A. M., Ekundayo, O. Y., Busayo, E. T., & Nel, W. (2019). Spatial assessment of drought severity in Cape Town area, South Africa. *Heliyon*, *5*(7). https://doi.org/10.1016/j.heliyon.2019.e02148, e02148.

Orimoloye, I. R., Zhou, L., & Kalumba, A. M. (2021). Drought disaster risk adaptation through ecosystem services-based solutions: Way forward for South Africa. *Sustainability*, *13*(8), 4132. https://doi.org/10.3390/su13084132.

Osinowo, A. A., Okogbue, E. C., Eresanya, E. O., et al. (2017). Evaluation of wind potential and its trends in the mid-Atlantic. *Modeling Earth Systems and Environment*, *3*, 1199–1213. https://doi.org/10.1007/s40808-017-0399-4.

Palmer, W. C. (1965). *Meteorologic drought* (p. 58). US Department of Commerce, Weather Bureau. Research Paper No. 45.

Palmer, W. C. (1968). Keeping track of crop moisture conditions, nationwide: The new crop moisture index. *Weatherwise*, *21*, 156–161.

Rahmstorf, S., & Coumou, D. (2011). Increase of extreme events in a warming world. *Proceedings of the National Academy of Sciences*, *108*(44), 17905–17909. https://doi.org/10.1073/pnas.1101766108.

Richard, R. H., Jr. (2015). An overview of weather and climate extremes – Products and trends. Weather and climate. *Extremes*, *10*, 1–9. https://doi.org/10.1016/j.wace.2015.11.001.

Ropo, O. I., Perez, M. S., Werner, N., & Enoch, T. I. (2017). Climate variability and heat stress index have increasing potential ill-health and environmental impacts in the East London. South Africa. *International Journal of Applied Engineering Research*, *12*(17), 6910–6918.

Sarhadi, A., & Soulis, E. D. (2017). Time-varying extreme rainfall intensity-duration-frequency curves in a changing climate. *Geophysical Research Letters*, *44*(5), 2454–2463.

Seneviratne, S. I., Nicholls, N., Easterling, D., Goodess, C. M., Kanae, S., Kossin, J., … Zhang, X. (2012). Changes in climate extremes and their impacts on the natural physical environment. In C. B. Field, V. Barros, T. F. Stocker, D. Qin, D. J. Dokken, K. L. Ebi, … P. M. Midgley (Eds.), *A special report of Working Groups I and II of the Intergovernmental Panel on Climate Change (IPCC). Managing the risks of extreme events and disasters to advance climate change adaptation* (pp. 109–230). Cambridge, UK: Cambridge University Press.

Shafer, B. A., & Dezman, L. E. (1982). Development of a surface water supply index (SWSI) to assess the severity of drought conditions in snowpack runoff areas. In *Preprints, Western SnowConf., Reno, NV* (pp. 164–175). Colorado State University.

Sillmann, J., Thorarinsdottir, T., Keenlyside, N., Schaller, N., Alexander, L. V., Hegerl, G., … Zwiers, F. W. (2017). Understanding, modeling and predicting weather and climate extremes: Challenges and opportunities. *Weather and Climate Extremes*, *18*, 65–74. https://doi.org/10.1016/j.wace.2017.10.003.

Sintayehu, D. W. (2018). Impact of climate change on biodiversity and associated key ecosystem services in Africa: A systematic review. *Ecosystem Health and Sustainability*, *4*(9), 225–239.

Stewart, I. T., Rogers, J., & Graham, A. (2020). Water security under severe drought and climate change: Disparate impacts of the recent severe drought on environmental flows and water supplies in Central California. *Journal of Hydrology X*, *7*. https://doi.org/10.1016/j.hydroa.2020.100054, 100054.

Stott, P. A., Christidis, N., Otto, F. E., Sun, Y., Vanderlinden, J. P., van Oldenborgh, G. J., … Zwiers, F. W. (2016). Attribution of extreme weather and climate-related events. *Wiley Interdisciplinary Reviews: Climate Change*, *7*(1), 23–41. https://doi.org/10.1002/wcc.380.

Svoboda, M., Fuchs, B. A., Poulsen, C., & Nothwehr, J. R. (2015). The drought risk atlas: Enhancing decision support for drought risk management in the United States. *Journal of Hydrology*, *526*, 274–286. https://doi.org/10.1016/j.jhydrol.2015.01.006.

Swain, D. L., Singh, D., Touma, D., & Diffenbaugh, N. S. (2020). Attributing extreme events to climate change: A new frontier in a warming world. *One Earth*, *2*(6), 522–527. https://doi.org/10.1016/j.oneear.2020.05.011.

Takane, Y., Ohashi, Y., Grimmond, C. S. B., Hara, M., & Kikegawa, Y. (2020). Asian megacity heat stress under future climate scenarios: Impact of air-conditioning feedback. *Environmental Research Communications*, *2*(1), 015004.

Thomas, C. D., Cameron, A., Green, R. E., Bakkenes, M., Beaumont, L. J., Collingham, Y. C., & Williams, S. E. (2004). Extinction risk from climate change. *Nature*, *427*(6970), 145–148. https://doi.org/10.1038/nature02121.

Trenberth, K. E. (2011). Attribution of climate variations and trends to human influences and natural variability. *Wiley Interdisciplinary Reviews: Climate Change*, *2*(6), 925–930. https://doi.org/10.1002/wcc.142.

Turco, M., Llasat, M. C., von Hardenberg, J., et al. (2014). Climate change impacts on wildfires in a Mediterranean environment. *Climatic Change*, *125*, 369–380. https://doi.org/10.1007/s10584-014-1183-3.

Twumasi, Y., Merem, E., Ayala-Silva, T., Osei, A., Petja, B., & Alexander, K. (2017). Techniques of remote sensing and GIS as tools for visualizing impact of climate change-induced flood in the southern African region. *American Journal of Climate Change*, *6*, 306–327. https://doi.org/10.4236/ajcc.2017.62016.

UN Secretariat General. (1994). *United Nations convention to combat drought and desertification in countries experiencing serious droughts and/or desertification, particularly in Africa*. Paris: UN Secretariat General. https://treaties.un.org/Pages/ViewDetails.aspx?src=TREATY&mtdsg_no=XXVII-10&chapter=27&clang=_en. (Accessed 14 September 2021).

UNISDR, C. (2015). *The human cost of natural disasters: A global perspective*. https://reliefweb.int/report/world/human-cost-natural-disasters-2015-global-perspective. (Accessed 26 September 2021).

Van Rooy, M. P. (1965). A rainfall anomaly index (RAI), independent of time and space. *Notos*, *14*, 43–48.

Vicente-Serrano, S. M., Beguería, S., & López-Moreno, J. I. (2010). A multiscalar drought index sensitive to global warming: The standardized precipitation evapotranspiration index. *Journal of Climate*, *23*(7), 1696–1718.

Vicente-Serrano, S. M., McVicar, T. R., Miralles, D. G., Yang, Y., & Tomas-Burguera, M. (2020). Unraveling the influence of atmospheric evaporative demand on drought and its response to climate change. *Wiley Interdisciplinary Reviews: Climate Change*, *11*(2), e632.

Wallace, J. M., & Hobbs, P. V. (1977). *Atmospheric science: An introductory survey* (p. 305). London: Academic Press.

White, I. (2013). *Water and the City: Risk, resilience and planning for a sustainable future*. Abingdon, UK: Routledge. https://doi.org/10.4324/9780203848319.

Wilhite, D. A., & Glantz, M. H. (1985). Understanding the drought phenomenon: the role of definitions. *Water International*, *10*(3), 111–120. https://doi.org/10.1080/02508068508686328.

Williams, A. P., Seager, R., Macalady, A. K., Berkelhammer, M., Crimmins, M. A., Swetnam, T. W., & Rahn, T. (2015). Correlations between components of the water balance and burned area reveal new insights for predicting forest fire area in the Southwest United States. *International Journal of Wildland Fire*, *24*(1), 14–26. https://doi.org/10.1071/WF14023.

WMO. (2020). *The State of the Global Climate 2020: Unpacking the indicators*. https://public.wmo.int/en/our-mandate/climate/wmo-statement-state-of-global-climate. (Accessed 21 September 2021).

Wolf, J., Adger, W. N., Lorenzoni, I., Abrahamson, V., & Raine, R. (2010). Social capital, individual responses to heat waves and climate change adaptation: An empirical study of two UK cities. *Global Environmental Change*, *20*(1), 44–52.

World Meteorological Organization Proceeding of the world climate conference-A conference of experts on climate and mankind. World Meteorological Organization (WMO)-WMO-No. 537, 1979. https://library.wmo.int/index.php?lvl=author_see&id=5288#.YUJP8J0zaUk.

World Meteorological Organization (WMO). (1975). *Drought and agriculture. Technical note no. 138, Report of the CAgM Working Group on Assessment of Drought* (p. 127). Geneva, Switzerland: WMO.

Wu, Y., Miao, C., Sun, Y., AghaKouchak, A., Shen, C., & Fan, X. (2021). Global observations and CMIP6 simulations of compound extremes of monthly temperature and precipitation. *GeoHealth*, *5*. https://doi.org/10.1029/2021GH000390, e2021GH000390.

Zhang, X., Hegerl, G., Seneviratne, S., Stewart, R., Zwiers, F., & Alexander, L. (2013). *WCRP grand challenge: Understanding and predicting weather and climate extremes.* https://www.wcrp-climate.org/images/documents/grand_challenges/GC_Extremes_v2.pdf. (Accessed 26 September 2021).

Zwiers, F. W., Lisa, V. A., Gabriele, C. H., Thomas, R. K., James, P. K., Phillippe, N., … Zhang, X. (2013). Climate extremes: Challenges in estimating and understanding recent changes in the frequency and intensity of extreme climate and weather events. In G. Asrar, & J. Hurrell (Eds.), *Climate Science for Serving Society.* Dordrecht: Springer. https://doi.org/10.1007/978-94-007-6692-1_13.

Further reading

Klein Tank, A. M. G., Zwiers, F. W., & Zhang, X. (2009). *Guidelines on analysis of extremes in a changing climate in support of informed decisions for adaptation. WMO/TD - no. 1500; WCDMP - no. 72.* Geneva, Switzerland: World Meteorological Organization.

Thomas, R. K., & David, R. E. (1999). Climate extremes: Selected review and future research directions. *Climate Change, 42*, 309–325.

William, R. T. (2014). Weather and climate extremes: Pacemakers of adaptation? *Weather and Climate Extremes, 5-6*, 29–39. https://doi.org/10.1016/j.wace.2014.08.001.

World Meteorological Organization (WMO) and Global Water Partnership (GWP). (2016). In M. Svoboda, & B. A. Fuchs (Eds.), *Integrated Drought Management Programme (IDMP), Integrated Drought Management Tools and Guidelines Series 2. Handbook of drought indicators and indices.* Geneva.

CHAPTER

2

The role of climate datasets in understanding climate extremes

Malcolm N. Mistry

Department of Economics, Ca' Foscari University of Venice, Venice, Italy
Department of Public Health, Environments and Society, London School of Hygiene and Tropical Medicine (LSHTM), London, United Kingdom

1 Introduction

Following some prominent studies that have convincingly shown the extent of the human influence on the 20th century temperature (Beniston & Stephenson, 2004; Kaufmann & Stern, 1997) and precipitation trends (Zhang et al., 2007), and on specific extreme weather events [e.g., flooding (Pall et al., 2011) and heat waves (Rahmstorf & Coumou, 2011)], the Intergovernmental Panel on Climate Change (IPCC) (The IPCC established in 1988 by the World Meteorological Organization (WMO) and the United Nations Environment Programme (UNEP), is an intergovernmental body of the United Nations for assessing the science related to climate change. In addition to the assessment reports published by its working groups, the IPCC periodically publishes other reports on climate change and extremes, such as the Special Report on Managing the Risks of Extreme Events and Disasters to Advance Climate Change Adaptation (SREX) (IPCC Summary for Policymakers, 2012)). In its Fifth Assessment Report (AR5) (IPCC, 2013) for the first time attributed changes in extreme weather and climate events to a changing climate. A growing body of literature in recent years (e.g. Giorgi, Raffaele, & Coppola, 2019; Masson-Delmotte et al., 2018) have shown how increase in historical global temperatures leads to changes in atmospheric patterns, with a subsequent intensification and increase in the frequency of heat waves and precipitation extremes (Avila-Diaz, Benezoli, Justino, Torres, & Wilson, 2020; Schaller et al., 2016). In addition, the IPCC in its most recent sixth assessment report (AR6) has dedicated a chapter on weather extremes (Seneviratne et al., 2021), further emphasizing the scientific consensus establishing a robust link between human influence on extreme events.

Extremes in weather and climate (Formally a "climate extreme," at times also referred to as "extreme weather" or an "extreme climate event," is "the occurrence of a value of a weather or climate variable above (or below) a threshold value near the upper (or lower) ends of the range of observed values of the variable" (Chapter 3 of the Special Report on Managing the Risks of Extreme Events and Disasters to Advance Climate Change Adaptation (SREX) of Working Groups I and II of the Intergovernmental Panel on Climate Change (IPCC) (Seneviratne et al., 2012)). Following the approach adopted by the IPCC SREX, for the sake of clarity, both "extreme weather events" and "extreme climate events" are collectively referred to as "climate extremes" in the remainder of this chapter. For avid readers, Section 2 of Morss, Wilhelmi, Meehl, and Dilling (2011) provides an interesting overview of the different definitions of weather and climate extremes), such as floods, droughts, heat and cold waves, hailstorms, and cyclones. These extremes can have significant societal, ecological, and economic impacts at various spatial and temporal scales (Easterling et al., 2000; Meehl et al., 2000). Not only are these extreme events as in the case of changes in the mean state of the climate heterogeneous across space and time (Alexander, 2016; Wartenburger et al., 2017), the social vulnerability to these extremes is also variable (Otto et al., 2017). A combination of heterogeneous extreme events fueled further by climate change, contrasting adaptive capacity of societies, and complex transmission channels in the wider global and regional economies, can not only foster inequality, but also disproportionately harm the already disadvantaged people (Cutter & Finch, 2008; Oppenheimer et al., 2014).

Climate Impacts on Extreme Weather
https://doi.org/10.1016/B978-0-323-88456-3.00005-8

Despite the scientific and technological advances in understanding the earth's climate system and the complex interactions between physical and human systems, such extreme events continue to cause substantial monetary and environmental loss and damages. Moreover, while rapid strides have been made in hazard mitigation efforts, social vulnerability and adaptive capacity of societies remain paramount in shaping the magnitude and directions of climate impacts on exposed communities (Morss et al., 2011). The impacts of climate extremes on society, such as the loss of lives and destruction of property, famine and other socioeconomic spillover effects (e.g., drop in agricultural productivity), can also culminate in human displacement and migration, and in more dire cases result in conflicts (Reuveny, 2007; Vesco, Kovacic, Mistry, & Croicu, 2021).

In spite of the ongoing improvements in mitigation and adaptive measures to address climate extremes, economic losses from extremes in weather and climate in recent decades have not only gone up dramatically, but catastrophic weather disasters in particular have also increased in frequency (Changnon, Pielke, Changnon, Sylves, & Pulwarty, 2000; Morss et al., 2011). Not surprisingly, researchers have increasingly identified climate extremes as one of the key challenges in climate studies (Sillmann et al., 2017), and as a natural outcome of such contributions to the scientific literature, a significant number of climate extremes datasets have also been assembled in recent years (e.g., Donat et al., 2013b; Mistry, 2019a). As a matter of fact, extremes in climate are of growing interest to a broad spectrum of experts within the climate change community. For instance, focusing on projections of changes in extremes, physical scientists have led the efforts of the climate modeling community to a better understanding of the dynamic processes embedded within climate extremes (Sillmann, Kharin, Zhang, Zwiers, & Bronaugh, 2013; Sillmann, Kharin, Zwiers, Zhang, & Bronaugh, 2013). On the other hand, social, environmental, and sectoral impacts experts have achieved considerable improvements in examining the impacts of changes and the potential for systems to cope and adapt to extreme weather events (Morss et al., 2011).

Along similar lines, the role of anthropogenic greenhouse gases (GHGs) in contributing to the observed intensification of extremes in climate is a topic gaining a lot of attention in recent years. Examples of such studies include Stott, Stone, and Allen (2004) for temperature, Min, Zhang, Zwiers, and Hegerl (2011) and Paik et al. (2020) for precipitation, and Diffenbaugh et al. (2017) for both temperature and precipitation extremes (Readers are guided to Stott et al. (2016) for an excellent review of literature and attribution of extreme weather and climate-related events). Moreover, the climate modeling community has also improved its understanding of how climate extremes are likely to change and evolve in a moderate and severe warming climate of the 21st century (Alexander & Tebaldi, 2012; Dosio, 2016; Sillmann, Kharin, Zhang, et al., 2013; Sillmann, Kharin, Zwiers, et al., 2013). In general, extremes in future climate are not only projected to increase in their intensity and frequency (IPCC Summary for Policymakers, 2012), but are also likely to be heterogeneous in space and time, depending on the climate scenario (Angélil et al., 2014; Seneviratne et al., 2021) and the availability of moisture in the atmosphere (Tabari, 2020). Yet as noted by Sillmann et al. (2017), "detecting, and even predicting the changes in large-scale circulation is a major challenge to be overcome in order to better predict changes in the odds of extreme events."

Analyses of climate extremes (includes attribution, understanding the drivers and mechanisms of climate extremes, as well as the sectoral impact assessment of climate extremes) fundamentally depends on the availability of high-quality records of the required input meteorological variables at subdaily or daily timescales (Seneviratne et al., 2021), usually bundled as so called "Climate Datasets." It is worth emphasizing though that while the role of suitable climate datasets (Section 3) remains at the heart of any climate impacts assessment study, equally important for understanding the complexity of extremes (complexity in climates extremes here refer to the associated feedbacks and teleconnections with broader scale atmospheric-oceanic circulations, and the physical mechanisms driving concurrent occurrences of extremes) are the "measures" or "indicators" of climate extremes, more commonly referred to as Climate Extremes Indices (CEIs) (Karl, Nicholls, & Ghazi, 1999; Zhang et al., 2011; Section 2). Generally, a climate extremes index (CEI) is a representation of a particular aspect of some data pertaining to extremes in climate. Put more formally, CEIs usually derived using one or more commonly available meteorological fields (namely near-surface daily temperature and precipitation), are measures of moderate or severe climate extremes, capable of representing some extreme aspect of a long-term climate record, as opposed to a mean aspect (Karl et al., 1999; Sillmann, Kharin, Zhang, et al., 2013). The CEIs not only play an important role in analysis of global and regional scale climate extremes, but also help the climatic modeling community and decision-makers in sectoral impacts assessment, mitigation, adaptation, risk reduction, and policy formulation and implementation (Mistry, 2019b; Mysiak et al., 2018). Driven by a desire for a better understanding of the evolution of extremes in a warming climate, researchers have also applied CEIs as benchmarking tools in climate model validation exercises (Alexander & Arblaster, 2017; Sillmann, Kharin, Zhang, et al., 2013; Sillmann, Kharin, Zwiers, et al., 2013).

Motivated largely by: (i) the demand of the climate sectoral-impact modelers and the wider scientific community; and (ii) the need for a robust definition of CEIs, international collaboration and efforts notably by the Expert Team on

Climate Change Detection and Indices [ETCCDI (Karl et al., 1999)]. The ETCCDI was formed by the World Climate Research Program (WCRP) of the World Meteorological Organization (WMO) Commission for Climatology (CCl) in 1999. For a historical backdrop of the WMO led initiatives leading to the ETCCDI and a summary of their research activities, readers are guided to Peterson and Manton (2008) and Zhang et al. (2011) for guidelines and recommendations for the usage of a broad suite of CEIs derived using meteorological fields. The ETCCDI led the first efforts in defining a set of CEIs that provide a comprehensive overview of temperature and precipitation statistics (Donat et al., 2013a, 2013b; Karl et al., 1999), and have played a leading role in developing software tools and routines for assembling CEIs (Alexander & Herold, 2016; Section 2). These efforts have also resulted in development of global and regional datasets of CEIs catering to the wider scientific and sectoral impacts modeling community (Section 4), again central to which is the availability of climate datasets offering quality-controlled, homogeneous (homogeneity in input climate data essentially implies that the observation records are not distorted from any unnatural cause, or put another way the time-series is consistent across the spatial domain of observations. Section 2.2.2 is dedicated to the various factors that can result in an inhomogeneous data, known methods to address this issue, and usage caution when compiling CEIs) global historical time-series of climatic variables, preferably at a high spatial resolution.

The rest of the chapter is organized as follows: Section 2 provides a brief overview of the ETCCDI and other CEIs. A discussion on the methodology commonly applied in constructing the CEIs, such as the required input meteorological fields, necessary quality controls and homogeneity checks, and the available software routines are also described in detail. Publicly available global and regional scale quality-controlled climate datasets commonly used as input data sources for the construction of CEIs are presented and discussed in Section 3. A comparison of spatiotemporal dimensions and the available meteorological variables in these datasets, and necessary caution when constructing CEIs find special mention in this section. Section 4 is devoted to a detailed discussion on the strengths and limitations of existing datasets of CEIs derived from both observations and model simulated data. The novelty, potential scope, application, and limitations of the datasets are central to this section. Recent advancements in understanding climate extremes, ongoing works and recommendations for further research conclude the chapter as Section 5.

2 CEIs: Definition, data quality control and methodology

Characterizing extremes under past and projected future climate has generated rapid interest dating back to the publication of the third assessment report of the IPCC in 2000 (Alexander, 2016). The CEIs have found wide-ranging applications both in climate change attribution studies (e.g., Angélil et al., 2014; Changnon et al., 2000; Min et al., 2011; Paik et al., 2020; Pall et al., 2011), as well as in sectoral climate impacts assessment [e.g., Li et al. (2018), Vicente-Serrano, Beguería, and López-Moreno (2010), and Wing, De Cian, and Mistry (2021) in agriculture, Vesco et al. (2021) and Wischnath and Buhaug (2014) in conflict, De Cian, Pavanello, Randazzo, Mistry, and Davide (2019), Guan (2009), Levesque et al. (2018), and Randazzo, De Cian, and Mistry (2020) in energy, and Antonelli, Coromaldi, Dasgupta, Emmerling, and Shayegh (2020), Bezerra et al. (2021), Orlov, Sillmann, Aaheim, Aunan, and de Bruin (2019), and Schleypen, Mistry, Saeed, and Dasgupta (2021) in labor productivity and labor supply). This is largely due to their simplicity, yet their utility as robust tools for impacts modelers and researchers in climate science (Alexander & Herold, 2016). For instance, subdaily and daily climatic records that are commonly required for computing CEIs are known to preserve the more—moderate to severe—"extreme" climate information, compared to the monthly climatology that can mask out important information on "extremes" which could be useful for sectoral impacts assessments (Sillmann et al., 2017; Zhang et al., 2011). Moreover, a large suite of CEIs based on percentiles, are by definition meaningful across different climate regimes. Even for the nonpercentile-based CEIs, applying conventional standardization allows researchers to compare results across time periods, regions and source datasets, thus facilitating consistent analyses at a global scale (Zhang et al., 2011).

2.1 Brief historical background of ETCCDI core and noncore indices

The ETCCDI has played a leading role in developing software tools and routines for assembling CEIs, often a computationally intensive exercise when utilizing high-resolution gridded or spatiotemporal data. Together with its offspring—the Expert Team on Sector-specific Climate Indices (ET-SCI) (https://climpact-sci.org/about/project/)—the ETCCDI has been active in the last two decades in outlining protocols and bridging the gap between climate science and impacts assessment communities. Their primary goal has been to "promote the use of globally consistent, sector-specific climate indices to bring out variability and trends in climate of particular interest to socio-economic sectors, and to help characterize the climate sensitivity of various sectors" (https://www.wmo.int/pages/prog/wcp/ccl/opace/archive/ccl-15/opace4/expertteam.php). The cumulative efforts of ETCCDI/ET-SCI and their collaborative partners have resulted in a steadily growing datasets of quality-controlled CEIs (Section 4), derived using input meteorological data from both historical observations and climate model projections (Section 3).

One of the preliminary accomplishments of the ETCCDI has been an internationally coordinated set of core climate indices consisting of 27 descriptive indices for moderate weather extremes (Alexander et al., 2006; Alexander & Herold, 2016; Zhang et al., 2011). Put in another way, CEIs characterize one or more extreme aspects of climate record at a given location. As opposed to describing the mean aspect of climate that is readily representable by standard statistical measures, CEIs provide standardization and meaning of extreme meteorological events, especially across different climate regimes. In addition, CEIs when applied as metrics, not only assist in the analysis of global and regional extremes in meteorological events, but also aid climate modelers and policymakers in the assessment of sectoral impacts.

The indices defined and developed by the ETCCDI/ET-SCI enable a comprehensive characterization of the frequency, amplitude, and persistence of extremes, and can be broadly categorized as indices based on "percentiles," "absolute," "threshold," and "duration" (the classification of the CEIs discussed above is based on the statistical characteristics of the indices. A more generic classification is based purely on the type of climate variable used in the construction of the index (e.g., temperature or precipitation based extremes), or the lower/upper tails of the probability distribution that the index characterizes (e.g. very cold or very warm extremes, low or high precipitation extremes). In general, the characteristics of extremes (such as the frequency, intensity, and duration) are the basis on which the extremes are defined and classified. Readers are guided to Bezerra et al. (2021), Menne and Williams (2009), Mysiak et al. (2018), and Sillmann et al. (2017) for further reading). The preliminary set of these 27 indices referred to as the "core indices" (Table 2.1) were developed keeping the detection and attribution needs of the research community in mind (Alexander et al., 2006; Zhang et al., 2011). As the discipline progressed, these 27 core indices began to be considered limited in scope; especially with regard to their usage in assessing sectoral impacts. To address the rising needs of sectoral impacts community, additional sector-relevant indices were subsequently recommended and developed by the ET-SCI (Alexander & Herold, 2016), referred to as "noncore" indices (https://www.wcrp-climate.org/data-etccdi.) (Table 2.2). It is worth emphasizing that these core and noncore indices can be compiled using both observations and climate model simulated data, and are applicable for assessing climate extremes at global as well as regional scales.

2.2 Input meteorological variables and quality checks for assembling the ETCCDI/ET-SCI suite of indices

2.2.1 *Subdaily or daily meteorological data for computing CEIs*

Given that the monthly averages of climate data can smooth over a lot of important information that is relevant for sectoral impacts assessment (Zhang et al., 2011), the ETCCDI set a framework for deriving indices from higher (temporal) frequency data. At the time of conceiving the framework for CEIs in the late 1990s, daily weather observations from both station-based and climate reanalysis data products were gaining momentum. CEIs (Tables 2.1 and 2.2) drawn from daily weather observations are useful to answer questions concerning aspects of the climate system that affect many human and natural systems, with particular emphasis on extremes (Zhang et al., 2011).

Assembling a consistent set of CEIs described in Tables 2.1 and 2.2 requires the following quality-controlled time-series of selective meteorological variables at daily timescales: (i) maximum near-surface air temperature (TX) [Units: ° Celsius], (ii) minimum near-surface air temperature (TN) [Units: ° Celsius], and (iii) total near-surface precipitation (PR) [Units: $\mathrm{kg\,m^2\,day^{-1}}$ or $\mathrm{mm\,day^{-1}}$]. It is important to mention that the indices discussed here are restricted to the CEIs defined by the ETCCDI/ET-SCI, and thus do not account for variables other than TN, TX, and PR. Indices constructed using a wider set of meteorological parameters (e.g., wind, humidity, solar radiation) therefore fall out of scope of this chapter.

In recent years with the advent of improved data assimilation systems, and a rapid expansion of storage space and computational resources, TN, TX and PR at subdaily time scales (e.g., hourly, 3 hourly) are readily available as high spatial resolution gridded data fields in the Network Common Data Form (NetCDF) file format (https://www.unidata.ucar.edu/software/netcdf/docs/netcdf_data_model.html) (discussed further in Section 3). For such subdaily gridded input data variables, command line tools such as the Climate Data Operators (CDO) (Schulzweida, 2018) and

TABLE 2.1 32 CORE ET-SCI indices.

Short name	Long name	Definition	Plain language description	Units	Time scale	Sector(s)
FD	Frost days	Number of days when TN < 0°C	Days when minimum temperature is below 0°C	Days	Mon/ Ann	H, AFS
TNlt2	TN below 2°C	Number of days when TN < 2°C	Days when minimum temperature is below 2°C	Days	Mon/ Ann	AFS
TNltm2	TN below −2°C	Number of days when TN < −2°C	Days when minimum temperature is below −2°C	Days	Mon/ Ann	AFS
TNltm20	TN below −20°C	Number of days when TN < −20°C	Days when minimum temperature is below −20° C	Days	Mon/ Ann	H, AFS
ID	Ice Days	Number of days when TX < 0°C	Days when maximum temperature is below 0°C	Days	Mon/ Ann	H, AFS
SU	Summer days	Number of days when TX > 25°C	Days when maximum temperature exceeds 25°C	Days	Mon/ Ann	H
TR	Tropical nights	Number of days when TN > 20°C	Days when minimum temperature exceeds 20°C	Days	Mon/ Ann	H, AFS
GSL	Growing Season Length	Annual number of days between the first occurrence of 6 consecutive days with TM > 5°C and the first occurrence of 6 consecutive days with TM < 5°C	Length of time in which plants can grow	Days	Ann	AFS
TXx	Max TX	Warmest daily TX	Hottest day	°C	Mon/ Ann	AFS
TNn	Min TN	Coldest daily TN	Coldest night	°C	Mon/ Ann	AFS
WSDI	Warm spell duration indicator	Annual number of days contributing to events where 6 or more consecutive days experience TX > 90th percentile	Number of days contributing to a warm period (where the period must be at least 6 days long)	Days	Ann	H, AFS, WRH
WSDId	User-defined WSDI	Annual number of days contributing to events where *d* or more consecutive days experience TX > 90th percentile	Number of days contributing to a warm period (where the minimum length is user-specified)	Days	Ann	H, AFS, WRH
CSDI	Cold spell duration indicator	Annual number of days contributing to events where 6 or more consecutive days experience TN < 10th percentile	Number of days contributing to a cold period (where the period must be at least 6 days long)	Days	Ann	H, AFS
CSDId	User-defined CSDI	Annual number of days contributing to events where *d* or more consecutive days experience TN < 10th percentile	Number of days contributing to a cold period (where the minimum length is user-specified)	Days	Ann	H, AFS, WRH
TXgt50p	Fraction of days with above average temperature	Percentage of days where TX > 50th percentile	Fraction of days with above average temperature	%	Mon/ Ann	H, AFS, WRH
TX95t	Very warm day threshold	Value of 95th percentile of TX	A threshold where days above this temperature would be classified as very warm	°C	Daily	H, AFS

Continued

TABLE 2.1 32 CORE ET-SCI indices—cont'd

Short name	Long name	Definition	Plain language description	Units	Time scale	Sector(s)
TMge5	TM of at least 5°C	Number of days when TM >= 5°C	Days when average temperature is at least 5°C	Days	Mon/ Ann	AFS
TMlt5	TM below 5°C	Number of days when TM < 5°C	Days when average temperature is below 5°C	Days	Mon/ Ann	AFS
TMge10	TM of at least 10°C	Number of days when TM >= 10°C	Days when average temperature is at least 10°C	Days	Mon/ Ann	AFS
TMlt10	TM below 10°C	Number of days when TM < 10°C	Days when average temperature is below 10°C	Days	Mon/ Ann	AFS
TXge30	TX of at least 30°C	Number of days when TX >= 30°C	Days when maximum temperature is at least 30°C	Days	Mon/ Ann	H, AFS
TXge35	TX of at least 35°C	Number of days when TX >= 35°C	Days when maximum temperature is at least 35°C	Days	Mon/ Ann	H, AFS
TXdTNd	User-defined consecutive number of hot days and nights	Annual count of *d* consecutive days where both TX > 95th percentile and TN > 95th percentile, where 10 >= *d* >= 2	Total consecutive hot days and hot nights (where consecutive periods are user-specified)	Events	Ann	H, AFS, WRH
GDDgrown	Growing degree Days	Annual sum of TM - *n* (where *n* is a user-defined location-specific base temperature and TM > *n*)	A measure of heat accumulation to predict plant and animal developmental rates	Degree-days	Ann	H, AFS
CDD	Consecutive dry days	Maximum number of consecutive dry days (when PR < 1.0 mm)	Longest dry spell	Days	Mon/ Ann	H, AFS, WRH
R20mm	Number of very heavy rain days	Number of days when PR >= 20 mm	Days when rainfall is at least 20 mm	Days	Mon/ Ann	AFS, WRH
PRCPTOT	Annual total wet-day PR	Sum of daily PR >= 1.0 mm	Total wet-day rainfall	mm	Mon/ Ann	AFS, WRH
R95pTOT	Contribution from very wet days	100*r95p/PRCPTOT	Fraction of total wet-day rainfall that comes from very wet days	%	Ann	AFS, WRH
R99pTOT	Contribution from extremely wet days	100*r99p/PRCPTOT	Fraction of total wet-day rainfall that comes from extremely wet days	%	Ann	AFS, WRH
RXdday	User-defined consecutive days PR amount	Maximum *d*-day PR total	Maximum amount of rain that falls in a user-specified period	mm	Mon/ Ann	H, AFS, WRH
SPI	Standardized Precipitation Index	Measure of "drought" using the Standardized Precipitation Index on time scales of 3, 6, 12, 24, 36 and 48 months. Calculated using the R SPEI package (Begueria & Vicente-Serrano, 2013)	A drought measure specified as a precipitation deficit	Unitless	Custom	H, AFS, WRH
SPEI	Standardized Precipitation Evapotranspiration Index	Measure of "drought" using the Standardized Precipitation Evapotranspiration Index on time scales of 3, 6, 12, 24, 36 and 48 months. Calculated using the R SPEI package (Begueria & Vicente-Serrano, 2013)	A drought measure specified using precipitation and evaporation	Unitless	Custom	H, AFS, WRH

Bold indicates index is also an ETCCDI index. *TX*, daily maximum near-surface air temperature; *TN*, daily minimum near-surface air temperature; *PR*, daily near-surface total precipitation; *H*, health; *AFS*, Agriculture and Food Security; *WRH*, Water Resources and Hydrology.
From *Alexander, L., Herold, N. (2016).* ClimPACT2 indices and software (R software package). *Available from: https://htmlpreview.github.io/?https://raw.githubusercontent.com/ARCCSS-extremes/climpact2/master/user_guide/ClimPACT2_user_guide.htm (Accessed 11 December* 2020*).*

TABLE 2.2 39 noncore ET-SCI indices.

Short name	Long name	Definition	Plain language description	Units	Time scale	Sector(s)
TXbdTNbd	User-defined consecutive number of cold days and nights	Annual number of *d* consecutive days where both TX < 5th percentile and TN < 5th percentile, where 10 >= *d* >= 2	Total consecutive cold days and cold nights (where consecutive periods are user-specified)	Events	Ann	H, AFS, WRH
DTR	Daily temperature range	Mean difference between daily TX and daily TN	Average range of maximum and minimum temperature	°C	Mon/Ann	
TNx	Max TN	Warmest daily TN	Hottest night	°C	Mon/Ann	
TXn	Min TX	Coldest daily TX	Coldest day	°C	Mon/Ann	
TMm	Mean TM	Mean daily mean temperature	Average daily temperature	°C	Mon/Ann	
TXm	Mean TX	Mean daily maximum temperature	Average daily maximum temperature	°C	Mon/Ann	
TNm	Mean TN	Mean daily minimum temperature	Average daily minimum temperature	°C	Mon/Ann	
TX10p	Amount of cool days	Percentage of days when TX < 10th percentile	Fraction of days with cool day time temperatures	%	Ann	
TX90p	Amount of hot days	Percentage of days when TX > 90th percentile	Fraction of days with hot day time temperatures	%	Ann	
TN10p	Amount of cold nights	Percentage of days when TN < 10th percentile	Fraction of days with cold night time temperatures	%	Ann	
TN90p	Amount of warm nights	Percentage of days when TN > 90th percentile	Fraction of days with warm night time temperatures	%	Ann	
CWD	Consecutive wet days	Maximum annual number of consecutive wet days (when PR >= 1.0 mm)	The longest wet spell	Days	Ann	
R10mm	Number of heavy rain days	Number of days when PR >= 10 mm	Days when rainfall is at least 10 mm	Days	Mon/Ann	
Rnnmm	Number of customized rain days	Number of days when PR >= *nn*	Days when rainfall is at least a user-specified number of mm	Days	Mon/Ann	
SDII	Daily PR intensity	Annual total PR divided by the number of wet days (when total PR >= 1.0 mm)	Average daily wet-day rainfall intensity	mm/day	Ann	
R95p	Total annual PR from heavy rain days	Annual sum of daily PR > 95th percentile	Amount of rainfall from very wet days	mm	Ann	
R99p	Total annual PR from very heavy rain days	Annual sum of daily PR > 99th percentile	Amount of rainfall from extremely wet days	mm	Ann	
Rx1day	Max 1-day PR	Maximum 1-day PR total	Maximum amount of rain that falls in one day	mm	Mon/Ann	
Rx5day	Max 5-day PR	Maximum 5-day PR total	Maximum amount of rain that falls in five consecutive days	mm	Mon/Ann	

Continued

TABLE 2.2 39 noncore ET-SCI indices—cont'd

Short name	Long name	Definition	Plain language description	Units	Time scale	Sector(s)
HWN (EHF/ Tx90/ Tn90)	Heat wave number (HWN) as defined by either the Excess Heat Factor (EHF), 90th percentile of TX or the 90th percentile of TN	The number of individual heat waves that occur each summer (Nov.-Mar. in southern hemisphere and May-Sep in northern hemisphere). A heat wave is defined as 3 or more days where either the EHF is positive, TX > 90th percentile of TX or where TN > 90th percentile of TN. Where percentiles are calculated from base period specified by user. See Perkins and Alexander (2013) for more details	Number of individual heat waves	Events	Ann	H, AFS, WRH
HWF (EHF/ Tx90/ Tn90)	Heat wave frequency (HWF) as defined by either the Excess Heat Factor (EHF), 90th percentile of TX or the 90th percentile of TN	The number of days that contribute to heat waves as identified by HWN. See Perkins and Alexander (2013) for more details	Total number of days that contribute to individual heat waves	Days	Ann	H, AFS, WRH
HWD (EHF/ Tx90/ Tn90)	Heat wave duration (HWD) as defined by either the Excess Heat Factor (EHF), 90th percentile of TX or the 90th percentile of TN	The length of the longest heat wave identified by HWN. See Perkins and Alexander (2013) for more details	Length of the longest heat wave	Days	Ann	H, AFS, WRH
HWM (EHF/ Tx90/ Tn90)	Heat wave magnitude (HWM) as defined by either the Excess Heat Factor (EHF), 90th percentile of TX or the 90th percentile of TN	The mean temperature of all heat waves identified by HWN. See Perkins and Alexander (2013) for more details	Average temperature across all individual heat waves	°C (°C^2 for EHF)	Ann	H, AFS, WRH
HWA (EHF/ Tx90/ Tn90)	Heat wave amplitude (HWA) as defined by either the Excess Heat Factor (EHF), 90th percentile of TX or the 90th percentile of TN	The peak daily value in the hottest heat wave (defined as the heat wave with highest HWM). See Perkins and Alexander (2013) for more details	Hottest day of the hottest heat wave	°C (°C^2 for EHF)	Ann	H, AFS, WRH
CWN_ECF	Cold wave number (CWN) as defined by the Excess Cold Factor (ECF)	The number of individual "coldwaves" that occur each year. See Nairn and Fawcett (2015) for more information	Number of individual coldwaves	Events	Ann	H, AFS, WRH
CWF_ECF	Cold wave frequency (CWF) as defined by the Excess Cold Factor (ECF)	The number of days that contribute to "coldwaves" as identified by ECF_HWN. See Nairn and Fawcett (2015) for more information	Total number of days that contribute to individual coldwaves	Days	Ann	H, AFS, WRH
CWD_ECF	Cold wave duration (CWD) as defined by the Excess Cold Factor (ECF)	The length of the longest "coldwave" identified by ECF_HWN. See Nairn and Fawcett (2015) for more information	Length of the longest coldwave	Days	Ann	H, AFS, WRH
CWM_ECF	Coldwave magnitude (CWM) as defined by the Excess Cold Factor (ECF)	The mean temperature of all "coldwaves" identified by ECF_HWN. See Nairn and Fawcett (2015) for more information	Average temperature across all individual coldwaves	°C^2	Ann	H, AFS, WRH
CWA_ECF	Coldwave amplitude (CWA) as defined by the Excess Cold Factor (ECF)	The minimum daily value in the coldest "coldwave" (defined as the coldwave with lowest ECF_HWM). See Nairn and Fawcett (2015) for more information	Coldest day of the coldest coldwave	°C^2	Ann	H, AFS, WRH

Bold indicates index is also an ETCCDI index. Sectoral abbreviations same as in Table 2.1.
From *Alexander, L., Herold, N. (2016).* ClimPACT2 indices and software (R software package). *Available from: https://htmlpreview.github.io/?https://raw.githubusercontent.com/ARCCSS-extremes/climpact2/master/user_guide/ClimPACT2_user_guide.htm (Accessed 11 December 2020).*

the NetCDF Command Operators (NCO) (Zender, 2008) are generally applied to aggregate subdaily meteorological variables to daily fields prior computation of the CEIs.

2.2.2 *Quality control and homogenization of input data*

While the use of data recorded on (at least) daily timescales is a prerequisite for computing the CEIs, the data if not quality-controlled (e.g., checked for systematic errors such as negative values for precipitation, or large number of outliers or missing observations in the sample of the data, etc.), and not homogenized, can lead to spurious results. In fact as emphasized by Della-Marta and Wanner (2006) ("all climate data have to be scrutinized thoroughly before they can be used to assess long-term changes in variability"), any changes in the conditions of measurement that can result in artificial shifts in the observed climate time-series, need to be treated with sophisticated statistical techniques and used with caution in climate research (Caussinus & Mestre, 2004). A thorough understanding of homogeneity of data in particular is therefore required before moving further.

Homogeneity of long-term time-series of meteorological variables is perhaps the single-most crucial factor affecting the reliability and robustness of results pertaining to climate extremes. Homogeneity implies that the observation records do not suffer from artificial bias. Numerous factors can influence the homogeneity of the input climate data, thus affecting the robustness of developed indicators for climate impacts and risks assessment. Some of these factors include changes in site location of the measurement instrument, as well as and the time of the day when the measurements are recorded. Additionally, changes in the method used to calculate daily-averages of totals of temperature and precipitation respectively are other common causes. For a comprehensive discussion on the potential factors causing inhomogeneity in input datasets, readers are guided to an excellent review by Jones (2016) and Alexander (2016).

Several improvements have been made in homogenization techniques in recent years that correct the artificial shifts (or breakpoints) in the times-series of input climate data variables, as well as in the generated extreme indicators when such shifts are difficult to correct in the input data itself (Precipitation is generally more difficult to treat for inhomogeneity due to lack of reliable methods, as well as due to the fundamental complexity of its distribution in space and time (Brown, Bradley, & Keimig, 2010)). A detailed discussion on different approaches used in homogenization of climate data falls outside the scope of this chapter [interested readers are guided to Brown et al. (2010), Della-Marta and Wanner (2006), Menne and Williams (2009), and Squintu, van der Schrier, Štěpánek, Zahradníček, and Tank (2020)]. Nevertheless, it is important to highlight that while modern homogenization techniques do make the resulting dataset usable for examining long-term trends in the mean state of the climate at a location (Brown et al., 2010; Trewin, 2013), drawing robust conclusions about changes in climate extremes and variability becomes more challenging to address (Brown et al., 2010). Thus, even the most sophisticated homogenization techniques currently available are not necessarily a panacea for treating spurious data that usually arise from long time-series of global station records (Sections 3.1 and 3.2).

2.3 Available software routines for computing ETCCDI/ET-SCI defined indices

Before moving on to the commonly used input data sources in compiling the CEIs (Section 3) and discussing the readily available datasets of global and regional CEIs (Section 4), a brief overview of the available software routines used in the computation of the CEIs is provided here. The reasons are twofold. First, understanding the software tools capable of handling the input data variables (types/formats) is as important for a user to be aware, as are the potential limitations of the input datasets if any. Second, and more importantly, the inherent discrepancy in a particular index compiled using the same underlying input data source can emanate more than often due to the choice of the software routine. Knowing the right tool for the right task is therefore an integral component when assembling datasets of CEIs.

A growing body of software environments tailored for the climate modeling are available as open-source tools. While such resources are generally applicable for the modeling, analysis, and visualization of climate data, only a handful of such software environments facilitate computation of the CEIs by way of built-in libraries. In addition, users have to pay attention to the methodology used in the computation of the CEIs. For instance, for percentile-based indices in particular, the definition used by the software routine in the computation of an index may not necessarily adhere to the definition (Percentile-based indices (Tables 2.1 and 2.2) are documented to be most sensitive to the method of computation. See Mahlstein, Spirig, Liniger, and Appenzeller (2015) for further details.) recommended and adapted by the ETCCDI/ET-SCI. Table 2.3 summarizes the open-source software routines (Other available softwares such as Matlab are omitted in the discussion as the discussion is limited to open-source softwares. Moreover, the discussion is restricted to software environments with compiled libraries/routines designed for computing the CEIs. While in principle it is possible and even computationally faster to compute the CEIs in compiled programming languages such

TABLE 2.3 Available open-source software environments and routines facilitating computation of core and/or noncore CEIs described in Tables 2.1 and 2.2.

Software	Supported operating systems	Software libraries	Supported input data file formats	Indices directly implemented[a]	Features and known limitations	References/web resources
R[b]	Windows, Linux, MacOS	ClimPACT[c] ClimInd	Ascii (.txt, .csv), netCDF (.nc)	All CEIs listed in Tables 10.1 and 10.2	ClimPACT is the ETCCDI/ET-SCI recommended package. Extensive data quality checks implemented. netCDF is not supported on Windows version. ClimInd includes a large number of climate indices in addition to most ETCCDI indices	Alexander and Herold (2016)/https://climpact-sci.org/get-started/ Reig-Gracia, Vicente-Serrano, Dominguez-Castro, and Bedia-Jiménez (2021)/https://rdrr.io/cran/ClimInd/
Python[d]	Windows, Linux, MacOS	ICCLIM Xclim	NetCDF (.nc, .nc4)	Only core CEIs in Tables 10.1 and 10.2	Additional non ETCCDI/ET-SCI indices also implemented. Data quality checks not part of the libraries	ICCLIM: https://icclim.readthedocs.io/en/latest/ xclim: https://xclim.readthedocs.io/en/stable/indices.html
CDO	Windows (via Cygwin or Linux emulator), Linux, MacOS	Combination of inbuilt functions	NetCDF (.nc, .nc4), grib (.gb)	Only core CEIs in Tables 10.1 and 10.2	Command line tool. Additional non ETCCDI/ET-SCI indices also implemented. Data quality checks not part of the functions. Definition of percentile differs to recommended definition of ETCCDI[e]	Schulzweida (2018)/https://code.mpimet.mpg.de/projects/cdo/embedded/cdo_eca.pdf
Julia[f]	Windows, Linux, MacOS	ClimateTools.jl	NetCDF (.nc, .nc4)	Only core CEIs in Tables 10.1 and 10.2	Additional non ETCCDI/ET-SCI indices also implemented. Data quality checks not part of the libraries	https://juliaclimate.github.io/ClimateTools.jl/stable/indices/

Softwares and routines for computing CEIs.

[a] *Direct computation here implies existing functions in the software facilitate computation of the CEI without other combination of functions.*

[b] *https://www.r-project.org/.*

[c] *Supersedes "RClimDex." Implements a number of R packages, including "SPEI" and "climdex.pcic." More details in package documentation. A web-based application built using R shinny app for computing CEIs using station observations is also available at https://ccrc-extremes.shinyapps.io/climpact/.*

[d] *https://www.python.org/.*

[e] *The definition of percentiles implemented in CDO does not follow the bootstrapping approach adopted in the R ClimPACT package.*

[f] *https://julialang.org/.*

as C, C++ and FORTRAN, no known libraries exist for direct implementation of the CEIs.) available for compiling the CEIs, providing an overview of their strengths and limitations, and some guidance notes for users.

3 Historical climate data for assembling the CEIs

Central to the construction of a quality-controlled homogenous historical dataset of CEIs is the availability of a long-term, continuous, and consistent high-quality record of the required input meteorological variables at subdaily or daily timescales (Prettel, 2011). Apart from homogeneity in the data source, the choice of underlying data sources can influence the robustness of assembled CEIs (Karl et al., 1999). A nonexhaustive list of such characteristics in the data includes the temporal and spatial scale resolution of the dataset from which meteorological fields are drawn, whether the drawn variables are measurements from in situ observations, or from a blend of observations and model simulated data commonly referred to as data assimilation systems or reanalysis data products (Sections 3.1 and 3.3). Interested readers are also guided to Oppenheimer et al. (2014) for further details.

As discussed in the following subsections, the indicators defined and developed for representing climate extremes often require only three key measures of daily weather. However, the diurnal cycle of meteorological variables can differ greatly depending on the type, location (e.g., distance from the coast or height above sea-level) and geographic terrain or location (e.g., leeward side of a mountain). Even though in recent years there has been a rise in availability of publicly available meteorological data sources, understanding the finer details of how such records are assembled is therefore important before its usage.

Another vital point to remember when using any climate dataset for studying climate extremes relates to the high frequency (subdaily/daily) data required for compiling the CEIs. As presented in Tables 2.1 and 2.2, a large number of indices are based on thresholds, defined as "consecutive" (e.g., CDD, CWD, HWD) and/or also "percentiles" (e.g., CSDI, WSDI). While in general, the different types of datasets discussed in the following subsections agree on the average state of the weather recorded at a location (i.e., long-term climatology), the agreement in the timing, magnitude and frequency of extreme weather, and intra-seasonal variation in the recorded observations of meteorological fields are not always consistent with reality. Mahlstein et al. (2015) discusses the importance of accurate estimation of the thresholds in observations based datasets.

Generally, the input data variables from observations can emanate from seven different sources: (i) Ground station as single point location or network of locations; (ii) Gridded data derived from network of locations; (iii) Reanalysis data; (iv) Satellite- or Remote Sensing-derived data; (v) Reconstruction data; (vi) Merged data sources (gauge, satellite, and reanalysis), and (vii) Global-retrospective-meteorological forcing data.

Among these seven data sources, data sources "(i)-(iii)" are more commonly employed in constructing CEIs and are therefore discussed in detail in the following Sections 3.1 to 3.3. On the contrary, satellite-derived estimates of meteorological fields (data source "(iv)") resolve the data a high spatial resolution only from the early 2000s onward (from newer satellite sensors) (Dell, Jones, & Olken, 2014), are generally less reliable (particularly for precipitation over land), and are therefore less suited and rarely applied in examining climate extremes. Moreover, remotely sensed temperature are essentially indirect estimates of temperature that do not represent the true near-surface temperature which is generally measured at about 2 meters above ground level (Rayner et al., 2020; Fig. 2.1). Perhaps, the biggest limitation of a satellite derived measurement of temperature is its dependence on a clear sky. Presence of clouds over a location does not facilitate measurement of the surface temperature. While their scope in examining climate extremes may be limited, and thus omitted from the discussion herein, satellite-derived data do find applications as "ingested" data in other data sources (Sections 3.2 and 3.3). For avid readers, a recent project "European Union (EU) Surface Temperature for All Corners of Earth (EUSTACE) (www.eustaceproject.org)" (Rayner et al., 2020) provides further details on satellite derived products and protocols.

Also not included in the discussion here are data sources "(v)–(vii)." Reconstructions of past climate conditions ("v") derived from paleoclimatology proxies are not at daily timescales and hence omitted. Merged data ("vi") on the other hand are essentially a combination of different data sources "i–iv," derived by merging gauge, satellite, and reanalysis data products. They are relatively newer data products and only a handful such are currently made available to the wider research community (Though not discussed here in detail, an example of such a dataset ("MSWEP") is mentioned in Table 2.4.). Lastly, Global-retrospective-Meteorological Forcing (GMF) data ("vii"), which are derived using multiple observational data sources and post processing techniques (e.g., bias correction). Since GMF datasets (GMFs are widely used for driving Global Climate Models (GCMs) with a consistent set of bias-corrected input meteorological parameters, thus permitting a direct comparison of model outputs) are a combination of one or more data sources listed in Table 2.4 and not frequently applied in constructing CEIs, they are reserved for discussion in Section 5.2.

FIG. 2.1 Schematic of skin temperature typically measured by satellites, contrary to actual near-surface air temperature as a variable of interest. Reproduced with permission from the EU Horizon 2020 funded EU Surface Temperature for All Corners of Earth (EUSTACE) project; *https://www.eustaceproject.org/users/why-a-new-temperature-dataset/*.

TABLE 2.4 Key global datasets providing subdaily or daily input variables (TN, TX or PR) required for compiling CEIs.

Data source (version)	Type[a]	Variables	Period[b]	Spatial resolution (Lat × Lon)[c]	Brief description	Link/source of data
BEST	SO, GCD	TN, TX	1880–Present	1.0° × 1.0° (for GCD)	The daily data is referred to as "experimental" data, i.e., a new product in a late stage of development. While it is a homogeneous dataset, homogenization is not perfect (it is achieved with respect to the mean and not the variance). The procedure thus does not capture all changes in variability. Along with the gridding/kriging approach used for interpolation, the dataset while superior to GHCN (discussed below), should be used with caution when studying extremes and variability	Rohde, Muller, Jacobsen, Muller, and Wickham (2013)/ http://berkeleyearth.org/data/
CHIRPS (ver. 2)	GCD	PR	1981–Present	0.05° × 0.05°	A quasiglobal dataset spanning latitudes 50°S–50°N (and all longitudes). Rainfall estimates are from rain gauge and satellite observations using sophisticated interpolation algorithm	Funk et al. (2015)/https://www.chc.ucsb.edu/data/chirps
CPC	GCD	TN, TX, PR	1979–Present	0.5°×0.5°	Available as two separate datasets as "TN, TX" and "PR." PR is global daily unified gauge-based analysis of precipitation. Datasets Gridded using the Shepard algorithm. Observations are recorded at 0600 Coordinated Universal Time (UTC)	https://psl.noaa.gov/data/gridded/data.cpc.globalprecip.html https://psl.noaa.gov/data/gridded/data.cpc.globaltemp.html
ECMWF-ERA5	Reanal.	TN, TX, PR	1950–Present[d]	0.25° × 0.25°	Only reanalysis product (along with ERA5 Land) providing data at hourly timesteps. Precipitation is accumulated hourly, thus daily values need to be aggregated using the hourly values	Hersbach et al. (2020)/https://cds.climate.copernicus.eu/cdsapp#!/dataset/reanalysis-era5-single-levels?tab=overview
ECMWF-ERA5-Land	Reanal.	TN, TX, PR	1981–Present	0.09° × 0.09°	Global land-surface dataset, consistent with atmospheric data from the ERA5 reanalysis from 1950 onwards. For locations near water bodies (e.g., lakes, rivers, seas), the gridded data may report missing observations, thus requiring spatial interpolation. Back extension (1950–80) will be made available as in the case of ERA5. PR at 24h UTC is the accumulated PR in the preceding 24h	Muñoz-Sabater et al. (2021)/https://cds.climate.copernicus.eu/cdsapp#!/dataset/reanalysis-era5-land?tab=overview
EUSTACE (ver. 1.1)	SO, GCD	TN, TX	1850–2015	0.25° × 0.25°	Unlike dynamical reanalysis, temperature is independent from numerical weather prediction models and extends further back in time. Data are assembled from seven different sources, are quality-controlled and come with a large amount of additional information on homogeneity. Daily fields are presented as ensembles. Dataset is not planned to be kept updated	Brugnara, Good, Squintu, Schrier, and Brönnimann (2019) and Rayner et al. (2020)/https://catalogue.ceda.ac.uk/uuid/7925ded722d743fa8259a93acc7073f2
GHCN-D (ver. 3)	SO	TN, TX, PR	1880–Present	NA	Raw daily climate observations are integrated from nearly 30 different data sources and near 100,000 stations (in 180 countries and territories). Data are quality-controlled but not homogenized, i.e., localized biases like station moves, time of observation changes, and instrument changes are not systematically corrected using homogeneity routines. Because these issues can	Menne, Durre, Vose, Gleason, and Houston (2012)/ https://www.ncei.noaa.gov/access/metadata/landing-page/bin/iso?id=gov.noaa.ncdc:C00861

					introduce significant biases, the dataset is not suited for looking at long-term trends. Moreover, the differing temporal evolution of the station network results in a relatively small number of stations making daily summaries available prior 1890. PR is available across more station records compared to TN and TX	
GLDAS	Reanal.	TN, TX, PR	1948–Present[e]	0.25°×0.25°	Satellite- and ground-based observational data ingested using advanced land surface modeling and data assimilation techniques. Users are cautioned when combining the data from the two versions as the merged time-series may not be homogeneous (see Mistry, 2019a, 2019c, for details)	Rodell et al. (2004)/https://ldas.gsfc.nasa.gov/gldas
GPCC (ver. 2020)	GCD	PR	1982–2019	1.0° × 1.0°	Daily Land-Surface Precipitation from Rain-Gauge. The dataset provides a gridded quantification for the following errors: (i) systematic gauge-measuring error (due to under catch of the true PR); and (ii) stochastic sampling error (due to a sparse network density). Latency period is longer compared to CHIRPS	Schamm et al. (2014)/https://opendata.dwd.de/climate_environment/GPCC/html/fulldata-daily_v2020_doi_download.html
HadGHCND	GCD	TN, TX	1950–Present	2.5° × 3.75°	Gridded version of GHCN-D	Caesar, Alexander, and Vose (2006)/https://www.metoffice.gov.uk/hadobs/hadghcnd/
HadISD (ver. 3.1.2.202105p)	SO	TN, TX[f], PR	1931–Present	NA	Based on NOAA-NCEI's ISD dataset. Quality-controlled using automated algorithm with special attention to retaining extreme values. Though data is not homogenized, homogeneity assessment is provided as an annual update process (see discussion paper Dunn, Willett, Parker, and Mitchell (2016) for further details). Dataset is updated frequently	Dunn et al. (2012, 2016)/https://www.metoffice.gov.uk/hadobs/hadisd/
MERRA-2	Reanal.	TN, TX, PR	1980–Present	0.5° × 0.625°	Developed by NASA-GMAO. In addition to raw PR, bias-corrected PR is also available	Randles et al. (2017)/https://disc.gsfc.nasa.gov/datasets/M2SDNXSLV_5.12.4/summary
MSWEP[g] (ver. 2.8)	Merg.	PR	1979–Present	0.1° × 0.1°	Derived using rain gauge, satellite, and reanalysis data to provide the highest quality precipitation estimates. Near real-time estimates are available with a latency of ~3h	Beck et al. (2019)/http://www.gloh2o.org/mswep/
NCEP-NCAR reanalysis 1	Reanal.	TN, TX, PR	1948–Present	2.5° × 2.5°	The data assimilation pattern changes in postsatellite era (1979–present), thus making the reanalysis inconsistent, though considered scientifically valid. Moreover, data from 1948 to 1957 was ingested at different subdaily timescales. NCEP-NCAR recommends using this data for ancillary purpose and not as primary research dataset. The dataset is mentioned here as it has been widely used in earlier research	Kalnay et al. (1996)/https://psl.noaa.gov/data/reanalysis/reanalysis.shtml
NCEP DOE REANALYSIS 2	Reanal.	TN, TX, PR	1979–Present[h]	2.5°×2.5°	Data begins in 1979 coinciding with the start date of modern satellite weather era. Despite the main objective to correct known errors in the Reanalysis 1[i] and update the parameterizations of the physical processes, southern hemisphere is poorly modeled. The spatial resolution	Kanamitsu et al. (2002)/https://psl.noaa.gov/data/gridded/data.ncep.reanalysis2.html

Continued

TABLE 2.4 Key global datasets providing subdaily or daily input variables (TN, TX or PR) required for compiling CEIs—cont'd

Data source (version)	Type[a]	Variables	Period[b]	Spatial resolution (Lat × Lon)[c]	Brief description	Link/source of data
					remains the same as in Reanalysis 1. Overall considered as a "first generation" product. NCEP-NCAR recommends using this data for ancillary purpose and not as primary research dataset. The dataset is mentioned here as it has been widely used in earlier research	
NOAA/ NCEP-CFSR (ver. 1 and 2)	Reanal.	TN, TX, PR	1979–Present[j]	0.3125°× 0.3125° (ver. 1) 0.5°×0.5°	Third generation reanalysis product. Though the data assimilation is upgraded in version 2, the model used in both versions remains the same	Ver. 1: Saha et al. (2010)/https://rda.ucar.edu/datasets/ds093.0/ Ver. 2: Saha et al. (2014)/https://rda.ucar.edu/datasets/ds094.0/
NOAA-CIRES-DOE (ver. 3)	Reanal.	TN, TX, PR	1836–2015	1.0° × 1.0°	20th century Reanalysis (20CR) provides analysis and estimates of associated uncertainty. Along with the earlier version (v2c), 20CRv3 is the first reanalysis dataset providing subdaily global data spanning over 100 years[k]. Likely to be affected by changes in the observation and modeling system, thus potentially introducing spurious changes, especially in regions of sparse coverage like the Southern Hemisphere	Slivinski et al. (2019)/https://psl.noaa.gov/data/gridded/data.20thC_ReanV3.html
JRA-55	Reanal.	TN, TX, PR	1958–Present	1.25°×1.25°	Apart from JRA-55 which is a full observing system reanalysis, the following two other products are also available in the JRA-55 family of products: (i) Fixed observing system reanalysis, and (ii) Without observation assimilation	Kobayashi et al. (2015)/https://jra.kishou.go.jp/JRA-55/index_en.html

See Table A2.1 in Appendix for abbreviations and originating country of the data source.

[a] GCD, *gridded climate data;* Merg, *merged data;* Reanly, *reanalysis;* SO, *station observations.*

[b] *"–Present" indicates that the dataset is generally available with a latency time period of about few days to 3 months. Readers are recommended to refer to the individual dataset documentation for further details.*

[c] *Where the data product is available at more than one spatial resolution, the finest scale is reported here. Readers are referred to the documentation for availability of data at other resolutions.*

[d] *The preliminary ERA5 dataset from 1950 to 1978 is available on the Climate data store (CDS): https://cds.climate.copernicus.eu/cdsapp#!/search?text=ERA5%20back%20extension&type=dataset. Earlier versions of the ECMWF reanalysis data products, namely ERA-interim and ERA40 are not discussed here as these are superseded by ERA5.*

[e] *Version 2.0 covers 1948–2014. Version 2.1 covers 2000–Present.*

[f] *Temperature is provided as subdaily mean temperature, which can then be aggregated to TN, TX (daily) for computation of CEIs.*

[g] *Multisource Weather (MSWX) is another high-resolution data product from the same data source providing TN, TX and PR. The reference paper is currently in preparation. Both MSWEP and MSWX can also be referred to as meteorological forcing datasets (data source "vii"), discussed further in Section 5.2.*

[h] *Plan exists to go back to 1950s.*

[i] *National Center for Atmospheric Research Staff (Eds). Last modified 08 May 2021. "The Climate Data Guide: NCEP Reanalysis (R2)." Retrieved from https://climatedataguide.ucar.edu/climate-data/ncep-reanalysis-r2.*

[j] *Ver. 1 covers 1979–March 2011, Ver. 2 covers 2011–Present.*

[k] *ECMWF's Coupled Ocean-Atmosphere Reanalysis of the 20th Century (CERA-20C) (not discussed here), is another comparable data product spanning 1901–2010. However, it does not provide the best estimate of the climate during the postsatellite era (when more comprehensive observations are available), as no upper air and satellite observations are assimilated.*

FIG. 2.2 Example of a ground station-based observation part of the UK Met Office (UKMO) "HadUK-Grid" surface monitoring network (see also Fig. 2.3 for wider network of ground station-based observations). The instruments are generally located in an open field away sufficiently far from any potential artificial influence or obstructions that can contaminate measurements, e.g., turbulence from roadside traffic/airport runway, shadow of a building structure, etc. Image from *HadUK-Grid—UKMO (https://www.metoffice.gov.uk/research/climate/maps-and-data/data/haduk-grid/haduk-grid). Readers are also guided to http://www.jma.go.jp/jma/en/Activities/surf/surf.html for an overview of Surface Weather Observing System.*

The remainder of this section begins by providing an overview of the three data sources (i–iii) with a detailed description of their key strengths and limitations (Sections 3.1–3.3). Examples of various global and regional climate datasets based on these sources are discussed in Section 3.4.

3.1 Station-derived meteorological observations (point-based and network)

Station-level weather data also referred to as "Station Observations," "Station Records," "Ground Observations," or simply "In situ measurements," have its origins dating back to the early 1800s when systematic monitoring of weather first began in Europe (Auffhammer, Hsiang, Schlenker, & Sobel, 2013). Based on near-surface instrument measurements of subdaily or daily meteorological fields, most often temperature and rainfall (measured by way of in situ weather loggers, sensors, rain-gauge, and radar instrumentation, etc., Fig. 2.2), station observations are at the heart of a region's meteorological observations network.

Many countries have a national body (usually the National Meteorological Agency) overseeing the network of stations spanning its geographical boundaries. Examples of such national meteorological organizations include: The UK Met Office (UKMO), Indian Meteorological Department (IMD), Japanese Meteorological Agency (JMA), Kenya Meteorological Department (KMD). Apart from being responsible for archiving the instrumentation records at different locations, the national meteorological agencies together with their designated regional partners are also responsible for quality control checks and maintenance of measurement instruments to ensure that the gaps in missing observations are avoided. Station records in general have an advantage of providing an accurate measurement of the "actual" location's climate. In developed countries (e.g., United States and United Kingdom), the network of ground stations is usually dense and the instrumentation records well documented to assess the homogeneity of records. For such regions, the station based observations can facilitate a fairly disaggregated level of analysis (Dell et al., 2014).

However, it is inevitable though that different measurement sites can still end up having differing lengths of observation records, as well as missing observations due to mechanical downtime, and inconsistency in time-series, e.g., due to relocation of an instrument to a different site, etc.; drawbacks generally requiring caution when applied for studying extremes in climate. Such gaps in observations are also routinely filled up using other so called "proxy" observational sources, generally by way of satellite/radar estimates, with the resulting time-series referred to as "reconstructed" time-series. Users therefore need to pay special attention in understanding how such blend of data sources can cause a potential bias in measured records, and affect the homogeneity of observation records. Moreover, even when the observations are complete and measured by the same instrument throughout the life of the weather observatory, measurement errors are still possible, for example due to inclement weather, such as strong winds affecting the true accumulation of rainfall in a rain-gauge (Goodison, Louie, & Yang, 1998).

3.2 Gridded climate data

Even when the instrumentation records across the network of observations are complete and reliable, geographical proximity of the measurements to the location of interest is often not guaranteed, especially in developing countries

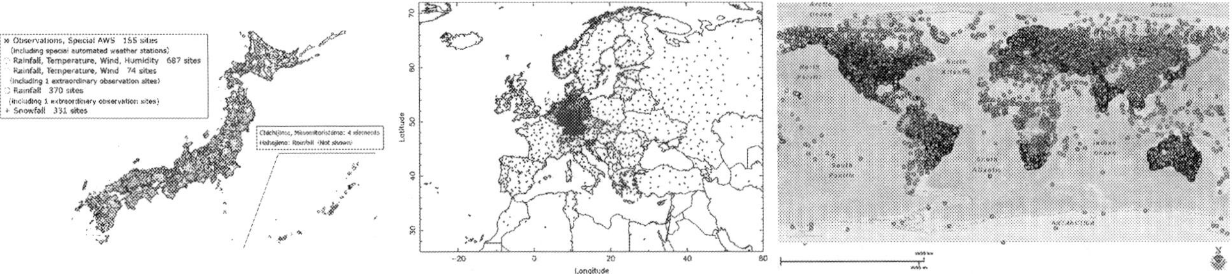

FIG. 2.3 Location of stations used in three different GCDs. (A) Japan Meteorological Agency. (B) European Climate Assessment & Dataset (ECA&D) gridded product E-OBS. (C) Global Historical Climatological Network-Daily (GHCN-Daily). Image *(A)* from *https://www.jma.go.jp/jma/en/Activities/observations.html; (B)* from van der Schrier, G., van den Besselaar, E. J. M., Klein Tank, A. M. G., and Verver, G. (2013), Monitoring European average temperature based on the E-OBS gridded data set. Journal of Geophysical Research: Atmospheres, 118, 5120–5135, doi:10.1002/jgrd.50444; *(C)* from NCEI-NOAA: *https://www.ncei.noaa.gov/access/metadata/landing-page/bin/iso?id=gov.noaa.ncdc%3AC00861.*

where infrastructure and funding limitations can result in sparse coverage. In such instances, especially in regions with sparse rain-gauge or radar network, satellite derived estimates for rainfall are also commonly used as "merged satellite gauge rainfall" daily data. It is worthy noting that in contrast to filling the temporal gaps in observations as discussed above for station observations, the data filling here is in the spatial dimension.

In addition, drawing from the limitations of sparse in situ measurements, the national meteorological divisions and the wider global climate community. For global datasets, both station networks and the gridded versions based on them, the underlying data are provided by national meteorological and hydrological services, as well as WMO partner organizations. have in recent years developed several "Gridded Climate Datasets" (GCDs), also referred to as "Gridded Meteorological Datasets." The underlying objective of such GCDs is to address the key shortcoming of spatial coverage of ground station observations, both within countries and spanning global scales. Using sophisticated statistical interpolation algorithms and bias correction techniques, these GCD spatially resolve the observations, or put simply, fill gaps between in situ measurements (Fig. 2.3A–C). The resulting data typical in NetCDF or GeoTiff file formats are recorded at each grid-cell (pixel) spanning the spatial extent of the regional or global station observations network, and have consistent start and end dates in the time-series at each grid-cell (i.e., a balanced panel of observations). This is often the single biggest advantage vis-à-vis the network of station observations discussed in Section 3.1, where in the station entry and exit dates can vary.

The various research groups making GCDs of surface temperature and precipitation available in principle utilize similar global network of land and sea surface observations. The differences across products are driven mainly by different approaches (interpolation, kriging, etc.) to develop the gridded products (Jones, 2016). Since a GCD is essentially derived using interpolation schemes, its accuracy thus depends on the spatial coverage of the station data (network density), as well as the choice of the interpolation algorithm. There is no single algorithm that fits best for all variables, as well as topography (For a detailed discussion on commonly used interpolation techniques and comparison of results, readers are referred to Hofstra, Haylock, New, Jones, and Frei (2008)). Accuracy in precipitation which varies spatially more compared to temperature, is generally more affected when the resulting gridded (interpolated) data is at a coarser gridded resolution (e.g., 2.5°×2.5°, which roughly translates to 275 km × 275 km at the equator).

In addition, compared to the station observations that record data daily and are updated in their respective data dissemination channels in near-real time, the same benefit cannot be availed in most GCDs with a latency time period much longer (generally a few months to a year). Apart from the infrequent updates in a typical GCD, a change in the interpolation method when implemented only in a new version covering recent time periods, can make merging the different versions of the same GCD problematic. Users are therefore recommended to pay attention to the dataset version documents when stacking panels from different versions of a GCD.

Other potential limitations pertaining to "Global" or even wider "Regional" GCDs, is the combination of individual country's station records in assembling a global dataset. Considering the heterogeneity in observation protocols and instrumentation usage across national meteorological bodies, merging of such data could introduce artificial noise in measurements. For example, users need to pay attention if the end-of-the-day (EOD) accumulation of daily precipitation are consistent across different versions, though this problem can as well arise in Reanalysis data as discussed in Section 3.3. Equally important when applying a regional or a global GCD is the density of stations which is rarely uniform in space. The grid-cells in a GCD constructed over a geographical spare network is likely to suffer from a

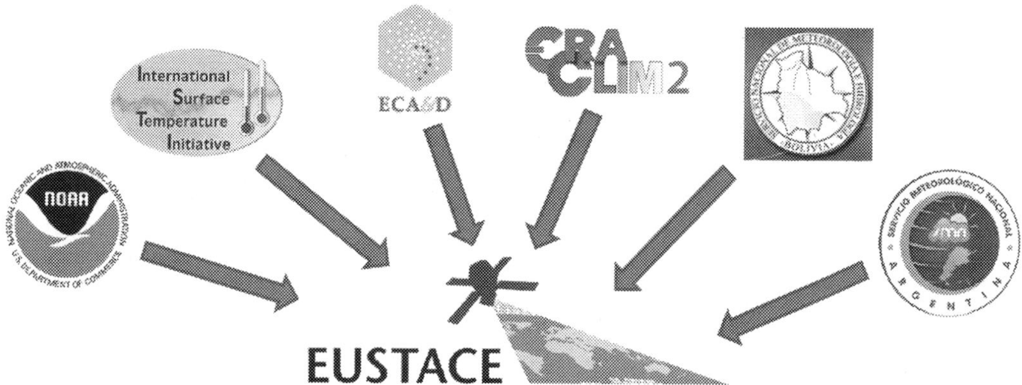

FIG. 2.4 Different sources of national, continental and global station observations utilized by EUSTACE for assembling a quality control consistent time-series of selective ECVs. Quality control algorithms automatically compare data series from multiple sources, accounting for rounding errors, duplication of records, conversion of units, EOD accumulation and temporal shifts among others. Data quality of each observation is assessed through a series of quality tests. In addition, a battery of homogeneity tests is applied to each temperature record to detect artificial (non-climatic) signals, often introduced by merging data from different stations. *Image reproduced with permission from the EU Horizon 2020 funded EU Surface Temperature for All Corners of Earth (EUSTACE) project; https://www.eustaceproject.org/*

larger uncertainty in measurement compared to its counterpart drawn over regions with a high station density. Generally for developing countries that have limited monitoring networks (Auffhammer et al., 2013), the choice of the underlying GCD used for studying extremes in climate can reveal different statistics.

A few if not all the issues discussed above, especially pertaining to inconsistencies in instrumentation records, data gathering and homogeneity, are common to both station observation and GCD. These issues are being largely addressed as part of wider initiatives, such as the Global Historical Climate Network (GHCN) led by the National Oceanic and Atmospheric Administration (NOAA) in the United States (Table 2.4), and EUMETNET (A network of 31 European National Meteorological Services based in Brussels, Belgium. https://www.eumetnet.eu/.). A primary objective of such initiatives is to create and maintain continental or even global data collections for several Essential Climate Variables (ECVs) [As defined by the WMO, an ECV is identified based on selective criteria and is defined as a "...physical, chemical or biological variable or a group of linked variables that critically contributes to the characterization of Earth's climate" (See https://public.wmo.int/en/programmes/global-climate-observing-system/essential-climate-variables for further details)], with special emphasis on daily temperature and precipitation (Fig. 2.4).

3.3 Reanalysis data

Despite a vast and rapidly growing global station-based observations network, there remains a number of locations where even ECVs (Both temperature and precipitation discussed in this chapter are defined by the WMO as ECVs.) are either not recorded or the density of network is too sparse to perform spatial interpolation for constructing a reliable long time-series. In addition, satellite-derived estimates of temperature and precipitation (mentioned earlier) are not particularly suitable for constructing CEIs and for the analyses of the extremes. It is however possible to use them with near-surface measurements to create a more complete record of observations (Rayner et al., 2020), a principle used in the dynamical reanalysis systems.

Reanalysis data products, also referred to as "retrospective analysis" have found wide applications in climate sciences dating back to the early 1990s (Compo et al., 2011; Hersbach et al., 2020). They are created using a so called "frozen" data assimilation scheme in conjunction with dynamical-physical-coupled numerical models and wide array of measurements (Auffhammer et al., 2013; Dee, Fasullo, Sheah, Walsh, & NCAR Staff, 2016; Dell et al., 2014; Fig. 2.5). Put simply, data assimilation is a way of combining subdaily observational data with numerical models, with a primary purpose to fill in the spatiotemporal gaps in true observations (Auffhammer et al., 2013). Thus, contrary to GCDs that use statistical interpolation between observations, reanalysis datasets use climate models to achieve the same objective. Among the input variables ingested by reanalysis data products, varying raw data sources ranging from ground station and ocean buoy measurements to upper atmospheric measurements [radiosonde (Freezing is for the entire time span of a reanalysis, thus inoculating them from changes in software.)] and satellite derived estimates are utilized to generate the past and the current state of the climate (Fig. 2.5).

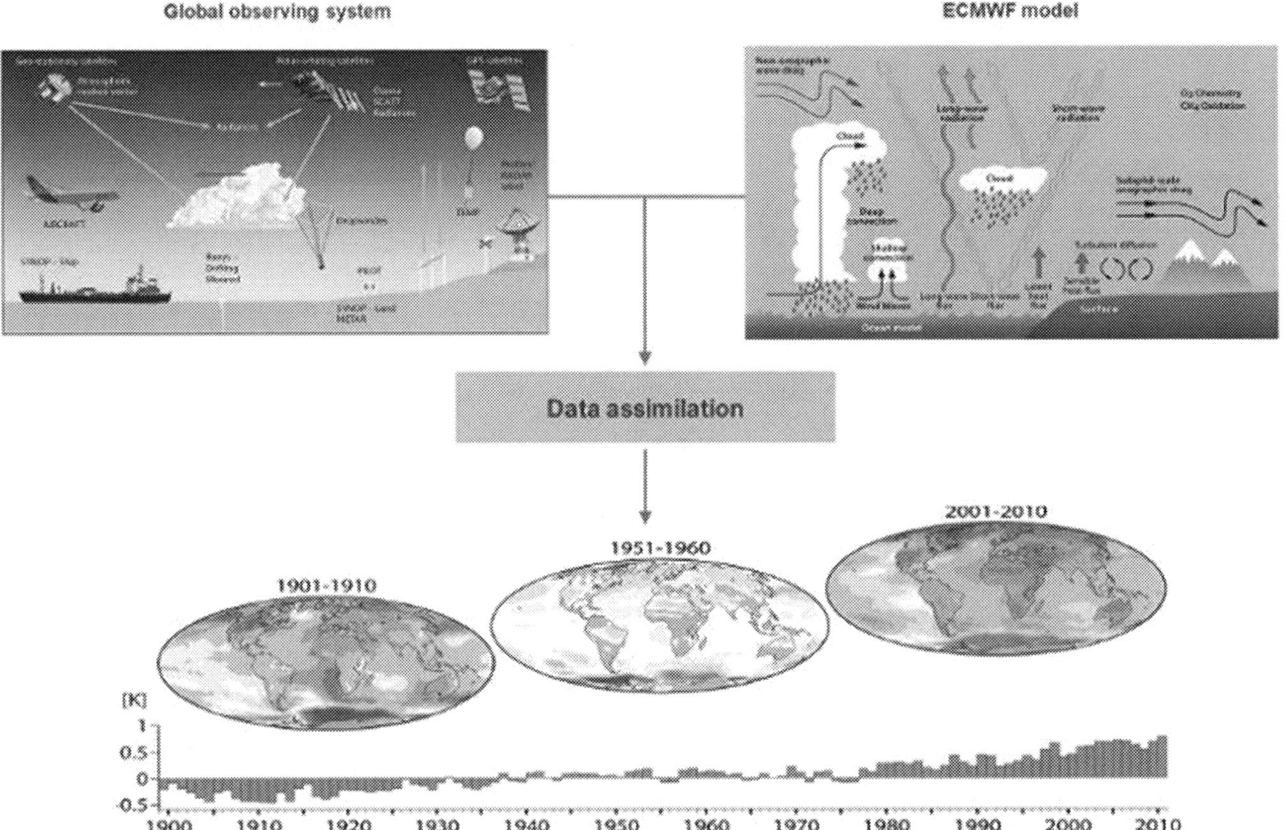

FIG. 2.5 Schematic of the ECMWF ERA5 global reanalysis. *From https://ecmwf.int/en/about/media-centre/news/2018/ecmwfs-era5-reanalysis-soon-extend-back-1979.*

The reanalysis paradigm uses a systematic approach to produce "complete" data sets suited for climate monitoring, impacts assessment and attribution studies among others, thus offering an immediate advantage over its station observations by offering a consistent past and current record of observations, covering both land and oceans. In general, the spatiotemporal features of temperature show high correlation between GCDs and reanalysis datasets. In contrast, the correlation of precipitation is generally poor in regions where ground station data is sparse, or the terrain is more rugged (A number of global reanalysis products have data starting from 1958 (see Table 2.4), which also coincides with the dawn of regular radiosonde observations globally.).

It is worthwhile mentioning that reanalysis are essentially observationally constrained model output (Mistry, 2019a) and not actual or perfect observations per se. In spite of the advantages mentioned above, variables such as precipitation that are primarily determined by the forecast model are generally poorly resolved.

Moreover, while the data assimilation scheme can be considered as "unchanging," the ingested variables inevitably form the "varying component" (Auffhammer et al., 2013). For a typical reanalysis data product spanning circa 50 years, the changes in the observation sources can thus potentially result in artificial variability, spurious trends and/or systematic biases (Bengtsson, Hagemann, & Hodges, 2004), though most recent state-of-the-art reanalysis data products such as the European Center of Medium-Range and Forecasts (ECMWF) Atmospheric Reanalysis v5 (ERA5) (Hersbach et al., 2020) have been well evaluated (Crossett, Betts, Dupigny-Giroux, & Bomblies, 2020; Jiang et al., 2021; Tarek, Brissette, & Arsenault, 2020) and are increasing applied in studying historical extremes in climate.

3.4 Existing global datasets suitable for computation of the CEIs

While there is no dearth of regional or global records of observed daily TN, TX and PR based on the above data sources, the selective datasets discussed in this section are chosen based on the following four criteria (For additional reading on the typical characteristics of climate datasets that are necessary for an adequate analysis of changes or trends in extremes, readers are guided to Heim (2015) and Observations: Atmosphere and Surface (2014):

Temporal resolution: The required three input variables (TN, TX and PR) are at either subdaily or daily timescales.

Long time-series: A consistent daily time-series for at least 30 years to facilitate computation of percentile thresholds as recommended by the ETCCDI [Also referred to as base period, the WMO and ETCCDI recommend at least a 30-year period for computation of percentiles. While 1961–1990 has generally been chosen for assembling percentile based CEIs, 1981–2010 has been more commonly applied in recent years (discussed in Section 4)].

Quality control: The times-series of the input variables is at least quality-controlled if not homogenous. When homogenization of the dataset is not performed, the dataset documentation highlights the limitations if any in usage notes.

Availability: The dataset is easily accessible without purchase requirements and/or restrictions on noncommercial usage.

It is important to reemphasize that the type of the input data source (Sections 3.1–3.3) and software routines (Table 2.3) used in assembling the CEIs, can result in differing estimates of magnitude, frequency, and long-term trends/statistics of climate extremes. Knowing the strengths of limitations of the underlying input data sources, as well the spatiotemporal coverage is therefore important when using these datasets for compiling CEIs, as well as for studying extremes in climate. The Tables 2.4 and 2.5 list the prominent global and regional dataset based on the different types of input data sources discussed above.

TABLE 2.5 Key regional datasets providing subdaily or daily input variables (TN, TX or PR) required for compiling CEIs.

Data source	Type[a]	Variables	Period	Spatial coverage	Spatial resolution (Lat × Lon)	Brief description	Link/source of data
APHRODITE (ver. 1901)	GCD	PR[b]	1998–2015	Monsoon Asia	0.25° × 0.25°	Quality control and EOD analysis automated using satellite derived algorithm. Daily fields contain information on the ratio of the station value to climatology (or ratio of grid-box containing rain gauges). The interpolation algorithm better represents extreme values. The shorter time-series of Ver. 1901 makes it unsuitable for percentile based indices. The earlier versions provide data from 1961 to 2007, though the data is inherently unsuitable for examining extremes (see ver. 1101_R2 documentation)	Yatagai et al. (2012)/http://aphrodite.st.hirosaki-u.ac.jp/products.html http://aphrodite.st.hirosaki-u.ac.jp/product_readme/Readme_V1901.pdf
DAYMET (ver. 4)	GCD	TN, TX, PR	1950–Present	North America	1 km × 1 km	Timespan varies, with North America and Hawaii from 1980 to present, and for Puerto Rico from 1950 to present, updated annually	Thornton et al. (2020)/https://daymet.ornl.gov/files/Thornton_Daymet_V4_submitted_2021-01-20.pdf (in review) https://daac.ornl.gov/DAYMET/guides/Daymet_Daily_V4.html
ECMWF-CERRA	GCD	TN, TX, PR	1961–2019	Europe	5.5 km × 5.5 km	Uses a full set of in situ observations and satellite information. High density surface observations will be	https://climate.copernicus.eu/copernicus-regional-reanalysis-europe-cerra

Continued

TABLE 2.5 Key regional datasets providing subdaily or daily input variables (TN, TX or PR) required for compiling CEIs—cont'd

Data source	Type	Variables	Period	Spatial coverage	Spatial resolution (Lat × Lon)	Brief description	Link/source of data
						added for the special surface analysis (called CERRA-Land), i.e., the regional reanalysis equivalent of ERA5-Land is also under preparation	
ECA&D	SO	TN, TX, PR	1920–Present[c]	Europe	NA	Observations are gathered across circa 20,000 meteorological stations from national weather centers (65 countries across Europe and Southern Mediterranean[d]). Data is quality-controlled. Both blended[e] and nonblended time-series are available. TN and TX are also available as homogenous blended series for selective stations	Klein Tank et al. (2002)/ https://www.ecad.eu/dailydata/predefinedseries.php#
E-OBS (ver. 23.0e)	GCD	TN, TX, PR	1920–Present	Europe	0.1° × 0.1°	Provides an ensemble dataset, though the general limitations of ECA&D data would apply	Cornes, van der Schrier, van den Besselaar, and Jones (2018) and Haylock et al. (2008)/https://surfobs.climate.copernicus.eu/dataaccess/access_eobs.php
NARR	GCD	TN, TX, PR	1979–Present	North America	~ 0.3°	The NARR dataset is an extension of the NCEP Global Reanalysis which is run over the North American Region. Different versions depending on temporal range. PR data comes from a variety of sources	Mesinger et al. (2006)/ https://www.ncdc.noaa.gov/data-access/model-data/model-datasets/north-american-regional-reanalysis-narr https://rda.ucar.edu/datasets/ds608.0/#!description (for updated version)
NLDAS (ver. 2)	GCD	TN, TX, PR	1979–Present	North America	0.125° × 0.125°	The data assimilation system incorporated is not complete. Remotely sensed estimates of land-surface states (e.g., soil moisture and snowpack) are not yet assimilated as part of the system	Mitchell et al. (2004)/ https://ldas.gsfc.nasa.gov/nldas; Xia et al. (2019) an overview of regional and global land data assimilation systems
TAMSAT (ver. 3.1)	GCD	PR	1983–Present	Africa	0.0375°× 0.0375°	Rainfall record is temporally complete with a record of "filled" days provided	Maidment et al. (2017)/ http://www.tamsat.org.uk/data

See Table A2.1 in Appendix for abbreviations and originating country of the data source.

[a] *Abbreviations as used in Table 2.4.*

[b] *Daily mean temperature is also made available as an additional dataset, though not relevant here for the computation of CEIs.*

[c] *Data for 1920 to 1949 have a research status and not meant to be updated with more station records or improvement in quality control. Present day data are generally available with a time latency of approx. 1 year.*

[d] *Some series in Scandinavian countries are missing as they are not provided by the corresponding national meteorological services.*

[e] *Missing gaps in time-series are filled by interpolating observations from nearby stations.*

3.4.1 *Global datasets*

Table 2.4 lists the prominent global datasets providing either of the three input climate variables (TX, TN or PR) required for assembling the CEIs. The datasets based on station observations, gridded climate and reanalysis, are the most commonly used and extensively studied datasets in the broader climate literature. Though not an exhaustive list, the global datasets discussed here have also been used for assembling global datasets of CEIs (discussed in Section 4).

3.4.2 *Regional datasets*

Often the focus of study is restricted to a specific country or a wider region (e.g., continent). In such instances, regional datasets, either station observed or reanalysis, can be a suitable alternative source. To conserve space, the regional datasets discussed in Table 2.5 are restricted to those that cover a wider geographical domain, and not individual countries.

4 Existing datasets of CEIs: Strengths and limitations

As discussed in earlier sections, climate indices are very useful tools to describe the past and current climate (Mahlstein et al., 2015), as well as for analyses of climate extremes (Zhang et al., 2011). Their appeal and application in climate change and climate extremes research stem from the simplicity, robustness and usage scope at wider spatiotemporal scales and climate regimes. Moreover, representation of broad scale weather patterns that are known and experienced by the wider public, are often easier to proxy by way of simple or compound indicators. Because most of the climate and climate extreme indices aggregate information in a nonlinear way, the consequent information these can reveal about possible impacts are more meaningful than simple temperature averages for instance.

Drawing data from a number of climate data sources, the research community have in recent years assembled a suite of temperature and precipitation based extreme indices at both global and regional scales. Table 2.6 summarizes

TABLE 2.6 Key global and regional datasets of CEIs.

Data source writea	Type	Period	Spatial coverage	Spatial resolution (Lat × Lon)	Brief description	Ref. paper/web link
ECA&D	SO	1920–Present	Europe	NA	Includes 76 indices comprising of a core set of 26 ETCCDI defined indices, and an additional set of 50 indices highlighting particular characteristics of climate change in Europe (including snow depth, sunshine duration, etc.). See ECA&D (Table 2.5) for general characteristics of input data variables	https://www.ecad.eu/indicesextremes/# (Dashboard available through same link)
E-OBS	GCD	1950–Present	Europe	0.1° × 0.1°	Median and uncertainty estimates (2.5 and 97.5 percentiles) for a total of 45 indices based on E-OBS Ver. 22.0e. See E-OBS (Table 2.5) for general characteristics of input data variables	Cornes et al. (2018) and Haylock et al. (2008)/https://surfobs.climate.copernicus.eu/dataaccess/access_eobs_indices.php
GHCNDEX	GCD	1951–Present	Global	2.5° × 2.5°	26 core ETCCDI indices. The dataset is produced by gridding CEIs calculated at the station locations onto a global grid. See GHCN-D (Table 2.4) for general characteristics of input data variables	Donat et al. (2013a)/https://www.climdex.org/learn/datasets/ (Dashboard)
GLDAS (vers. 2.1 and 2.2)	Reanly	1970–2016	Global	0.25° × 0.25°	The indices for years 2010 to 2016 are computed using GLDAS Ver. 2.2 data which could result in a break in time-series at some locations. Users are referred to discussion paper for further details on usage. Updated version of the dataset available from the author upon request. See GLDAS (Table 2.4) for general characteristics of input data variables	Mistry (2019a, 2019b)

Continued

TABLE 2.6 Key global and regional datasets of CEIs—cont'd

Data source write	Type	Period	Spatial coverage	Spatial resolution (Lat × Lon)	Brief description	Ref. paper/web link
GLDAS (vers. 2.1 and 2.2)	Reanly.	1970–2018	Global	0.25° × 0.25°	The indices for years 2010 to 2018 are computed using GLDAS Ver. 2.2 data which could result in a break in time-series at some locations. Users are referred to discussion paper for further details on usage. Includes only Cooling (CDD)- and Heating (HDD)-Degree-Days. Updated version of the dataset available from the author upon request. See GLDAS (Table 2.4) for general characteristics of input data variables	Mistry (2019c, 2019d)
HadEX3 (Ver 3.0.3)	GCD	1901–2018	Global	1.25° × 1.875°	29 core ETCCDI indices assembled utilizing nearly quality-controlled 7000 (17,000) station observations for temperature (precipitation). As in GHCNDEX, the dataset is produced by gridding CEIs calculated at the station locations onto a global grid. However, compared to GHCNDEX that uses only GHCN-D as input, HadEX3 (and its predecessor HadEX2[b]) utilizes a mix of large collections of data and country-based inputs (Dunn, Donat, & Alexander, 2014). Both GHCNDEX and HadEX3 are therefore based on largely independent sets of input stations, though the gridding method is identical. However, the HadEX family of datasets is less frequently updated compared to GHCNDEX which is updated monthly or annually	Dunn et al. (2020)/ https://www.metoffice.gov.uk/hadobs/hadex3/index.html, https://www.climdex.org/learn/datasets/ (Dashboard)
HadGHCND	GCD	1950–2011	Global	2.5° × 3.75°	Only temperature-based indices are made available. Dataset is not updated	https://www.metoffice.gov.uk/hadobs/hadghcnd/download.html
INDECIS	GCD	1950–2019	Europe		A gridded dataset of 125 climate indices derived using E-OBS and ERA5. In addition to temperature and precipitation, indices based on biometeorology and aridity, among others are also included. Available through interactive dashboard	Dominguez-Castro, Reig, Noguera, et al. (2020) and Domínguez-Castro, Reig, Vicente-Serrano, et al. (2020)/ https://indecis.csic.es/#map_name=aci_month#map_position=839 (Dashboard)
S-14	GMF	1950–2019	Global	0.5° × 0.5°	27 core ETCCDI indices are made available using five different GMF datasets. In addition, the historical and future period CEIs from bias-corrected GCMs' simulated data are also included	Iizumi, Takikawa, Hirabayashi, Hanasaki, and Nishimori (2017)/ http://h08.nies.go.jp/s14/

See Table A2.1 in Appendix for abbreviations.

[a] *CEIs datasets based on earlier versions of legacy reanalysis data products (ECMWF-ERA-Interim and ERA40, and NCEP-NCAR and NCEP DOE—Table 2.4) can be found at https://crd-data-donnees-rdc.ec.gc.ca/CCCMA/products/CLIMDEX/REANALYSES/historical/.*

[b] *HadEX was the first global land-based gridded dataset of temperature and precipitation extremes covering the second half of the 20th century (Donat et al., 2013a).*

the characteristics, strengths and limitations of key historical datasets of CEIs. Though the legacy datasets of CEIs assembled using older reanalysis data products are not mentioned here, readers are guided to the relevant references for the sake of completeness. In addition, a few recent data dashboards facilitating download and analysis of CEIs are also mentioned where applicable. The dashboards depending on the functionality commonly find appeal within the climate sectoral impacts modeling community, whose primary aim is to perform rapid spatiotemporal aggregation, trend analysis or conversion to a user-friendly file format.

5 Conclusion and recommendations

5.1 Recent advancements in understanding climate extremes

With the advent of quality-controlled global scale high-resolution historical observations of climate, supercomputing infrastructures, big data storage, improved parameterizations of small-scale atmospheric processes and state-of-the-art climate modeling simulations, the scientific community have made rapid strides to understand the underlying mechanisms fueling high-frequency climate extremes (Alexander, 2016; Pfahl, O'Gorman, & Fischer, 2017). In particular, the spatiotemporal variations in past and current climate extremes both at different geographic scales (Alexander et al., 2006; Alexander & Arblaster, 2017), as well as the projected changes of extremes under different climate change scenarios (Dosio, 2016; Fotso-Nguemo et al., 2018; Liao, Xu, Zhang, Li, & Tian, 2019; Ongoma, Chen, Gao, Nyongesa, & Polong, 2018; Sillmann, Kharin, Zwiers, et al., 2013), are some of the recent advancements made by researchers in understanding climate extremes.

In addition, noteworthy progress has been made in recent decades in our understanding of the detection, attribution, modeling, and prediction of climate extremes. Led by the ETCCDI and their collaborative partners, the development of a broad suite of CEIs (Tables 2.1 and 2.2) derived from readily available subdaily/daily temperature and precipitation has been one such commendable contribution. The result has been a steadily growing inventory of high-resolution quality-controlled datasets of CEIs (Table 2.6), covering both historical and future periods, spanning global and regional domains. In addition, sophisticated data assimilation and statistical approaches have also led to improved measurements and homogenization of input data variables used for assembling the CEIs, thereby reducing the inherent bias and uncertainties emanating from observations and estimation techniques.

Equally commendable improvements are evident in development of software routines (Table 2.3) used for computation and evaluation of a broad suite of CEIs. With increasing availability of high-resolution global observational records often spanning more than a hundred years (Table 2.4), the climate research and modeling community often deal with terabytes of data when analyzing global scale extremes. Apart from the need to handle memory intensive operations and storage requirements when assembling the CEIs, validating the generated datasets of indices against benchmark observations also requires special approaches (See Avila-Diaz, Bromwich, Wilson, Justino, & Wang, 2021; Sippel et al., 2015; Thorarinsdottir, Sillmann, Haugen, Gissibl, & Sandstad, 2020 for further reference).

Understanding how dynamical models represent climate extremes in their simulations is another research domain gaining rapid strides. For instance, convection-permitting modeling of the current and future climate now available at continental-scales (Liu et al., 2017) are expected to facilitate better understanding of precipitation-based extremes. In addition, results from ongoing global and regional climate model simulations such as those developed by the consortium of GCMs participating in the Coupled Model Intercomparison Project-Phase 6 (CMIP6, Eyring et al. (2016)) (Established in 1995 by the WCRP's Working Group of Coupled Modeling (WGCM), the CMIP project has coordinated climate model experiments involving multiple international modeling teams worldwide), PRIMAVERA (https://www.primavera-h2020.eu/), HighResMIP (Haarsma et al., 2016), and CORDEX (The WCRP Coordinated Regional Downscaling Experiment (CORDEX, http://wcrp-cordex.ipsl.jussieu.fr/) provides an internationally coordinated framework to improve regional climate scenarios.) (Giorgi, Jones, & Asrar, 2009) are expected to throw further insights on both the likely scale of changes in global and regional climate extremes, as well as on the underlying mechanisms exacerbating such changes. A few such recent studies (e.g., Seneviratne & Hauser, 2020) included in the recent IPCC AR6 (Seneviratne et al., 2021) have already provided evidence that regional climate sensitivity is a distinct feature of climate models and a key determinant of projected regional impacts.

5.2 Scope and recommendation for future research

In spite of the progress made toward a better understanding of the extreme aspects in our climate system, robust findings of which have made a significant contribution to the IPCC Assessment Reports (Sillmann et al., 2017), scope for further work remains in a number of avenues. For instance, although scientific interest in regional and global climate extremes has risen in the past two decades, understanding and modeling high-frequency weather extremes (e.g., subdaily or daily intense precipitation events at localized scales resulting in flash floods) is yet in its nascent stage (A few examples of recent studies include Ávila, Justino, Wilson, Bromwich, and Amorim (2016) and Ávila, Guerrero,

Escobar, and Justino (2019). Readers are also guided to a discussion on "Short-duration extreme events" in Sillmann et al. (2017)). Improved in-situ ground based and remote-sensed measurements of key meteorological parameters, and availability of quality-controlled high spatiotemporal resolution datasets are expected to spur further interest in this field.

Assessments of climate extremes derived from different climate data sources (covered in Section 3), are also an ongoing research area worthy of mention. Such comparative assessments (see Donat et al., 2014; Dunn et al., 2014, for examples) throw insights into the strengths and limitations of different data products in representation of climate extremes, and potential reasons for uncertainties and biases if any. (One such example is regridding or interpolation employed in data source 'ii' (Section 3.2) and its by-products, data sources "vi–vii.") However, such comparative exercises have generally not kept pace with the availability of newer regional and global input climate data sources (Tables 2.4 and 2.5). The research community could therefore undertake a more systematic evaluation of different datasets in evaluating climate extremes at a global scale.

Still pertaining to climate data sources, two further potential avenues of research are promising and require further investigation. First, multiple data sources are increasingly being used as "reference climatology" in the bias-correction (Bias-correction is a statistical adjustment to match the output of model or reanalysis data closer to reference climatology. Though unavoidable (Iizumi et al., 2017), bias-correction if not correctly applied, can (in addition to the actual bias) result in removing or smoothing the important variations in the climate variable (Mahlstein et al., 2015), especially in the tails of the distribution.) of subdaily/daily data from GCMs and reanalysis data. Such data sources referred to as "GMF" (Readers are guided to Iizumi et al. (2017), Weedon et al. (2014), Lange (2018), and Cucchi et al. (2020) for further details and examples of prominent forcing datasets.) when used with appropriate bias-correction methods, help to improve the inconsistencies in models' and reanalysis data products (Iizumi et al., 2017). However, currently, a limited number of studies examine the contribution of different bias-correction methods and input data sources to the uncertainty in past and future (projected) extremes in climate (Iizumi et al., 2017; Mahlstein et al., 2015). Second, and as emphasized by Sillmann et al. (2017), a large number of existing datasets already facilitate computation of the CEIs and analysis of a broad suite of extremes. Nevertheless, new data sources such as satellite or remote sensing, merged and GMF (Section 3 data sources "v," "vi," and "vii," respectively), should be exploited to examine their suitability in understanding climate extremes.

Finally, another recent research area that remains not only an open and challenging research question, but also a need of the hour, pertains to "Compound Extreme Events" (Mysiak et al., 2018). Also sometimes referred to as "Connected Extreme Events" or "Concurrent Extremes," these are simultaneous occurrences of same or different types of climate extremes at two or more different locations, or the back-to-back occurrence of different climate extremes at the same location (Raymond et al., 2020; Zscheischler et al., 2018, 2020). For example, a simultaneous occurrence of warm and cold spells in Central Africa and North America respectively, or high intensity precipitation event followed by flooding would be termed as compound climate extreme events. Such simultaneous extreme events typically have a low probability of occurrence, but generally end up causing a higher impact than a univariate extreme event. Here again, CEIs are being increasingly used as tools for understanding "Compound Extreme Events" (Irannezhad, Moradkhani, & Kløve, 2018), but a thorough evaluation of climate datasets and CEIs capable of representing the teleconnections across wider geographic scales in past, current and future climates remains a promising research avenue.

TABLE A2.1 List of datasets abbreviations used in Tables 2.4–2.6.

APHRODITE	Asian precipitation—highly-resolved observational data integration toward evaluation (Japan)
BEST	Berkley Earth Surface Temperatures (United States)
CERRA	Copernicus regional reanalysis for Europe (Europe)
CFSR	Climate Forecast System Reanalysis
CHIRPS	Climate Hazards Group InfraRed Precipitation with Station data (United States)
CPC	Climate Prediction Center (United States)
ECA&D/E-OBS	European Climate Assessment & Dataset, and its gridded version E-OBS (Europe)
ECMWF	European Center for Medium Range Weather Forecasts (Europe)
ERA5	5th Generation ECMWF Reanalysis (Europe)
EUSTACE	European Union (EU) Surface Temperature for All Corners of Earth (Europe)

TABLE A2.1 List of datasets abbreviations used in Tables 2.4–2.6—cont'd

APHRODITE	Asian precipitation—highly-resolved observational data integration toward evaluation (Japan)
GHCN-D	Global Historical Climate Network-Daily
GLDAS	Global Land Data Acquisition System (United States)
GPCC	Global Precipitation Climatology Centre (Germany)
HadEX3	Hadley Centre Global Climate Extremes Index 3 (United Kingdom)
HadGHCND	Hadley Centre (derived) GHCN-D (United Kingdom)
HadISD	Met Office Hadley Centre (United Kingdom) ISD (For ISD see below)
INDECIS	N.A.
ISD	Integrated Surface Database (United States)
JRA-55	Japanese 55-year Reanalysis (developed by Japan Meteorological Agency—JMA)
MERRA	Modern-Era Retrospective analysis for Research and Applications (United States)
MSWEP	Multisource Weighted-Ensemble Precipitation
NARR	North American Regional Reanalysis (United States)
NASA-GMAO	National Aeronautics and Space Administration-Global Modeling and Assimilation Office
NCEI	National Centers for Environmental Information (United States)
NCEP	National Centers for Environmental Prediction (formerly known as the National Climatic Data Center, NCDC) (United States)
NCAR	National Center for Atmospheric Research (United States)
NLDAS	North American Land Data Assimilation System
NOAA-CIRES-DOE	National Oceanic and Atmospheric Administration-Cooperative Institute for Research in Environmental Sciences-(the US) Department of Energy (United States)
S-14	N.A.
TAMSAT	Tropical Applications of Meteorology using SATellite data and ground-based observations (United Kingdom)

The country/region where the institution is based is mentioned in braces where applicable.

Appendix

Acknowledgments

The author was funded by a grant from the European Research Council (ERC) under the European Unions Horizon 2020 research and innovation program, under grant agreement no. 756194 (ENERGYA, PI Prof. Enrica De Cian). The author is grateful to Enrico Scoccimarro (CMCC), Sepehr Marzi (CMCC), and Paola Vesco (Uppsala Universitet) for constructive comments that helped to improve the structure of the chapter. Any remaining errors are those of the author.

Conflict of interest

The author declares no conflict of interest.

References

Alexander, L., & Herold, N. (2016). *ClimPACT2 indices and software (R software package)*. Available from: https://htmlpreview.github.io/?https://raw.githubusercontent.com/ARCCSS-extremes/climpact2/master/user_guide/ClimPACT2_user_guide.htm. (Accessed 11 December 2019).

Alexander, L., & Tebaldi, C. (2012). Climate and weather extremes: Observations, modelling, and projections. In A. Henderson-Sellers, & K. McGuffie (Eds.), *The future of the world's climate* (2nd ed., pp. 253–288). Boston: Elsevier, ISBN:978-0-12-386917-3 (chapter 10).

Alexander, L. V. (2016). Global observed long-term changes in temperature and precipitation extremes: A review of progress and limitations in IPCC assessments and beyond. *Weather and Climate Extreme*, *11*, 4–16.

Alexander, L. V., & Arblaster, J. M. (2017). Historical and projected trends in temperature and precipitation extremes in Australia in observations and CMIP5. *Weather and Climate Extreme, 15*, 34–56.

Alexander, L. V., Zhang, X., Peterson, T. C., Caesar, J., Gleason, B., Klein Tank, A. M. G., et al.…. (2006). Global observed changes in daily climate extremes of temperature and precipitation. *Journal of Geophysical Research: Atmospheres, 111*(D5). https://agupubs.onlinelibrary.wiley.com/action/showCitFormats?doi=10.1029%2F2005JD006290.

Angélil, O., Stone, D. A., Tadross, M., Tummon, F., Wehner, M., & Knutti, R. (2014). Attribution of extreme weather to anthropogenic greenhouse gas emissions: Sensitivity to spatial and temporal scales. *Geophysical Research Letters, 41*, 2150–2155.

Antonelli, C., Coromaldi, M., Dasgupta, S., Emmerling, J., & Shayegh, S. (2020). Climate impacts on nutrition and labor supply disentangled—An analysis for rural areas of Uganda. In *Environment and development economics* (pp. 1–26). Cambridge University Press.

Auffhammer, M., Hsiang, S. M., Schlenker, W., & Sobel, A. (2013). Using Weather data and climate model output in economic analyses of climate change. *Review of Environmental Economics and Policy, 7*, 181–198.

Ávila, Á., Guerrero, F., Escobar, Y., & Justino, F. (2019). Recent precipitation trends and floods in the Colombian Andes. *Water, 11*, 379.

Ávila, A., Justino, F., Wilson, A., Bromwich, D., & Amorim, M. (2016). Recent precipitation trends, flash floods and landslides in southern Brazil. *Environmental Research Letters, 11*, 114029.

Avila-Diaz, A., Benezoli, V., Justino, F., Torres, R., & Wilson, A. (2020). Assessing current and future trends of climate extremes across Brazil based on reanalyses and earth system model projections. *Climate Dynamics, 55*, 1403–1426.

Avila-Diaz, A., Bromwich, D. H., Wilson, A. B., Justino, F., & Wang, S.-H. (2021). Climate extremes across the north American Arctic in modern reanalyses. *Journal of Climate, 34*, 2385–2410.

Beck, H. E., Wood, E. F., Pan, M., Fisher, C. K., Miralles, D. G., van Dijk, A. I. J. M., … Adler, R. F. (2019). MSWEP V2 global 3-hourly 0.1° precipitation: Methodology and quantitative assessment. *Bulletin of the American Meteorological Society, 100*, 473–500.

Begueria, S., & Vicente-Serrano, S. M. (2013). *SPEI: Calculation of the standardised precipitation-evapotranspiration index. R package version 1.6*. http://CRAN.R-project.org/package=SPEI.

Bengtsson, L., Hagemann, S., & Hodges, K. I. (2004). Can climate trends be calculated from reanalysis data? *Journal of Geophysical Research: Atmospheres, 109*(D11). https://agupubs.onlinelibrary.wiley.com/doi/10.1029/2004JD004536.

Beniston, M., & Stephenson, D. B. (2004). Extreme climatic events and their evolution under changing climatic conditions. *Global and Planetary Change, 44*, 1–9.

Bezerra, P., da Silva, F., Cruz, T., Mistry, M., Vasquez-Arroyo, E., Magalar, L., … Schaeffer, R. (2021). Impacts of a warmer world on space cooling demand in Brazilian households. *Energy and Buildings, 234*, 110696.

Brown, P. J., Bradley, R. S., & Keimig, F. T. (2010). Changes in extreme climate indices for the Northeastern United States, 1870–2005. *Journal of Climate, 23*, 6555–6572.

Brugnara, Y., Good, E., Squintu, A. A., Schrier, G., & Brönnimann, S. (2019). The EUSTACE global land station daily air temperature dataset. *Geoscience Data Journal, 6*, 189–204.

Caesar, J., Alexander, L., & Vose, R. (2006). Large-scale changes in observed daily maximum and minimum temperatures: Creation and analysis of a new gridded data set. *Journal of Geophysical Research, 111*, D05101.

Caussinus, H., & Mestre, O. (2004). Detection and correction of artificial shifts in climate series. *Journal of the Royal Statistical Society: Series C: Applied Statistics, 53*, 405–425.

Changnon, S. A., Pielke, R. A., Changnon, D., Sylves, R. T., & Pulwarty, R. (2000). Human factors explain the increased losses from weather and climate extremes. *Bulletin of the American Meteorological Society, 81*, 437–442.

Compo, G. P., Whitaker, J. S., Sardeshmukh, P. D., Matsui, N., Allan, R. J., Yin, X., Gleason, B. E., Vose, R. S., Rutledge, G., Bessemoulin, P., et al. (2011). The twentieth century reanalysis project. *Quarterly Journal of the Royal Meteorological Society, 137*, 1–28.

Cornes, R. C., van der Schrier, G., van den Besselaar, E. J. M., & Jones, P. D. (2018). An ensemble version of the E-OBS temperature and precipitation data sets. *Journal of Geophysical Research: Atmospheres, 123*, 9391–9409.

Crossett, C. C., Betts, A. K., Dupigny-Giroux, L.-A. L., & Bomblies, A. (2020). Evaluation of daily precipitation from the ERA5 global reanalysis against GHCN observations in the Northeastern United States. *Climate, 8*, 148.

Cucchi, M., Weedon, G. P., Amici, A., Bellouin, N., Lange, S., Müller Schmied, H., … Buontempo, C. (2020). WFDE5: Bias-adjusted ERA5 reanalysis data for impact studies. *Earth System Science Data, 12*, 2097–2120.

Cutter, S. L., & Finch, C. (2008). Temporal and spatial changes in social vulnerability to natural hazards. *Proceedings of the National Academy of Sciences, 105*, 2301–2306.

De Cian, E., Pavanello, F., Randazzo, T., Mistry, M., & Davide, M. (2019). Households' adaptation in a warming climate. Air conditioning and thermal insulation choices. *Environmental Science & Policy, 100*, 136–157.

Dee, D., Fasullo, J., Sheah, D., Walsh, J., & NCAR Staff (Eds.). (2016). *The climate data guide: Atmospheric reanalysis: Overview & comparison tables*. Available from: https://climatedataguide.ucar.edu/climate-data/atmospheric-reanalysis-overview-comparison-tables.

Dell, M., Jones, B. F., & Olken, B. A. (2014). What do we learn from the Weather? The New climate-economy literature. *Journal of Economic Literature, 52*, 740–798.

Della-Marta, P. M., & Wanner, H. (2006). A method of homogenizing the extremes and mean of daily temperature measurements. *Journal of Climate, 19*, 4179–4197.

Diffenbaugh, N. S., Singh, D., Mankin, J. S., Horton, D. E., Swain, D. L., Touma, D., Charland, A., Liu, Y., Haugen, M., Tsiang, M., et al. (2017). Quantifying the influence of global warming on unprecedented extreme climate events. *Proceedings of the National Academy of Sciences, 114*, 4881–4886.

Domínguez-Castro, F., Reig, F., Noguera, I., van der Schrier, G., Vicente-Serrano, S., Peña-Angulo, D., … Kenawy, A. M. (2020). *El INDECIS*.

Domínguez-Castro, F., Reig, F., Vicente-Serrano, S. M., Aguilar, E., Peña-Angulo, D., Noguera, I., … El Kenawy, A. M. (2020). A multidecadal assessment of climate indices over Europe. *Scientific Data, 7*, 125.

Donat, M. G., Alexander, L. V., Yang, H., Durre, I., Vose, R., & Caesar, J. (2013a). Global land-based datasets for monitoring climatic extremes. *Bulletin of the American Meteorological Society, 94*, 997–1006.

Donat, M. G., Alexander, L. V., Yang, H., Durre, I., Vose, R., Dunn, R. J. H., Willett, K. M., Aguilar, E., Brunet, M., Caesar, J., et al. (2013b). Updated analyses of temperature and precipitation extreme indices since the beginning of the twentieth century: The HadEX2 dataset. *Journal of Geophysical Research: Atmospheres, 118*, 2098–2118.

Donat, M. G., Sillmann, J., Wild, S., Alexander, L. V., Lippmann, T., & Zwiers, F. W. (2014). Consistency of temperature and precipitation extremes across various global gridded in situ and reanalysis datasets. *Journal of Climate, 27*, 5019–5035.

Dosio, A. (2016). Projections of climate change indices of temperature and precipitation from an ensemble of bias-adjusted high-resolution EURO-CORDEX regional climate models. *Journal of Geophysical Research: Atmospheres, 121*, 5488–5511.

Dunn, R. J. H., Alexander, L. V., Donat, M. G., Zhang, X., Bador, M., Herold, N., et al.…. (2020). Development of an updated global land in situ-based data set of temperature and precipitation extremes: HadEX3. *Journal of Geophysical Research: Atmospheres, 125*(16). https://agupubs.onlinelibrary.wiley.com/doi/full/10.1029/2019JD032263.

Dunn, R. J. H., Donat, M. G., & Alexander, L. V. (2014). Investigating uncertainties in global gridded datasets of climate extremes. *Climate of the Past, 10*, 2171–2199.

Dunn, R. J. H., Willett, K. M., Parker, D. E., & Mitchell, L. (2016). Expanding HadISD: Quality-controlled, sub-daily station data from 1931. *Geoscientific Instrumentation, Methods and Data Systems*, (5), 473–491.

Dunn, R. J. H., Willett, K. M., Thorne, P. W., Woolley, E. V., Durre, I., Dai, A., … Vose, R. S. (2012). HadISD: A quality-controlled global synoptic report database for selected variables at long-term stations from 1973–2011. *Climate of the Past, 8*, 1649–1679.

Easterling, D. R., Meehl, G. A., Parmesan, C., Changnon, S. A., Karl, T. R., & Mearns, L. O. (2000). Climate extremes: Observations, modeling, and impacts. *Science, 289*, 2068–2074.

Eyring, V., Bony, S., Meehl, G. A., Senior, C. A., Stevens, B., Stouffer, R. J., & Taylor, K. E. (2016). Overview of the coupled model intercomparison project phase 6 (CMIP6) experimental design and organization. *Geoscientific Model Development, 9*, 1937–1958.

Fotso-Nguemo, T. C., Chamani, R., Yepdo, Z. D., Sonkoué, D., Matsaguim, C. N., Vondou, D. A., & Tanessong, R. S. (2018). Projected trends of extreme rainfall events from CMIP5 models over Central Africa. *Atmospheric Science Letters, 19*, e803.

Funk, C., Peterson, P., Landsfeld, M., Pedreros, D., Verdin, J., Shukla, S., Husak, G., Rowland, J., Harrison, L., Hoell, A., et al. (2015). The climate hazards infrared precipitation with stations—A new environmental record for monitoring extremes. *Scientific Data, 2*, 150066.

Giorgi, F., Jones, C., & Asrar, G. (2009). Addressing climate information needs at the regional level: The CORDEX framework. *World Meteorological Organization. Bulletin, 58*, 175–183.

Giorgi, F., Raffaele, F., & Coppola, E. (2019). The response of precipitation characteristics to global warming from climate projections. *Earth System Dynamics, 10*, 73–89.

Goodison, B., Louie, P. Y. T., & Yang, D. (1998). *WMO solid precipitation measurement intercomparison. Final reports* (p. 318).

Guan, L. (2009). Preparation of future weather data to study the impact of climate change on buildings. *Building and Environment, 44*, 793–800.

Haarsma, R. J., Roberts, M. J., Vidale, P. L., Senior, C. A., Bellucci, A., Bao, Q., Chang, P., Corti, S., Fučkar, N. S., Guemas, V., et al. (2016). High resolution model Intercomparison project (HighResMIP∼v1.0) for CMIP6. *Geoscientific Model Development, 9*, 4185–4208.

Haylock, M. R., Hofstra, N., Klein Tank, A. M. G., Klok, E. J., Jones, P. D., & New, M. (2008). A European daily high-resolution gridded data set of surface temperature and precipitation for 1950–2006. *Journal of Geophysical Research: Atmospheres, 113*(D20). https://agupubs.onlinelibrary.wiley.com/doi/10.1029/2008JD010201.

Heim, R. R. (2015). An overview of weather and climate extremes—Products and trends. *Weather and Climate Extreme, 10*, 1–9.

Hersbach, H., Bell, B., Berrisford, P., Hirahara, S., Horányi, A., Muñoz-Sabater, J., Nicolas, J., Peubey, C., Radu, R., Schepers, D., et al. (2020). The ERA5 global reanalysis. *Quarterly Journal of the Royal Meteorological Society, 146*, 1999–2049.

Hofstra, N., Haylock, M., New, M., Jones, P., & Frei, C. (2008). Comparison of six methods for the interpolation of daily, European climate data. *Journal of Geophysical Research: Atmospheres, 113*, D21110.

Iizumi, T., Takikawa, H., Hirabayashi, Y., Hanasaki, N., & Nishimori, M. (2017). Contributions of different bias-correction methods and reference meteorological forcing data sets to uncertainty in projected temperature and precipitation extremes. *Journal of Geophysical Research: Atmospheres, 122*, 7800–7819.

Intergovernmental Panel on Climate Change (Ed.). (2014). *Observations: Atmosphere and surface. Climate change 2013—The physical science basis: Working group I contribution to the fifth assessment report of the intergovernmental panel on climate change* (pp. 159–254). Cambridge: Cambridge University Press, ISBN:9781107057999.

IPCC. (2013). *Climate change 2013: The physical science basis. Contribution of working group I to the fifth assessment report of the intergovernmental panel on climate change*. Cambridge, United Kingdom and New York, NY, USA: Cambridge University Press, ISBN:978-1-107-66182-0.

IPCC Summary for Policymakers. (2012). Managing the risks of extreme events and disasters to advance climate change adaptation. In *A special report of working groups I II intergovernmental panel on climate change* (pp. 3–21).

Irannezhad, M., Moradkhani, H., & Kløve, B. (2018). Spatiotemporal variability and trends in extreme temperature events in Finland over the recent decades: Influence of northern hemisphere teleconnection patterns. *Advances in Meteorology, 2018*, 7169840.

Jiang, Q., Li, W., Fan, Z., He, X., Sun, W., Chen, S., … Wang, J. (2021). Evaluation of the ERA5 reanalysis precipitation dataset over Chinese mainland. *Journal of Hydrology, 595*, 125660.

Jones, P. (2016). The reliability of global and hemispheric surface temperature records. *Advances in Atmospheric Sciences, 33*, 269–282.

Kalnay, E., Kanamitsu, M., Kristler, R., Collins, W., Deaven, D., Gnadin, L., Iredell, M., Saha, S., White, G., Woollen, J., et al. (1996). NCEP NCAR renanalysis. *Bulletin of the American Meteorological Society, 77*, 437–471.

Kanamitsu, M., Ebisuzaki, W., Woollen, J., Yang, S. K., Hnilo, J. J., Fiorino, M., & Potter, G. L. (2002). NCEP–DOE AMIP-II reanalysis (R-2). *Bulletin of the American Meteorological Society, 83*, 1631–1643.

Karl, T. R., Nicholls, N., & Ghazi, A. (1999). CLIVAR/GCOS/WMO workshop on indices and indicators for climate extremes workshop summary. In T. R. Karl, N. Nicholls, & A. Ghazi (Eds.), *Weather and climate extremes: Changes, variations and a perspective from the insurance industry* (pp. 3–7). Dordrecht: Springer Netherlands, ISBN:978-94-015-9265-9.

Kaufmann, R. K., & Stern, D. I. (1997). Evidence for human influence on climate from hemispheric temperature relations. *Nature, 388*, 39–44.

Klein Tank, A. M. G., Wijngaard, J. B., Können, G. P., Böhm, R., Demarée, G., Gocheva, A., Mileta, M., Pashiardis, S., Hejkrlik, L., Kern-Hansen, C., et al. (2002). Daily dataset of 20th-century surface air temperature and precipitation series for the European climate assessment. *International Journal of Climatology, 22*, 1441–1453.

Kobayashi, S., Ota, Y., Harada, Y., Ebita, A., Moriya, M., Onoda, H., Onogi, K., Kamahori, H., Kobayashi, C., Endo, H., et al. (2015). The JRA-55 reanalysis: General specifications and basic characteristics. *Journal of the Meteorological Society of Japan, II*(93), 5–48.

Lange, S. (2018). Bias correction of surface downwelling longwave and shortwave radiation for the EWEMBI dataset. *Earth System Dynamics, 9*, 627–645.

Levesque, A., Pietzcker, R. C., Baumstark, L., De Stercke, S., Grübler, A., & Luderer, G. (2018). How much energy will buildings consume in 2100? A global perspective within a scenario framework. *Energy, 148*, 514–527.

Li, G., Zhang, X., Cannon, A. J., Murdock, T., Sobie, S., Zwiers, F., … Qian, B. (2018). Indices of Canada's future climate for general and agricultural adaptation applications. *Climatic Change, 148*, 249–263.

Liao, X., Xu, W., Zhang, J., Li, Y., & Tian, Y. (2019). Global exposure to rainstorms and the contribution rates of climate change and population change. *Science of the Total Environment, 663*, 644–653.

Liu, C., Ikeda, K., Rasmussen, R., Barlage, M., Newman, A. J., Prein, A. F., Chen, F., Chen, L., Clark, M., Dai, A., et al. (2017). Continental-scale convection-permitting modeling of the current and future climate of North America. *Climate Dynamics, 49*, 71–95.

Mahlstein, I., Spirig, C., Liniger, M. A., & Appenzeller, C. (2015). Estimating daily climatologies for climate indices derived from climate model data and observations. *Journal of Geophysical Research: Atmospheres, 120*, 2808–2818.

Maidment, R. I., Grimes, D., Black, E., Tarnavsky, E., Young, M., Greatrex, H., Allan, R. P., Stein, T., Nkonde, E., Senkunda, S., et al. (2017). A new, long-term daily satellite-based rainfall dataset for operational monitoring in Africa. *Scientific Data, 4*, 170063.

Masson-Delmotte, V., et al. (Eds.). (2018). *IPCC summary for policymakers Global warming of 1.5 deg C. An IPCC special report on the impacts of global warming of 1.5 deg C above pre-industrial levels and related global greenhouse gas emission pathways, in the context of strengthening the global.* Geneva, Switzerland: World Meteorological Organization.

Meehl, G. A., Karl, T., Easterling, D. R., Changnon, S., Pielke, R., Changnon, D., Evans, J., Groisman, P. Y., Knutson, T. R., Kunkel, K. E., et al. (2000). An introduction to trends in extreme weather and climate events: Observations, socioeconomic impacts, terrestrial ecological impacts, and model projections. *Bulletin of the American Meteorological Society, 81*, 413–416.

Menne, M. J., Durre, I., Vose, R. S., Gleason, B. E., & Houston, T. G. (2012). An overview of the global historical climatology network-daily database. *Journal of Atmospheric and Oceanic Technology, 29*, 897–910.

Menne, M. J., & Williams, C. N. (2009). Homogenization of temperature series via pairwise comparisons. *Journal of Climate, 22*, 1700–1717.

Mesinger, F., DiMego, G., Kalnay, E., Mitchell, K., Shafran, P. C., Ebisuzaki, W., Jović, D., Woollen, J., Rogers, E., Berbery, E. H., et al. (2006). North American regional reanalysis. *Bulletin of the American Meteorological Society, 87*, 343–360.

Min, S.-K., Zhang, X., Zwiers, F. W., & Hegerl, G. C. (2011). Human contribution to more-intense precipitation extremes. *Nature, 470*, 378–381.

Mistry, M. N. (2019a). A high-resolution global gridded historical dataset of climate extreme indices. *Data, 4*, 41. https://doi.org/10.3390/DATA4010041.

Mistry, M. N. (2019b). A high-resolution (0.25 degree) historical global gridded dataset of climate extreme indices (1970-2016). In *PANGAEA*. https://doi.pangaea.de/10.1594/PANGAEA.898014.

Mistry, M. N. (2019c). *A high-resolution (0.25 degree) historical global gridded dataset of monthly and annual cooling and heating degree-days (1970-2018) based on GLDAS data.* https://doi.org/10.1594/PANGAEA.903123.

Mistry, M. N. (2019d). Historical global gridded degree-days: A high-spatial resolution database of CDD and HDD. *Geoscience Data Journal, 6*, 214–221.

Mitchell, K. E., Lohmann, D., Houser, P. R., Wood, E. F., Schaake, J. C., Robock, A., et al..... (2004). The multi-institution north American land data assimilation system (NLDAS): Utilizing multiple GCIP products and partners in a continental distributed hydrological modeling system. *Journal of Geophysical Research: Atmospheres, 109*(D7). https://agupubs.onlinelibrary.wiley.com/doi/10.1029/2003JD003823.

Morss, R. E., Wilhelmi, O. V., Meehl, G. A., & Dilling, L. (2011). Improving societal outcomes of extreme weather in a changing climate: An integrated perspective. *Annual Review of Environment and Resources, 36*, 1–25.

Muñoz-Sabater, J., Dutra, E., Agustí-Panareda, A., Albergel, C., Arduini, G., Balsamo, G., Boussetta, S., Choulga, M., Harrigan, S., Hersbach, H., et al. (2021). ERA5-land: A state-of-the-art global reanalysis dataset for land applications. *Earth System Science Data Discussions, 2021*, 1–50.

Mysiak, J., Torresan, S., Bosello, F., Mistry, M., Amadio, M., Marzi, S., … Sperotto, A. (2018). Climate risk index for Italy. *Philosophical Transactions. Series A, Mathematical, Physical, and Engineering Sciences, 376*, 20170305.

Nairn, J. R., & Fawcett, R. J. B. (2015). The excess heat factor: A metric for heatwave intensity and its use in classifying heatwave severity. *International Journal of Environmental Research and Public Health, 12*, 227–253.

Ongoma, V., Chen, H., Gao, C., Nyongesa, A. M., & Polong, F. (2018). Future changes in climate extremes over equatorial East Africa based on CMIP5 multimodel ensemble. *Natural Hazards, 90*, 901–920.

Oppenheimer, M., Campos, M., Warren, R., Birkmann, J., Luber, G., O'Neil, B., … Van, V. D. (2014). In C. B. Field, V. R. Barros, D. J. Dokken, et al. (Eds.), *IPCC climate change 2014: Impacts, adaptation, and vulnerability. Part A: Global and sectoral aspects. Contribution of working group II to the fifth assessment report of the intergovernmental panel on climate.* Cambridge, United Kingdom and New York, NY, USA: IPCC.

Orlov, A., Sillmann, J., Aaheim, A., Aunan, K., & de Bruin, K. (2019). Economic losses of heat-induced reductions in outdoor worker productivity: A case study of Europe. *Economics of Disasters and Climate Change, 3*, 191–201.

Otto, I. M., Reckien, D., Reyer, C. P. O., Marcus, R., Le Masson, V., Jones, L., … Serdeczny, O. (2017). Social vulnerability to climate change: A review of concepts and evidence. *Regional Environmental Change, 17*, 1651–1662.

Paik, S., Min, S.-K., Zhang, X., Donat, M. G., King, A. D., & Sun, Q. (2020). Determining the anthropogenic greenhouse gas contribution to the observed intensification of extreme precipitation. *Geophysical Research Letters, 47*. e2019GL086875.

Pall, P., Aina, T., Stone, D. A., Stott, P. A., Nozawa, T., Hilberts, A. G. J., … Allen, M. R. (2011). Anthropogenic greenhouse gas contribution to flood risk in England and Wales in autumn 2000. *Nature, 470*, 382–385.

Perkins, S. E., & Alexander, L. V. (2013). On the measurement of heat waves. *Journal of Climate, 26*, 4500–4517.

Peterson, T. C., & Manton, M. J. (2008). Monitoring changes in climate extremes: A tale of international collaboration. *Bulletin of the American Meteorological Society, 89*, 1266–1271.

Pfahl, S., O'Gorman, P. A., & Fischer, E. M. (2017). Understanding the regional pattern of projected future changes in extreme precipitation. *Nature Climate Change, 7*, 423.

Prettel, L. E. (2011). Impact of weather and climate extremes. In *Impact of weather and climate extremes* (pp. 1–269). Hauppauge, NY: Nova Science Publishers. ISBN: 9781607414216 https://primoa.library.unsw.edu.au/primo-explore/fulldisplay/UNSW_ALMA21143088760001731/UNSWS.

Rahmstorf, S., & Coumou, D. (2011). Increase of extreme events in a warming world. *Proceedings of the National Academy of Sciences, 108*, 17905–17909.

Randazzo, T., De Cian, E., & Mistry, M. N. (2020). Air conditioning and electricity expenditure: The role of climate in temperate countries. *Economic Modelling, 90*, 273–287.

Randles, C. A., da Silva, A. M., Buchard, V., Colarco, P. R., Darmenov, A., Govindaraju, R., Smirnov, A., Holben, B., Ferrare, R., Hair, J., et al. (2017). The MERRA-2 aerosol reanalysis, 1980 onward. Part I: System description and data assimilation evaluation. *Journal of Climate*, *30*, 6823–6850.

Raymond, C., Horton, R. M., Zscheischler, J., Martius, O., AghaKouchak, A., Balch, J., Bowen, S. G., Camargo, S. J., Hess, J., Kornhuber, K., et al. (2020). Understanding and managing connected extreme events. *Nature Climate Change*, *10*, 611–621.

Rayner, N. A., Auchmann, R., Bessembinder, J., Brönnimann, S., Brugnara, Y., Capponi, F., Carrea, L., Dodd, E. M. A., Ghent, D., Good, E., et al. (2020). The EUSTACE project: Delivering global, daily information on surface air temperature. *Bulletin of the American Meteorological Society*, *101*, E1924–E1947.

Reig-Gracia, F., Vicente-Serrano, S. M., Dominguez-Castro, F., & Bedia-Jiménez, J. (2021). *Clim Ind: Climate indices*. Available from https://CRAN.R-project.org/package=ClimInd (accessed on 10 November 2020).

Reuveny, R. (2007). Climate change-induced migration and violent conflict. *Political Geography*, *26*, 656–673.

Rodell, M., Houser, P. R., Jambor, U., Gottschalck, J., Mitchell, K., Meng, C.-J., Arsenault, K., Cosgrove, B., Radakovich, J., Bosilovich, M., et al. (2004). The global land data assimilation system. *Bulletin of the American Meteorological Society*, *85*, 381–394.

Rohde, R., Muller, A., Jacobsen, R., Muller, E., & Wickham, C. (2013). A new estimate of the average earth surface land temperature spanning 1753 to 2011. *Geoinformatics Geostatistics An Overview*, *1*(2). https://doi.org/10.4172/2327-4581.1000103.

Saha, S., Moorthi, S., Pan, H.-L., Wu, X., Wang, J., Nadiga, S., Tripp, P., Kistler, R., Woollen, J., Behringer, D., et al. (2010). The NCEP climate forecast system reanalysis. *Bulletin of the American Meteorological Society*, *91*, 1015–1058.

Saha, S., Moorthi, S., Wu, X., Wang, J., Nadiga, S., Tripp, P., Behringer, D., Hou, Y.-T., Chuang, H., Iredell, M., et al. (2014). The NCEP climate forecast system version 2. *Journal of Climate*, *27*, 2185–2208.

Schaller, N., Kay, A. L., Lamb, R., Massey, N. R., van Oldenborgh, G. J., Otto, F. E. L., Sparrow, S. N., Vautard, R., Yiou, P., Ashpole, I., et al. (2016). Human influence on climate in the 2014 southern England winter floods and their impacts. *Nature Climate Change*, *6*, 627–634.

Schamm, K., Ziese, M., Becker, A., Finger, P., Meyer-Christoffer, A., Schneider, U., … Stender, P. (2014). Global gridded precipitation over land: A description of the new GPCC first guess daily product. *Earth System Science Data*, *6*, 49–60.

Schleypen, J. R., Mistry, M. N., Saeed, F., & Dasgupta, S. (2021). Sharing the burden: Quantifying climate change spillovers in the European Union under the Paris agreement. *Spatial Economic Analysis*, *17*(1), 67–82.

Schulzweida, U. (2018). *Max P.I. for M. Climate data operators (CDO) user guide, version 1.9.0.*

Seneviratne, S., Nicholls, N., Easterling, D., Goodess, C., Kanae, S., Kossin, J., Luo, Y., Marengo, J., Mcinnes, K., Rahimi, M., et al. (2012). Changes in climate extremes and their impacts on the natural physical environment. In *Managing the risks of extreme events and disasters to advance climate change adaptation*, ISBN:9781107607804 (chapter 3).

Seneviratne, S. I., & Hauser, M. (2020). Regional climate sensitivity of climate extremes in CMIP6 versus CMIP5 multimodel ensembles. *Earth's Future*, *8*. e2019EF001474.

Seneviratne, S. I., Zhang, X., Adnan, M., Badi, W., Dereczynski, C., Di Luca, A., … Climate Extreme Events in a Changing Climate. (2021). In V. Masson-Delmotte, A. Pirani, S. L. Connors, C. Péan, S. Berger, N. Caud, … O. Yelekçi (Eds.), *Climate change 2021: The physical science basis. Contribution of working group I to the sixth assessment report of the intergovernmental panel on climate change* Cambridge University Press (in press).

Sillmann, J., Kharin, V. V., Zhang, X., Zwiers, F. W., & Bronaugh, D. (2013). Climate extremes indices in the CMIP5 multimodel ensemble: Part 1. Model evaluation in the present climate. *Journal of Geophysical Research: Atmospheres*, *118*, 1716–1733.

Sillmann, J., Kharin, V. V., Zwiers, F. W., Zhang, X., & Bronaugh, D. (2013). Climate extremes indices in the CMIP5 multimodel ensemble : Part 2. Future climate projections. *Journal of Geophysical Research: Atmospheres*, *118*, 2473–2493.

Sillmann, J., Thorarinsdottir, T., Keenlyside, N., Schaller, N., Alexander, L. V., Hegerl, G., … Zwiers, F. W. (2017). Understanding, modeling and predicting weather and climate extremes: Challenges and opportunities. *Weather and Climate Extremes*, *18*, 65–74.

Sippel, S., Zscheischler, J., Heimann, M., Otto, F. E. L., Peters, J., & Mahecha, M. D. (2015). Quantifying changes in climate variability and extremes: Pitfalls and their overcoming. *Geophysical Research Letters*, *42*, 9990–9998.

Slivinski, L. C., Compo, G. P., Whitaker, J. S., Sardeshmukh, P. D., Giese, B. S., McColl, C., Allan, R., Yin, X., Vose, R., Titchner, H., et al. (2019). Towards a more reliable historical reanalysis: Improvements for version 3 of the twentieth century reanalysis system. *Quarterly Journal of the Royal Meteorological Society*, *145*, 2876–2908.

Squintu, A. A., van der Schrier, G., Štěpánek, P., Zahradníček, P., & Tank, A. K. (2020). Comparison of homogenization methods for daily temperature series against an observation-based benchmark dataset. *Theoretical and Applied Climatology*, *140*, 285–301.

Stott, P. A., Christidis, N., Otto, F. E. L., Sun, Y., Vanderlinden, J.-P., van Oldenborgh, G. J., Vautard, R., von Storch, H., Walton, P., Yiou, P., et al. (2016). Attribution of extreme weather and climate-related events. *WIREs Climate Change*, *7*, 23–41.

Stott, P. A., Stone, D. A., & Allen, M. R. (2004). Human contribution to the European heatwave of 2003. *Nature*, *432*, 610–614.

Tabari, H. (2020). Climate change impact on flood and extreme precipitation increases with water availability. *Scientific Reports*, *10*, 13768.

Tarek, M., Brissette, F. P., & Arsenault, R. (2020). Evaluation of the ERA5 reanalysis as a potential reference dataset for hydrological modelling over North America. *Hydrology and Earth System Sciences*, *24*, 2527–2544.

Thorarinsdottir, T. L., Sillmann, J., Haugen, M., Gissibl, N., & Sandstad, M. (2020). Evaluation of CMIP5 and CMIP6 simulations of historical surface air temperature extremes using proper evaluation methods. *Environmental Research Letters*, *15*, 124041.

Thornton, M. M., Shrestha, R., Wei, Y., Thornton, P. E., Kao, S., & Wilson, B. E. (2020). *Daymet: Daily surface weather data on a 1-km grid for North America, version 4.*

Trewin, B. (2013). A daily homogenized temperature data set for Australia. *International Journal of Climatology*, *33*, 1510–1529.

Vesco, P., Kovacic, M., Mistry, M., & Croicu, M. (2021). Climate variability, crop and conflict: Exploring the impacts of spatial concentration in agricultural production. *Journal of Peace Research*, *58*, 98–113.

Vicente-Serrano, S. M., Beguería, S., & López-Moreno, J. I. (2010). A multiscalar drought index sensitive to global warming: The standardized precipitation evapotranspiration index. *Journal of Climate*, *23*, 1696–1718.

Wartenburger, R., Hirschi, M., Donat, M. G., Greve, P., Pitman, A. J., & Seneviratne, S. I. (2017). Changes in regional climate extremes as a function of global mean temperature: An interactive plotting framework. *Geoscientific Model Development*, *10*, 3609–3634.

Weedon, G. P., Balsamo, G., Bellouin, N., Gomes, S., Best, M. J., & Viterbo, P. (2014). The WFDEI meteorological forcing data set: WATCH forcing data methodology applied to ERA-interim reanalysis data. *Water Resources Research*, *50*, 7505–7514.

Wing, I. S., De Cian, E., & Mistry, M. N. (2021). Global vulnerability of crop yields to climate change. *Journal of Environmental Economics and Management*, 102462.

Wischnath, G., & Buhaug, H. (2014). On climate variability and civil war in Asia. *Climatic Change, 122*, 709–721.

Xia, Y., Hao, Z., Shi, C., Li, Y., Meng, J., Xu, T., … Zhang, B. (2019). Regional and global land data assimilation systems: Innovations, challenges, and prospects. *Journal of Meteorological Research, 33*, 159–189.

Yatagai, A., Kamiguchi, K., Arakawa, O., Hamada, A., Yasutomi, N., & Kitoh, A. (2012). APHRODITE: Constructing a long-term daily gridded precipitation dataset for Asia based on a dense network of rain gauges. *Bulletin of the American Meteorological Society, 93*, 1401–1415.

Zender, C. S. (2008). Analysis of self-describing gridded geoscience data with netCDF operators (NCO). *Environmental Modelling and Software, 23*, 1338–1342.

Zhang, X., Alexander, L., Hegerl, G. C., Jones, P., Tank, A. K., Peterson, T. C., … Zwiers, F. W. (2011). Indices for monitoring changes in extremes based on daily temperature and precipitation data. *Wiley Interdisciplinary Reviews: Climate Change, 2*, 851–870.

Zhang, X., Zwiers, F. W., Hegerl, G. C., Lambert, F. H., Gillett, N. P., Solomon, S., … Nozawa, T. (2007). Detection of human influence on twentieth-century precipitation trends. *Nature, 448*, 461–465.

Zscheischler, J., Martius, O., Westra, S., Bevacqua, E., Raymond, C., Horton, R. M., van den Hurk, B., AghaKouchak, A., Jézéquel, A., Mahecha, M. D., et al. (2020). A typology of compound weather and climate events. *Nature Reviews Earth & Environment, 1*, 333–347.

Zscheischler, J., Westra, S., van den Hurk, B. J. J. M., Seneviratne, S. I., Ward, P. J., Pitman, A., AghaKouchak, A., Bresch, D. N., Leonard, M., Wahl, T., et al. (2018). Future climate risk from compound events. *Nature Climate Change, 8*, 469–477.

CHAPTER

3

Advances in weather and climate extremes

Victor Nnamdi Dike[a,b], Zhao-Hui Lin[a,c], Chenglai Wu[a,c], and Colman Chikwem Ibe[b]

[a]International Center for Climate and Environment Sciences, Institute of Atmospheric Physics, Chinese Academy of Sciences, Beijing, China

[b]Energy, Climate, and Environment Sciences Group, Imo State Polytechnic Umuagwo, Ohaji, Imo State, Nigeria

[c]China-Pakistan Joint Research Center on Earth Sciences, Chinese Academy of Sciences-Higher Education Commission (CAS-HEC), Islamabad, Pakistan

1 Introduction

Weather and climate extremes have profound impacts on humans and society; and, against the current background of climate change, such extremes may render many parts of the world inhabitable. These extremes include extreme hot and cold temperatures, precipitation extremes, and extreme convective events such as hurricanes, tornadoes, and hailstorms. The impact of these weather and climate extremes is daunting on the livelihood of the people. The increase in the frequency and intensity of weather and climate extremes has continued to inflict exponential huge human and financial costs on society (IPCC, 2012, 2013, 2021a). For instance, in 2016, the overall worldwide losses from meteorological, hydrological, or climatological disasters amounted to US$127bn, with 2016 being the fifth costliest year for insured losses since 1980. In the United States alone, there were about 16 events with damages of at least US$1 bn in 2017, resulting in the deaths of 282 people (Smith, 2018). The frequency of extreme events continued to increase, with 22 separate weather and climate disaster events across the United States in 2020, breaking the previous annual record of 16 events that occurred in 2017 and 2011. It is estimated that the average overall damage/costs of weather and climate extremes per annum have risen to a new record of $120.0 billion/year (NCEI, 2021).

Nonetheless, the impact of weather and climate extremes is more devastating in developing countries given that the population is largely dependent on seasonal climate to drive critical sectors of the economy (IPCC, 2013; Niang et al., 2014). For instance, the Sahel drought in the 1970s/early 1980s caused millions of deaths, hampered development and economic growth throughout the region, as agrarian activities in the region are largely dependent on summer precipitation (Agnew & Chappell, 1999; Held, Delworth, Lu, Findell, & Knutson, 2005; Zeng, 2003). Also, the 2013 summer heatwaves over eastern China were the hottest on record. Economic losses associated with the accompanying drought were estimated at $8.4 bn (Sun et al., 2014). Similarly, the impact of heatwaves has continued to ravage the Middle East and North Africa (Khan, Shahid, Ismail, Ahmed, & Nawaz, 2019; Zittis, Hadjinicolaou, Almazroui, et al., 2021). Particularly, the India-Pakistan heatwave of 2015 resulted in more than 2500 and 2000 deaths in India and Pakistan, respectively (Wehner, Stone, Krishnan, AchutaRao, & Castillo, 2016). Apart from the above-listed events, there are many more episodes of extreme weather and climate extremes with far-reaching impacts on people. The most recent events include the 2021 heatwave in North America (Schiermeier, 2021), Zhengzhou flood disaster in China (Normile, 2021), and Germany and Belgium flood disasters (Cornwall, 2021), among others. The plight of the affected population has continued to attract the attention of international aid organizations as well as the scientific community, which has encouraged research activities intended to unravel the characteristics of the extreme event in terms of causal mechanisms and prospects.

Arguably, the investigation of extreme events and their relation to climate change and variability is one of the most challenging areas in climate research. Consequently, the World Climate Research Programme (WCRP) recognized

Climate Impacts on Extreme Weather
https://doi.org/10.1016/B978-0-323-88456-3.00007-1

weather and climate extremes as one of the grand challenges (Zhang et al., 2014) and accordingly identified the importance of "understanding and predicting weather and climate extremes" (Alexander et al., 2016; Zhang et al., 2014). The WCRP promotes the extreme grand challenges through community-organized workshops, conferences, and strategic planning meetings to identify exciting and high-priority research that requires international partnership and coordination, and that yields "actionable information" for decision-makers (Seneviratne & Zwiers, 2015). Nonetheless, the "extremes grand challenge" is focused on the four most prevalent aspects of extreme events, namely heavy precipitation, heatwave, drought, and storm. This is organized around four overarching themes, which are documentation (focusing on observational requirements), understanding (focusing on the relative roles of different spatial scales and their interactions), simulation (focusing on model reliability and improvement), and attribution (focusing on unraveling the contributors to extreme events). Overall, research questions addressed under this grand challenge are related to the development of observational products on extremes based both on in situ and satellite data, the understanding of large-scale versus regional-scale mechanisms leading to the occurrence of extremes, the prediction, and simulation of extremes in climate models, as well as the possible attribution of extreme events to anthropogenic climate forcing (Alexander et al., 2016; Seneviratne & Zwiers, 2015).

As expected, these concerted efforts have led to notable advances in weather and extreme research. These include expansion of observation and numerical model simulations to enhance our understanding of weather and climate extremes. Thus, the ensuing publications have contributed immensely to the Intergovernmental Panel on Climate Change (IPCC) reports intended to provide policymakers with regular scientific assessments on climate change, its implications, and potential future risks, as well as to put forward adaptation and mitigation options. Sillmann, Daloz, Schaller, and Schwingshackl (2021) highlighted an overview of the current scientific understanding and approaches of analyzing weather and climate extremes in the context of observed climate change. Here, we focus on a broad aspect of the advances in the science of weather and climate extremes, including advances in the documentation, simulation, attribution, and quantification of these extremes and the projection of their future occurrences.

2 Evolution of definitions of weather and climate extremes

At this point, it is imperative to highlight the evolution of the definition of weather and climate extremes. The extremes are by definition rare, often difficult to define, and not well observed, which pose considerable challenges in their characterization (IPCC, 2012). However, in practice, there are many definitions of weather and climate extremes that tend to characterize the rarity and severity of the events. According to IPCC reports, weather and climate extremes are rare events at a particular place and time of year (IPCC, 2012, 2013). These events fall outside the realm of normal patterns, exceed local averages, or set a record. The characteristics of extreme events may vary from place to place. This is in line with the World Meteorological Organization (WMO) description of extreme weather and climate events as the occurrence of a value of a weather variable above (or below) a threshold value near the upper (or lower) ends of the range of its observed values in a specific region. Moreover, individual studies often choose the annual maxima or arbitrary thresholds such as the 95th, 99th, and 99.9th percentiles above which to define an extreme value. However, there can be unintended consequences of using such thresholds and the associated interpretation of results (Alexander et al., 2016). Hence, as part of the effort to address the grand challenge of weather and climate extremes, WCRP constituted the Expert Team on Climate Change Detection and Indices (ETCCDI) to coordinate the definition of extreme climate indices to facilitate intercomparison (Karl, Nicholls, & Ghazi, 1999a; Peterson, 2005; Peterson & Manton, 2008; Zhang et al., 2011).

In pursuance of its mandate to address the need for the objective measurement and characterization of climate variability and change, the joint CCl/CLIVAR/JCOMM Expert Team (ET) on ETCCDI defined a suite of 27 core indices (Karl, Nicholls, & Ghazi, 1999b; Peterson, 2005; Peterson & Manton, 2008). Zhang and Yang (2004) and Zhang et al. (2005) provide a detailed list of the indices. These extreme climate indices are relevant for climate change detection, and the comparison of modeled data and observations. The extreme indices fall broadly under four categories: "duration," "thresholds," "absolute," and "percentiles," accounting for both moderate and severe extremes. In particular, the indices are tailored to be representative of droughts, precipitation extremes, heat waves, and cold waves, characterizing spells and durations, as well as magnitude and intensity (Sillmann, Kharin, Zhang, Zwiers, & Bronaugh, 2013).

One of the disadvantages of the ETCCDI indices is that few of the indices are specifically sector-relevant. Hence, it was realized that it was important to get sectors involved in the development of the sector-specific indices so that more application-relevant indices could be developed to better support adaptation. Thus, the Expert Team on Climate Risk and Sector-specific Indices (ETCRSCI) was commissioned, and later on Expert Team on Sector-Specific Climate Indices (ET-SCI), with terms of reference centered on climate information for adaptation and risk management across sectors.

A set of 34 core indices was agreed which includes the 27 ETCCD indices. In particular, Perkins and Alexander (2013) and Perkins, Alexander, and Nairn (2012) added heat wave indices that constitute most of the ET-SCI extreme climate indices. This is thus intended to improve decision-making for planning, operations, risk management, and for adaptation to both climate change and variability (covering time scales from seasonal to centennial), tailored to meet sector-specific needs. Specifically, ET-SCI will work with sector-based agencies and experts, including those of relevant WMO Technical Commissions, particularly the Commission for Climatology for health, the Commission for Hydrology (CHy) for water, and the Commission for Agricultural Meteorology (CAgM) for agriculture and food security, to facilitate the use of climate information in users' decision-support systems for climate risk management and adaptation strategies. It should be noted that index development is an ongoing activity as additional sector needs arise and other sectors are considered within the terms of reference and deliverables of ET-SCI. Nevertheless, both ETCCDI and ET-SCI commissioned the development of RClimDEX (Zhang & Yang, 2004) and ClimPACT2 (Alexander & Herold, 2016) to produce an easy and consistent way of calculating the indices.

Numerous observational and numerical model studies have used these indices to study variability and changes in climate extremes across timescales (Ayugi et al., 2021; Diatta, Mbaye, & Sambou, 2020; Dike, Lin, & Ibe, 2020; Dike, Lin, Wang, & Nnamchi, 2019; Khan et al., 2019; Perkins et al., 2012; Perkins & Alexander, 2013; Peterson & Manton, 2008; Seneviratne & Zwiers, 2015; Sillmann, Kharin, Zhang, et al., 2013; Sun et al., 2014). Furthermore, considering the fact that most daily observation datasets are not openly available (Dike et al., 2018), the ETCCDI facilitated several regional workshops, where representatives of National Meteorological and Hydrological Services (NHMSs) could attend and calculate the climate extremes indices from their daily data without needing to release those data publicly (Peterson & Manton, 2008). This effort has yielded three generations of a land-surface dataset of climate extremes indices (HadEX) (Caesar, Alexander, & Vose, 2006; Donat et al., 2013a, 2013b; Dunn et al., 2020). The datasets provide a tool for understanding climate extremes, especially in developing countries (Dunn et al., 2020).

More recently, an international effort has been focused on global subdaily rainfall extremes, through the INTENSE (INTElligent use of climate models for adaptatioN to non-Stationary hydrological Extremes) project (Blenkinsop et al., 2018). The INTENSE project has contributed substantially to the research efforts of the Global Energy and Water Exchanges (GEWEX) Hydroclimatology Panel (GHP) and the WCRP "Grand Challenge" on extremes (Alexander et al., 2016). Following deliberations with the climate observations and modeling communities, the project identified a new set of indices that describe important attributes of subdaily extremes (Table 3.1; Lewis et al., 2018). These indices are intentionally described to correspond with the existing daily ETCCDI indices. For example, indices of monthly maxima of hourly and multihourly rainfall will be provided along with monthly counts of threshold exceedances, as well as indices reflecting the diurnal cycle (e.g., monthly index of the wettest hour). These indices are consistent with the naming and methodological conventions of the ETCCDI daily indices for easy application by users and the data will be made freely accessible for research purposes (Alexander et al., 2019; Lewis, Guerreiro, Blenkinsop, & Fowler, 2018).

TABLE 3.1 Definitions of the proposed set of subdaily precipitation indices.

Type of index	Index
Monthly maximum indices	Rx1h monthly maximum 1h precipitation
	Rx3h monthly maximum 3h precipitation
	Rx6h monthly maximum 6h precipitation
	Rx1hP percent of daily total that fell in the Monthly maximum 1h precipitation
Diurnal cycle indices	LW1H monthly likely wettest hour within a day
	LD1H monthly likely driest hour within a day
	DLW1H dispersion around monthly likely wettest hour within a day
	S1HII simple hourly precipitation intensity index
	CW1H maximum length of wet spell
Frequency/threshold indices	R10mm1h monthly count of hours when PRCP $\geq$10mm
	R20mm1h monthly count of hours when PRCP $\geq$20mm
	Rxmm1h annual count of hours when PRCP $\geq$ nnmm, nn is a user defined threshold
General indices	PRCPTOT1h annual total precipitation in wet hours

3 Spatiotemporal scales of measurement of weather and climate extremes

The measurement and statistics of extreme values of weather and climate parameters have progressed from a general art and science to a highly sophisticated body of knowledge and technology that may be applied to a wide range of practical activities (Radinović & Ćurić, 2014). This progress includes advances in the measurement of frequency, duration, and intensity among others on both weather and climate scales (El Kenawy, López-Moreno, & Vicente-Serrano, 2011; IPCC, 2013; Niang et al., 2014; Perkins & Alexander, 2013; Peterson, 2005; Peterson & Manton, 2008; Radinović & Ćurić, 2012, 2014). Many aspects of weather and climate extremes are well represented by daily means (Perkins & Alexander, 2013; Zhang et al., 2011). Nonetheless, some aspects of weather extremes are better represented by subdaily datasets (Alexander et al., 2019; Lewis et al., 2018). Both observational and numerical model studies have analyzed weather and climate at different timescales (Fischer, Beyerle, & Knutti, 2013; Förster & Thiele, 2020; Kirchmeier-Young & Zhang, 2020; Li, Guo, Sha, & Yang, 2021; Myhre, Alterskjær, Stjern, et al., 2019; Sillmann et al., 2017; Sillmann, Kharin, Zhang, et al., 2013; Yao et al., 2021). Notably, it has been found that the spatial scale of weather and climate extremes are nonlinear (Angélil, Stone, Perkins-Kirkpatrick, et al., 2018) and not homogeneous (Hu et al., 2019).

Obviously, the spatiotemporal scales of quantifying weather and climate extremes are largely dependent on the availability of datasets to test the stochastic hypothesis and different scales. Thus, the advancement in spatiotemporal scales of measurement can further be discussed based on the spatiotemporal improvements in both observational and model datasets. Interestingly, spatiotemporal scales in the measurement of weather and climate extremes have continued to improve over the years. For instance, the spatial resolution of observational and model datasets has improved from coarse to finer resolutions. Particularly, improved observations and sampling of under-sampled areas have enhanced both spatial and temporal scales of observational datasets needed for monitoring and detection of weather and climate extremes. The progress in HadEX is one such example. HadEX was the first global data set to contain all 27 indices defined by the ETCCDI (Alexander et al., 2006), which covered the period from 1951 to 2003. HadEX2 extended the length of the data set, from 1901 to 2010 and the number of land-surface stations contributing to HadEX2 was larger than that in HadEX. Given the spatial distribution of the station datasets, both HadEX and HadEX2 were made available on $3.75° \times 2.5°$ longitude-latitude grid resolution. It is also important to mention the Global Historical Climatology Network daily climate extremes (GHCNDEX; Donat et al., 2013a) and (HadGHCND; Caesar et al., 2006), which includes temperature extremes that are similar efforts interpolated into the same grid used by HadEX. Moreover, most of ETCCDI estimated using RClimdex were calculated on an annual scale. The advent of ET-SCI provided the opportunity for monthly estimation of most of the extreme climate indices. This measurement of climate extremes implemented in CLIMPACT2 enabled seasonal scale analysis of these extremes which serves the needs of specific sectors. Thus, HadEX3 consolidated on the advances in HadEX, and HadEX2 by providing a finer resolution dataset of 1.875×1.25 longitude-latitude grid and covering the temporal scale from 1901 to 2018 (Dunn et al., 2020). However, progress is still limited and there are key regions of the globe such as Africa where the issue of very limited access to data remains (Alexander et al., 2019). Hence, data availability in recent decades remains a challenge in Africa (Dike et al., 2018).

Nonetheless, it is becoming increasingly clear that indices based on daily data will miss or mask some of the most intense events that could lead to flash flooding. Consequently, the focus has shifted to the use of indices for subdaily data. However, access to and availability of subdaily station data is even more limited than that of daily data (Alexander et al., 2019). Interestingly, some datasets have been or are being developed as part of the INTENSE project (Blenkinsop et al., 2018) intended to monitor and detect subdaily rainfall extremes. For example, the Global Subdaily Rainfall dataset (GSDR; Lewis et al., 2019) has been developed as the first major international effort to focus on global subdaily rainfall extremes. These dataset will be useful for the estimation of subdaily precipitation extremes. Hopefully, this effort will address the case of small-scale convective extremes which are often masked by daily means.

Additionally, model simulation has aided our ability to understand future changes in weather and climate extremes. However, the range of changes reported in the climate modeling literature is very large, sometimes leading to contradictory results for a given extreme weather and climate event. Much of this uncertainty stems from the incomplete understanding of the physics of extreme weather processes, lack of representation of mesoscale processes in coarse-resolution climate models, and the effect of natural climate variability at multidecadal time scales (Moore, Matthews, Simmons, & Leduc, 2015). Interestingly, the progress in model development has continued to address these problems to a commendable extent. One good example is the consistent progress in the Coupled Model Intercomparison Project (CMIP; Eyring et al., 2016; Meehl et al., 2007; Meehl, Covey, McAvaney, Latif, & Stouffer, 2005; Taylor, Stouffer, & Meehl, 2012) and the regionally downscaled version Coordinated Regional Climate Downscaling experiment (CORDEX; Gutowski Jr. et al., 2016). Hence, the model outputs have been used to investigate variations and changes in weather and climate extremes across scales.

4 Changes in frequency, intensity, duration of climate extremes

Extreme climate indices are defined to capture the frequency, intensity, and duration of climate extremes. Many studies have used these indices to investigate changes in weather and climate extremes (IPCC, 2021a; Sillmann, Kharin, Zwiers, Zhang, & Bronaugh, 2013; Sun et al., 2014; Zittis et al., 2021). There is a seeming consensus that the changes in these extremes are invigorated by global warming (IPCC, 2012, 2013, 2018, 2021a, 2021b). Analytically, the present-day change can simply be estimated as the long-term trend of a time series which can be assessed using a wide variety of datasets. Meanwhile, a clear change can also be established from the year-to-year variations of climate extremes, estimated as a departure from the long-term mean. It is imperative to note that the results may show strong interannual variability rather than a long-term change. For instance, Sun et al. (2014) obtained a strong interannual variability before 1990 and a clear change thereafter, the study thus focused on the rapid increase of the anomalous temperature extremes after the 1990s. Meanwhile (Mishra, Ganguly, Nijssen, & Lettenmaier, 2015) showed a clear change. This may also hold for projected change in climate extremes. However, results from studies vary depending on the choice of baseline and time slide selected for the future projection (Liersch et al., 2020). To foster intercomparison of results, the WMO recommends using the 30-year period of 1961 to 1990 as the climate normal when comparing with future periods and that this should be maintained as a reference for monitoring long-term climate variability and change (WMO, 2014). Later on, the 30-year baseline period was updated, 1981–2010, to give a more recent context for understanding weather and climate extremes (WMO, 2017). The IPCC used the 20-year period 1986–2005 as the baseline in many diagrams in the Fifth Assessment Report (IPCC, 2014) and the Sixth Assessment Report used 1995–2014 as the baseline period (IPCC, 2021a, 2021b; Liersch et al., 2020). Meanwhile, the relative change between a specific baseline and a corresponding future period is computed as shown in Eq. (3.1):

$$\Delta FC_{i,j} = \frac{FC_{\text{future},i} - FC_{\text{base},j}}{FC_{\text{base},j}} \tag{3.1}$$

where FC_{base} is the average of the annual values of the baseline period and FC_{future} is the average of a future period. The index i refers to different future periods. The index j represents the baselines.

Several studies have used these methods to investigate the future changes in the frequency, intensity, and duration of temperature and precipitation extremes (Freychet, Hegerl, Mitchell, & Collins, 2021; Ge, Zhu, Luo, Zhi, & Wang, 2021; King et al., 2018; Sillmann, Kharin, Zwiers, et al., 2013; Westra et al., 2014; Xu, Gao, Giorgi, et al., 2018).

For instance, we present in Fig. 3.1 is the projected changes in summer precipitation extremes under a high greenhouse gas emission scenario, Fig. 3.1a shows that the projected summer changes in total precipitation amount are regionally dependent in monsoon regions, such that CMIP6 models project a significant and robust increase in PRCPTOT over east Asia, north Africa, North America, India and South Asian monsoon regions toward the end of the 21st century. Nonetheless, the projected changes in total summer precipitation amount are seemingly linked to the projected changes in the frequency (RR1) and the intensity (RX5day) of precipitation extremes. In the next section, we discuss in detail the changes in frequency, intensity, and duration of climate extremes based on studies that have quantified the changes in climate extremes.

4.1 Changes in frequency of weather and climate extremes

On the scale of weather events, it has been suggested that human-induced climate change has invigorated more frequent extreme weather events (Robinson, 2021). However, certain types of extreme weather events have become more frequent while other types have become less frequent (Moore et al., 2015; Robinson, 2021). In particular, heat waves have become more frequent while extreme cold days are less frequent (Meehl et al., 2000). Nonetheless, both observational and numerical model studies also suggest an increase in the frequency of extreme precipitation (Moore et al., 2015; Myhre et al., 2019; Robinson, 2021). It is however, important to note that the reported increase in the frequency of extreme precipitation is regionally dependent; hence, they are not homogenous across the globe. Additionally, Tabari, Hosseinzadehtalaei, AghaKouchak, and Willems (2019) indicate that uncertainties in the projected changes are highly heterogeneous across latitudes. Such that the tropical and subtropical regions are identified as the global uncertainty hotspots, with the Sahara desert and the southern part of the Middle East being the local hotspots. This heterogeneity on regional scales reflects the important role of local/regional processes and feedbacks, which can substantially impact the effects of climate change at regional levels (Sillmann et al., 2017, 2021). Nonetheless, the changes in the frequency of weather extremes are attributable to climate change resulting from anthropogenic forcing (Mann, Rahmstorf, Kornhuber, et al., 2017; Ornes, 2018).

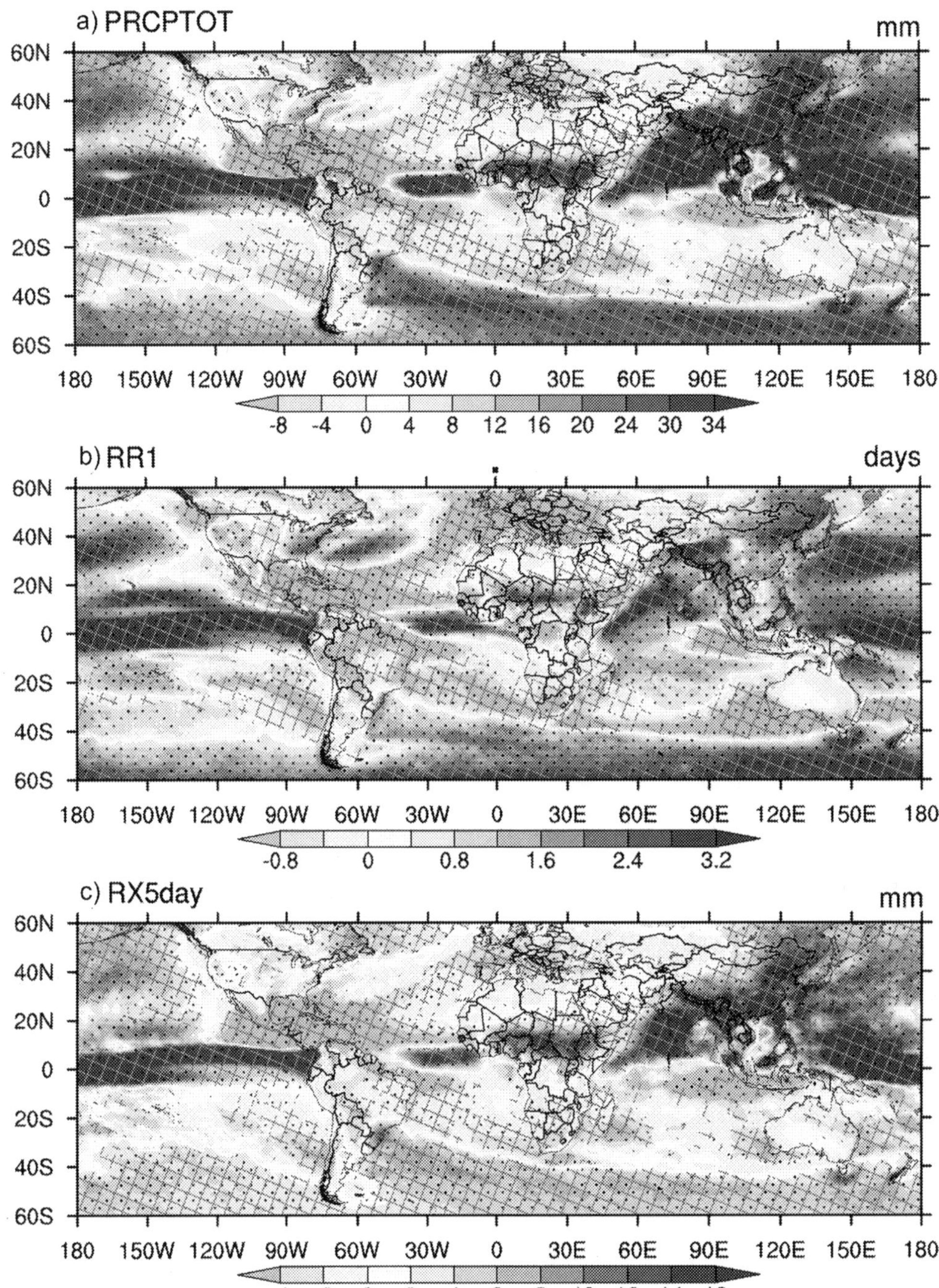

FIG. 3.1 Spatial distribution of the projected changes in precipitation extremes under the SSP5-8.5 emission scenario, 2081-2100 relative to reference period 1995–2014 during the summer July-August-September (JJA) season. (a) Total precipitation amount (PRCPTOT), (c) Frequency of rainy days estimated by the number of wet-days >1 mm (RR1). (c), maximum consecutive 5-day precipitation (RX5day). Shadings are the multimodel ensemble mean change. The *black dots* indicate statistically significant changes at the 95% confidence level and the slopped *green* (*gray* in print version) hatching indicates grid points where at least 70% of the GCMs agree on the sign of the change in ensemble mean.

Although some extreme climate indices represent the frequency of extreme events, many studies have investigated the frequency of extreme events as a function of their return periods (Herold et al., 2021). Return periods are derived from the inverse of the occurrence probability of an event, with lower return periods indicating a higher frequency of an event (Herold et al., 2021; Kharin, Zwiers, Zhang, & Hegerl, 2007). Similarly, it is found that rarer extreme events will experience greater changes in frequency in China, especially under higher warming (Li, Bai, You, Hou, & Li, 2021). Similar results are obtained for subdaily precipitation over Europe as Hosseinzadehtalaei, Tabari, and Willems (2020) indicated that the frequency of subdaily extreme precipitation events of 50- and 100-year return periods will be tripled under the high-end RCP8.5 scenario. In addition, the frequency change in weather and climate extremes can also be assessed as the change in the number of events above the threshold given by the reference period (see Myhre et al., 2019; Sun et al., 2014 for details). Interestingly, it is found that the frequency of climate extremes have increased in most land areas and this is projected to increase further due to the impact of global warming (Myhre et al., 2019). For instance, Myhre et al. (2019) studied the changes in precipitation extremes throughout Europe and found that the frequency of extreme precipitation has increased extensively under global warming. Besides, Fig. 3.1b shows that the changes in the projected frequency of summer wet-days are regionally dependent in monsoon regions, such that CMIP6 models project a significant and robust increase in RR1 over east Asia, north Africa, North America, India and South Asian monsoon regions toward the end of the 21st century. Relatedly, Sun et al. (2014) demonstrated that the frequency of hot summers has increased in China and will continue to increase toward the end of the 21st century. Similar inferences are obtained from studies that used extreme temperature indices that represent the frequency of high-temperature extremes throughout the region (Chen & Dong, 2021; Shi, Jiang, Chen, & Li, 2018). In a warmer world, the frequency of extreme heat waves would double over most of the globe under 2°C global warming (Dosio, Mentaschi, Fischer, & Wyser, 2018). In Africa, Iyakaremye et al. (2021) demonstrated that warm days' frequency is expected to upsurge under shared socioeconomic pathways (SSP), thus by SSP2-4.5 (26%–59%) and SSP5-8.5 (30%–69%) relative to the recent climate and highlighted the increase population exposure the high temperature extremes. The changes in frequency of heatwaves have also been linked to global warming. Moreover, there is seeming consensus on the global scale increase in high temperature extremes and this persistent increase in high temperature extremes is attributed to the increase in greenhouse gas emissions and the consequential increase in global warming (Chen & Dong, 2021; Shi et al., 2018; Sun et al., 2014).

4.2 Changes in intensity of weather and climate extremes

In a warming climate, saturation vapor pressure increases exponentially with atmospheric temperature according to the Clausius-Clapeyron relationship (Lenderink, Barbero, Loriaux, & Fowler, 2017; Skliris, Zika, Nurser, Josey, & Marsh, 2016). This increase in atmospheric water vapor, in turn, invigorates heavier than normal precipitation (Liu, Liu, Shiu, Li, & Zhang, 2016; Trenberth, Dai, Rasmussen, & Parsons, 2003). Observational studies have shown the intensification of precipitation extremes (Dike et al., 2020; Guerreiro et al., 2018). Moreover, Westra et al. (2014) found evidence that extreme rainfall intensity has been increasing on the global scale in recent years, potentially leading to an increase in the magnitude and frequency of flash floods. Interestingly, more intense but less frequent precipitation extremes have been reported in some regions (Bichet & Diedhiou, 2018; Luong, Dasari, & Hoteit, 2020). In fact, the intense weather and climate extremes are characterized by record-shattering severe extremes for both precipitation (Fischer, Sippel, & Knutti, 2021) and temperature extremes (Nangombe et al., 2018). Evidence shows that human-induced increases in greenhouse gases have contributed to the observed intensification of heavy precipitation events (Min, Zhang, Zwiers, & Hegerl, 2011). The intense weather and climate extremes have also been projected to increase significantly in the future (Fischer et al., 2013; Nangombe et al., 2018; Westra et al., 2014) according to the Clausius-Clapeyron relationship (Lenderink et al., 2017; Moustakis, Papalexiou, Onof, & Paschalis, 2021; Skliris et al., 2016). Moreover, Fig. 3.1c illustrates that the accumulated changes in total summer precipitation is significantly linked to the projected changes in intense precipitation extremes over the aforementioned regions.

Moreover, Iyakaremye et al. (2021) found that the intensity of high-temperature extremes is anticipated to escalate between 0.25°C and 1.8°C and 0.6°C to 4 °C under SSP2-4.5 and SSP5-8.5, respectively, in mid-21st century over Africa. They concluded that ~353.6 million person-days under SSP2-4.5 and ~401.4 million person-days under SSP5-8.5 will be exposed to intense high-temperature extremes mainly due to the interaction effect. In a 1.5°C world, 13.8% of the world population will be exposed to severe heat waves at least once every 5 years. This fraction becomes nearly three times larger (36.9%) under 2°C warming, i.e., a difference of around 1.7 billion people across the globe (Dosio et al., 2018). Already vulnerable regions like Africa will be hard-hit, as significant warming across Africa is projected at the 1.5°C warming world and is amplified at the 2°C world, exceeding the mean global warming rate (Nangombe, Zhou, Zhang, Zou, & Li, 2019). Extreme heat can have adverse impacts on human health, leading to decreased worker productivity, causing increased morbidity and, more severely, resulting in premature death (Dosio et al., 2018; Sillmann et al., 2021). Not only will the human population be affected by the intensification of high-temperature extremes, but the impact on the ecosystem will be daunting in the future (Stillman, 2019). Nonetheless, heat waves are also often interlinked with drought conditions and importantly, they can mutually exacerbate each other (Sillmann et al.,

2021; Trenberth et al., 2014). As such, the intensification of dry conditions has also been documented (El Kenawy et al., 2020; Fathian et al., 2020; Zhang et al., 2018) and is projected to intensify in the future (Ukkola, De Kauwe, Roderick, Abramowitz, & Pitman, 2020).

4.3 Changes in timing, duration seasonality of weather, and climate extremes

Global warming is intensifying extreme weather and climate events arising from fundamental properties of the thermodynamics of air (Trenberth et al., 2003). Particularly, the intensification of extreme precipitation will likely affect ecosystems and flooding around the world. It has also been observed that the changes in the climate systems have also altered the timing, duration, and indeed the seasonality of precipitation (Kumar, 2013; Li, Yu, & Sun, 2013; Mondal & Mujumdar, 2015; Moustakis et al., 2021; Persiano et al., 2020; Tan, Wu, Liu, & Chen, 2020). For instance, statistically significant changes in the seasonality and duration of precipitation extremes with shorter spell duration that are uniformly distributed throughout the year are projected over the United States (Moustakis et al., 2021). Furthermore, Feng, Porporato, and Rodriguez-Iturbe (2013) also reported a shift in seasonal magnitude, timing, and duration of precipitation in the tropics. Similar changes have also been reported over central Asia (Dike, Lin, Kece, Langendijk, & Nath, 2022), and the Alps (Brönnimann et al., 2018). Dike et al. (2022) found a seeming future change in the seasonality of the projected precipitation extremes over Central Asia with a likelihood of drier (wetter) spring over the southern (northern) central Asia and conversely wetter (drier) summer over northern (southern) central Asia which is departure from norm over the subregions. Moreover, Brönnimann et al. (2018) demonstrated that RX1day events have become less frequent in late summer and more frequent in early summer and early autumn, when it is cooler. Hence, the seasonality shift in RX1day extreme is perhaps related to dry and warm summer. As such the accompanying compound extreme events may have a far-reaching impact on the people. It has also been suggested that extreme events will likely intensify in the future, while their duration and seasonal timing will shift in many regions in a warmer future (Moustakis et al., 2021; Persiano et al., 2020). Arguably, changes in the seasonality precipitation extremes may be more important than the changes in mean precipitation (Chou et al., 2013; Lan, Lo, Chen, & Yu, 2019), due to more significant impact induced on critical sectors like agriculture and ecology. Interestingly, studies have investigated the mechanism associated with the changes in precipitation seasonality (Chou et al., 2013; Lan et al., 2019). It is found that changes in seasonal precipitation are associated with thermodynamic changes due to the increase in water vapor and the dynamic response due to atmospheric circulation changes (Chou et al., 2013; Geng, Xie, Zheng, & Wang, 2020; Lan et al., 2019).

4.4 Large-scale features associated with weather and climate extremes

Compelling scientific evidence has demonstrated that the observed and projected changes in weather and climate extremes are largely invigorated by global warming resulting from human activities (IPCC, 2021a, 2021b; Sun et al., 2014; Tabari, Madani, & Willems, 2020). Particularly, process-based understanding of the mechanism associated with the changes in the extremes links the increasing anthropogenic greenhouse gas emissions to global warming, which creates instability in the climate system (Trenberth et al., 2003). Moreover, extreme events are modulated by a complex system of large-scale atmospheric interactions. These interactions stem from instability in the climate system, comprising ocean, land, and atmosphere. Arguably, natural variabilities in the ocean are known to induce extreme weather and climate events at different time scales (Joshi, Rai, Kulkarni, & Kucharski, 2020; Wang & Yan, 2011). For example, a Madden-Julian Oscillation (MJO) at subseasonal timescale, El Niño-Southern Oscillation (ENSO) at interannual timescale, the Pacific Decadal Oscillation (PDO), and Atlantic multidecadal oscillation at decadal timescale. The underlying mechanisms have been noted in many studies (Joshi et al., 2020; Matsueda & Takaya, 2015; Muhammad, Lubis, & Setiawan, 2021; Stephenson et al., 2014; Wei, Yan, & Li, 2021). These factors drive dynamic conditions like monsoon systems, atmospheric rivers, low-level jets, tropical and extratropical cyclones, and subtropical highs, among others, which invigorate precipitation and temperature extremes (Hatsuzuka, Sato, & Higuchi, 2021; Matsueda & Takaya, 2015). Moreover, the characteristics of these natural variabilities are been altered by global warming (IPCC, 2021a, 2021b), as these dynamic processes are strongly interconnected with thermodynamic factors (Liu, Tan, Gan, et al., 2020; Wang & Yan, 2011) through the Clausius-Clapeyron relationship (Trenberth et al., 2003).

In the context of global warming, both atmospheric dynamics and thermodynamic conditions have been undergoing significant changes (Pfahl, O'Gorman, & Fischer, 2017; Witze, 2018), leading to changes in the atmospheric moisture sources, transport pathways, and their development and evolution that are highly associated with occurrences of extreme precipitation events (Zhou, Huang, Wang, & Cheng, 2018). Changes in the moisture in the air depend on temperature. When the air is heated by 1°C, it can hold approximately 7% more water (Hatsuzuka et al., 2021; Liu et al., 2020; Trenberth et al., 2003). The severity of extreme precipitation events is often determined by the amount of water vapor in the atmosphere and an increase in atmospheric water vapor due to an increase in its temperature, and the amount of precipitation is expected to increase (Liu et al., 2020; Pfahl et al., 2017; Trenberth et al., 2003). In addition to thermodynamic mechanisms for water vapor, extreme precipitation events also result from dynamic mechanisms of multiscale weather systems like atmospheric rivers, low-level jets, and tropical cyclones, subtropical highs, among others. Their complex interactions drive atmospheric moisture transport and provide water vapor for precipitation and its extremes (Liu et al., 2020; Zhou et al., 2018). The dynamics of each of these weather systems are typically restricted by some regional synoptic circulations (Liu et al., 2020). As the dynamic response of the extremes is associated with the changes in the spatial distribution of atmospheric circulations, hence, the changes in weather and climate extremes are not regionally and seasonally homogenous (Dike et al., 2022). Accordingly, changes in synoptic circulations could impact the persistent water vapor sources and pathways for extreme precipitation that are provided by these weather systems. Moreover, the underlying mechanism causing the air to rise in a mesoscale convective system, precipitation is often classified as convective precipitation due to atmospheric instability and dynamic precipitation arising from large-scale lifting or orographic lifting (see, Liu et al., 2020), and references therein. It has been suggested that the synoptic occurrence of precipitation extremes is largely determined by large-scale atmospheric circulations, while thermodynamic aspects of the atmosphere determine the intensity of precipitation extremes (Busuioc, Dobrinescu, Birsan, Dumitrescu, & Orzan, 2015; O'Gorman & Schneider, 2009; Pfahl et al., 2017).

5 Innovations in weather and climate extremes prediction

Subseasonal to seasonal prediction of extreme weather events is one of the major challenges to the scientific community (Vitart et al., 2019). The prediction of extreme weather events 2 to 6 weeks ahead has immense social-economic benefits for hazard prevention and risk management as well as economic planning (Wang & Moon, 2018). Considering this challenge and the importance of subseasonal to seasonal prediction, the World Weather Research Program in alliance with the World Climate Research Program established subseasonal to seasonal prediction project (S2S) project in 2013 (Vitart & Robertson, 2018). Since the inception of the S2S project, giant strides have been made to facilitate seamless subseasonal to seasonal prediction of extreme weather and climate events. This includes enhanced understanding of the physical basis for subseasonal prediction which is primarily rooted in the intrinsic predictability of large-scale circulation associated with MJO which modulates extreme weather events in the tropics and extratropics (Wang & Moon, 2018). Interestingly, recent advances model physics and the implementation of deep learning bias correction method have significantly improved the accurate prediction of tropical MJO and its teleconnections (Hirons, Inness, Vitart, & Bechtold, 2013; Kim, Ham, Joo, & Son, 2021; Wang & Moon, 2018). It is therefore expected that in the near future, S2S framework for forecasts of extreme events will be fully operational at community level to provide an early warning for extreme events a few weeks in advance.

It is also imperative to state that the innovations that artificial intelligence signifies a beginning of a new era in extreme weather and climate prediction. Recent studies have shown that deep learning approach can skillfully represent subgrid processes in climate models (Rasp, Pritchard, & Gentine, 2018). Moreover, it has been found that deep learning multiyear ENSO forecasts are more skillful than those of current state-of-the-art dynamical forecast systems (Ham, Kim, & Luo, 2019). In addition, artificial intelligence is used to reconstruct missing climate information (Kadow, Hall, & Ulbrich, 2020) and parameterization of climate models (Rasp et al., 2018). This suggests that model uncertainties and biases which have impeded the accuracy of climate models in the past can be corrected by artificial intelligence (Baño-Medina, Manzanas, & Gutiérrez, 2020; Reichstein et al., 2019). As such, it is expected that in the future artificial intelligence will enhance extreme weather and climate predictions for disaster risk reduction (WMO, 2022).

6 Conclusion

Extreme weather and climate events undoubtedly constitute a huge threat to human existence and the ecosystem. It is important to characterize weather and climate extremes for the development of mitigation and adaptation strategies. Several global concerted efforts commissioned the scientific community to pursue new research frontiers in order to shape our understanding of weather and climate extremes. Over the years of continuous work, substantial evidence has emerged that clearly shows that the occurrence of extreme weather and climate events are invigorated by global warming. Furthermore, studies have shown that global warming is mainly driven by the increase in greenhouse gas emissions (IPCC, 2012, 2013, 2021a). Consequently, record-breaking extreme weather and climate events have continued to ravage different parts of the world, and its impact is daunting and inimical to sustainable development. Moreover, the scientific community has continued to make progress in research frontiers aimed at providing credible information needed to alleviate the impact of weather and climate extremes around the globe. Specifically, the IPCC reports are handy with a summarized report for policymakers which provides a high-level summary of the understanding of the current state of the climate, including how it is changing and the role of human influence (Tabari et al, 2020). IPCC reports also provide a high-level summary on the state of knowledge about possible future climate change (especially those relevant to key regions and sectors), as well as the mitigation of human-induced climate change (IPCC, 2021b).

This chapter has captured in fair terms some of the progress made so far toward better documentation, understanding, simulation, and the attribution of weather and climate extremes. One notable challenge to the understanding of weather and climate extremes was the lack of datasets needed for the monitoring and detection of extreme events. Hence, expansion of observation capability provides more reliable information on the state of the atmosphere and ocean surface from land-based and space-based instruments. This allows seamless analysis of extreme weather and climate events using climate extreme indices which particularly describe temperature and precipitation extremes, and are tailored to meet sector-specific needs. These indices have encouraged analysis of extreme climate events at global and regional scales using high resolution in situ and satellite datasets. Later on, the subdaily precipitation indices were defined to account for extreme weather events. As short-duration extreme precipitation can cause devastating flooding that puts lives, infrastructure, and natural ecosystems at risk. However, a significant barrier to answering extant scientific questions is the lack of availability subdaily rainfall data on the global scale. Interestingly, ongoing efforts have led to the collation of several global subdaily rainfall datasets based on gauge observations. The dataset has a wide range of applications, including improving our understanding of the nature and drivers of subdaily rainfall extremes, improving and validating high-resolution climate models, and developing a high-resolution subdaily subdaily rainfall dataset of indices. In addition to improved observations, continuous progress in climate model development has also contributed significantly to our understanding of how extreme weather and climate have changed in the past and may change in the future. In particular, the Coupled Model Intercomparison Project and its diagnostic MIPs like Detection and Attribution Model Intercomparison Project (DAMIP) and CORDEX, among others, provide experiments to quantify the role of greenhouse gases.

Interestingly, numerous studies have used these indices and datasets to investigate weather and climate extremes and concluded that weather and climate extremes have increased in both intensity and frequency (IPCC, 2012, 2013, 2021a). However, less frequent and more intense precipitation extremes have been observed in some regions (Bichet & Diedhiou, 2018; Luong et al., 2020). Nonetheless, future projections indicate that one of the most visible consequences of a warming world is an increase in the intensity and frequency of extreme weather events (IPCC, 2021a). Thus, efforts to pursue low global warming might be of considerable benefit. Although the 2.0°C limit above preindustrial levels has been under discussion for years as a benchmark for dangerous impacts, many small island countries highly susceptible to the dangers of climate change have raised concerns that a 2°C target might not be enough to shield them from the risk associated with global warming. Therefore, they welcomed the Paris agreement centered on limiting global warming to 1.5°C with great expectations (IPCC, 2018). In fact, the latest IPCC report opined that in the next 20 years global temperature is expected to reach or exceed 1.5°C of warming (IPCC, 2021a, 2021b), unless there are rapid, sustained, and large-scale reductions in climate change-causing greenhouse gas emissions, including CO_2, methane, and others. The report further notes that efforts to reduce greenhouse gas emissions over the past decades have been wholly insufficient, and charged nationally determined contributions to net-zero emissions (IPCC, 2021b).

Although significant efforts have been made to document, understand, simulate, and attribute weather and climate extremes, the investigation of extreme events and their relation to climate change and variability remains one of the most challenging areas in climate research. For progress to be sustained, the scientific community needs early-career scientists' deep empathy for event-driven losses and casualties, who will use advanced statistical tools to explore the scale of interaction of extreme events (Carlson, 2015). Interestingly, recent innovations in artificial intelligence present a new frontier that holds the potential to take extreme weather and climate research to a new dimension, especially in model physics and parameterizations needed to reduce biases and uncertainties in climate model simulations. In addition to its potentials in revolutionizing subseasonal prediction of weather extremes needed for early warning system as severe extreme weather persists.

Acknowledgments

The authors thankfully acknowledge the funding provided by the CAS-TWAS Network for Sustainable Development and the support from the NSFC research fund for international young scientists (Grant No. 42150410394).

References

Agnew, C. T., & Chappell, A. (1999). Drought in the Sahel. *GeoJournal, 48*(4), 299–311.

Alexander, L., et al. (2016). *Implementation plan for WCRP grand challenge on understanding and predicting weather and climate extremes—The "Extremes Grand Challenge"*. Version 30th June 2016 https://www.wcrp-climate.org/images/documents/grand_challenges/WCRP_Grand_Challenge_Extremes_Implementation_Plan_v20160708.pdf. Accessed 10 September 2021.

Alexander, L., & Herold, N. (2016). *ClimPACTv2 indices and software. A document prepared on behalf of the Commission for Climatology (CCl) Expert Team on Sector-Specific Climate Indices (ET-SCI), Sydney*. WMO Publications. https://epic.awi.de/id/eprint/49274/1/ClimPACTv2_manual.pdf.

Alexander, L. V., et al. (2006). Global observed changes in daily climate extremes of temperature and precipitation. *Journal of Geophysical Research: Atmospheres, 111*, D05109. https://doi.org/10.1029/2005JD006290.

Alexander, L. V., et al. (2019). On the use of indices to study extreme precipitation on sub-daily and daily timescales. *Environmental Research Letters, 14*(12), 125008. https://doi.org/10.1088/1748-9326/ab51b6.

Angélil, O., Stone, D., Perkins-Kirkpatrick, S., et al. (2018). On the nonlinearity of spatial scales in extreme weather attribution statements. *Climate Dynamics, 50*, 2739–2752. https://doi.org/10.1007/s00382-017-3768-9.

Ayugi, B., Zhihong, J., Zhu, H., Ngoma, H., Babaousmail, H., Rizwan, K., et al. (2021). Comparison of CMIP6 and CMIP5 models in simulating mean and extreme precipitation over East Africa. *International Journal of Climatology, 41*(15), 6474–6496. https://doi.org/10.1002/joc.7207.

Baño-Medina, J., Manzanas, R., & Gutiérrez, J. M. (2020). Configuration and intercomparison of deep learning neural models for statistical downscaling. *Geoscientific Model Development, 13*(4), 2109–2124.

Bichet, A., & Diedhiou, A. (2018). Less frequent and more intense rainfall along the coast of the Gulf of Guinea in West and Central Africa (1981–2014). *Climate Research, 76*(3), 191–201.

Blenkinsop, S., Fowler, H. J., Barbero, R., Chan, S. C., Guerreiro, S. B., Kendon, E., … Tye, M. R. (2018). The INTENSE project: Using observations and models to understand the past, present and future of sub-daily rainfall extremes. *Advances in Science and Research, 15*, 117–126. https://doi.org/10.5194/asr-15-117-2018.

Brönnimann, S., Rajczak, J., Fischer, E. M., Raible, C. C., Rohrer, M., & Schär, C. (2018). Changing seasonality of moderate and extreme precipitation events in the Alps. *Natural Hazards and Earth System Sciences, 18*, 2047–2056. https://doi.org/10.5194/nhess-18-2047-2018.

Busuioc, A., Dobrinescu, A., Birsan, M.-V., Dumitrescu, A., & Orzan, A. (2015). Spatial and temporal variability of climate extremes in Romania and associated large-scale mechanisms. *International Journal of Climatology, 35*(7), 1278–1300.

Caesar, J., Alexander, L., & Vose, R. (2006). Large-scale changes in observed daily maximum and minimum temperatures: Creation and analysis of a new gridded data set. *Journal of Geophysical Research: Atmospheres, 111*, D05101. https://doi.org/10.1029/2005JD006280.

Carlson, D. (2015). Foreword. *Weather and Climate Extremes, 9*, 1. https://doi.org/10.1016/j.wace.2015.08.004.

Chen, W., & Dong, B. (2021). Projected near-term changes in temperature extremes over China in the mid-twenty-first century and underlying physical processes. *Climate Dynamics, 56*(5), 1879–1894.

Chou, C., Chiang, J., Lan, C. W., Chung, C. H., Liao, Y. C., & Lee, C. J. (2013). Increase in the range between wet and dry season precipitation. *Nature Geoscience, 6*(4), 263–267. https://doi.org/10.1038/ngeo1744.

Cornwall, W. (2021). Europe's deadly floods leave scientists stunned. *Science, 373*(6553), 372–373. https://doi.org/10.1126/science.abl5271.

Diatta, S., Mbaye, M. L., & Sambou, S. (2020). Evaluating hydro-climate extreme indices from a regional climate model: A case study for the present climate in the Casamance river basin, southern Senegal. *Scientific African, 10*, e00584. https://doi.org/10.1016/j.sciaf.2020.e00584.

Dike, V. N., Lin, Z., Kece, F., Langendijk, G. S., & Nath, D. (2022). Evaluation and multi-model projection of seasonal precipitation extremes over Central Asia based on CMIP6 simulations. *International Journal of Climatology*. https://doi.org/10.1002/joc.7641 (accepted author manuscript).

Dike, V. N., Lin, Z., Wang, Y., & Nnamchi, H. (2019). Observed trends in diurnal temperature range over Nigeria. *Atmospheric and Oceanic Science Letters, 12*(2), 131–139.

Dike, V. N., Lin, Z.-H., & Ibe, C. C. (2020). Intensification of summer rainfall extremes over Nigeria during recent decades. *Atmosphere, 11*(10), 1084. https://doi.org/10.3390/atmos11101084.

Dike, V. N., et al. (2018). Obstacles facing Africa's young climate scientists. *Nature Climate Change, 8*(6), 447–449.

Donat, M. G., et al. (2013a). Global land-based datasets for monitoring climatic extremes. *Bulletin of the American Meteorological Society, 94*(7), 997–1006.

Donat, M. G., et al. (2013b). Updated analyses of temperature and precipitation extreme indices since the beginning of the twentieth century: The HadEX2 dataset. *Journal of Geophysical Research: Atmospheres, 118*(5), 2098–2118.

Dosio, A., Mentaschi, L., Fischer, E. M., & Wyser, K. (2018). Extreme heat waves under 1.5 °C and 2 °C global warming. *Environmental Research Letters, 13*(5), 054006. https://doi.org/10.1088/1748-9326/aab827.

Dunn, R. J. H., et al. (2020). Development of an updated global land in situ-based data set of temperature and precipitation extremes: HadEX3. *Journal of Geophysical Research: Atmospheres, 125*(16). e2019JD032263.

El Kenawy, A., López-Moreno, J. I., & Vicente-Serrano, S. M. (2011). Recent trends in daily temperature extremes over northeastern Spain (1960–2006). *Natural Hazards and Earth System Sciences, 11*(9), 2583–2603.

El Kenawy, A. M., et al. (2020). Evidence for intensification of meteorological droughts in Oman over the past four decades. *Atmospheric Research, 246*, 105126. https://doi.org/10.1016/j.atmosres.2020.105126.

Eyring, V., et al. (2016). Overview of the Coupled Model Intercomparison Project Phase 6 (CMIP6) experimental design and organization. *Geoscientific Model Development, 9*(5), 1937–1958.

Fathian, F., et al. (2020). Assessment of changes in climate extremes of temperature and precipitation over Iran. *Theoretical and Applied Climatology, 141*(3), 1119–1133.

Feng, X., Porporato, A., & Rodriguez-Iturbe, I. (2013). Changes in rainfall seasonality in the tropics. *Nature Climate Change, 3*(9), 811–815.

Fischer, E. M., Beyerle, U., & Knutti, R. (2013). Robust spatially aggregated projections of climate extremes. *Nature Climate Change, 3*(12), 1033–1038.

Fischer, E. M., Sippel, S., & Knutti, R. (2021). Increasing probability of record-shattering climate extremes. *Nature Climate Change, 11*(8), 689–695.

Förster, K., & Thiele, L.-B. (2020). Variations in sub-daily precipitation at centennial scale. *npj Climate and Atmospheric Science, 3*(1), 13. https://doi.org/10.1038/s41612-020-0117-1.

Freychet, N., Hegerl, G., Mitchell, D., & Collins, M. (2021). Future changes in the frequency of temperature extremes may be underestimated in tropical and subtropical regions. *Communications Earth & Environment, 2*(1), 28. https://doi.org/10.1038/s43247-021-00094-x.

Ge, F., Zhu, S., Luo, H., Zhi, X., & Wang, H. (2021). Future changes in precipitation extremes over Southeast Asia: Insights from CMIP6 multi-model ensemble. *Environmental Research Letters*, *16*(2), 024013. https://doi.org/10.1088/1748-9326/abd7ad.

Geng, Y.-F., Xie, S.-P., Zheng, X.-T., & Wang, C.-Y. (2020). Seasonal dependency of tropical precipitation change under global warming. *Journal of Climate*, *33*(18), 7897–7908.

Guerreiro, S. B., et al. (2018). Detection of continental-scale intensification of hourly rainfall extremes. *Nature Climate Change*, *8*(9), 803–807.

Gutowski, W. J., Jr., et al. (2016). WCRP coordinated regional downscaling experiment (CORDEX): A diagnostic MIP for CMIP6. *Geoscientific Model Development*, *9*(11), 4087–4095.

Ham, Y.-G., Kim, J.-H., & Luo, J.-J. (2019). Deep learning for multi-year ENSO forecasts. *Nature*, *573*(7775), 568–572.

Hatsuzuka, D., Sato, T., & Higuchi, Y. (2021). Sharp rises in large-scale, long-duration precipitation extremes with higher temperatures over Japan. *npj Climate and Atmospheric Science*, *4*(1), 29. https://doi.org/10.1038/s41612-021-00184-9.

Held, I. M., Delworth, T. L., Lu, J., Findell, K. L., & Knutson, T. R. (2005). Simulation of Sahel drought in the 20th and 21st centuries. *Proceedings of the National Academy of Sciences of the United States of America*, *102*(50), 17891–17896.

Herold, N., et al. (2021). Projected changes in the frequency of climate extremes over southeast Australia. *Environmental Research Communications*, *3*(1), 011001. https://doi.org/10.1088/2515-7620/abe6b1.

Hirons, L. C., Inness, P., Vitart, F., & Bechtold, P. (2013). Understanding advances in the simulation of intraseasonal variability in the ECMWF model. Part I: The representation of the MJO. *Quarterly Journal of the Royal Meteorological Society*, *139*(675), 1417–1426.

Hosseinzadehtalaei, P., Tabari, H., & Willems, P. (2020). Climate change impact on short-duration extreme precipitation and intensity–duration–frequency curves over Europe. *Journal of Hydrology*, *590*(125), 249. https://doi.org/10.1016/j.jhydrol.2020.125249.

Hu, Z., et al. (2019). "Dry gets drier, wet gets wetter": A case study over the arid regions of central Asia. *International Journal of Climatology*, *39*(2), 1072–1091.

IPCC. (2012). Summary for policymakers: Managing the risks of extreme events and disasters to advance climate change adaptation. In C. B. Field, V. Barros, T. F. Stocker, Q. Dahe, D. J. Dokken, K. L. Ebi, … S. K. Allen (Eds.), *Planning for climate change. A special report of working groups I and II of the intergovernmental panel on climate change (IPCC)* (pp. 111–128). Cambridge, UK, New York, NY, USA: Cambridge University Press.

IPCC. (2013). Climate change 2013: The physical science basis. In T. F. Stocker, et al. (Eds.), *Contribution of working group I to the fifth assessment report of the intergovernmental panel on climate change*Cambridge University Press.

IPCC. (2014). Climate change 2014: Synthesis report. In Core Writing Team, R. K. Pachauri, & L. A. Meyer (Eds.), *Contribution of working groups I, II and III to the fifth assessment report of the intergovernmental panel on climate change*. Geneva, Switzerland: IPCC. 151 pp.

IPCC. (2018). Summary for policymakers. In V. Masson-Delmotte, P. Zhai, H. O. Pörtner, D. Roberts, J. Skea, P. R. Shukla, … T. Waterfield (Eds.), *Global warming of 1.5 °C. An IPCC special report on the impacts of global warming of 1.5 °C above pre-industrial levels and related global greenhouse gas emission pathways, in the context of strengthening the global response to the threat of climate change, sustainable development, and efforts to eradicate poverty* (p. 32). Geneva, Switzerland: World Meteorological Organization.

IPCC. (2021a). Climate change 2021: The physical science basis. In V. Masson-Delmotte, P. Zhai, A. Pirani, S. L. Connors, C. Péan, S. Berger, … B. Zhou (Eds.), *Contribution of working group I to the sixth assessment report of the intergovernmental panel on climate change* Cambridge University Press (In press).

IPCC. (2021b). Summary for policymakers. In V. Masson-Delmotte, P. Zhai, A. Pirani, S. L. Connors, C. Péan, S. Berger, … B. Zhou (Eds.), *Climate change 2021: The physical science basis. Contribution of working group I to the sixth assessment report of the intergovernmental panel on climate change*Cambridge University Press (In press).

Iyakaremye, V., et al. (2021). Increased high-temperature extremes and associated population exposure in Africa by the mid-21st century. *Science of the Total Environment*, *790*, 148162. https://doi.org/10.1016/j.scitotenv.2021.148162.

Joshi, M. K., Rai, A., Kulkarni, A., & Kucharski, F. (2020). Assessing changes in characteristics of hot extremes over India in a warming environment and their driving mechanisms. *Scientific Reports*, *10*(1), 2631. https://doi.org/10.1038/s41598-020-59427-z.

Kadow, C., Hall, D. M., & Ulbrich, U. (2020). Artificial intelligence reconstructs missing climate information. *Nature Geoscience*, *13*(6), 408–413.

Karl, T. R., Nicholls, N., & Ghazi, A. (1999a). CLIVAR/GCOS/WMO workshop on indices and indicators for climate extremes workshop summary. In T. R. Karl, N. Nicholls, & A. Ghazi (Eds.), *Weather and climate extremes: Changes, variations and a perspective from the insurance industry* (pp. 3–7). Netherlands, Dordrecht: Springer.

Karl, T. R., Nicholls, N., & Ghazi, A. (1999b). Clivar/GCOS/WMO workshop on indices and indicators for climate extremes workshop summary. *Climatic Change*, *42*(1), 3–7.

Khan, N., Shahid, S., Ismail, T., Ahmed, K., & Nawaz, N. (2019). Trends in heat wave related indices in Pakistan. *Stochastic Environmental Research and Risk Assessment*, *33*(1), 287–302.

Kharin, V. V., Zwiers, F. W., Zhang, X., & Hegerl, G. C. (2007). Changes in temperature and precipitation extremes in the IPCC ensemble of global coupled model simulations. *Journal of Climate*, *20*(8), 1419–1444.

Kim, H., Ham, Y. G., Joo, Y. S., & Son, S. W. (2021). Deep learning for bias correction of MJO prediction. *Nature Communications*, *12*(1), 3087. https://doi.org/10.1038/s41467-021-23406-3.

King, A. D., et al. (2018). On the linearity of local and regional temperature changes from 1.5°C to 2°C of global warming. *Journal of Climate*, *31*(18), 7495–7514.

Kirchmeier-Young, M. C., & Zhang, X. (2020). Human influence has intensified extreme precipitation in North America. *Proceedings of the National Academy of Sciences*, *117*(24), 13308–13313.

Kumar, P. (2013). Seasonal rain changes. *Nature Climate Change*, *3*(9), 783–784.

Lan, C.-W., Lo, M.-H., Chen, C.-A., & Yu, J.-Y. (2019). The mechanisms behind changes in the seasonality of global precipitation found in reanalysis products and CMIP5 simulations. *Climate Dynamics*, *53*(7), 4173–4187.

Lenderink, G., Barbero, R., Loriaux, J. M., & Fowler, H. J. (2017). Super-Clausius-Clapeyron scaling of extreme hourly convective precipitation and its relation to large-scale atmospheric conditions. *Journal of Climate*, *30*(15), 6037–6052.

Lewis, E., Fowler, H., Alexander, L., Dunn, R., McClean, F., Barbero, R., … Blenkinsop, S. (2019). GSDR: A global sub-daily rainfall dataset. *Journal of Climate*, *32*(15), 4715–4729.

Lewis, E., Guerreiro, S., Blenkinsop, S., & Fowler, H. J. (2018). *Quality control of a global sub-daily precipitation dataset*. American Geophysical Union. Fall Meeting 2018, abstract #H41H-22.

Li, J., Yu, R., & Sun, W. (2013). Duration and seasonality of hourly extreme rainfall in the central eastern China. *Acta Meteorologica Sinica*, *27*(6), 799–807.

Li, Y., Bai, J., You, Z., Hou, J., & Li, W. (2021). Future changes in the intensity and frequency of precipitation extremes over China in a warmer world: Insight from a large ensemble. *PLoS One*, *16*(5), e0252133. https://doi.org/10.1371/journal.pone.0252133.

Li, Z., Guo, L., Sha, Y., & Yang, K. (2021). Knowledge map and global trends in extreme weather research from 1980 to 2019: A bibliometric analysis. *Environmental Science and Pollution Research*, *28*, 49755–49773. https://doi.org/10.1007/s11356-021-13825-6.

Liersch, S., et al. (2020). One simulation, different conclusions—The baseline period makes the difference! *Environmental Research Letters*, *15*(10), 104014. https://doi.org/10.1088/1748-9326/aba3d7.

Liu, B., Tan, X., Gan, T. Y., et al. (2020). Global atmospheric moisture transport associated with precipitation extremes: Mechanisms and climate change impacts. *WIREs Water*, *7*, e1412. https://doi.org/10.1002/wat2.1412.

Liu, R., Liu, S. C., Shiu, C.-J., Li, J., & Zhang, Y. (2016). Trends of regional precipitation and their control mechanisms during 1979–2013. *Advances in Atmospheric Sciences*, *33*(2), 164–174.

Luong, T. M., Dasari, H. P., & Hoteit, I. (2020). Extreme precipitation events are becoming less frequent but more intense over Jeddah, Saudi Arabia. Are shifting weather regimes the cause? *Atmospheric Science Letters*, *21*, e981. https://doi.org/10.1002/asl.981.

Mann, M., Rahmstorf, S., Kornhuber, K., et al. (2017). Influence of anthropogenic climate change on planetary wave resonance and extreme weather events. *Scientific Reports*, *745*, 242. https://doi.org/10.1038/srep45242.

Matsueda, S., & Takaya, Y. (2015). The global influence of the madden Julian oscillation on extreme temperature events. *Journal of Climate*, *28*(10), 4141–4151.

Meehl, G. A., Covey, C., McAvaney, B., Latif, M., & Stouffer, R. J. (2005). Overview of the coupled model intercomparison project. *Bulletin of the American Meteorological Society*, *86*(1), 89–93.

Meehl, G. A., et al. (2000). Trends in extreme weather and climate events: Issues related to modeling extremes in projections of future climate change. *Bulletin of the American Meteorological Society*, *81*(3), 427–436.

Meehl, G. A., et al. (2007). The WCRP CMIP3 multimodel dataset: A new era in climate change research. *Bulletin of the American Meteorological Society*, *88*(9), 1383–1394.

Min, S.-K., Zhang, X., Zwiers, F. W., & Hegerl, G. C. (2011). Human contribution to more-intense precipitation extremes. *Nature*, *470*(7334), 378–381.

Mishra, V., Ganguly, A. R., Nijssen, B., & Lettenmaier, D. P. (2015). Changes in observed climate extremes in global urban areas. *Environmental Research Letters*, *10*(2), 024005. https://doi.org/10.1088/1748-9326/10/2/024005.

Mondal, A., & Mujumdar, P. P. (2015). Modeling non-stationarity in intensity, duration and frequency of extreme rainfall over India. *Journal of Hydrology*, *521*, 217–231.

Moore, T. R., Matthews, H. D., Simmons, C., & Leduc, M. (2015). Quantifying changes in extreme weather events in response to warmer global temperature. *Atmosphere-Ocean*, *53*(4), 412–425.

Moustakis, Y., Papalexiou, S. M., Onof, C. J., & Paschalis, A. (2021). Seasonality, intensity, and duration of rainfall extremes change in a warmer climate. *Earth's Future*, *9*. https://doi.org/10.1029/2020EF001824. e2020EF001824.

Muhammad, F. R., Lubis, S. W., & Setiawan, S. (2021). Impacts of the Madden–Julian oscillation on precipitation extremes in Indonesia. *International Journal of Climatology*, *41*(3), 1970–1984.

Myhre, G., Alterskjær, K., Stjern, C. W., et al. (2019). Frequency of extreme precipitation increases extensively with event rareness under global warming. *Scientific Reports*, *916*, 063. https://doi.org/10.1038/s41598-019-52277-4.

Nangombe, S., et al. (2018). Record-breaking climate extremes in Africa under stabilized 1.5 °C and 2 °C global warming scenarios. *Nature Climate Change*, *8*(5), 375–380.

Nangombe, S. S., Zhou, T., Zhang, W., Zou, L., & Li, D. (2019). High-temperature extreme events over Africa under 1.5 and 2 °C of global warming. *Journal of Geophysical Research: Atmospheres*, *124*(8), 4413–4428.

NCEI. (2021). *U.S. billion-dollar weather and climate disasters*. NOAA National Centers for Environmental Information. https://www.ncdc.noaa.gov/billions/. 10.25921/stkw-7w73.

Niang, I., et al. (2014). Africa. In V. R. Barros, C. B. Field, D. J. Dokken, M. D. Mastrandrea, K. J. Mach, T. E. Bilir, … L. L. White (Eds.), *Climate change: Impacts, adaptation, and vulnerability. Part B: Regional aspects. Contribution of working group II to the fifth assessment report of the intergovernmental panel on climate change* (pp. 1199–1265). Cambridge, United Kingdom and New York, NY, USA: Cambridge University Press.

Normile, D. (2021). Zhengzhou subway flooding a warning for other major cities. *Science*. https://doi.org/10.1126/science.abl6944.

O'Gorman, P. A., & Schneider, T. (2009). Scaling of precipitation extremes over a wide range of climates simulated with an idealized GCM. *Journal of Climate*, *22*(21), 5676–5685.

Ornes, S. (2018). Core concept: How does climate change influence extreme weather? Impact attribution research seeks answers. *Proceedings of the National Academy of Sciences*, *115*(33), 8232–8235.

Perkins, S. E., Alexander, L. A., & Nairn, J. R. (2012). Increasing frequency, intensity and duration of observed global heatwaves and warm spells. *Geophysical Research Letters*, *39*, L20714. https://doi.org/10.1029/2012GL053361.

Perkins, S. E., & Alexander, L. V. (2013). On the measurement of heat waves. *Journal of Climate*, *26*(13), 4500–4517.

Persiano, S., et al. (2020). Changes in seasonality and magnitude of sub-daily rainfall extremes in Emilia-Romagna (Italy) and potential influence on regional rainfall frequency estimation. *Journal of Hydrology: Regional Studies*, *32*, 100751. https://doi.org/10.1016/j.ejrh.2020.100751.

Peterson, T. C. (2005). Climate change indices. *WMO Bulletin*, *54*(2), 83–86.

Peterson, T. C., & Manton, M. J. (2008). Monitoring changes in climate extremes: A tale of international collaboration. *Bulletin of the American Meteorological Society*, *89*(9), 1197–1198.

Pfahl, S., O'Gorman, P. A., & Fischer, E. M. (2017). Understanding the regional pattern of projected future changes in extreme precipitation. *Nature Climate Change*, *7*(6), 423–427.

Radinović, D., & Ćurić, M. (2012). Criteria for heat and cold wave duration indexes. *Theoretical and Applied Climatology*, *107*(3), 505–510. https://doi.org/10.1007/s00704-011-0495-8.

Radinović, D., & Ćurić, M. (2014). Measuring scales for daily temperature extremes, precipitation and wind velocity. *Meteorological Applications*, *21*(3), 461–465.

Rasp, S., Pritchard, M. S., & Gentine, P. (2018). Deep learning to represent subgrid processes in climate models. *Proceedings of the National Academy of Sciences*, *115*(39), 9684–9689.

Reichstein, M., et al. (2019). Deep learning and process understanding for data-driven earth system science. *Nature, 566*(7743), 195–204.

Robinson, W. A. (2021). Climate change and extreme weather: A review focusing on the continental United States. *Journal of the Air & Waste Management Association, 71*, 1186–1209. https://doi.org/10.1080/10962247.2021.1942319.

Schiermeier, Q. (2021). Climate change made North America's deadly heatwave 150 times more likely. *Nature*. https://doi.org/10.1038/d41586-021-01869-0.

Seneviratne, S. I., & Zwiers, F. W. (2015). Attribution and prediction of extreme events: Editorial on the special issue. *Weather and Climate Extremes, 9*, 2–5.

Shi, C., Jiang, Z.-H., Chen, W.-L., & Li, L. (2018). Changes in temperature extremes over China under 1.5 °C and 2 °C global warming targets. *Advances in Climate Change Research, 9*(2), 120–129.

Sillmann, J., Daloz, A. S., Schaller, N., & Schwingshackl, C. (2021). Extreme weather and climate change. In T. M. Letcher (Ed.), *Climate change* (3rd ed., pp. 359–372). Elsevier (chapter 16).

Sillmann, J., Kharin, V. V., Zhang, X., Zwiers, F. W., & Bronaugh, D. (2013). Climate extremes indices in the CMIP5 multimodel ensemble: Part 1. Model evaluation in the present climate. *Journal of Geophysical Research: Atmospheres, 118*(4), 1716–1733.

Sillmann, J., Kharin, V. V., Zwiers, F. W., Zhang, X., & Bronaugh, D. (2013). Climate extremes indices in the CMIP5 multimodel ensemble: Part 2. Future climate projections. *Journal of Geophysical Research: Atmospheres, 118*(6), 2473–2493.

Sillmann, J., et al. (2017). Understanding, modeling and predicting weather and climate extremes: Challenges and opportunities. *Weather and Climate Extremes, 18*, 65–74.

Skliris, N., Zika, J. D., Nurser, G., Josey, S. A., & Marsh, R. (2016). Global water cycle amplifying at less than the Clausius-Clapeyron rate. *Scientific Reports, 6*(1), 38752. https://doi.org/10.1038/srep38752.

Smith, A. (2018). *2017 U.S. billion-dollar weather and climate disasters: A historic year in context*. NOAA National Centers for Environmental Information. https://doi.org/10.13140/RG.2.2.17249.53609.

Stephenson, T. S., et al. (2014). Changes in extreme temperature and precipitation in the Caribbean region, 1961–2010. *International Journal of Climatology, 34*(9), 2957–2971.

Stillman, J. H. (2019). Heat waves, the new normal: Summertime temperature extremes will impact animals, ecosystems, and human communities. *Physiology, 34*(2), 86–100.

Sun, Y., et al. (2014). Rapid increase in the risk of extreme summer heat in Eastern China. *Nature Climate Change, 4*(12), 1082–1085.

Tabari, H., Hosseinzadehtalaei, P., AghaKouchak, A., & Willems, P. (2019). Latitudinal heterogeneity and hotspots of uncertainty in projected extreme precipitation. *Environmental Research Letters, 14*(12), 124032. https://doi.org/10.1088/1748-9326/ab55fd.

Tabari, H., Madani, K., & Willems, P. (2020). The contribution of anthropogenic influence to more anomalous extreme precipitation in Europe. *Environmental Research Letters, 15*(10), 104077. https://doi.org/10.1088/1748-9326/abb268.

Tan, X., Wu, Y., Liu, B., & Chen, S. (2020). Inconsistent changes in global precipitation seasonality in seven precipitation datasets. *Climate Dynamics, 54*(5), 3091–3108.

Taylor, K. E., Stouffer, R. J., & Meehl, G. A. (2012). An overview of CMIP5 and the experiment design. *Bulletin of the American Meteorological Society, 93*(4), 485–498.

Trenberth, K. E., Dai, A., Rasmussen, R. M., & Parsons, D. B. (2003). The changing character of precipitation. *Bulletin of the American Meteorological Society, 84*(9), 1205–1218.

Trenberth, K. E., et al. (2014). Global warming and changes in drought. *Nature Climate Change, 4*(1), 17–22.

Ukkola, A. M., De Kauwe, M. G., Roderick, M. L., Abramowitz, G., & Pitman, A. J. (2020). Robust future changes in meteorological drought in CMIP6 projections despite uncertainty in precipitation. *Geophysical Research Letters, 47*(11). e2020GL087820.

Vitart, F., & Robertson, A. W. (2018). The sub-seasonal to seasonal prediction project (S2S) and the prediction of extreme events. *npj Climate and Atmospheric Science, 1*(1), 3. https://doi.org/10.1038/s41612-018-0013-0.

Vitart, F., et al. (2019). Sub-seasonal to seasonal prediction of weather extremes. In A. W. Robertson, & F. Vitart (Eds.), *Sub-seasonal to seasonal prediction* (pp. 365–386). Elsevier (chapter 17).

Wang, B., & Moon, J.-Y. (2018). Subseasonal prediction of extreme weather events. In *Bridging science and policy implication for managing climate extremes. World Scientific Series on Asia-Pacific Weather and Climate* (pp. 33–48). World Scientific. https://doi.org/10.1142/9789813235663_0003.

Wang, Y., & Yan, Z. (2011). Changes of frequency of summer precipitation extremes over the Yangtze River in association with large-scale oceanic-atmospheric conditions. *Advances in Atmospheric Sciences, 28*(5), 1118. https://doi.org/10.1007/s00376-010-0128-7.

Wehner, M., Stone, D., Krishnan, H., AchutaRao, K., & Castillo, F. (2016). The deadly combination of heat and humidity in India and Pakistan in summer 2015. *Bulletin of the American Meteorological Society, 97*(12), S81–S86.

Wei, W., Yan, Z., & Li, Z. (2021). Influence of Pacific Decadal Oscillation on global precipitation extremes. *Environmental Research Letters, 16*(4), 044031. https://doi.org/10.1088/1748-9326/abed7c.

Westra, S., et al. (2014). Future changes to the intensity and frequency of short-duration extreme rainfall. *Reviews of Geophysics, 52*(3), 522–555.

Witze, A. (2018). Why extreme rains are gaining strength as the climate warms. *Nature, 563*, 458–460.

WMO. (2014). Scientists urge more frequent updates of 30-year climate baselines to keep pace with rapid climate change. WMO-No., 997. *WMO Publications*.

WMO. (2017). *WMO guidelines on the calculation of climate normals*. Tech. Rep. WMO- No. 1203.

WMO. (2022). Artificial intelligence for disaster risk reduction: Opportunities, challenges, and prospects. *WMO Bulletin, 71*(1).

Xu, Y., Gao, X., Giorgi, F., et al. (2018). Projected changes in temperature and precipitation extremes over China as measured by 50-yr return values and periods based on a CMIP5 ensemble. *Advances in Atmospheric Sciences, 35*, 376–388. https://doi.org/10.1007/s00376-017-6269-1.

Yao, J., et al. (2021). Intensification of extreme precipitation in arid Central Asia. *Journal of Hydrology, 598*. 125760.

Zeng, N. (2003). Drought in the Sahel. *Science, 302*(5647), 999–1000. https://doi.org/10.1126/science.1090849.

Zhang, D., Zhang, Q., Qiu, J., Bai, P., Liang, K., & Li, X. (2018). Intensification of hydrological drought due to human activity in the middle reaches of the Yangtze River, China. *The Science of the Total Environment, 637–638*, 1432–1442.

Zhang, X., Alexander, L., Hegerl, G. C., Jones, P., Tank, A. K., Peterson, T. C., … Zwiers, F. W. (2011). Indices for monitoring changes in extremes based on daily temperature and precipitation data. *WIREs Climate Change, 2*, 851–870. https://doi.org/10.1002/wcc.147.

Zhang, X., Hegerl, G., Seneviratne, S., Stewart, R., Zwiers, F., & Alexander, L. (2014). *WCRP grand challenge: Science underpinning the prediction and attribution of extreme events*. https://www.wcrp-climate.org/documents/GC_Extremes.pdf. Accessed 10 September 2021.

Zhang, X., & Yang, F. (2004). *ClimDex (1.0)—User manual*. Climate research branch environment Canada Downs view, Ontario.

Zhang, X., et al. (2005). Trends in Middle East climate extreme indices from 1950 to 2003. *Journal of Geophysical Research: Atmospheres, 110*(D22). https://doi.org/10.1029/2005JD006181.

Zhou, X., Huang, G., Wang, X., & Cheng, G. (2018). Future changes in precipitation extremes over Canada: Driving factors and inherent mechanism. *Journal of Geophysical Research: Atmospheres, 123*(11), 5783–5803.

Zittis, G., Hadjinicolaou, P., Almazroui, M., et al. (2021). Business-as-usual will lead to super and ultra-extreme heatwaves in the Middle East and North Africa. *npj Climate and Atmospheric Science, 4*, 20. https://doi.org/10.1038/s41612-021-00178-7.

CHAPTER

4

Uncertainties in daily rainfall over West Africa: Assessment of gridded products and station gauges

Imoleayo E. Gbode[a,b], *Joseph D. Intsiful*[c], *Akintomide Afolayan Akinsanola*[d,e], *Akintayo T. Abolude*[f], *and Kehinde O. Ogunjobi*[b,g]

[a]West African Science Service Center on Climate Change and Adapted Land Use, Federal University of Technology Akure, Akure, Nigeria

[b]Department of Meteorology and Climate Science, Federal University of Technology Akure, Akure, Nigeria

[c]Division of Mitigation and Adaptation, Green Climate Fund, Songdo Incheon, Republic of Korea

[d]Department of Geography, University of Georgia, Athens, GA, United States

[e]Environmental Science Division, Argonne National Laboratory, Lemont, IL, United States

[f]Nouveau Projects Limited, Lagos, Nigeria

[g]West African Science Service Center on Climate Change and Adapted Land Use, Competence Center, Ouagadougou, Burkina Faso

1 Introduction

West African rainfall varies at both spatial and temporal scales, which pose a significant challenge for assessing and understanding climate change over the region. This variation is partly due to the heterogeneous nature of the region's climate and presence of geographic features, such as the Saharan deserts, mountains and plateaus, sizeable lakes, land-sea contrast, and sea surface temperature of the surrounding Atlantic Ocean (Akinsanola, Ogunjobi, Gbode, & Ajayi, 2015; Sylla, Giorgi, Coppola, & Mariotti, 2013). These features influence not only the distribution and statistics of daily rainfall and extremes, but also the onset and evolution of the monsoon rainy season (Sylla, Dell'Aquila, Ruti, & Giorgi, 2010). This effect may exert huge impact on natural systems and human activities than average rainfall distribution (Parry, Canziani, Palutikof, Van Der Linden, & Hanson, 2007). Sectors such as agriculture, health, and water resources require useful information on how the weather and extreme events affect their operations. This information is readily accessible from meteorological data but cannot be directly applied to specific sectors. Transforming this data to directly applicable forms, thus, there is need for an assessment of long-term historical and projected climate records over the West African region (Akinsanola & Zhou, 2019). In this case, such assessment is a careful evaluation of a product's ability to reproduce values of hydroclimatic variables of the West African monsoon systems (Gbode, Adeyeri, et al., 2019).

Regional Climate Models (RCMs) are widely used to simulate the future changes in weather and climate. The outputs of RCMs however contain uncertainties introduced by different sources including the forcing data, parameterization, resolution, and algorithm (Foley, 2010; Gbode, Dudhia, Ogunjobi, & Ajayi, 2019). Although these models can be verified with available observational data, the associated uncertainties dominate subject of discourse in atmospheric science research. In recent times, RCMs are run at convection-permitting scales (i.e., grid size less than 4 km), which allows the model to explicitly resolve convection and reduce the uncertainties introduced by cumulus parameterization (Prein et al., 2015). In a data sparse region like West Africa with very limited observation networks, gridded products are used as alternative datasets to evaluate and verify RCM simulations over the region (Akinsanola et al., 2017). Some

Climate Impacts on Extreme Weather
https://doi.org/10.1016/B978-0-323-88456-3.00003-4

commonly used gridded rainfall products include: Climate Hazards Group InfraRed Precipitation with Station data (CHIRPS); Tropical Rainfall Measuring Mission (TRMM); Global Precipitation Climatology Project (GPCP); Global Precipitation Climatology Center (GPCC); Climate Prediction Center (CPC); CPC MORPHing technique (CMORPH); Global Satellite Mapping of Precipitation (GSMAP); Multi-Source Weighted-Ensemble Precipitation (MSWEP); and African Rainfall Climatology (ARC). These gridded products show differences in distribution among themselves because not only do they differ by algorithm and or platforms, but also by input data (Sylla et al., 2013). Also, the construction of gridded dataset for model validation, especially in data sparse regions with high rainfall variability, is quite challenging. Thus, the need for assessing substitute products that can best replicate the intended observed climatological feature of observed rainfall (Akinsanola et al., 2017). As a result of these differences and disagreement between products, consolidated research efforts must be directed toward intercomparison of datasets at various temporal timescales.

Model validation and or intercomparison of datasets is generally not a new subject globally (Maraun, 2016; Tapiador et al., 2017) and regionally (Akinsanola et al., 2015, 2017). Roca et al. (2010) compared satellite and ground rainfall data of 2006 and reported that over West Africa the satellite products analyzed describe rainfall variability similarly to ground measurements. The study reported that most satellite products possess high regional and seasonal skills at 10-day scale and that they are sensitive to the passage of African easterly wave, but their skills vary on a daily basis. Other products like TRMM and GPCP show good agreement in representing the observed seasonal rainfall pattern compared with station data at different spatial resolutions over Africa (Herrmann & Mohr, 2011). Using hydrological approach, Gosset, Viarre, Quantin, and Alcoba (2013), evaluated the performance of eight products relative to gauge networks from Benin and Niger between June and September from 2003 to 2010. Results reveal that CMORPH among two others had the highest correlation with the observed from the daily mean rainfall and rainfall accumulation indices analyzed. Different gridded datasets provide quite different representations of daily precipitation behavior as shown in Sylla et al. (2013). The study analyzed different indices including precipitation intensity, simple daily intensity, frequency of wet days, and length of wet and dry spells from three products, including the Famine Early Warning System dataset, over Africa for the period 1998–2007. The observation products analyzed exhibit substantial systematic biases in mean rainfall, especially in higher order daily precipitation statistics. Akinsanola et al. (2017) reported that ARC v.2 was inconsistent in capturing the dry(wet) conditions associated with the El Niño (La Niña) over West Africa when analyzing ARC, UDEL, GPCC, CRU, and TRMM. The study recommended the latter three for use. Maidment et al. (2017) used different versions of Tropical Application of Meteorology Using Satellite Data and Ground-Based Observations (TAMSAT v.2.0 & v.3.0) to compare ground-based observations from five African countries and concluded that TAMSAT does reliably well in detecting rainy days, although with less skill in capturing rainfall amount. More recently, Satgé et al. (2020) assessed the reliability of gridded products over West Africa using 2000–03 data and concluded that of the 23 datasets considered the best performing were MSWEP v.2.2 (CHIRPS v.2) for daily(monthly) timescale.

A research gap however exists in the place of exploring the daily characteristics of rainfall distribution. The skills of various gridded products can be further assessed using designated hydroclimatological indices such as those defined by the World Meteorological Organization (WMO) Expert Team on Sector-Specific Climate Indices (ET-SCI). Also, previous efforts could be improved by assessing monsoon onset, peak, and retreat with respect to the evaluated indices, which is critical to inform planning agricultural production and food security, and other rainfall-dependent activities. To fill these gaps, this chapter presents an intercomparison across different gridded (merged satellite-gauge rainfall) products in terms of characteristics of daily rainfall and extreme events to assess uncertainties present in them over West Africa region. The uncertainties are assessed by comparing the gridded precipitation products with station-gauges over selected stations in Nigeria. Precisely, a range of daily precipitation statistics was examined for three seasons: March-May (MAM); July-August (JJA); and September-November (SON), which encapsulate the monsoon period as well as its onset and cessation.

The remaining part of this chapter is structured as follows: Section 2 describes the observational datasets and methods used, while Section 3 presents results and discussions and Section 4 gives the summary and conclusion.

2 Data and methods

2.1 Station gauges

Daily precipitation from 24 synoptic stations in Nigeria is used as a reference data to evaluate the uncertainties in the gridded products. The stations are distributed across the three climatic subregions of West Africa (see red dots on Fig. 4.1) namely: Guinea Coast, Savannah, and Sahel. Nigeria makes a fair representation for understanding the West African climatic zone since all three subregions are present in the country. Fig. 4.1 shows the distribution of the stations

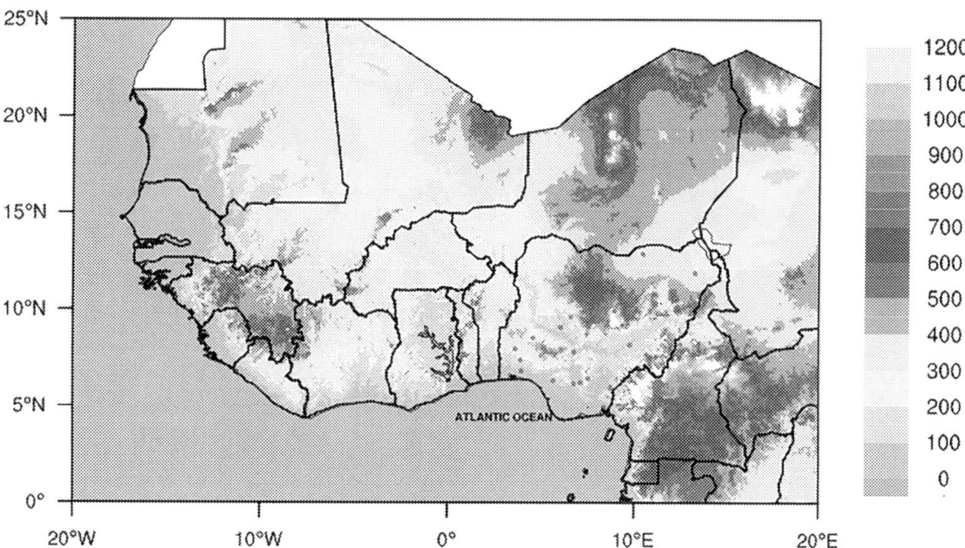

FIG. 4.1 Map of West Africa showing elevation at 100m intervals. Station locations are indicted by *red* (*gray* in print version) *dots* over Nigeria.

used in Nigeria and their topographic characteristics. The density and spread of the selected stations is similar to those used in previous studies (Akinsanola & Ogunjobi, 2017; Gbode, Adeyeri, et al., 2019).

The gauge data were extracted from the archives of the Nigerian Meteorological Agency (NiMet) and cover the period 1998–2013 (16 years). This period is selected to ensure a consistent and equal time slice for all gauge data and gridded products used. Quality control of the data is done based on standard guidelines presented in previous studies (Gbode, Adeyeri, et al., 2019; Gbode, Akinsanola, & Ajayi, 2015). During quality control, erroneous data, such as negative precipitation or days with precipitation greater than 200 mm, minimum temperature greater than maximum temperature, and minimum and maximum temperatures greater than 6 standard deviation (SD) from the long-term mean (Gbode, Adeyeri, et al., 2019; Gbode et al., 2015), were replaced with −999.9, which is recognized as missing data by climPACT2 in order to avoid erroneous computation. The reason for using 6 SD is to reduce the number of outliers in the dataset, which may not necessarily be an outlier mostly due to some extreme events in the highly variable precipitation data.

2.2 Gridded products

Consolidating on the results of referenced literatures, six of the best performing precipitation products over West Africa were selected for this work with the following details. Climate Hazards Group InfraRed Precipitation with Station data version 2 (CHIRPS v2), a 0.25° horizontal resolution of merged satellite and rain gauge precipitation estimate produced from 1981 to near-present (Funk et al., 2015); Tropical Rainfall Measuring Mission (TRMM), 0.25° resolution 3B42 product is one of the reliable sources for merged high-quality satellite-derived precipitation estimates (Huffman et al., 2007); Global Precipitation Climatology Project (GPCP) version 1.3, a 1° resolution satellite-derived dataset from 1996 to present (Huffman, Adler, Bolvin, & Gu, 2009; Huffman, Bolvin, & Adler, 2016); Global Precipitation Climatology Center (GPCC) version 1, also available at 1° resolution (Schamm et al., 2015); Climate Prediction Center (CPC) 0.5° resolution unified (gauge-based) precipitation (Xie, Chen, & Shi, 2010); and CPC MORPHing technique (CMORPH) version 1 raw precipitation at 0.25° resolution (Joyce, Janowiak, Arkin, & Xie, 2004).

CHIRPS v2 is a rainfall estimate from rain gauge and satellite observations created to support drought early warning and environmental monitoring (Funk et al., 2015). This product enables the assessment of historical rainfall variability in order to identify and evaluate severe rainfall deficits in a timely manner. CHIRPS applied satellite-based precipitation fields to provide improved climatology and reduced systematic bias dataset. The TRMM applied the Version 7 TRMM Multi-Satellite Precipitation Analysis (TMPA) algorithm to produce the dataset (Huffman et al., 2014). The dataset was derived from the three-hourly combined microwave-IR estimates with gauge adjustment computed on quasiglobal grids about 2 months after the end of each month starting in January 1998. The GPCP v1.3 daily analysis is a globally complete precipitation estimates at a spatial resolution of 1° latitude-longitude and daily time scale from October 1996 to the present (Adler et al., 2017). This product is an improved version of GPCP v1.2 and a companion to the monthly analysis generated through the World Climate Research Program (WCRP) and Global

Energy and Water Exchanges (GEWEX) activities. GPCC provides daily global data of land-surface precipitation based on precipitation data provided by national meteorological and hydrological services, regional and global data collections as well as WMO Global Telecommunication System (GTS)-data (Schamm et al., 2013). This 1° resolution data contains the daily totals on a rectilinear grid with temporal resolution ranging from January 1988 to present. CPC provides gauge-based global precipitation analysis over land (Xie, 2010). The temporal resolution of the data ranges from 1979 to present times and it is suitable for global monsoon applications. CMORPH applies techniques exclusively derived from low orbiter satellite microwave observations, whose features are transported via spatial propagation information that is obtained entirely from geostationary satellite IR data to produce global precipitation analyses at very high spatial and temporal resolution. Though this method is not a precipitation estimation algorithm, but a means by which estimates from existing microwave rainfall algorithms can be combined, making it extremely flexible to incorporate any precipitation estimates from any microwave satellite source (Joyce, Janowiak, Arkin, & Xie, 2004).

After computing the daily statistics in terms of climate indices, the gridded precipitation products were interpolated to $1° \times 1°$ horizontal resolution using the first-order conservative remapping method (Jones, 1999). This is important to bring the gridded observations to the same grid size so as to directly intercompare the products by grid-to-grid point.

2.3 Methodology

Firstly, an assessment of mean precipitation as well as the time-latitude precipitation cross section over West Africa is presented to examine onset, peak, and retreat of monsoon rainy season over the region (Hourdin et al., 2010; Le Barbe, Lebel, & Tapsoba, 2002). As mentioned earlier, a 16-year period is considered in this analysis to ensure a consistent and equal time slice for all gauge data and gridded products used. Although the time slice may not be enough to deduce robust conclusions for climate applications, it, however, provides useful insight on how the chosen datasets describe the characteristics of the West African monsoon rainfall. Furthermore, different hydroclimatologic indices based on the WMO Expert Team on Sector-Specific Indices (ET-SCI) are used to assess rainfall characteristics calculated from daily climate data including total wet-day, maximum 1-day precipitation, very heavy precipitation days, and consecutive dry(wet) days. ET-SCI was reestablished at the 17th session of the Commission for Climatology (CCl) to develop a number of sector-specific climate indices in close collaboration with experts from health, agriculture, and water sectors. Also, ET-SCI worked closely with the disbanded CCl Expert Team on Climate Change Detection and Indices (ET-CCDI), which developed RClimDex software used for calculating climate indices. Detailed description of the indices used inclusive of the definition and estimation are provided in Table 4.1. These indices were generated using the WMO ET-SCI climPACT version 2 software and are relevant to characterize the climate sensitivity of various sectors, such as, health, agriculture, food security, water resources and hydrology. ET-SCI increased the number of climate indices provided by ET-CCDI and further group them into categories for applications in different priority sectors. Extensive details on the indices are available from the climPACT website (ClimPACT, n.d.).

For detailed analysis of precipitation uncertainty, we evaluate quantitative measures of mean difference and pattern correlation calculated from the mean rainfall climatology over the West African region. Also, we intercompare the gridded products among each other and with the station gauge data.

TABLE 4.1 Description of calculated precipitation indices based on the Expert Team on Sector-Specific Indices (ET-SCI).

ID	Indicator name	Definitions	Units	Sector[a]
PRCPTOT	Total wet-day precipitation	Total precipitation from wet days (RR[a] $\geq$1 mm)	mm	AFS, WRH
RX1day	Maximum 1-day precipitation total	The value of daily precipitation amount which is the maximum precipitation in 1 day (max RR)	mm	H, AFS, WRH
R20mm	Number of very heavy precipitation days	Total count of days when RR $\geq$20 mm	days	AFS, WRH
CDD	Consecutive dry days	Maximum number of consecutive days with RR < 1 mm	days	H, AFS, WRH
CWD	Consecutive wet days	Maximum number of consecutive days with RR $\geq$1 mm	days	H, AFS, WRH

[a] RR, *rainfall;* H, *health;* AFS, *agriculture and food security;* WRH, *water resources and hydrology.*

3 Results and discussion

3.1 Mean precipitation

The mean precipitation of rain days (i.e., rainfall ≥ 1 mm) from station gauges (color filled circles) and the gridded products for the three seasons MAM, JJA and SON over West Africa is presented in Fig. 4.2.

During the premonsoon season (MAM), rainfall is mostly experienced in the Guinea coast subregion, south of 10°N. The maximum values are observed along the southern coast of the region. Generally, the gradient of rainfall decreases with increasing latitudes. In JJA, the Inter-Tropical Convergence Zone (ITCZ), which defines the region of main moisture flux convergence, is situated in the northern hemisphere and displaced northward. The ITCZ over land defines the Inter-Tropical Discontinuity (ITD), which determines the amount of atmospheric moisture in the region. The ITD reaches its peak northernmost latitudinal position of 22° and the ITCZ attains a quasiequilibrium position around 10°N in August. At this time, the monsoon is fully developed, and the associated rainfall is widely distributed across the West Africa region except for Sahara Desert (Nicholson, 2013). Also, the climatological rainfall maxima are clearly seen along the monsoon belt and over complex terrains of Cameroun Mountains, Jos Plateau, and Fouta-Djallon highlands, which extends further to the western coast of West Africa. The SON season marks the end of the monsoon as the ITD retreats southward resulting to a notable decrease in the rainfall amount of this season, especially in the northern part.

Further, Tables 4.2 and 4.3 present the statistics of the seasonal precipitation biases for the three seasons and the pattern correlation, respectively, between station gauges and neighboring grid points over the area of interest. Varying degree of differences can be found across the gridded products in terms of the spatial extent and magnitude of rainfall (Table 4.2). In general, there is a good agreement between station gauges and gridded observations in JJA within the 5–15°N latitudinal band and along the southern coast, virtually, in all seasons. CMORPH and CPC show higher mean precipitation values compared to other products, especially in the orographic areas and along the monsoon belt. Compared to gauge data, CMORPH and CPC show relatively large differences as well. CHIRPS, TRMM, GPCP, and GPCC all comparatively produced better monsoon belt defined by the precipitation pattern relative to each other and the in-situ measurements in almost all seasons. The statistical analyses presented in Tables 4.3 show an overall higher correlation coefficient (above 0.9) in most of the observation datasets except for CPC and CMORPH, where values of 0.8 are obtained. These values show a strong linear relationship in the observed precipitation patterns among the gridded products. However, significant differences are evident in the magnitude of precipitation.

CMORPH in general produces the highest rainfall amount while CPC shows the lowest amount. CHIRPS, TRMM, GPCP, and GPCC all have intermediate values, with TRMM rainfall being the lowest in all seasons and GPCP higher than the CHIRPS and GPCC in some seasons.

Furthermore, the biases between the gauges and gridded products vary with season and individual product. In general, some of the products including GPCP, GPCC, and CMORPH show overestimation of precipitation as indicated by predominance of positive biases while others slightly over and underestimates precipitation at different seasons. Consistent with Fig. 4.2, there is a good agreement between gauges and products, most especially in MAM and SON seasons. Nevertheless, CMORPH overestimates precipitation amount at regions south of 11°N latitude while CPC underestimates the observed magnitude. The inability of CPC to agree well with the gauges, mostly in JJA, is partly associated with its inadequate description of the rain belt and the maximum over complex terrains. In other products, the precipitation amount relative to the gauges is quite similar within 5–10°N but vary significantly outside this latitude band. The gauge-product correlations in Table 4.3 show lesser variability across CHIRPS, TRMM, GPCP and GPCC, with correlation ranging from 0.86 to 0.98. For datasets with ubiquitous biases (i.e., CPC and CMORPH), the correlation ranges from 0.51 to 0.93. In general, the high precipitation correlation quantitatively corroborates that the observations agree well with the spatial distribution pattern of gauges at seasonal scale (Fig. 4.2).

Effectively, the characteristics of mean precipitation discussed here shows that CMORPH (CPC) had the maximum (minimum) amount while CHIRPS had the closest values relative to station gauge data. The results of the performance of the gridded products found in this research also corroborate the findings of (Sylla et al., 2013).

3.2 Evolution of mean daily monsoon

We analyze the observed evolution of monsoon over the West African region based on daily rainfall data. This analysis is presented in Fig. 4.3, which shows latitude-time characteristics of average daily rainfall between longitudes 10°W and 10°E of West Africa for the (a) CHIRPS, (b) TRMM, (c) GPCP, (d) GPCC, (e) CPC and (f) CMORPH for the period 1998–2017 and throughout the year. All the gridded observations show the monsoon onset in

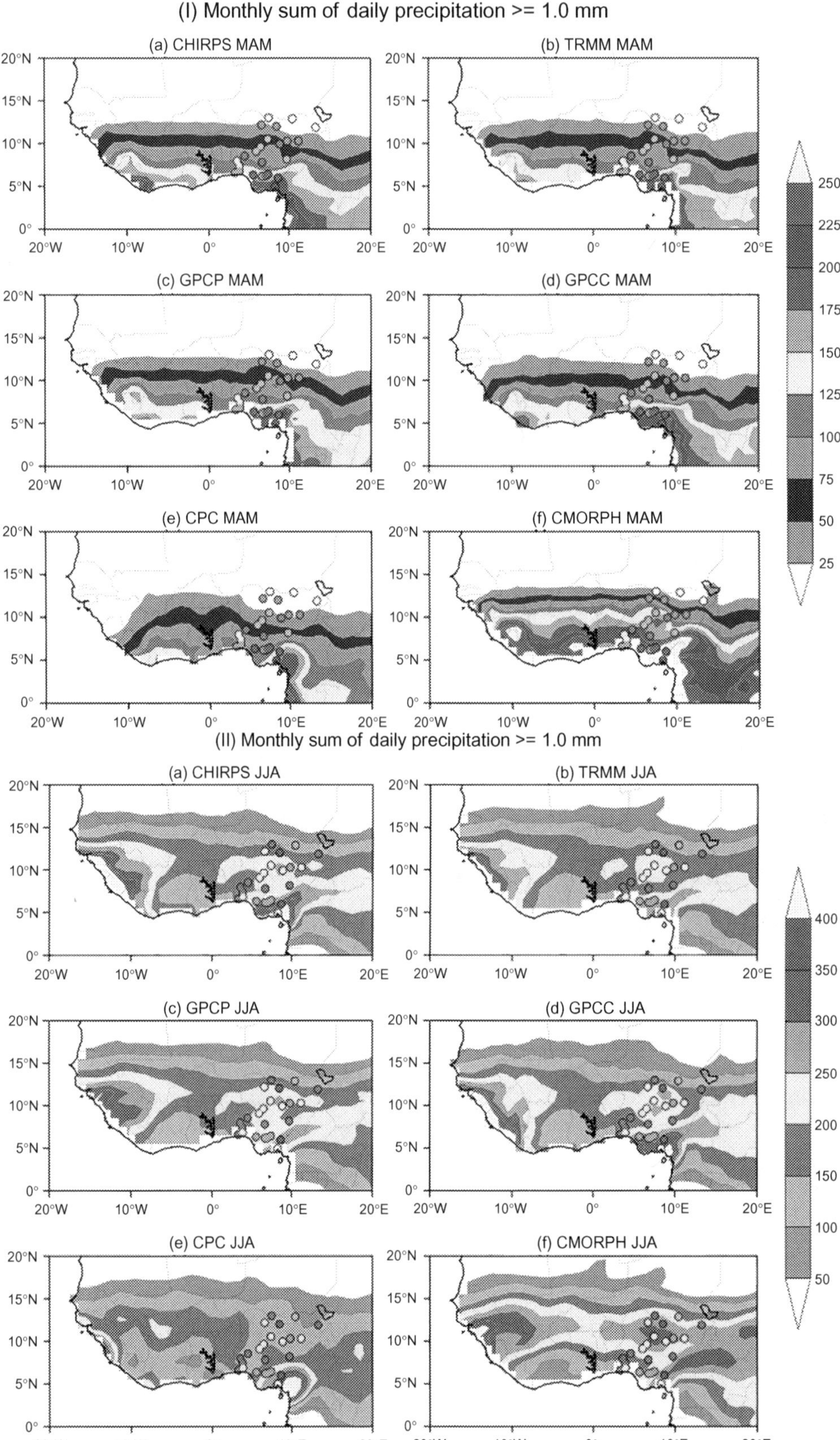

FIG. 4.2 (I)–(III) Average precipitation amount for MAM, JJA and SON in (a) CHIRPS, (b) TRMM, (c) GPCP, (d) GPCC, (e) CPC and (f) CMORPH. The station gauge data are plotted as filled *color circles* at the same scale with the gridded observations.

(Continued)

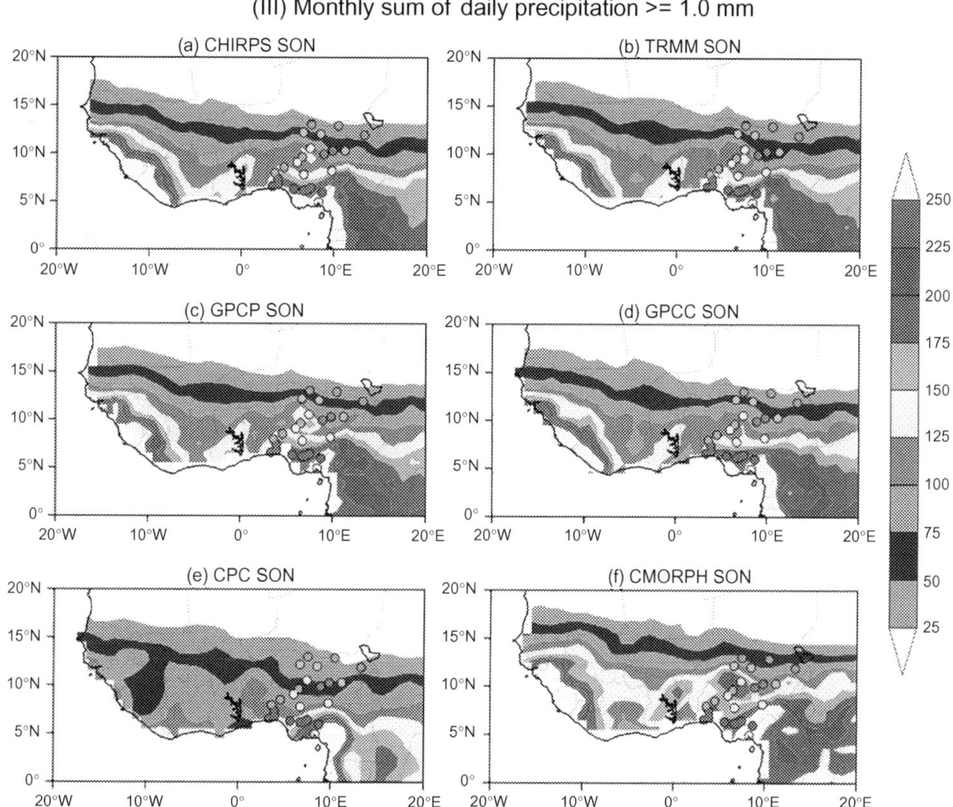

FIG. 4.2, CONT'D

TABLE 4.2 Differences in average seasonal precipitation between the different gridded rainfall products over latitudes 5–15°N and longitudes 10°W–10°E domain and grid points closest to selected station gauges in Nigeria.

	TRMM	GPCP	GPCC	CPC	CMORPH	OBS
CHIRPS	1.5	−2.5	1.0	16.4	−56.4	−1.4
	10.4	1.7	−2.5	37.8	−24.2	2.2
	6.1	−2.0	2.6	27.6	−23.0	1.2
TRMM		−4.0	−0.5	14.9	−57.9	0.1
		−8.7	−12.9	27.4	−34.6	−14.4
		−8.0	−3.5	21.5	−29.1	−7.2
GPCP			3.5	18.9	−53.9	5.2
			−4.2	36.1	−25.9	−6.0
			4.6	29.5	−21.1	11.1
GPCC				15.5	−57.4	3.1
				40.3	−21.7	9.6
				25.0	−25.6	2.3
CPC					−72.8	−25.1
					−61.9	−62.5
					−50.6	−7.2
CMORPH						74.8
						17.9
						36.1

The values of the rows are subtracted from the columns in the order of MAM, JJA, and SON seasons. The unit of all values is expressed in millimeters (mm).

TABLE 4.3 Pattern correlation coefficients of average seasonal precipitation across different rainfall products calculated over the whole West African region (0–20°N and 20°W–20°E) and at grid points closest to gauge station.

	TRMM	GPCP	GPCC	CPC	CMORPH	OBS
CHIRPS	0.99	0.99	0.98	0.94	0.97	0.97
	0.99	0.97	0.99	0.91	0.93	0.86
	0.99	0.99	0.99	0.91	0.94	0.96
TRMM		0.99	0.99	0.92	0.97	0.94
		0.99	0.99	0.92	0.96	0.88
		0.99	0.99	0.90	0.96	0.93
GPCP			0.98	0.93	0.98	0.96
			0.97	0.91	0.97	0.89
			0.98	0.89	0.97	0.94
GPCC				0.92	0.94	0.98
				0.92	0.93	0.94
				0.91	0.94	0.98
CPC					0.90	0.93
					0.87	0.68
					0.83	0.93
CMORPH						0.90
						0.58
						0.79

The values are computed column and row match in the order of MAM, JJA and SON seasons.

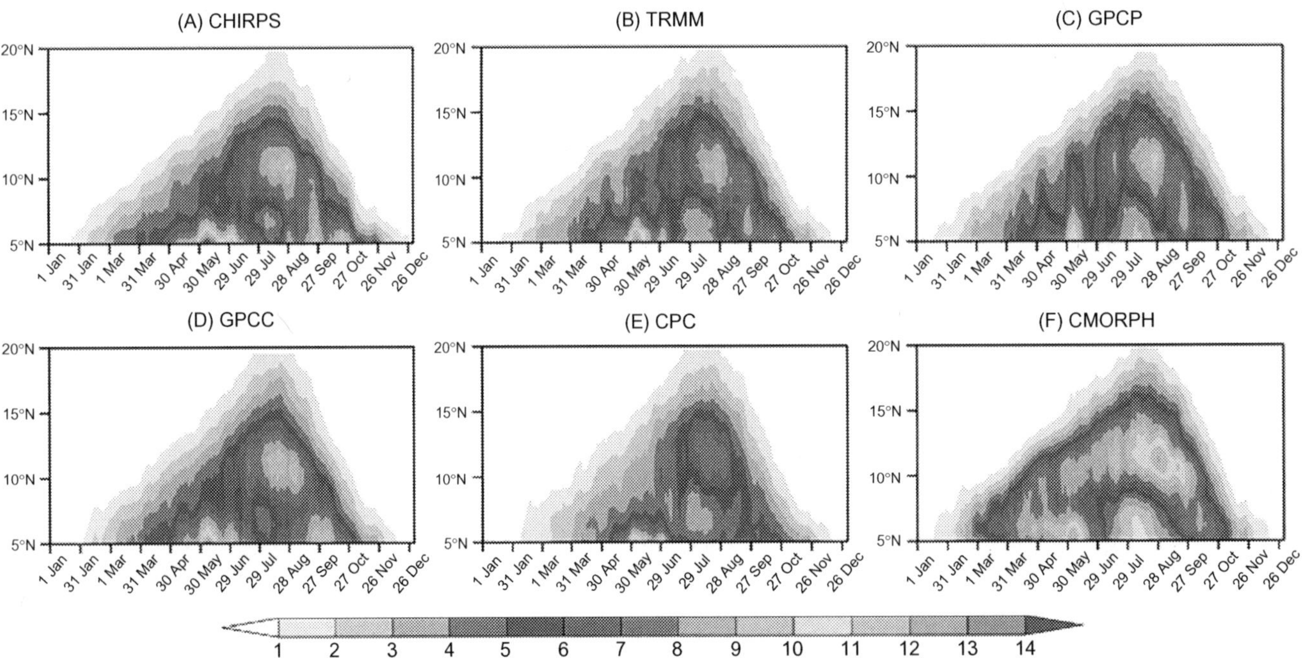

FIG. 4.3 Latitude-time cross-section of precipitation (mm day^{-1}) for (A) CHIRPS, (B) TRMM, (C) GPCP, (D) GPCC, (E) CPC, and (F) CMORPH. The data is averaged along 10°W to 10°E longitudes.

April-May over the Guinea coast except for CMORPH, where the onset is defined in March. The rain band extends northward from latitude 5°N, where the maximum rainfall amount is received. The "monsoon jump" (Hagos & Cook, 2007; Sultan & Janicot, 2000) ensues in early June when the monsoon is fully developed, and the core of the rain band displaced to a quasistationary position of latitude 10°N. This produces the maximum rainfall received around the same latitude and gives rise to rainfall onset in the Sahel area as well as the appearance of rainfall break (little dry season) in the Guinea coast. Last, a gradual retreat of the rains in early September that marks the cessation of the monsoon through significant decrease in intensity and southward movement of the rain band. All the gridded products show the observed heavy intensity rainfall centered in 5°N in May through to June. The CPC show lower intensity of peak rainfall over the Sahel relative to other products and was also unable to appropriately describe the rainfall evolution pattern. The CMORPH show the highest peak in Sahel, with a near linear transit of the rain band. All the other gridded products considered showed relatively larger differences in heavy rainfall characteristics in the Sahel region.

3.3 Extreme precipitation indices

The results of the maximum 1-day rainfall (RX1day) for each of the seasons (i.e., MAM, JJA, and SON) are presented in Fig. 4.4. All products agreed with the gauge stations data and replicated the latitudinal gradient with a northward decrease in observed gauge rainfall. The seasonal mean for maximum 1-day rainfall was lowest in CHIRPS and highest in CMORPH for all seasons considered. The region along the Guinea coast showed the largest biases among the products for MAM between CHIRPS and CMORPH relative to TRMM, GPCP, GPCC, and CPC. In JJA, the magnitude of observed differences between gridded products along the Guinea coast is considerably lesser than in MAM, while in the central part of Nigeria, TRMM and CMORPH had better agreement with the gauges. All the six gridded products show decreased values during the SON, with a more consistent performance relative to gauge stations around the central and coastal regions of Nigeria.

Fig. 4.5 shows the average monthly count of days with very heavy precipitation for MAM, JJA and SON for each gridded product. On average, the observed R20mm have higher values above 7 days during the JJA. Higher values of very heavy precipitation events occur over the highlands of West Africa. The gauge revealed higher R20mm along the Guinea coast near Cameroun mountains and areas around 10°N. Again, there are some differences across the gridded products. There is a noticeable lower (higher) count in CHIRPS (CMORPH) for MAM relative to other products. A degree of agreement exists for MAM among TRMM, GPCP, and GPCC, mostly in the middle belt of Nigeria. In the same season, CMORPH contains on the average more than 4 days of R20mm compared to other products. During JJA, the areas around the Jos Plateau, Cameroun mountains, and Fouta Djallon highlands through to the edge of western coast of the region have higher occurrence of heavy rainfall activities. The GPCP rainfall agrees quite well with gauges while others including CHIRPS, TRMM, GPCC, and CPC perform better in the southern parts of Nigeria. All products apart from CMORPH portray, in general, lower counts of R20mm events. The situation is not too different for SON, but there is, however, a decrease in heavy rainfall activities in all datasets in addition to similar characteristics among TRMM, GPCP, and GPCC.

Other paramount sets of daily precipitation characteristics for assessing possible climate impacts on core sector-specific areas are the consecutive wet days (CWD) and dry days (CDD). Fig. 4.6 shows the average consecutive wet spell length (i.e., maximum number of consecutive days with rainfall ≥ 1 mm), for in-situ measurements and gridded products and for the three seasons under consideration. The gauges show longer maximum wet day spell south of 10°N in MAM. CHIRPS, GPCP, GPCC, and CMORPH all show a gradient of longer maximum wet spell that extends from the south coast to about 10°N. The spatial extent of longer CWD in TRMM is limited not too far from the southern coast of the region and similar gradient is restricted further to the south in CPC. The spatial extent of CWD increases in JJA when the longer wet spell is averagely well above 6 (9) days for gauges (gridded products). GPCC compared to GPCP, CMORPH, CHIRPS and TRMM show longer maximum wet day spells above 9 days, especially over the complex terrains of Jos Plateau, Fouta Djallon highlands and Cameroun mountains along the ITD. The average counts of CWD reduce to about 6 days when the ITD retreats southward marking the end of the monsoon in SON. TRMM and CPC agree well with the gauges, mostly over the Sahel. On the average, all the observational datasets have longer CWD than the gauges.

Fig. 4.6 presents the daily precipitation characteristic of consecutive dry days. The figure shows the average consecutive dry spell length (i.e., maximum number of consecutive days with rainfall <1 mm) for station gauge and gridded products. The gauge shows that consecutive dry spell is apparently dominant in the Sahel and less in the southern parts of the considered area. Black circles represent cases where missing values exist in some months in the Sahel region. This is only observed in MAM and SON when there is longer period of CDD, that is, 64 and 12 days,

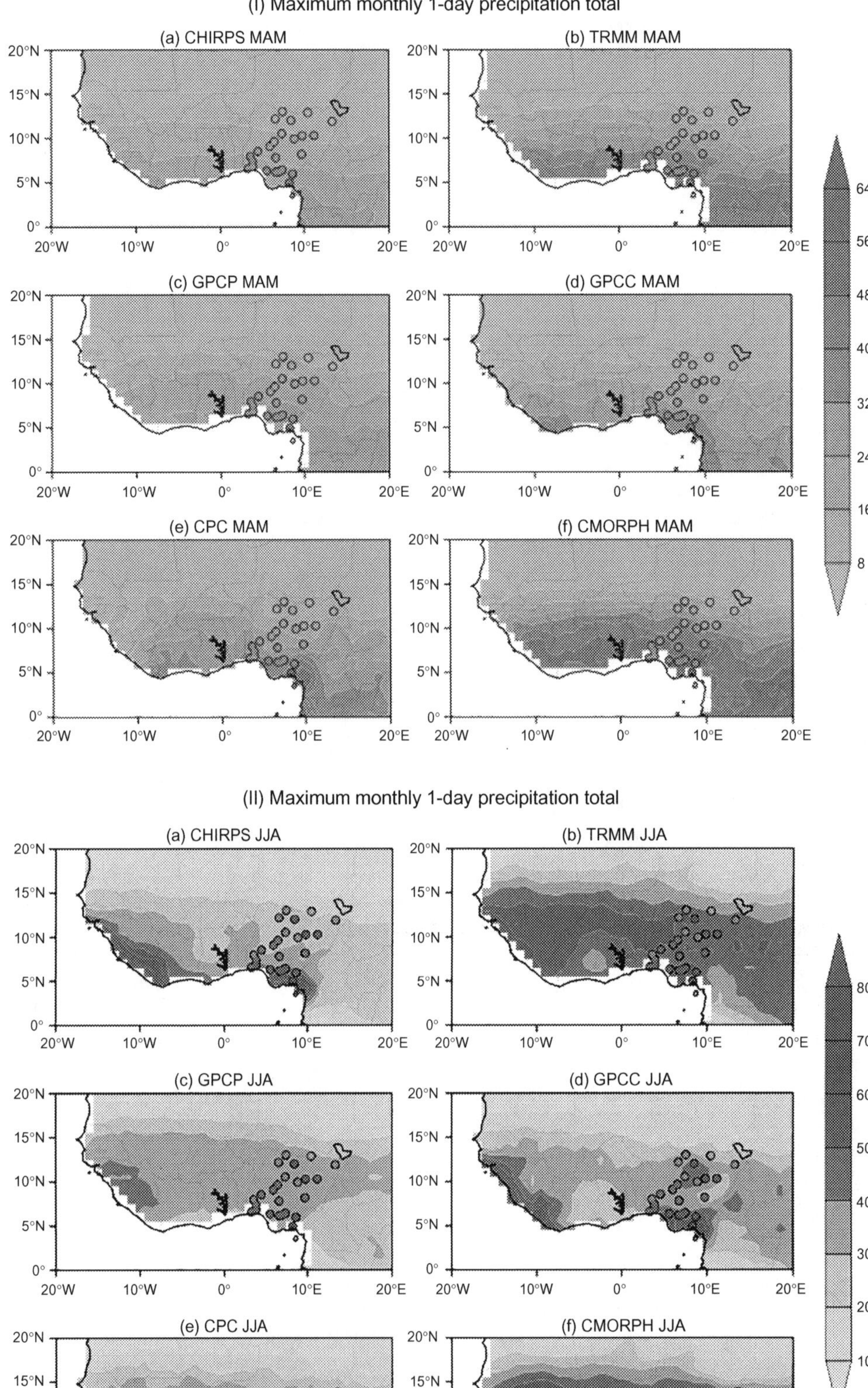

FIG. 4.4 (I)–(III) Average consecutive 1-day precipitation amount for MAM, JJA and SON in (a) CHIRPS, (b) TRMM, (c) GPCP, (d) GPCC, (e) CPC, and (f) CMORPH. The station gauge data are plotted as filled *color circles* at the same scale with the gridded observations.

(Continued)

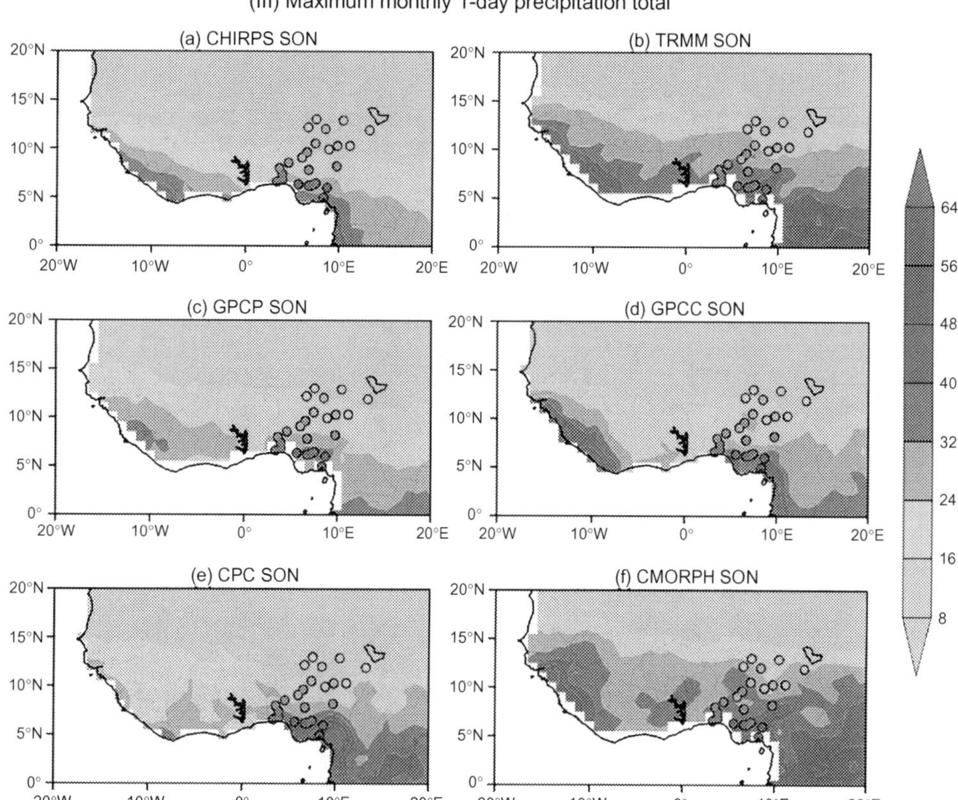

FIG. 4.4, CONT'D

respectively. Generally, in MAM, a belt of CDD is present around 10°N in the observational datasets. During JJA, the spatial distribution of CDD is highly variable in the gauges. Furthermore, the observed pattern of short CDD coincides with areas of complex terrains and consistent with the areas of longer CWD found in Fig. 4.7. When the monsoon activity ends in SON, the CDD increases in the Sahel and extends toward the southern coast areas of West Africa.

4 Summary and conclusion

The relevance of rainfall to core sectors of the economy like health, water resources and hydrology, and agriculture and food security cannot be overemphasized. Although rainfall is the major source of water in West Africa for a variety of purposes, sparsity of sufficient long-term station gauge data (from observations) makes it challenging to fully understand its temporal and spatial variability over the region. Therefore, gridded products are now been used as alternatives for studying the underlying characteristics of rainfall as well as evaluating the performance of climate model outputs. These gridded products are a blend of satellite and in situ station observations from available data, although these data do contain some measure of uncertainties. Therefore, evaluating the product performance before application is essential.

This chapter compared different gridded products including CHIRPS, TRMM, GPCP, GPCC, CPC, and CMORPH, and evaluated their performance against rainfall gauge station data across Nigeria within the context of the observation uncertainties. We focused on the mean precipitation distribution from rainy days as well as some sector-specific indices calculated from daily rainfall. The analysis revealed substantial differences among the different products in some of the metrics evaluated. Mostly, the products can vary significantly in terms of mean precipitation amount with CHIRPS, TRMM, GPCP and GPCC having the least and CPC and CMORPH the most variation. These characteristics also vary seasonally although with different magnitudes. The differences are more pronounced in a broad array of estimates derived from higher order statistics, such as frequency, intensity, and duration of rain events. For example,

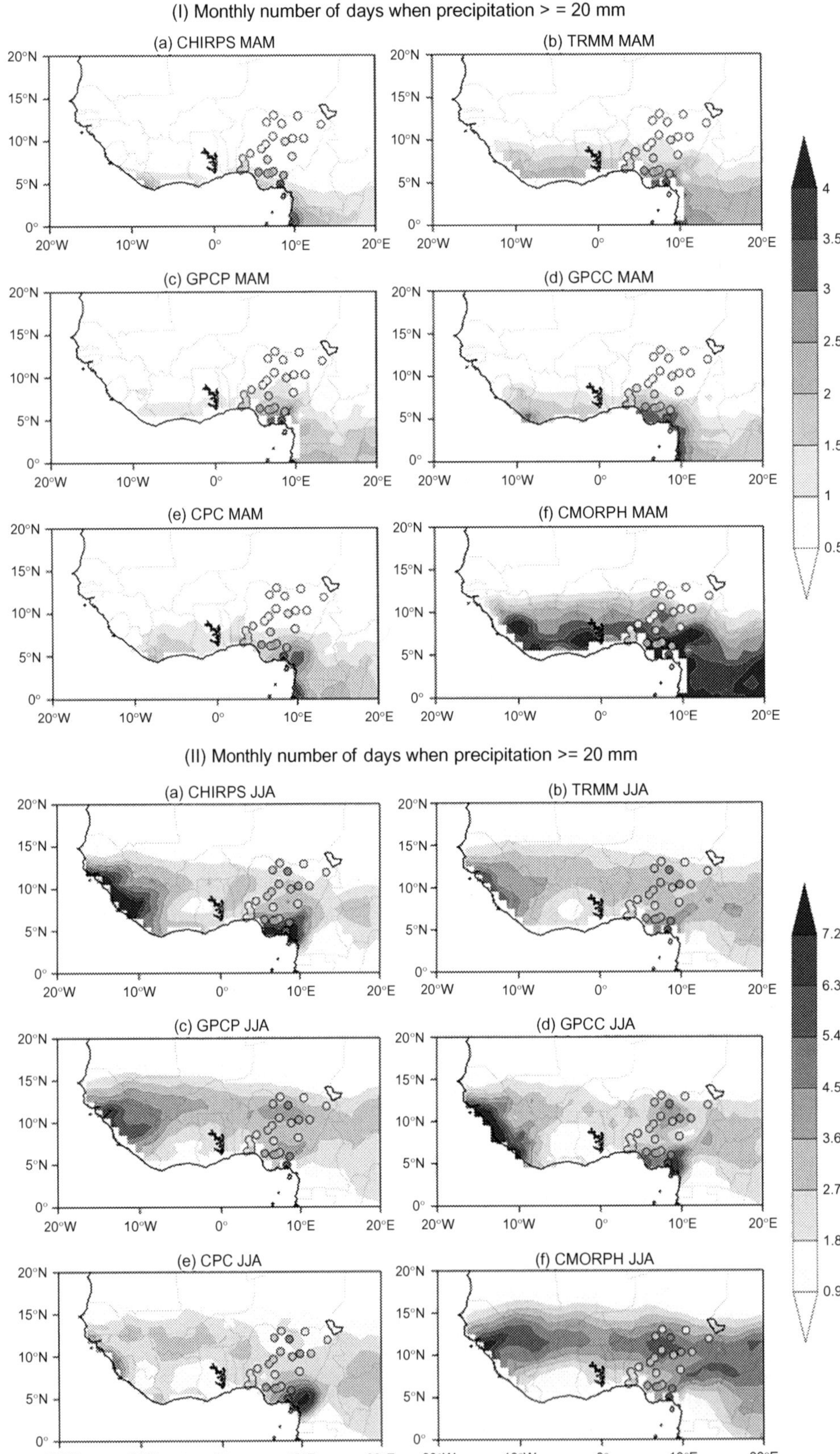

FIG. 4.5 (I)–(III) Same as in Fig. 4.4 but for average monthly count of days with very heavy precipitation (i.e., count of days with precipitation ≥20 mm).

(Continued)

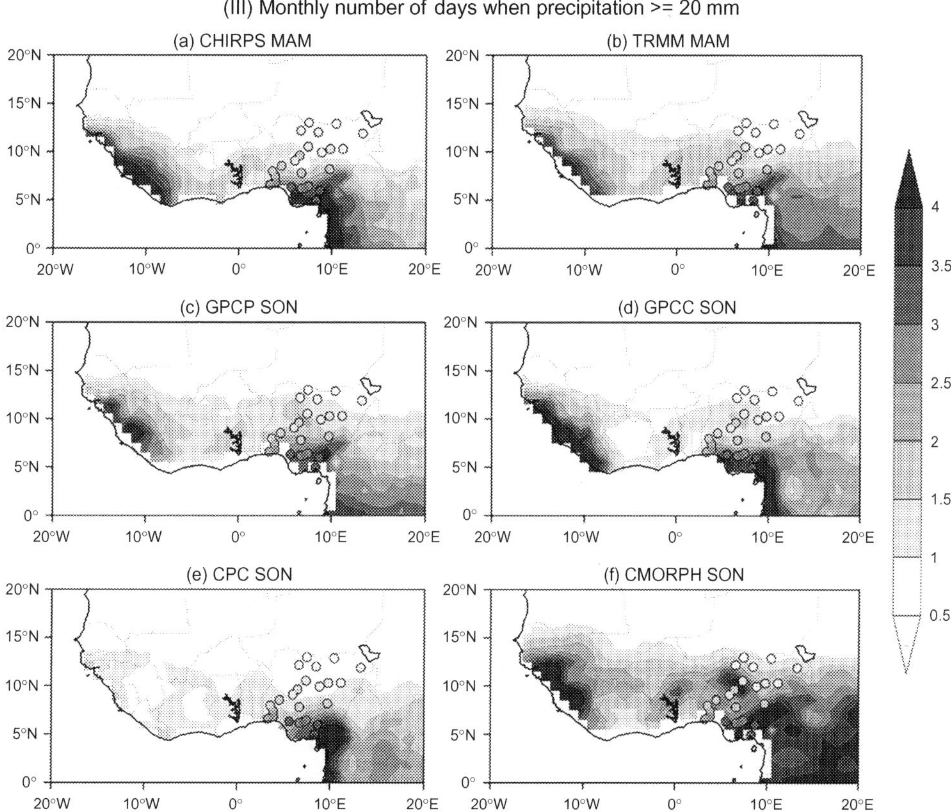

FIG. 4.5, CONT'D

the maximum length of wet spells is more in GPCC but it has less precipitation intensity relative to CMORPH, mostly in JJA. CMORPH displayed good agreement in reproducing RX1-day in Sahel region but less in the Guinea Coast.

The CWD show patterns (shown in Fig. 4.7) similar to the mean precipitation amount where wetter conditions are observed in the southern coast and drier conditions in the northern part of the region, especially in the desert region. Although there are differences between gauge and the gridded observations, both products agree more in regions north of 10°N in all seasons.

The quantitative assessment of the gridded products relative to station gauges reflects the underlying uncertainties. These uncertainties are clearly seen in a wide range of positive and negative biases. For example, CHIRPS, TRMM, GPCP and GPCP have lower values of precipitation bias compared to CPC (negative bias) and CMORPH (positive bias). The spatial pattern of precipitation over West Africa as well as the monsoon evolution is mainly captured in some of the observations.

Generally, the analysis shows that significant uncertainties are present among gridded observational rainfall datasets that significantly impact the characteristics of their daily rainfall distributions over the West Africa. This result is consistent with previous studies (Sylla et al., 2013). Though the observational products to some extent agree with the gauge measurements, the variances among them pose a great challenge regarding the analysis and the choice of reference for regional climate model evaluation over the monsoon region. In terms of rainfall amount, CHIRPS and GPCC, respectively, have better agreement relative to the gauge rainfall, highlighting the benefits of using rain gauge for their generation. Performance of other gridded products aside from CPC and CMORPH also compares favorably well relative to gauge data, therefore making their application equally suitable for impact studies. This result is, however, subject to the available station data used, which is only limited to Nigeria. However, it is critical to emphasize that users should be careful when drawing conclusions from the use of any of these gridded observational products, most especially when applying derived results to inform decision-making process in climate-sensitive sectors of the economy such as health, water resources and hydrology, and agriculture and food security. Also, a beneficial approach to

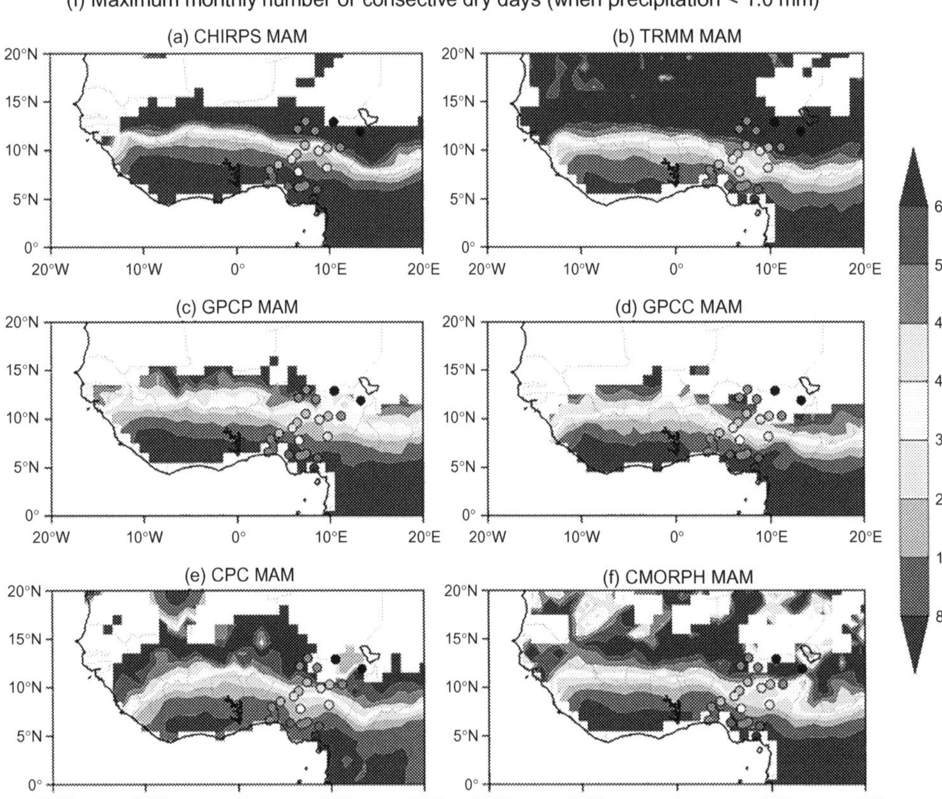

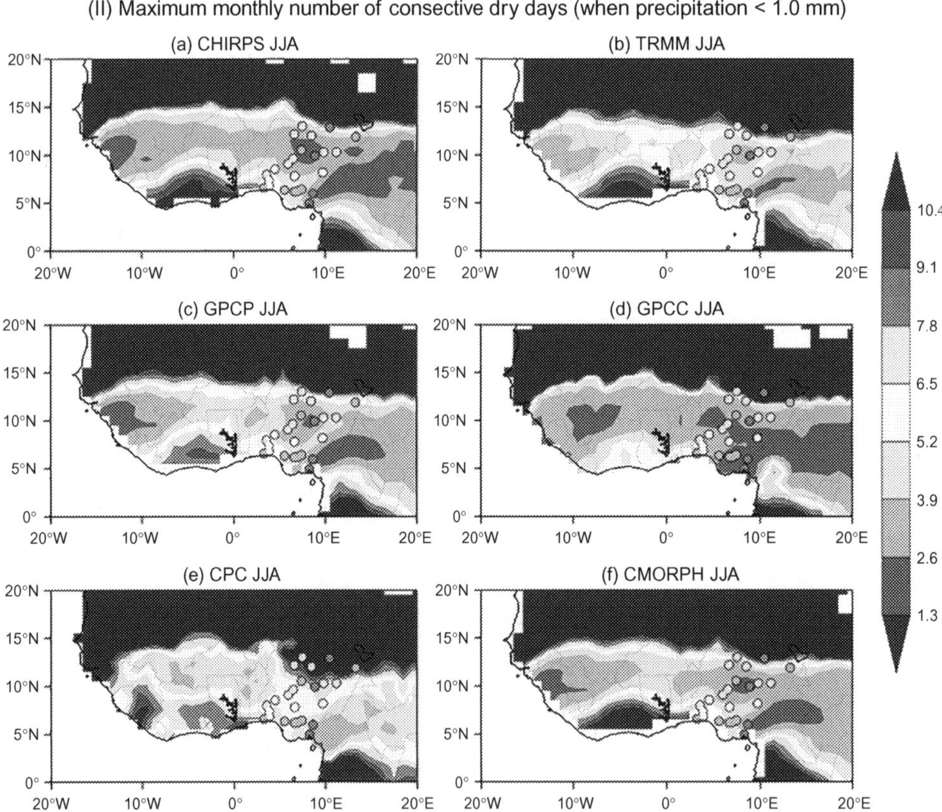

FIG. 4.6 (I)–(III) Same as in Fig. 4.4 but for average monthly consecutive dry days (i.e., count of days with precipitation ≥1 mm). Some stations are represented with *black circles* because of missing values in some months.

(Continued)

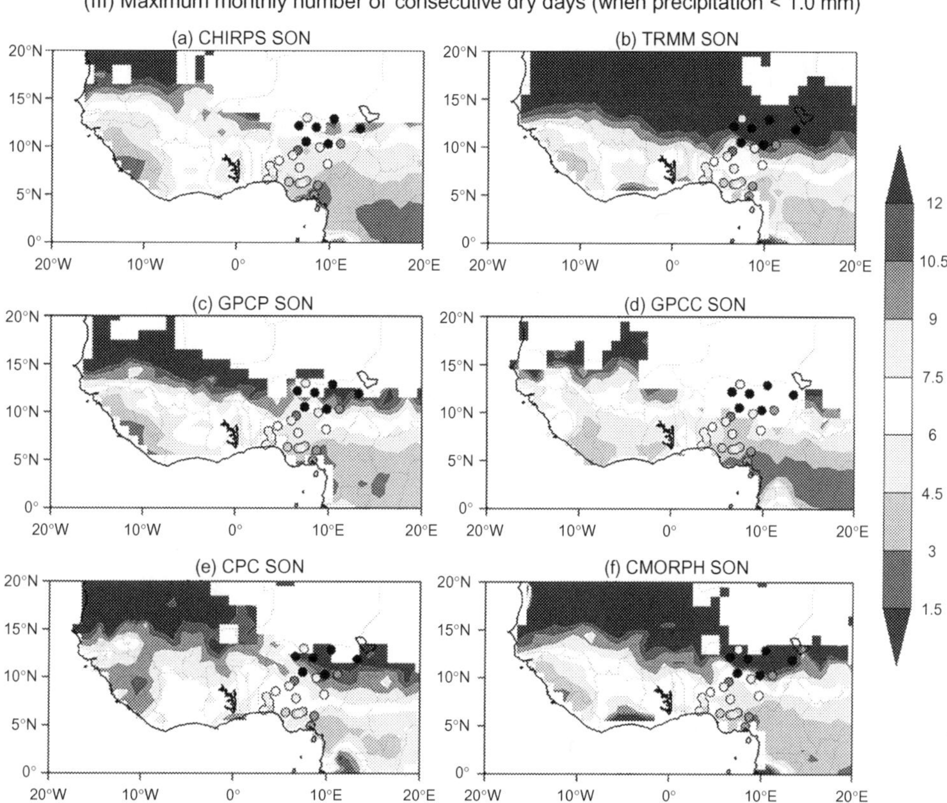

FIG. 4.6, CONT'D

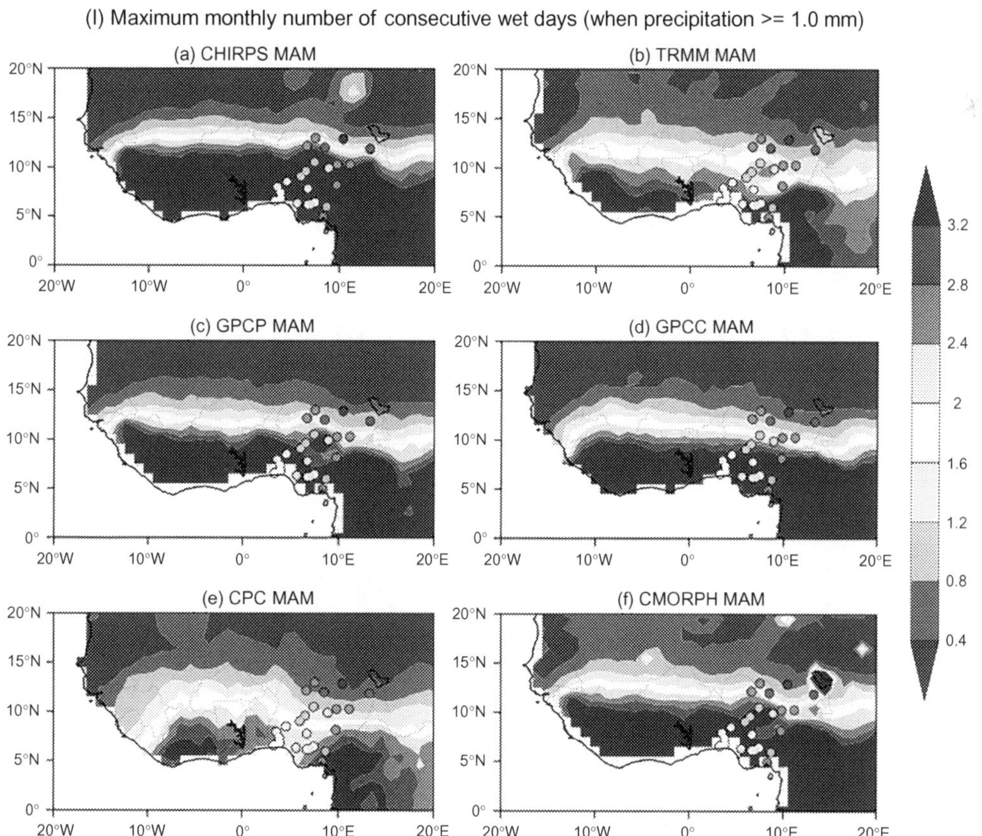

FIG. 4.7 (I)–(III) Same as in Fig. 4.4 but for average monthly consecutive wet days (i.e., count of days with precipitation ≥1 mm).

(Continued)

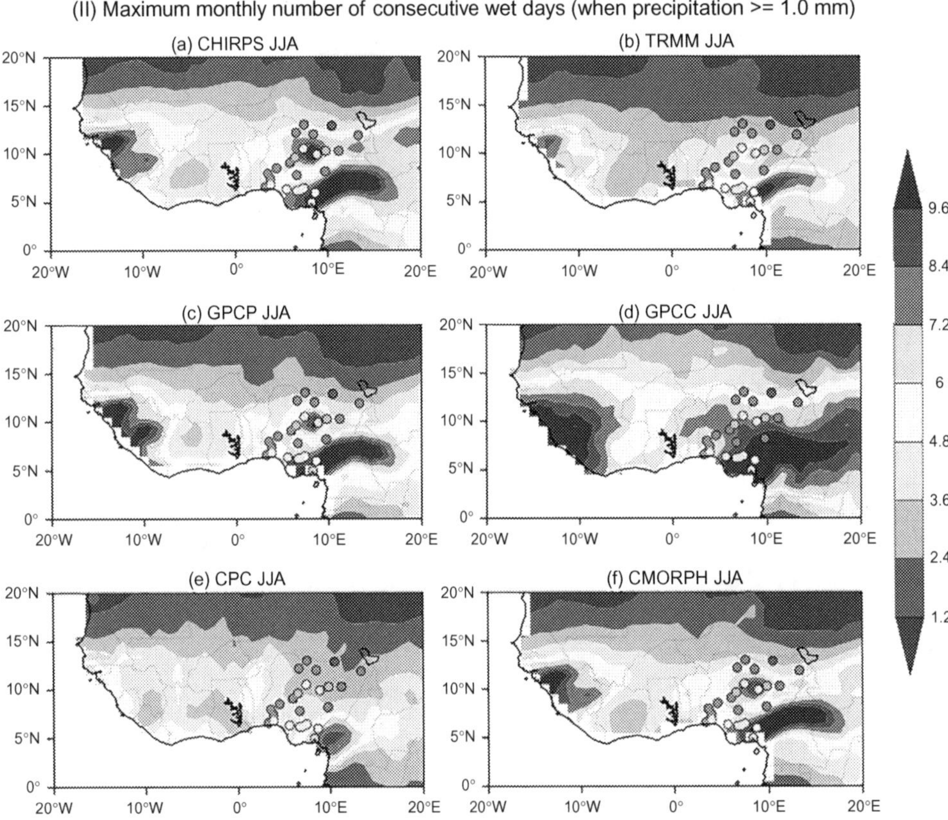
(II) Maximum monthly number of consecutive wet days (when precipitation >= 1.0 mm)
(a) CHIRPS JJA
(b) TRMM JJA
(c) GPCP JJA
(d) GPCC JJA
(e) CPC JJA
(f) CMORPH JJA
20°N
15°N
10°N
5°N
0°
20°W
10°W
0°
10°E
20°E
9.6
8.4
7.2
6
4.8
3.6
2.4
1.2

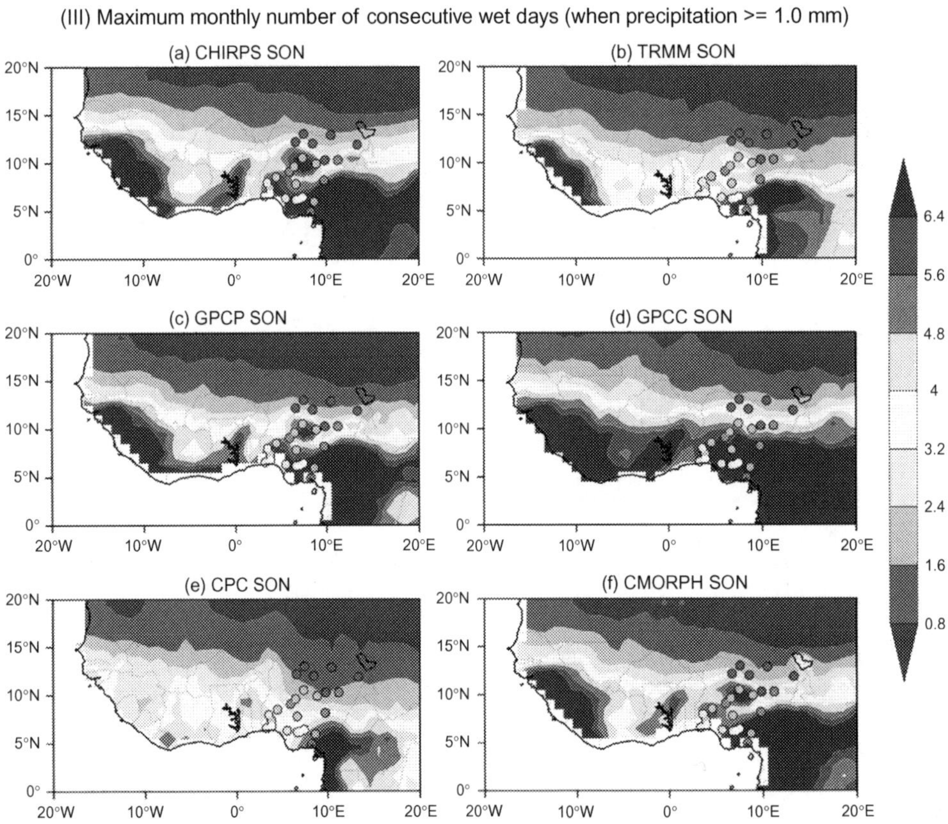
(III) Maximum monthly number of consecutive wet days (when precipitation >= 1.0 mm)
(a) CHIRPS SON
(b) TRMM SON
(c) GPCP SON
(d) GPCC SON
(e) CPC SON
(f) CMORPH SON
20°N
15°N
10°N
5°N
0°
20°W
10°W
0°
10°E
20°E
6.4
5.6
4.8
4
3.2
2.4
1.6
0.8

FIG. 4.7, CONT'D

increase the confidence of using these products is to perform bias correction based on increased number of station data. Future studies can expand on the sector-specific indices and/or explore a gridded products ensemble mean.

Conflict of interest

The authors declare no conflict of interest as regards the publication of this manuscript.

References

Adler, R., Wang, J., Sapiano, M., Huffman, G., Bolvin, D., Nelkin, E., et al. (2017). *Global precipitation climatology project (GPCP) climate data record (CDR), version 1.3 (daily)*. National Centers for Environmental Information. https://doi.org/10.7289/V5RX998Z (Accessed March 2019).

Akinsanola, A. A., & Ogunjobi, K. O. (2017). Recent homogeneity analysis and long-term spatio-temporal rainfall trends in Nigeria. *Theoretical and Applied Climatology*, *128*(1–2), 275–289.

Akinsanola, A. A., Ogunjobi, K. O., Ajayi, V. O., Adefisan, E. A., Omotosho, J. A., & Sanogo, S. (2017). Comparison of five gridded precipitation products at climatological scales over West Africa. *Meteorology and Atmospheric Physics*, *129*(6), 669–689.

Akinsanola, A. A., Ogunjobi, K. O., Gbode, I. E., & Ajayi, V. O. (2015). Assessing the capabilities of three regional climate models over CORDEX Africa in simulating West African summer monsoon precipitation. *Advances in Meteorology*, *2015*. https://doi.org/10.1155/2015/935431. Article ID 935431, 13 pages.

Akinsanola, A. A., & Zhou, W. (2019). Projection of West African summer monsoon rainfall in dynamically downscaled CMIP5 models. *Climate Dynamics*, *53*, 81–95. https://doi.org/10.1007/s00382-018-4568-6.

ClimPACT. https://climpact-sci.org/indices (Accessed May 2020).

Foley, A. M. (2010). Uncertainty in regional climate modelling: A review. *Progress in Physical Geography*, *34*(5), 647–670.

Funk, C., Peterson, P., Landsfeld, M., Pedreros, D., Verdin, J., Shukla, S., et al. (2015). The climate hazards infrared precipitation with stations—A new environmental record for monitoring extremes. *Scientific Data*, *2*, 150066. https://doi.org/10.1038/sdata.2015.66.

Gbode, I. E., Adeyeri, O. E., Menang, K. P., Intsiful, J. D., Ajayi, V. O., Omotosho, J. A., et al. (2019). Observed changes in climate extremes in Nigeria. *Meteorological Applications*, *26*(4), 642–654. https://doi.org/10.1002/met.1791.

Gbode, I. E., Akinsanola, A. A., & Ajayi, V. O. (2015). Recent changes of some observed climate extreme events in Kano. *International Journal of Atmospheric Sciences*. https://doi.org/10.1155/2015/298046.

Gbode, I. E., Dudhia, J., Ogunjobi, K. O., & Ajayi, V. O. (2019). Sensitivity of different physics schemes in the WRF model during a West African monsoon regime. *Theoretical and Applied Climatology*, *136*(1–2), 733–751. https://doi.org/10.1007/s00704-018-2538-x.

Gosset, M., Viarre, J., Quantin, G., & Alcoba, M. (2013). Evaluation of several rainfall products used for hydrological applications over West Africa using two high-resolution gauge networks. *Quarterly Journal of the Royal Meteorological Society*, *139*(673), 923–940.

Hagos, S. M., & Cook, K. H. (2007). Dynamics of the west African monsoon jump. *Journal of Climate*, *20*(21), 5264–5284.

Herrmann, S. M., & Mohr, K. I. (2011). A continental-scale classification of rainfall seasonality regimes in Africa based on gridded precipitation and land surface temperature products. *Journal of Applied Meteorology and Climatology*, *50*(12), 2504–2513.

Hourdin, F., Musat, I., Guichard, F., Ruti, P. M., Favot, F., Filiberti, M. A., et al. (2010). AMMA-model intercomparison project. *Bulletin of the American Meteorological Society*, *91*, 95–104. https://doi.org/10.1175/2009BAMS2791.1.

Huffman, G., Bolvin, D., Braithwaite, D., Hsu, K., Joyce, R., & Xie, P. (2014). *Integrated multi-satellite retrievals for GPM (IMERG), version 4.4*. NASA's Precipitation Processing Center. ftp://arthurhou.pps.eosdis.nasa.gov/gpmdata/ (Accessed March 2019).

Huffman, G. J., Adler, R. F., Bolvin, D. T., & Gu, G. (2009). Improving the global precipitation record: GPCP version 2.1. *Geophysical Research Letters*, *36*(17). https://doi.org/10.1029/2009GL040000.

Huffman, G. J., Bolvin, D. T., & Adler, R. F. (2016). *GPCP version 1.2 one-degree daily precipitation data set*. Research data archive at the National Center for Atmospheric Research, Computational and Information Systems Laboratory. https://doi.org/10.5065/D6D50K46.

Huffman, G. J., Bolvin, D. T., Nelkin, E. J., Wolff, D. B., Adler, R. F., Gu, G., et al. (2007). The TRMM multisatellite precipitation analysis (TMPA): Quasi-global, multiyear, combined-sensor precipitation estimates at fine scales. *Journal of Hydrometeorology*, *8*(1), 38–55.

Jones, P. W. (1999). First- and second-order conservative remapping schemes for grids in spherical coordinates. *Monthly Weather Review*, *127*(9), 2204–2210.

Joyce, R. J., Janowiak, J. E., Arkin, P. A., & Xie, P. (2004). CMORPH: A method that produces global precipitation estimates from passive microwave and infrared data at high spatial and temporal resolution. *Journal of Hydrometeorology*, *5*, 487–503. https://www.cpc.ncep.noaa.gov/products/janowiak/cmorph_description.html (Accessed March 2019).

Le Barbe, L., Lebel, T., & Tapsoba, D. (2002). Rainfall variability in West Africa during the years 1950–1990. *Journal of Climate*, *15*(2), 187–202.

Maidment, R. I., Grimes, D., Black, E., Tarnavsky, E., Young, M., Greatrex, H., et al. (2017). A new, long-term daily satellite-based rainfall dataset for operational monitoring in Africa. *Scientific Data*, *4*, 170063.

Maraun, D. (2016). Bias correcting climate change simulations—A critical review. *Current Climate Change Reports*, *2*(4), 211–220.

Nicholson, S. E. (2013). The West African Sahel: A review of recent studies on the rainfall regime and its interannual variability. *ISRN Meteorology*, *2013*. https://doi.org/10.1155/2013/453521, 453521.

Parry, M. L., Canziani, O. F., Palutikof, J. P., Van Der Linden, P. J., & Hanson, C. E. (2007). *Contribution of working group II to the fourth assessment report of the intergovernmental panel on climate change*. Cambridge, UK and New York, NY: Cambridge University Press.

Prein, A. F., Langhans, W., Fosser, G., Ferrone, A., Ban, N., Goergen, K., et al. (2015). A review on regional convection-permitting climate modeling: Demonstrations, prospects, and challenges. *Reviews of Geophysics*, *53*(2), 323–361.

Roca, R., Chambon, P., Jobard, I., Kirstetter, P. E., Gosset, M., & Bergès, J. C. (2010). Comparing satellite and surface rainfall products over West Africa at meteorologically relevant scales during the AMMA campaign using error estimates. *Journal of Applied Meteorology and Climatology*, *49*(4), 715–731.

Satgé, F., Defrance, D., Sultan, B., Bonnet, M. P., Seyler, F., Rouché, N., et al. (2020). Evaluation of 23 gridded precipitation datasets across West Africa. *Journal of Hydrology*, *581*, 124412.

Schamm, K., Ziese, M., Becker, A., Finger, P., Meyer-Christoffer, A., Rudolf, B., et al. (2013). *GPCC first guess daily product at 1.0o: Near real-time first guess daily land-surface precipitation from rain-gauges based on SYNOP data*. Global Precipitation Climatology Centre (GPCC, http://gpcc.dwd.de/) at Deutscher Wetterdienst. https://doi.org/10.5676/DWD_GPCC/FG_D_100.

Schamm, K., Ziese, M., Raykova, K., Becker, A., Finger, P., Meyer-Christoffer, A., et al. (2015). *GPCC full data daily version 1.0 at 1.0: Daily land-surface precipitation from rain-gauges built on GTS-based and historic data*. https://doi.org/10.5676/DWD_GPCC/FD_D_V1_100.

Sultan, B., & Janicot, S. (2000). Abrupt shift of the ITCZ over West Africa and intraseasonal variability. *Geophysical Research Letters*, *27*(20), 3353–3356.

Sylla, M. B., Dell'Aquila, A., Ruti, P. M., & Giorgi, F. (2010). Simulation of the intraseasonal and the interannual variability of rainfall over West Africa with RegCM3 during the monsoon period. *International Journal of Climatology*, *30*, 1865–1883. https://doi.org/10.1002/joc.2029.

Sylla, M. B., Giorgi, F., Coppola, E., & Mariotti, L. (2013). Uncertainties in daily rainfall over Africa: Assessment of gridded gridded products and evaluation of a regional climate model simulation. *International Journal of Climatology*, *33*(7), 1805–1817.

Tapiador, F. J., Navarro, A., Levizzani, V., García-Ortega, E., Huffman, G. J., Kidd, C., et al. (2017). Global precipitation measurements for validating climate models. *Atmospheric Research*, *197*, 1–20. https://doi.org/10.1016/j.atmosres.2017.06.021.

Xie, P. (2010). CPC unified gauge-based analysis of global daily precipitation. In *24th Conference on hydrology*. https://www.cpc.ncep.noaa.gov/products/Global_Monsoons/gl_obs.shtml (Accessed March 2019).

Xie, P., Chen, M., & Shi, W. (2010). CPC unified gauge-based analysis of global daily precipitation. In *Vol. 2. Preprints, 24th conference on hydrology, Atlanta, GA, Amer. Meteor. Soc 2010*.

CHAPTER

5

Features of regional Indian monsoon rainfall extremes

Hamza Varikoden and M.J.K. Reji

Indian Institute of Tropical Meteorology, Ministry of Earth Sciences, Pune, India

1 Introduction

Globally, India constitutes 2.4% of the total land area, supporting around 16.7% of the total population. It is approximated that the country's 54.6% human resources are involved in agriculture and allied sectors (https://censusindia.gov.in), which contributed approximately 17.8% of the country's Gross Value Added (GVA) for the year 2019–20 (Government of India, 2021). Rain-fed agriculture employs approximately 67% of the net sown area, supplying 44% of the food grains, which backs 40% of the population (Venkateswarlu, 2011). The Indian subcontinent receives about 85% of its annual rainfall during the summer monsoon season from June to September (Oza & Kishtawal, 2014; Turner & Annamalai, 2012). For centuries monsoon has been conceptualized as a massive land-sea breeze, but later studies identified monsoon as the seasonal migration of intertropical convergence zone (ITCZ) (Gadgil, 2003).

1.1 The role of Indian monsoon variability in extreme rainfall events over India

The variability of Indian Summer Monsoon Rainfall (ISMR) significantly impacts crop yield and crop productivity (Gadgil & Kumar, 2006; Prasanna, 2014; Preethi & Revadekar, 2013), greatly affecting the developing economy of the country. The interannual and intraseasonal variability of ISMR influences the occurance and intensity of extreme rainfall events.

Interannual variability of ISMR is the year to year variability of mean ISMR which can be identified as flood year, drought year and normal years (Fig. 5.1). A particular year is considered as a flood (drought) year when the percentage departure of ISMR of that year is 10% above (below) as compared to the mean ISMR (Fig. 5.1). El Niño Southern Oscillation (ENSO) and the Indian Ocean Dipole (IOD) are considered as the major drivers of interannual variability of ISMR and particularly extreme rainfall events (Krishnaswamy et al., 2015). The ENSO is a coupled oscillation between ocean and atmosphere in the tropical pacific ocean and it is generally characterized by the shift of warm water pool in the western tropical pacific ocean (Sen Gupta, McNeil, Henderson-Sellers, & McGuffie, 2012). The ENSO bridges the ISMR through the equatorial Walker and local Hadley circulations (Ashok, Guan, & Yamagata, 2001; Ashok, Guan, Saji, & Yamagata, 2004; Goswami, 1998; Varikoden & Preethi, 2013). In general, the warm (cold) phase of ENSO suppresses (enhances) the ISMR (Rasmusson & Carpenter, 1983; Sikka, 1980). The IOD is the anomaly gradient of SST between west and east of the Equatorial Indian Ocean and positive (negative) IOD refers to the warming in the western(eastern) Indian Ocean (Saji, Goswami, Vinayachandran, & Yamagata, 1999). IODs have a dominant influence on ERE's as compared to the ENSO in recent decades particularly the frequency of extreme rainfall events (Krishnaswamy et al., 2015). The frequency and intensity of positive IOD events manifest an unanticipated rise in recent years (Behera, Doi, Ratnam, & Behera, 2021). Thus increasing trend of IOD events in recent decades advocate the possible increase in the ERE's in the near future.

Intraseasonal variability of ISMR is driven by intraseasonal oscillation which can be discerned as alternative active (wet) and break (dry) periods (Gadgil, 2003; Lawrence & Webster, 2001). The intensity of wet spells and frequency of dry spells are showing a statistically significant increasing trend, while the intensity of the dry spell manifesting a statistically significant decreasing trend (Singh, Horton, et al., 2014; Singh, Tsiang, Rajaratnam, & Diffenbaugh, 2014). Thus, interannual and intraseasonal variabilities of monsoon significantly affect ERE's over India.

Climate Impacts on Extreme Weather
https://doi.org/10.1016/B978-0-323-88456-3.00014-9

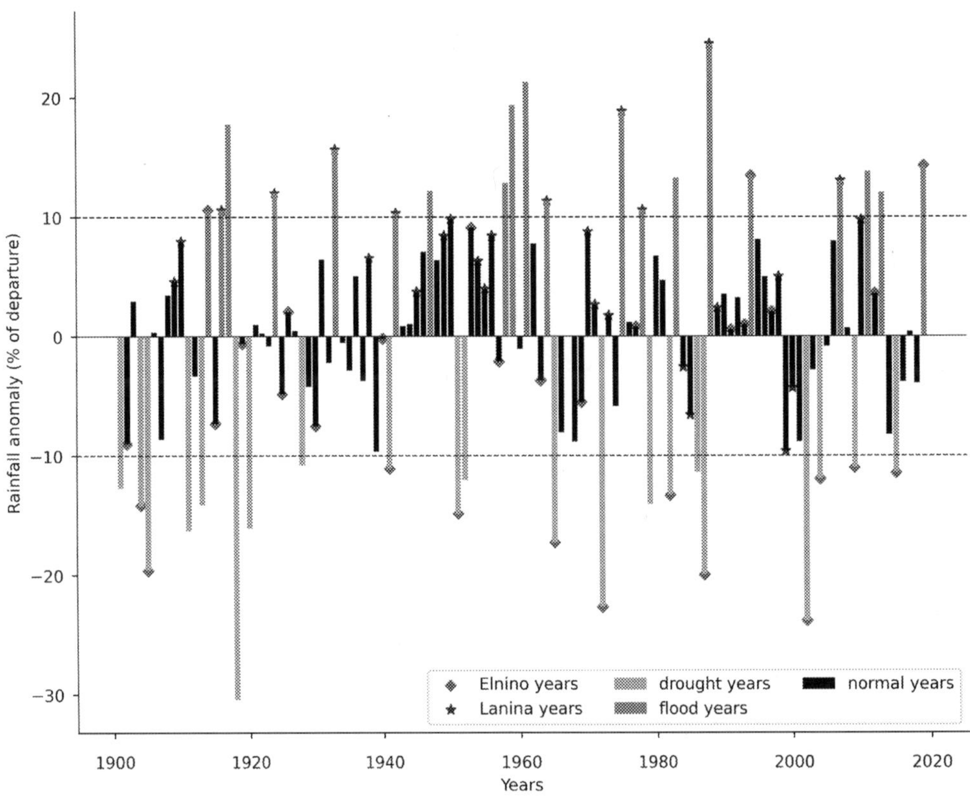

FIG. 5.1 Time series of percentile departures of ISMR anomalies for the 1901 to 2018 period Using daily gridded rainfall dataset from IMD (0.25° × 0.25° resolution). The *blue* (*gray* in print version) and *red* (*light gray* in print version) *bars* are wet and dry years, respectively, El Nino years are marked with stars and are La Nina years are marked with diamonds. The Niño 3.4 index for the period of 1901–2018 was retrieved from the climate prediction center, NOAA. *Niño 3.4 index data from http://www.cpc.ncep.noaa.gov.*

1.2 Impacts of global warming on extreme rainfall events over India

The frequency and duration of extreme rainfall events are highly influenced by global warming (Allan & Soden, 2008; Goswami, Venugopal, Sengupta, Madhusoodanan, & Xavier, 2006; Karmakar, Chakraborty, & Nanjundiah, 2017; Min, Zhang, Zwiers, & Hegerl, 2011). In the era of global warming, the atmosphere is warmer. Thus, it can hold high amounts of moisture (7% moisture per 1°C temperature rise) based on the Clausius-Clapeyron equation (Easterling, 2000; Trenberth, Dai, Rasmussen, & Parsons, 2003). The rate of increase in the intensity of long and short duration floods due to warming is consistent with the rate of increase in moisture (Fowler et al., 2021). The enhanced capacity of the atmosphere to hold moisture due to warming leads to an increase in both intensity and frequency of heavy precipitation (Burt, 2005; Clark, 2005; Coumou & Rahmstorf, 2012; Dai, 2006; Devrani, Singh, Mudd, & Sinclair, 2015; Joshi & Rajeevan, 2006; Simmons, Willett, Jones, Thorne, & Dee, 2010). In general, atmospheric warming affects the EREs at least in three different mechanisms. (1) The intensity of extreme rainfall events increases with the moisture enhancement if the atmospheric circulation remains unaltered. However, the circulation is changing and affecting the rainfall; thus, it leads to two other mechanisms. (2) Continuous warming of the atmosphere enhances the atmospheric stability, which in turn slows down the circulation, thus reducing the intensity of extreme rainfall events, and (3) manifests the strengthening of storms at a rate proportional to the release of latent heat and thus increasing the intensity of extreme rainfall events (Pendergrass, 2018). The effect of increasing moisture on extreme rainfall events remains significant. Still, the contribution is far less than due to climate change, and therefore, it can be concluded that climate change is the predominant driver of the extreme rainfall events (Lenderink & Fowler, 2017; Pendergrass & Hartmann, 2014; Trenberth, Fasullo, & Shepherd, 2015; Wang, Zhao, Yoon, Klotzbach, & Gillies, 2018).

1.3 Socio-economic impacts of extreme rainfall events over India

In addition to large scale floods and droughts, abnormal fluctuations in the form of localized floods and flash floods cause an enormous impact on human beings in terms of homelessness, destruction of infrastructure, loss of lives, and consequent disease outbreaks (Kshirsagar, Shinde, & Mehta, 2006; Mishra & Shah, 2018; Singh & Kumar, 2013). Irrespective of the decline in monsoon circulation, extreme rainfall events have been showing an increasing trend in the last few decades over various parts of the country (Goswami et al., 2006; Guhathakurta, Menon, Inkane, Krishnan, & Sable, 2017; Rajeevan, Bhate, & Jaswal, 2008; Roxy et al., 2017; Singh & Kumar, 2013). The global economic loss due to flooding is about USD 30 million per year, while that, for India, is reckoned as USD 3 million per year for the last few decades (Roxy et al., 2017). The Intergovernmental Panel on Climate Change (IPCC, 2018; Poloczanska, Mintenbeck, Portner, Roberts, & Levin, 2018) report elucidated the intensification of frequency, intensity, and amount of heavy and extreme rainfall globally, particularly over the Indo-Gangetic plain, Tibetan plateau, and Himalayan mountain ranges closer to India, the second-most populous country globally, has 1.39 billion people spread across the major states and union territories. According to Amarasinghe, Amarnath, Alahacoon, and Ghosh (2020), floods and droughts account for more than 50% of the natural hazards in India. In India, about USD 60 billion loss was solely due to regional floods, out of 99 billion US dollars lost from extreme weather events (such as floods, tropical cyclones, heat waves, cold waves, lightning, etc.) in the previous 50 years. Furthermore, among the mortality from all other extreme weather events, floods alone contribute 46.1% of the mortality in India, as shown in Fig. 5.2. Floods and heavy rains are significant causes of mortality due to extreme weather events in all states except the eastern coastal states. The primary reason for the mortality rate in the eastern coastal states is cyclonic storms. The heat waves strongly impact the mortality rate in Telangana, Rajasthan, and Andhra Pradesh states. As per the Global Flood Inventory (GFI) developed by Adhikari et al. (2010), the majority of the flooding (64%) is formed by short-duration heavy rains and subsequently by torrential rainfall that contributed to more than 11% of the total rainfall.

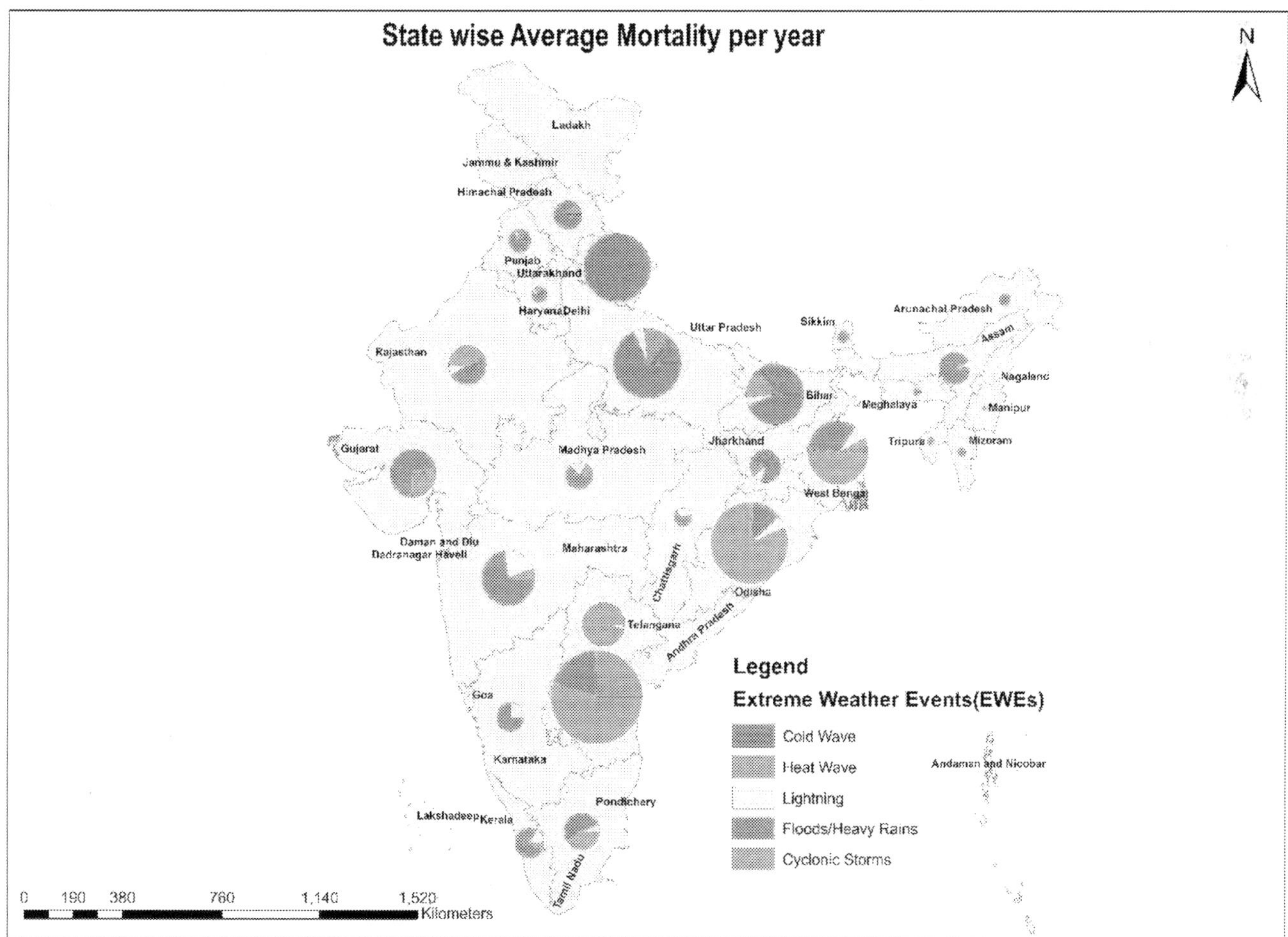

FIG. 5.2 Schematic representation of the distribution of mortality from various extreme weather events in different states of India during 1970–2019. The size of the *circle* represents the average mortality of each state. *From Ray, K., Giri, R. K., Ray, S. S., Dimri, A. P. & Rajeevan, M. (2021). An assessment of long-term changes in mortalities due to extreme weather events in India: A study of 50 years' data, 1970–2019.* Weather and Climate Extremes, 32, *100315. https://doi.org/10.1016/j.wace.2021.100315.*

All the extreme rainfall events share a few common characteristics: the dynamical lifting of moist air, continuous supply of moisture, and static instability (Doswell, Brooks, & Maddox, 1996). However, the magnitude and intensity of extreme rainfall events depend on the geographical characteristics and causative mechanisms. This work investigates the extreme rainfall events evolution and characteristics throughout India and addresses the impact of synoptic systems on the extreme rainfall events from a local to regional scale. This literature aims to develop an overall understanding of the Indian monsoon extreme rainfall events.

2 Data and methods

2.1 Data

High spatial resolution (0.25° × 0.25°) daily gridded rainfall data for the summer monsoon season (June-September) during the period 1901–2019 from India Meteorological Department (IMD) is utilized in the study for rainfall analysis. The data are retrieved from 6955 rain gauge stations in India. The density of the rain gauge stations was about 1500 for the first few years, but the number reached around 4000 by the end of the 20th century (Pai, Sridhar, Badwaik, & Rajeevan, 2015; Rajeevan et al. (2008). In addition to the gridded data set, Tropical Rainfall Measuring Mission (TRMM) and Multi-satellite Precipitation Analysis (TMPA) for research [TRMM 3B42, version 7 (TRMM 3B42-V7)] (Huffman et al., 2007) with a horizontal resolution of 0.25° × 0.25° were also used in the study.

The Niño 3.4 index is the Sea Surface Temperature (SST) anomaly in the Niño 3.4 region, i.e., 5° N–5° S, 170°–120° W, which is considered for the identification of El Niño and La Niña events. The Niño 3.4 index for the period of 1901–2019 was retrieved from the Climate Prediction Center (CPC) of the National Oceanic and Atmospheric Administration (NOAA) (http://www.cpc.ncep.noaa.gov).

The zonal and meridional (u and v) wind fields at 200 hPa with high resolution (0.25° × 0.25°) were retrieved from the European Centre for Medium-Range Weather Forecasts' fifth set of reanalysis data (ERA5) for July 2005. ERA5 provides a better representation of tropospheric circulations as compared to other data sets. Additionally, horizontal wind fields and specific humidity for the period November 1, 2015 to December 2, 2015, were obtained from National Aeronautics and Space Administration Modern-Era Retrospective-Analysis for Research and Applications (MERRA) with a horizontal resolution of 0.5° × 0.66° and 6-h temporal resolution were also used to characterize the circulation features during the Chennai rainfall extremes.

2.2 Methods

2.2.1 *Percentile departure of rainfall*

The percentile departure of summer monsoon rainfall (PDR) from its mean is calculated so that the wet years (>10%) and deficient years (<10%) can be identified. It is calculated as shown in Eq. (5.1)

$$PDRi = \frac{Ri - R_{\text{mean}}}{R_{\text{mean}}} * 100 \tag{5.1}$$

Ri is the seasonal mean rainfall for the specific year, and R_{mean} represents the mean rainfall of summer monsoon rainfall for the period 1901–2019. Wet years are identified as flood years and dry years are identified as drought years and this simple figure deciphers the interannual variability of ISMR in a single glance.

2.2.2 *Moisture transport*

The vertically integrated horizontal water vapor transport (IVT) based on Lavers, Villarini, Allan, Wood, and Wade (2012) is utilized for moisture transport assessment. The IVT is calculated (kg m^{-1} s^{-1}) using specific humidity (q), zonal and meridional (u and v) wind fields (Eq. 5.2) at different atmospheric levels from 1000 to 300 hPa

$$IVT = \sqrt{\left(\frac{1}{g}\int_{1000}^{300} qu dp\right)^2 + \left(\frac{1}{g}\int_{1000}^{300} qv dp\right)^2} \tag{5.2}$$

where g is the acceleration due to gravity (9.8 m s^{-2}), and p is pressure. Calculation of moisture transport is of great importance for the identification of occurrence and intensity of extreme rainfall events.

2.2.3 Probable maximum precipitation

The probable maximum precipitation (PMP) is theoretically the greatest depth of rain for a given duration that is physically possible over a given size of storm area in a particular region over a specific time of year (Ozone, 1985). The PMP is mainly used to design reservoirs and other major flood protection works. Ideally, PMP is used to estimate the largest flood that could occur in a hydrological basin.

2.2.4 Estimation of extreme events

In many studies (e.g., Jamshadali, Reji, Varikoden, & Vishnu, 2021; Roy & Balling, 2004; Varikoden, Preethi, & Revadekar, 2012), the extreme rainfall event is estimated based on the calculated percentile values based on daily rainfall data for the climatological period. If the percentile value is more than the 95th percentile, then the events are considered as extreme rainfall events. This percentile value may not be fixed for all the studies. This value may be changed from 90th to 99.9th percentiles, depending on the time slice considered, regions selected, and the study's objective. In some studies, a threshold value is also added to the percentile value to avoid the discrepancies arising from the regions which receive climatologically low rainfall (Goswami et al., 2006; Roxy et al., 2017).

The extreme rainfall events can also be estimated based on rainfall indices such as 1-day (Rx1 day), 2-day (Rx2 day), and 5-day (Rx5 day) maximum rainfall. These indices are computed at each latitude-longitude grid point based on the daily rainfall data. In these indices, a rainy day is considered when the daily rainfall is equal or exceeds 1 mm. An extreme rainfall event at one place may be treated as a normal rainfall event at another location based on its climatological values. Additionally, some studies estimate the EREs merely based on the threshold values such as 7 cm day^{-1} (Sinha Ray & Srivastava, 2000), 10 and 15 cm day^{-1} (Goswami et al., 2006).

3 Results

3.1 The history of extreme events

The study of extreme rainfall events gained adequate recognition at the beginning of the 21st century. The definition was not restricted to particular parameters and included various extreme metrics with many spatial, temporal and magnitude scales. The extreme rainfall studies in India embarked in the 1970s (Dhar & Kamte, 1971; Dhar, Rakhecha, & Sangam, 1975) with the preparation of PMP charts, which were then considered as the definition of extreme rainfall events. The development of PMP charts initially intended to adopt the best possible design for the development of infrastructures such as nuclear power plants, dams and other reservoirs (Dhar et al., 1975; Dhar & Kamte, 1971). Hydro-meteorologists continue to explore PMP with improved statistical methods such as Clavet-Gaumont and Hershfield PMP calculation for developmental designs (Deshpande & Gore, 1999; Kulkarni, Nandargi, & Mulye, 2010; Rakhecha & Clark, 2000; Sarkar & Maity, 2020). The PMP charts are best suitable for small drainage basins but might not be good for large diversified regions (World Meteorological Organization, 2009). The PMP studies fail to decipher the exact concept of extreme rainfall events over the diversified geography of the Indian subcontinent. Thus, the PMP and extreme rainfall studies developed as two separate branches of hydrometeorology.

The literature gradually adapted the term *extreme rainfall*, but its delineation remained ambiguous. The studies classified extreme rainfall based on the highest amount of rainfall received over a region during a specified period (1, 2, and 3 days maximum), which is similar to the concept of PMP (Rakhecha & Soman, 1994; Soman, Krishna, & Singh, 1988). One of the pioneer studies on extreme rainfall using the highest rainfall during the period (1-, 2-, 3-, ... 10-day period) was carried out by Soman et al. (1988) for the Kerala region and found that southern Peninsular India, especially the southern hilly regions of Kerala, manifested a decreasing trend in extreme rainfall at 95% confidence level. Roy and Balling (2004) found an overall increase in extreme precipitation events over India during the period 1910–2000 based on seven different indices of annual rainfall (annual rainfall, largest 1-, 5-, and 30-day event rainfall, extreme events of higher than 90th, 95th, and 97.5th percentile values).

Later, researchers identified the necessity of an index for extreme rainfall rather than explaining it concerning maximum rain for a particular duration. As an outset to develop a unified index, Sinha Ray and Srivastava (2000) identified heavy rainfall based on a single threshold value ($\geq$7 cm day^{-1}) for all the 151 stations widely distributed all over India. The study manifested a decreasing trend in heavy rainfall at most stations during the winter season; however, an increasing trend in heavy rain is registered over the west coast stations during the summer monsoon season. Goswami et al. (2006) selected a different threshold value to identify the extreme rainfall events and the threshold value was 15 cm day^{-1} or above over the central Indian region. The threshold value of this index is made based on the climatological conditions prevailing in that region. This study concluded an increasing trend in the frequency and magnitude of the extreme rainfall event over central India. However, Rajeevan et al. (2008) found that the magnitude of a rising trend in extreme rainfall events over central India is less compared to Goswami et al. (2006), and they attributed this dampening of the increasing trend to the multidecadal variation associated with sea surface temperature anomalies over the tropical Indian Ocean. Roxy et al. (2017) identified the homogeneity of extreme events over central India, extending from Gujarat to the east coast of Odisha (more than 500,000 km^2) based on a 15 cm day^{-1} or more threshold and compared this value with a 99.5th percentile value over the region. The trend and variability obtained as per the fixed threshold index and percentile-based index were strongly related to a correlation coefficient of 0.97, which is highly statistically significant. The study manifested a 75% increase in the frequency of extreme events throughout central India, which accounts for an increase of at least 13 events $decade^{-1}$. Francis and Gadgil (2006) also used the same threshold value and found that extreme rainfall events over the west coast of India are frequent from mid-June to mid-August, and these are associated with the active spells of Indian summer monsoon rainfall. (Ajayamohan & Rao, 2008) considered a threshold value of more than 10 cm day^{-1} (99.6th percentile of mean rainfall) for extreme rainfall event over the Ganges-Mahanadi basin in central India and identified an increasing probability of ERE. The increasing likelihood of extreme rainfall events is attributed to the recent abnormal Indian Ocean warming.

In addition to studies based on the area average rainfall of certain regions, spatial analysis on individual grid points revealed that the trends in the extreme rainfall events are nonuniform throughout the nation, especially in central India (Ghosh, Luniya, & Gupta, 2009). Ghosh et al. (2009) identified a significant increasing trend in the spatial heterogeneity of extreme rainfall throughout India using the extreme value theory. This spatial heterogeneity is attributed to the spatial variability of the convective constituents of rainfall, and thus it explicitly communicates the driver of extreme events (Ghosh et al., 2016). The extreme rainfall events, which manifest homogeneity in rainfall characteristics over large regions, are driven by synoptic systems (Nikumbh, Chakraborty, & Bhat, 2019; Nikumbh, Chakraborty, Bhat, & Frierson, 2020; Revadekar, Varikoden, Preethi, & Mujumdar, 2016; Roxy et al., 2017).

The definition and characterization of extreme rainfall event influence the conclusion drawn by the study (Pendergrass, 2018). The number of extreme events in a year or season is considered as the number of grid points at which the rainfall exceeded a threshold value (Goswami et al., 2006; Rajeevan et al., 2008; Roxy et al., 2017). The latter studies recognized extreme events occurring concurrently over several adjacent grids as a single event (Nikumbh et al., 2019, 2020). The trend value of the number of extreme events in central India based on the former method is 0.34 $year^{-1}$, while the trend value based on the latter approach is 0.13 $year^{-1}$ (Nikumbh et al., 2019). Thus, the characterization of the extreme event does cause considerable differences in the result. Hence, while considering the result of a study, the researchers should give equal importance to the method followed by the specific research.

3.2 Regional rainfall extremes

The geographical diversity of the Indian subcontinent plays a significant role in regional extreme events. Even though extreme rainfall events occur randomly throughout India for a multitude of reasons, studies have identified the west coast, central parts of India and northeast India as the preferred locations for the occurrence of extreme rainfall events (Pattanaik & Rajeevan, 2010). Due to the increasing spatial variability of extreme rainfall events throughout India, a regional-specific study is necessary to identify the trend and drivers of extreme events (Ghosh, Das, Kao, & Ganguly, 2012).

3.2.1 Peninsular India extreme rainfall events

Peninsular India is mainly the region of South India, which consists of various topographical and climatic patterns. Peninsular India has an inverted triangular shape covered by the Arabian Sea, the Bay of Bengal, and the mainland on the west, east, and north. The extreme rainfall events in Kerala, especially the extreme rainfall received over the hilly region, exhibit a decreasing trend (Soman et al., 1988). This trend is attributed to deforestation and possible changes in the southwest monsoon circulation over the west coast (Singh, Kumar, & Soman, 1989). The mean monsoon rainfall in July also showed a consistent decreasing trend over Kerala in addition to the extreme rainfall events (Guhathakurta & Rajeevan, 2008; Krishnakumar, Prasada Rao, & Gopakumar, 2009; Kumar, Pant, Parthasarathy, & Sontakke, 1992; Varikoden, Revadekar, Kuttippurath, & Babu, 2019). Southern regions of the Western Ghats (WG) exhibited a significant (95% confidence level) decreasing trend in extreme rainfall events. However, a significant (95% confidence level) increasing trend in the extreme rainfall event was observed in the north of 12°N of the WG and at some stations to the east of the WG in the central parts of the Peninsula of India (Rakhecha & Soman, 1994). The regions between 14°–16° N and near 19° N were identified as the region having a maximum probability of receiving intense or extreme rainfall during the summer monsoon period in the WG (Francis & Gadgil, 2006). Most of the extreme rainfall events experienced in this region (62%) are associated with synoptic or larger systems. The decreasing trend of southwest monsoon rainfall in Kerala (southern parts of WG) and increasing trend of extreme rainfall centered around 15° N is associated with the northward shift of the Low-Level Jetstream (LLJ) from 10° N to 15° N, which is related to abnormal warming of the northern Arabian Sea and adjoining continents (Rajendran, Kitoh, Srinivasan, Mizuta, & Krishnan, 2012; Sandeep & Ajayamohan, 2015; Varikoden et al., 2019).

Moreover, the city Chennai in the Tamil Nadu state in peninsular India has been more prone to intense floods in the past (e.g., 1943, 1976, 1985, 1996, 1998, 2005, 2010, and 2015) during the northeast monsoon period (Boyaj, Ashok, Ghosh, Devanand, & Dandu, 2018; Gupta & Nair, 2011). In general, these intense downpours were mainly associated with depressions and cyclonic storms in the Bay of Bengal. The 2002 flood (16–20 cm) in October was associated with a low-pressure system in the Gulf of Mannar (southwest Bay of Bengal). Similarly, the 2005 flood during the northeast monsoon (42 cm in around 40 h) was attributed to a deep depression over the Bay of Bengal (Gupta & Nair, 2010). Several floods were also reported during 2006, 2007 and 2008 (Gupta & Nair, 2011). Boyaj et al. (2018) studied the major flood event in southeast India centered on Chennai during November-December. Boyaj et al. (2018) found that the concurrent warming of the southern Bay of Bengal and strong El Niño conditions in the Pacific led to the modulation of the intensity of the synoptic events in the Bay of Bengal and thus the heavy rainfall events.

3.2.2 *Central India extreme rainfall events*

The magnitude and frequency of extreme rainfall events in central India manifest an increasing trend (Goswami et al., 2006; Pattanaik & Rajeevan, 2010; Rajendran & Kitoh, 2008; Roxy et al., 2017). The ERE's in central India is coupled with a multitude of factors such as Indian ocean warming, El Niño events, Increasing frequency of IOD and low-pressure system (LPS) (Ajayamohan, Merryfield, & Kharin, 2010; Nikumbh et al., 2019; Pattanaik & Rajeevan, 2010; Suthinkumar, Babu, & Varikoden, 2019). The extreme rainfall events over central India are regulated by the cool SST anomalies over the south-eastern equatorial Indian Ocean (Karuna Sagar, Rajeevan, & Vijaya Bhaskara Rao, 2017). The warming trend of the Indian Ocean associated with the increasing frequency of IOD years in recent periods advocates the rising probability of extreme rainfall events to occur over central India in coming years. The Indian Ocean warming partially contributed to the deterioration of the monsoon circulation (Gao et al., 2018; Krishnan et al., 2016; Roxy et al., 2015) and the contribution of aerosol increases and land-use changes (Krishnan et al., 2016). The increasing trend of localized extreme rainfall events is attributed to the deterioration of monsoon circulation compared to the impacts of climate change. The moisture surges associated with the variability of monsoon westerlies due to continuous warming of the Indian Ocean resulted in a threefold increase in extreme events over central India (Roxy et al., 2017). Suthinkumar et al. (2019) studied the extreme rainfall event in the 2017 monsoon period and they found that the central and north-central Indian regions are vulnerable to heavy precipitation and this is observed during the active phase of the monsoon. Further, they attributed these extreme events to the presence of deep convective cloud bands along with low-level cyclonic disturbances.

The LPS plays an important role in the development of extreme events in central India (Ajayamohan et al., 2010; Krishnamurthy & Ajayamohan, 2010; Rajeevan et al., 2008; Revadekar et al., 2016). Although the frequency of LPS is increasing (Ajayamohan et al., 2010), the studies manifest a reduction in cyclonic storms and depressions (Cohen & Boos, 2014; Hunt, Turner, Inness, Parker, & Levine, 2016; Patwardhan et al., 2020; Prajeesh, Ashok, & Rao, 2013; Sandeep, Ajayamohan, Boos, Sabin, & Praveen, 2018; Vishnu, Francis, Shenoi, & Ramakrishna, 2016). Revadekar et al. (2016) explored the role of the LPS in the spatial distribution of the extreme rainfall event throughout India and they found that a major portion of the country exhibits a positive relationship between the extreme rainfall event and the frequency of cyclonic disturbances. Further, they found that the 99th and 95th percentile values of rainfall show that the spatial extent and strength of the positive relationship decreases with an increase in threshold values. But later it was identified that the cooccurrence of LPS and secondary cyclonic circulation together at 500 hPa resulted in the development of large-sized extreme rainfall events (area $\geq 70 \times 103\,km^2$) over central India (Nikumbh et al., 2019, 2020).

3.2.3 *Himalayan extreme rainfall events*

The Himalayas are the main natural water resource (water tower) of the three major river systems of the Indian subcontinent. The Intergovernmental Panel on Climate Change report (IPCC, 2018) pointed out the critical role of the Himalayas in the provision of water to the continental monsoon and clearly stated its vulnerability to climate change. Therefore, the study of extremes in understanding the region's vulnerability and its influence on the neighboring areas is imperative. The extreme rainfall event occurring in northern India, especially regions adjacent to the Himalayan region, is regulated by an extended network of teleconnections, primarily via Rossby waves. This teleconnection manifests a lead-lag relationship between extreme events occurring concurrently in Africa, East Asia and Europe (Boers et al., 2019). The tropical-extratropical interaction developed as a result of the monsoon circulation and midlatitude westerly trough enhances the probability of extreme rainfall events (Vellore et al., 2016; Priya,

Krishnan, Mujumdar, & Houze, 2017). The Pakistan flood in 2010 is also attributed to the interaction between tropical large scale surges and an extra-tropical disturbance downstream of the European blocking high (Hong, Hsu, Lin, & Chiu, 2011; Lau & Kim, 2012; Mujumdar et al., 2012; Wang et al., 2018). The anomalous propagation of monsoon LPS developed over the Bay of Bengal is also associated with the Pakistan flood in 2010 (Houze, Rasmussen, Medina, Brodzik, & Romatschke, 2011). The presence of high pressure on the Tibetan plateau favors the interaction between the southward penetrating westerly trough (dry-cold air with high potential vorticity) located east of the blocking and the monsoon LPS system (moist air with lower-level convergence) facilitating the occurrence of extreme rainfall events in the Himalayan regions (Hong et al., 2011; Joseph et al., 2015; Martius et al., 2013). Nandargi and Dhar (2011) studied maximum 1-day rainfall as the extreme rainfall event over 475 stations in the Himalayan region during the 1871–2007 period. They concluded that the 1-day extreme rainfall event is mainly located between the Siwalik and the Great Himalayan ranges. Stations with higher elevations have recorded fewer extremes, and stations with altitudes <1500 m have recorded more extreme rainfall events. An increase in the frequency of extreme rainfall event has been registered from the 1951–60 decade onward, with a decrease in the frequency of EREs in the recent period of 2001–07, in which the prevailing monsoon was weak. It is also noted that a reduction in monsoon disturbance frequency leads to a reduction in extreme rainfall events.

3.2.4 *Extreme rainfall events in the north east*

The northeast region of India has seven states covering a region with an area of about 262,230 km^2, and its altitude varies from 1 to 7 km above sea level. The region's agriculture is primarily rain-fed, and the seasonal mean rainfall over the region is approximately 200 cm (Kuttippurath et al., 2021; Parthasarathy & Yang, 1995). Moreover, this region shows a weak out-of-phase relationship with the other areas in India during the southwest monsoon period (Shukla & Mooley, 1987; Varikoden & Babu, 2015).

These regions are prone to flooding and flash floods associated with extreme rainfall events that pose a significant threat to the region, damaging crops, livestock and property (Goswami et al., 2006). Since the northeast regions are climatologically high rainy areas, the rainfall from extreme events contributes about 30% to the total rainfall (Suthinkumar et al., 2019). Interaction between large-scale circulation and the local topography plays a crucial role in determining the weather and climate (Varikoden & Revadekar, 2019). In northeast India, the Brahmaputra valley is engulfed by relatively tall mountain ranges on three sides and therefore, this river basin is one of the worst food affected river basins in India due to its large catchment area and relative position (Kale, 2003; Prasad, Vinay Kumar, Singh, & Singh, 2006). The Brahmaputra river basin receives almost 90% of its annual rainfall during the southwest monsoon season. The physiography, relative geomorphic position of the Brahmaputra river basin and dynamic monsoon favor the flood risk in Assam. During June 2012, Assam received 28% more rainfall than average, which led to the breach of embankments in 43 locations of Brahmaputra tributaries (Shivaprasad Sharma, Roy, Chakravarthi, & Srinivasa Rao, 2018). The extreme rainfall events in this region are mainly cand consequent landslides resulted as Goswami, Mukhopadhyay, Mahanta, and Goswami (2010) explained. Therefore, the extremes show a decrease in its frequency based on 1975–2006. However, (Varikoden and Revadekar (2019)) found a significant increase in extreme rainfall event (more than 95th and 99th percentile values) based from 2000 to 2015. They found that the average longevity of extreme rainfall events is about 3 days with a standard deviation of 2 days. Intensification of low-level southerly wind from the Bay of Bengal and convergence over the regions is conducive to the rain peak.

3.3 Selected case studies

3.3.1 *Kerala flood 2018*

The Kerala flood in 2018 is considered one of the worst floods in south India, which received around 164% higher rainfall than normal between August 1st and 19th, 2018 (Kerala Post-Disaster Needs Assessment, 2018). The mean summer monsoon rainfall in Kerala shows a decreasing trend (Mishra & Shah, 2018). Before 2018, 1924 and 1961 were the wettest years in the history of Kerala, with rainfall being around 3600 mm. The rainfall in 2018 was far less compared to 1924 and 1961 till the first week of August. However, 2018 experienced an anomalous wet condition with heavy rainfall between 8 and 18 of August 2018 throughout the entire state. After the extreme rainfall event, the state experienced below normal rainfall, as evidenced in Fig. 5.3. It is clear that before the extreme event (first week of August), the rainfall distribution was below average in all the places of peninsular India (Fig. 5.3A). The rainfall suddenly came to above normal within a few days. The above-normal rainfall has been extensively distributed throughout entire states, especially in the hilly areas of southern parts of Kerala (Fig. 5.3B). Kerala experienced more than 100 mm day^{-1}, particularly in the southern parts of Kerala during these events. This special isolation of the rainfall aggravated the intensity of the flood and subsequent loss to the population and other infrastructure. After 11 days of incessant rainfall, the entire state went to below normal

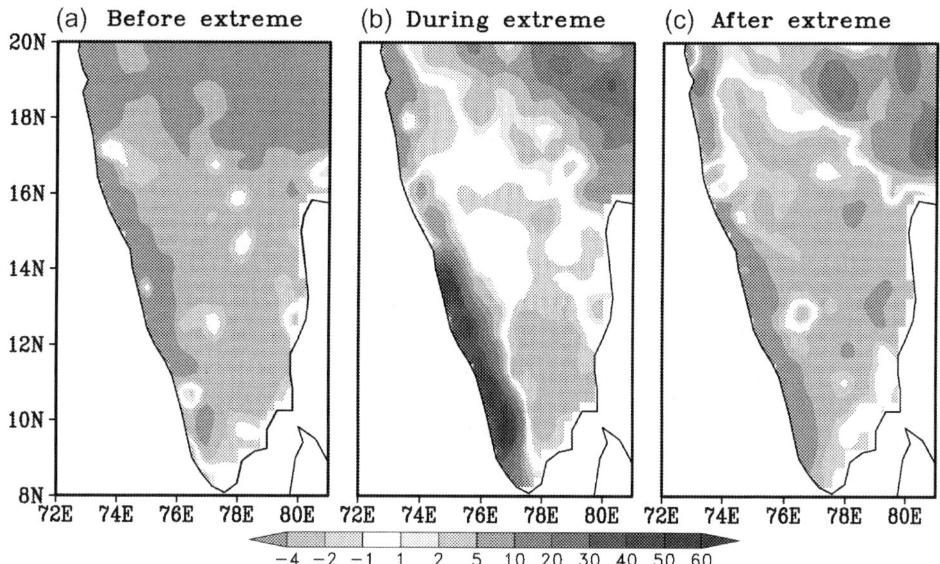

FIG. 5.3 Spatial distribution of rainfall anomalies using daily gridded rainfall dataset from IMD (0.25° × 0.25° resolution). (A) before: August 1–5, 2018, (B) during the event: August 8–18, 2018 and (C) after: August 25–30, 2018, the extreme rainfall event. The base period for the climatology is considered as 1985 to 2005.

condition again (Fig. 5.3C). Major reservoirs (6 out of 7) reached almost their total capacity by the first spell of the extreme event (8 to 10 August). Thus, the reservoirs experience a threat of demolition unless it drains excess rainfall received during the second part of extreme events (August 14–18). This threat of destruction of reservoirs aggravated the situation and caused the entire state to flood and related calamities (Mishra et al., 2018).

The extreme rainfall event and consequent landslides resulted in the death of 483 people and caused an enormous economic loss (Indian Express, 2018; Martha, Roy, Khanna, Mrinalni, & Vinod Kumar, 2019). Multiple studies attempted to explore the cause of this intense flood event (Hunt & Menon, 2020; Mishra et al., 2018; Mishra & Shah, 2018; Mohandas et al., 2020; Ramasamy, Gunasekaran, Rajagopal, Saravanavel, & Kumanan, 2019; Sudheer et al., 2019; Viswanadhapalli et al., 2019). The land-use land-cover changes, reservoir storage and operations, intense rainfall and other natural factors together caused the extreme flood of Kerala (Boyaj, Dasari, Hoteit, & Ashok, 2020; Mishra & Shah, 2018). Viswanadhapalli et al. (2019) found this extreme rainfall event was due to southwestward tilt in the depression over the Bay of Bengal and a strong low level westerly over the Arabian Sea. The modeling studies by Sudheer et al. (2019) and Hunt and Menon (2020) manifested that the 2018 flood was inevitable even if the authorities initiated the discharge of excess water before the flood. The Idukki, Kakki, and Periyar reservoirs upstream of the catchment received extreme rainfall for 1 to 15 days with a return period of 500 years (Mishra et al., 2018), and reservoirs required 34% more capacity to accommodate the excess precipitation received during the flood (Hunt & Menon, 2020).

3.3.2 *Chennai flood in 2015*

The northwest monsoon contributes only about 11% of the total annual rainfall in India. Few areas of the East coast of peninsular India receive approximately 50% of their rainfall during this period, particularly the eastern coast of Tamil Nadu (George, Charlotte, & Ruchith, 2011; Rao & Jagannathan, 1953). In 2015, Chennai experienced incessant rainfall for more than 1 month from November 1st to December 5th with a cumulative rainfall of 1416.8 mm, which is more than three-fold its normal value of 408.4 mm (Kamaljit et al., 2016; Mishra, 2016). The study of dynamics of the 2015 Chennai flood event by Kamaljit et al. (2016) unravels the anomalous southward extension of the subtropical westerly trough at mid-tropospheric level induced strong convection and development of deep convective clouds, which led to the downpour of 150 to 200 mm rainfall per hour over Chennai. It caused flooding throughout Chennai and left 1.8 million people destitute, which accounted for a monetary loss of about 3 billion US dollars (Kamaljit et al., 2016; Mishra & Shah, 2018; Narasimhan, Bhallamudi, Mondal, Ghosh, & Mujumdar, 2016; Selvaraj, Pandiyan, Yoganandan, & Agoramoorthy, 2016).

The year 2015 was one of the most vital El Nino years (Hu & Fedorov, 2017), and during that year, the Bay of Bengal experienced above normal SST. Boyaj et al. (2018) manifesting the increase in the propensity of above-normal rainfall over southeast India during October-December is associated with the warming of the southern Bay. The concurrent warming of the South Bay of Bengal and the El Nino in the Pacific leads to low-level convergence in the regions of extreme rainfall. Dhana Lakshmi and Satyanarayana (2019) argued that this extreme event occurred due to the pumping of moisture atmospheric rivers toward the core region of the extreme rainfall event based on different observational

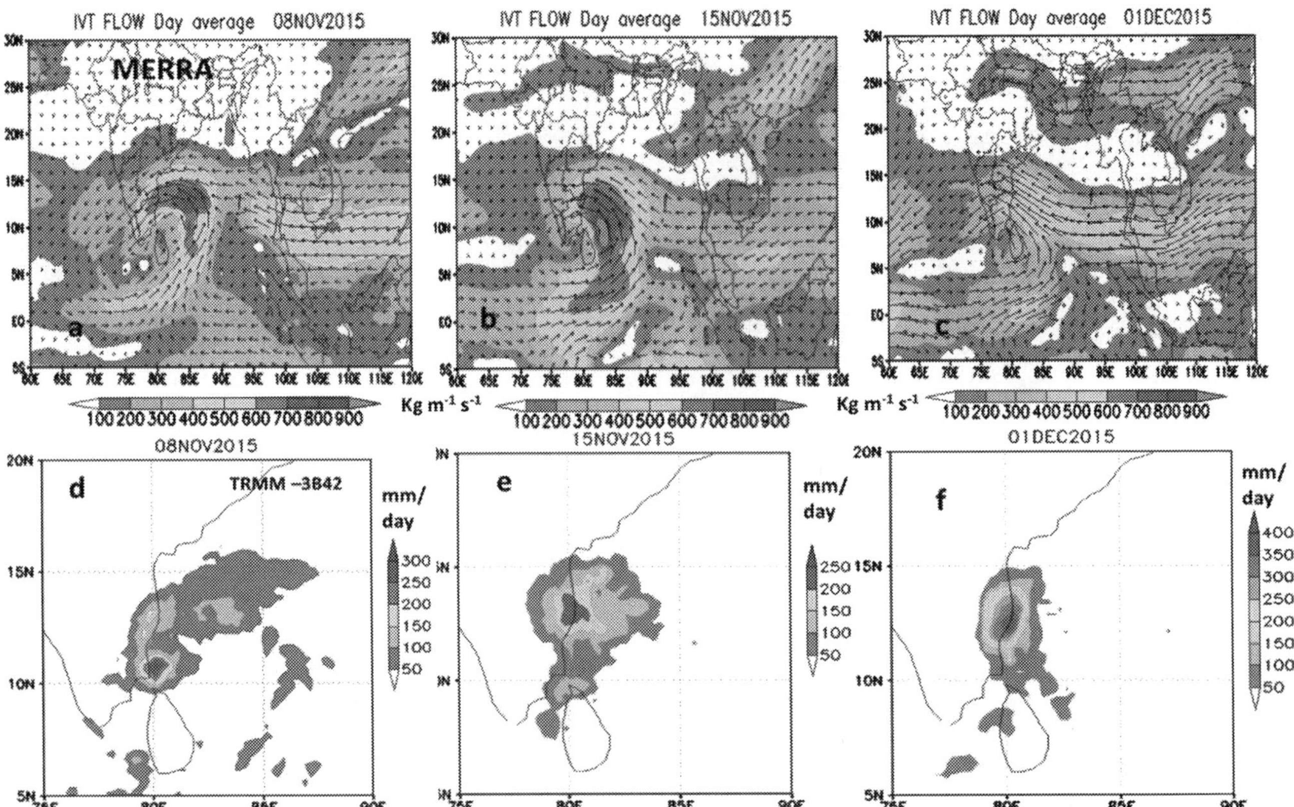

FIG. 5.4 Spatial distribution of integrated moisture transport (inner panels) and extreme rainfall from the TRMM (bottom panels) during 8, 15, November and 1 December 2015. *From Dhana Lakshmi, D. & Satyanarayana, A. N. V. (2019). Influence of atmospheric rivers in the occurrence of devastating flood associated with extreme precipitation events over Chennai using different reanalysis data sets.* Atmospheric Research, 215, *12–36. https://doi.org/10.1016/j.atmosres.2018.08.016, with permission.*

data sets. The distribution of vertically integrated moisture transport from MERRA reanalysis (Fig. 5.4 upper panel) and corresponding daily rainfall (Fig. 5.4 bottom panel) from the TRMM-3B42 product on November 8th and 15th and December 1st, 2015, show an in-phase relationship. The rainfall has shown a maximum peak of 300, 250, and 400 mm day^{-1} for November 8 and 15 and December 1, 2015, respectively. The spatial structure of fields with integrated moisture also shows a similar distribution throughout the regions of extreme rainfall event.

3.3.3 *Mumbai flood 2005*

Mumbai city lies on the west coast of peninsular India with 157 km^2 with 3 million people (https: //mumbaicity.gov.in/) experienced a flash flood on July 26th, 2005 (Gupta, 2007). As per the Indian Meteorological Department, the annual mean rainfall over the Santa Cruz observatory of Mumbai Airport is 2300 mm. Mumbai witnessed an unprecedented rainfall of 994 mm recorded at the Santa Cruz observatory of Mumbai airport for 24 h on July 26th, 2005 (Jenamani, Bhan, & Kalsi, 2006). The flood led to an economic loss of about 1.7 billion USD with a mortality of about 500 people (Hallegatte et al., 2010). Ranger et al. (2011) study's with SRES A2 scenario, an "upper bound climate scenario," unravels the likelihood of doubling 2005 like rainfall event and tripling monetary loss over Mumbai by 2080 due to climate change.

The studies on the Mumbai flood found that an anomalous strong westerly wind powered by a low-pressure system with the aid of orography and a weak offshore trough developed a mesoscale convective system comprising strong thunderstorm cells over Mumbai. This system led to the extreme rainfall event over Santa Cruz on July 26th (Boyaj et al., 2018; Gupta & Nair, 2011; Jenamani et al., 2006; Mohapatra, Kumar, & Bandyopadhyay, 2005; Shyamala & Bhadram, 2006; Vaidya & Kulkarni, 2007). In addition to the impact of LPS, an upper tropospheric westerly trough intrusion is visible during extreme rainfall event in Fig. 5.5. Even though Mumbai experienced the worst effects of extreme rainfall event, the other regions in central India (15°–19°N) also experienced heavy rainfall during this period (Fig. 5.5). Thus, the ERE at Mumbai was a result of organized interaction between synoptic and large-scale systems.

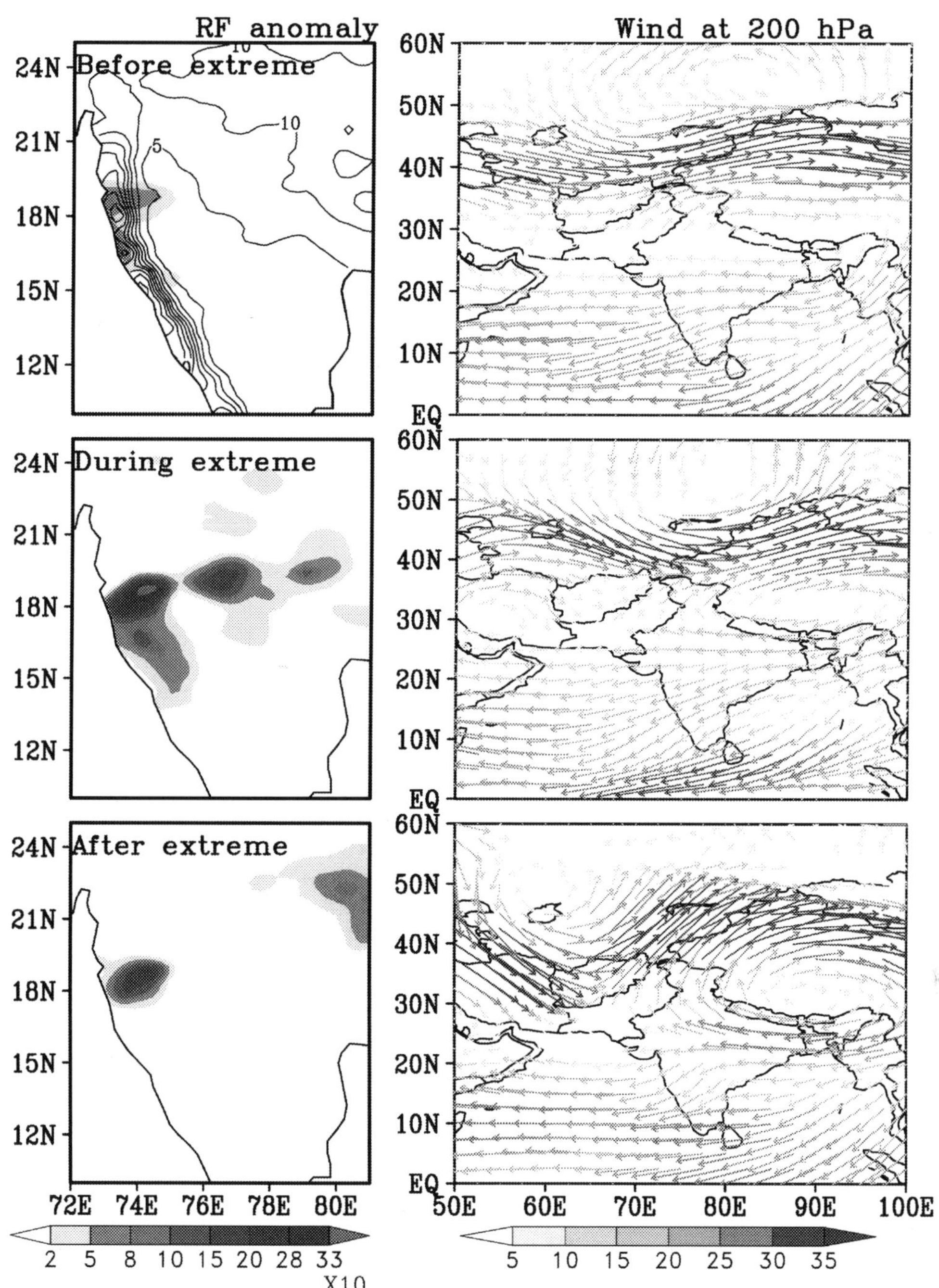

FIG. 5.5 Spatial distribution of rainfall anomalies (left panels) and upper-level circulations (right panels) before (22 July 2005), during (26 July 2005) and after (31 July 2005) the ERE in the Mumbai region. The zonal and meridional wind fields at 200hPa with high resolution (0.25° × 0.25°) from the European Centre for Medium-Range Weather Forecasts' fifth set of reanalysis data is used for identifying upper air circulation during before, during and after the event. The base period considered for calculating the anomalies is from 1985 to 2015.

3.3.4 *Uttarakhand flood 2013*

The state of Uttarakhand lies on the southern slope of the Himalayas and is prone to natural calamities like heavy rainfall, floods, earthquakes and landslides. Flooding is a routine disaster in Uttarakhand, which occurs either during July, August or September, but it seldom occurs during June. Nevertheless, the rainfall during the June 14–17, 2013, and subsequent flooding over Uttarakhand and the nearby locations of Himachal Pradesh and Uttar Pradesh was never experienced before. Dehradun, the closest meteorological station, recorded 370mm of rainfall during the extreme rainfall event period (Das, 2013). Some regions of the state, especially the capital city, received approximately 847% more rainfall than usual (Wake, 2013). A monsoon-like circulation prevailed during this period which is most unlikely to occur in the early phase of causing heavy rainfall over the valley of hills (Dobhal, Gupta, Mehta, & Khandelwal, 2013; Dube et al., 2014; Kotal, Roy, & Roy Bhowmik, 2014; Mishra & Srinivasan, 2013). Upper and lower tropospheric synoptic conditions during June remain conducive for the development of heavy to very heavy rainfall with a monsoon-like circulation over Uttarakhand. The interaction of the mid-tropospheric westerly trough, which extended southward over north India and LPS, resulted in the development of a convectively unstable atmosphere (Singh, Horton, et al., 2014; Singh, Tsiang, et al., 2014; Wang, Davies, Huang, & Gillies, 2011; Ullah & Shouting, 2013; Joseph et al., 2015). The low-level convergence and Somali jet strengthening accompanied the advection of moist warm air (Joseph et al., 2015). The orographic lift induced by the Himalayan terrain and the conducive unstable atmosphere triggered mesoscale convective clouds and heavy rainfall (Houze et al., 2011).

The heavy rainfall (64.5–124.4 mm) to very heavy rainfall (124.5–244.4 mm) received over a short duration melted the snow and Chorabari glacier (Sati & Gahalaut, 2013), and it triggered around 4873 landslides to ensue (Martha et al., 2015). Unlike the normal years, where the thawing of snow occurs before monsoon onset (Dube et al., 2014; Joseph et al., 2015), about 30% of snow remained frozen in the river beds (Durga Rao, Venkateshwar Rao, Dadhwal, & Diwakar, 2014). The heavy precipitation thawed the snow in the river beds and bursting out of the river eventually led to massive floods and excess mass wastage (Andermann et al., 2012; Dubey, Shukla, Ningreichon, & Usham, 2013; Durga Rao et al., 2014; Siderius et al., 2013). As per the joint rapid damage and needs assessment report submitted by the World Bank in partnership with Uttarakhand, the Asian Development Bank and the Global Facility for Disaster Reduction and Recovery (GFDRR), 580 people died and over 5400 people were missing. The reconstruction cost was estimated at INR 1504.8 million (USD 25.08 million) (Dani & Motwani, 2013).

3.4 Projections of extreme rainfall events over India

The Global climate models (GCMs) fail to provide a reliable projection of the intensity, frequency, and spatial distribution of the extreme rainfall events over vast regions due to coarse resolution of the global models (Allan & Soden, 2008; Min et al., 2011; O'Gorman & Schneider, 2009; Sun, Solomon, Dai, & Portmann, 2006), which advocates replacing global projections with regional scale projection through dynamical or statistical downscaling, particularly over India. However, this remains challenging for scientists due to the uncertainty and variability of the Indian climate (Sengupta & Rajeevan, 2013; Woo, Singh, Oh, & Lee, 2019). Model projection studies often stumble over the tropics and subtropics due to the inadequate representation of organized convective systems (Toreti et al., 2013). The model resolution plays an essential role in the simulation of extreme rainfall intensity over a region (Ramu, Chowdary, Ramakrishna, & Kumar, 2018; Varikoden et al., 2018). The high-resolution model provides better projection over India (Rajendran, Sajani, Jayasankar, & Kitoh, 2013). Thus projection study using very high-resolution (~20 km grid size) Meteorological Research Institute (MRI) atmospheric GCM (AGCM) by Mizuta et al. (2006) manifested the intensification of mean seasonal rainfall as well as extreme precipitation over the interior areas of the Indian subcontinent, whereas a significant reduction in the orographic and extreme rainfall events over the southern parts of the WG region (Abish & Arun, 2019; Jayasankar, Rajendran, & Surendran, 2018; Rajendran et al., 2012, 2013; Rajendran & Kitoh, 2008).

A study by Kim, Oh, Woo, and Kripalani (2019) using the Coupled Model Intercomparison Project Phase 5 (CMIP5) simulation data found that the extreme rainfall events over the Indian subcontinent are recognized in high-resolution model simulations. In contrast, the extreme rainfall events in northeast India are better manifested by the medium resolution model simulations. The projection study based on Regional Climate Model version4 (RegCM4) with a spatial resolution of 50 km predicted an increase of 0.56 mm day^{-1} for seasonal mean rainfall, and the frequency of maximum rainfall received in 1 day (RX1 day) is projected to increase (0.27/decade) as compared to mean rainfall (0.01/decade) over India (Rai, Singh, & Dash, 2020). The high emissions scenario (RCP 8.5) of CMIP5 identified a decrease in the light rainfall, increasing extreme rain. Thus, the Indian summer monsoon rainfall is projected to increase at 0.74 ± 0.36 mm/day (Jayasankar, Surendran, & Rajendran, 2015). The frequency of extreme rainfall events is projected to increase over the south and central India by the mid and end of the 21st century as per RCP 8.5 using simulation data of CMIP5 and Climate of the 20th century plus (C20C+) detection and attribution (D&A) project (Mukherjee, Aadhar, Stone, & Mishra, 2018).

4 Conclusions

Being a developing country with an agrarian-based society, India struggles to cope with extreme rainfall events. The extreme rainfall events study in India took a long time to evolve and progress, prompted by climate change. A multitude of rainfall indices developed for the identification of extreme rainfall events made it difficult to put forward a unified result for the exploration of causative mechanisms and investigation of the forecasting potential of extreme rainfall events. However, almost all studies have identified an increasing trend in the frequency and intensity of ERE's over India. Although there is a growing trend in the spatial heterogeneity of extreme rainfall events on the local scale, the large-scale extreme rainfall events are primarily homogeneous and are associated with synoptic as well as large scale atmospheric and oceanic systems.

The causative mechanism behind extreme rainfall events varies according to the regional characteristics. In the north and northeast Indian regions, the extreme rainfall events are mainly attributed to the interaction between extra-tropical westerly trough intrusion and LPS. However, the Himalayan topography aids the development of a mesoscale convective system for the development and sustenance of extreme rainfall events. The strength of the monsoon circulation also modifies the characteristics of extreme rainfall events in the north and northeast India. However, in the northeast regions, the extreme events are mainly associated with the break monsoon period, and synoptic-scale systems mainly cause them. It isn't easy to characterize and identify the physical and dynamical reason for the development of extreme rainfall events in central India. The extreme rainfall events in central India may be attributed to the LPS, cooccurrence of LPS and secondary cyclonic circulation, weakening of the monsoon circulation, IOD, warming of the Indian Ocean, El Nino and extra-tropical westerly trough intrusion. But it is evident that in most of the extreme rainfall events the presence of LPS or a combination of LPS and an extra-tropical westerly trough are there. The extreme rainfall events in Peninsular India, particularly on the west coast, are associated with the variations in LLJ. However, the presence of LPS and extra-tropical westerly troughs along with the offshore vortex during extreme rainfall events in peninsular India suggests a common ingredient for the existence of extreme rainfall events, even though they differ in their relative contribution.

Further studies to recognize the degree of influence caused by the interaction between the LPS and the extratropical westerly trough in extreme rainfall events on a large scale may answer the long-standing question regarding the causative mechanism and the potential for forecasting extreme events in a region. Further, the projection studies of rainfall extremes in India show a general increase in their frequency and intensity based on the different indices of extreme rainfall events.

Acknowledgment

The authors are thankful to the Director, Indian Institute of Tropical Meteorology (IITM), for providing the necessary facilities and to the Executive Director, Centre for Climate Change Research (CCCR), IITM, for encouragement. The second author is thankful for the financial assistance provided by the CSIR. The IITM is fully funded by the Ministry of Earth Sciences (MoES), Govt. of India.

References

Abish, B., & Arun, K. (2019). Resolving the weakening of orographic rainfall over India using a regional climate model RegCM 4.5. *Atmospheric Research*, *227*, 125–139. https://doi.org/10.1016/j.atmosres.2019.05.003.

Adhikari, P., Hong, Y., Douglas, K. R., Kirschbaum, D. B., Gourley, J., Adler, R., et al. (2010). A digitized global flood inventory (1998-2008): Compilation and preliminary results. *Natural Hazards*, *55*(2), 405–422. https://doi.org/10.1007/s11069-010-9537-2.

Ajayamohan, R. S., Merryfield, W. J., & Kharin, V. V. (2010). Increasing trend of synoptic activity and its relationship with extreme rain events over Central India. *Journal of Climate*, *23*(4), 1004–1013. https://doi.org/10.1175/2009JCLI2918.1.

Ajayamohan, R. S., & Rao, S. A. (2008). Indian Ocean dipole modulates the number of extreme rainfall events over India in a warming environment. *Journal of the Meteorological Society of Japan. Ser. II*, *86*(1), 245–252.

Allan, R. P., & Soden, B. J. (2008). Atmospheric warming and the amplification of precipitation extremes. *Science*, *321*(5895), 1481–1484. https://doi.org/10.1126/science.1160787.

Amarasinghe, U., Amarnath, G., Alahacoon, N., & Ghosh, S. (2020). How do floods and drought impact economic growth and human development at the sub-national level in India? *Climate*, *8*(11), 123. https://doi.org/10.3390/cli8110123.

Andermann, C., Longuevergne, L., Bonnet, S., Crave, A., Davy, P., & Gloaguen, R. (2012). Impact of transient groundwater storage on the discharge of Himalayan rivers. *Nature Geoscience*, *5*(2), 127–132. https://doi.org/10.1038/ngeo1356.

Ashok, K., Guan, Z., Saji, N. H., & Yamagata, T. (2004). Individual and combined influences of ENSO and the Indian Ocean Dipole on the Indian summer monsoon. *Journal of Climate*, *17*(16), 3141–3155. https://doi.org/10.1175/1520-0442(2004)017<3141:IACIOE>2.0.CO;2.

Ashok, K., Guan, Z., & Yamagata, T. (2001). Impact of the Indian Ocean dipole on the relationship between the Indian monsoon rainfall and ENSO. *Geophysical Research Letters*, *28*(23), 4499–4502. https://doi.org/10.1029/2001GL013294.

Behera, S. K., Doi, T., Ratnam, J. V., & Behera, S. K. (2021). *5—Air–sea interactions in tropical Indian Ocean: The Indian ocean dipole* (pp. 115–139). Elsevier. https://doi.org/10.1016/B978-0-12-818156-0.00001-0.

Boers, N., Goswami, B., Rheinwalt, A., Bookhagen, B., Hoskins, B., & Kurths, J. (2019). Complex networks reveal global pattern of extreme-rainfall teleconnections. *Nature*, *566*(7744), 373–377. https://doi.org/10.1038/s41586-018-0872-x.

Boyaj, A., Ashok, K., Ghosh, S., Devanand, A., & Dandu, G. (2018). The Chennai extreme rainfall event in 2015: The Bay of Bengal connection. *Climate Dynamics*, *50*(7–8), 2867–2879. https://doi.org/10.1007/s00382-017-3778-7.

Boyaj, A., Dasari, H. P., Hoteit, I., & Ashok, K. (2020). Increasing heavy rainfall events in South India due to changing land use and land cover. *Quarterly Journal of the Royal Meteorological Society*, *146*(732), 3064–3085. https://doi.org/10.1002/qj.3826.

Burt, S. (2005). Cloudburst upon Hendraburnick down: The Boscastle storm of 16 August 2004. *Weather*, *60*(8), 219–227. https://doi.org/10.1256/wea.26.05.

Clark, C. (2005). The cloudburst of 2 July 1893 over the Cheviot Hills, England. *Weather*, *60*(4), 92–97. https://doi.org/10.1256/wea.06.04.

Cohen, N. Y., & Boos, W. R. (2014). Has the number of Indian summer monsoon depressions decreased over the last 30 years? *Geophysical Research Letters*, *41*(22), 7846–7853. https://doi.org/10.1002/2014GL061895.

Coumou, D., & Rahmstorf, S. (2012). A decade of weather extremes. *Nature Climate Change*, *2*(7), 491–496. https://doi.org/10.1038/nclimate1452.

Dai, A. (2006). Recent climatology, variability, and trends in global surface humidity. *Journal of Climate*, *19*(15), 3589–3606. https://doi.org/10.1175/JCLI3816.1.

Dani, S., & Motwani, A. (2013). *India-Uttarakhand disaster: Joint rapid damage and needs assessment report* (No. 82643) (pp. 1–92). The World Bank.

Das, P. K. (2013). The Himalayan Tsunami'-cloudburst, flash flood & death toll: A geographical postmortem. *IOSR Journal of Environmental Science*, *7*(2). https://doi.org/10.9790/2402-0723345.

Deshpande, N. R., & Gore, A. P. (1999). On computation of probable maximum precipitation. *Calcutta Statistical Association Bulletin*, *49*(1–2), 113–122. https://doi.org/10.1177/0008068319990112.

Devrani, R., Singh, V., Mudd, S. M., & Sinclair, H. D. (2015). Prediction of flash flood hazard impact from Himalayan river profiles. *Geophysical Research Letters*, *42*(14), 5888–5894. https://doi.org/10.1002/2015GL063784.

Dhana Lakshmi, D., & Satyanarayana, A. N. V. (2019). Influence of atmospheric rivers in the occurrence of devastating flood associated with extreme precipitation events over Chennai using different reanalysis data sets. *Atmospheric Research*, *215*, 12–36. https://doi.org/10.1016/j.atmosres.2018.08.016.

Dhar, O. N., & Kamte, P. P. (1971). Estimation of extreme rainfall over North India. *Indian Journal of Meteorology and Geophysics*, *22*(3), 559–566.

Dhar, O. N., Rakhecha, P. R., & Sangam, R. B. (1975). Study of major rainstorms over and near Mahi basin up to Kadana dam site for evaluation of probable maximum design storm. *Indian Journal of Power & River Valley Development*, *25*, 29–35.

Dobhal, D. P., Gupta, A. K., Mehta, M., & Khandelwal, D. D. (2013). Kedarnath disaster: Facts and plausible causes. *Current Science*, *105*(2), 171–174. http://www.currentscience.ac.in/Volumes/105/02/0171.pdf.

Doswell, C. A., Brooks, H. E., & Maddox, R. A. (1996). Flash flood forecasting: An ingredients-based methodology. *Weather and Forecasting*, *560–581*. https://doi.org/10.1175/1520-0434(1996)011<0560:FFFAIB>2.0.CO;2.

Dube, A., Ashrit, R., Ashish, A., Sharma, K., Iyengar, G. R., Rajagopal, E. N., et al. (2014). Forecasting the heavy rainfall during Himalayan flooding-June 2013. *Weather and Climate Extremes*, *4*, 22–34. https://doi.org/10.1016/j.wace.2014.03.004.

Dubey, C. S., Shukla, D. P., Ningreichon, A. S., & Usham, A. L. (2013). Orographic control of the Kedarnath disaster. *Current Science*, *105*(11), 1474–1476. http://www.currentscience.ac.in/Volumes/105/11/1474.pdf.

Durga Rao, K. H. V., Venkateshwar Rao, V., Dadhwal, V. K., & Diwakar, P. G. (2014). Kedarnath flash floods: A hydrological and hydraulic simulation study. *Current Science*, *106*(4), 598–603. http://www.currentscience.ac.in/Volumes/106/04/0598.pdf.

Easterling, D. R. (2000). Climate extremes: Observations, modeling, and impacts. *Science*, 2068–2074. https://doi.org/10.1126/science.289.5487.2068.

Fowler, H. J., Lenderink, G., Prein, A. F., Westra, S., Allan, R. P., Ban, N., et al. (2021). Anthropogenic intensification of short-duration rainfall extremes. *Nature Reviews Earth and Environment*, *2*(2), 107–122. https://doi.org/10.1038/s43017-020-00128-6.

Francis, P. A., & Gadgil, S. (2006). Intense rainfall events over the west coast of India. *Meteorology and Atmospheric Physics*, *94*(1–4), 27–42. https://doi.org/10.1007/s00703-005-0167-2.

Gadgil, S. (2003). The Indian monsoon and its variability. *Annual Review of Earth and Planetary Sciences*, *31*(1), 429–467. https://doi.org/10.1146/annurev.earth.31.100901.141251.

Gadgil, S., & Kumar, K. R. (2006). The Asian monsoon—Agriculture and economy. In *The Asian monsoon* (pp. 651–683). Berlin, Heidelberg: Springer.

Gao, M., Ding, Y., Song, S., Lu, X., Chen, X., & McElroy, M. B. (2018). Secular decrease of wind power potential in India associated with warming in the Indian Ocean. *Science Advances*, *4*(12). https://doi.org/10.1126/sciadv.aat5256.

George, G., Charlotte, B. V., & Ruchith, R. D. (2011). Interannual variation of northeast monsoon rainfall over southern peninsular India. *Indian Journal of Marine Sciences*, *40*(1), 98–104. http://nopr.niscair.res.in/bitstream/123456789/11376/1/IJMS%2040%281%29%2098-104.pdf.

Ghosh, S., Das, D., Kao, S. C., & Ganguly, A. R. (2012). Lack of uniform trends but increasing spatial variability in observed Indian rainfall extremes. *Nature Climate Change*, *2*(2), 86–91. https://doi.org/10.1038/nclimate1327.

Ghosh, S., Luniya, V., & Gupta, A. (2009). Trend analysis of Indian summer monsoon rainfall at different spatial scales. *Atmospheric Science Letters*, *10*(4), 285–290. https://doi.org/10.1002/asl.235.

Ghosh, S., Vittal, H., Sharma, T., Karmakar, S., Kasiviswanathan, K. S., Dhanesh, Y., et al. (2016). Indian summer monsoon rainfall: Implications of contrasting trends in the spatial variability of means and extremes. *PLoS One*, *11*(7), e0158670. https://doi.org/10.1371/journal.pone.0158670.

Goswami, B. B., Mukhopadhyay, P., Mahanta, R., & Goswami, B. N. (2010). Multiscale interaction with topography and extreme rainfall events in the northeast Indian region. *Journal of Geophysical Research-Atmospheres*, *115*(12). https://doi.org/10.1029/2009JD012275.

Goswami, B. N. (1998). The physics of ENSO-monsoon connection. *Indian Journal of Marine Sciences*, *27*(1), 82–89.

Goswami, B. N., Venugopal, V., Sengupta, D., Madhusoodanan, M. S., & Xavier, P. K. (2006). Increasing trend of extreme rain events over India in a warming environment. *Science*, *314*(5804), 1442–1445. https://doi.org/10.1126/science.1132027.

Government of India. (2021). *Economic survey 2020-21*. https://www.indiabudget.gov.in/economicsurvey.

Guhathakurta, P., Menon, P., Inkane, P. M., Krishnan, U., & Sable, S. T. (2017). Trends and variability of meteorological drought over the districts of India using standardized precipitation index. *Journal of Earth System Science*, *126*(8). https://doi.org/10.1007/s12040-017-0896-x.

Guhathakurta, P., & Rajeevan, M. (2008). Trends in the rainfall pattern over India. *International Journal of Climatology*, *28*(11), 1453–1469. https://doi.org/10.1002/joc.1640.

Gupta, A. K., & Nair, S. S. (2010). Flood risk and context of land-uses: Chennai city case. *Journal of Geography and Regional Planning*, *3*(12), 365–372. https://doi.org/10.5897/JGRP.9000057.

Gupta, A. K., & Nair, S. S. (2011). Urban floods in Bangalore and Chennai: Risk management challenges and lessons for sustainable urban ecology. *Current Science*, *100*(11), 1638–1645. http://www.ias.ac.in/currsci/10jun2011/1638.pdf.

Gupta, K. (2007). Urban flood resilience planning and management and lessons for the future: A case study of Mumbai, India. *Urban Water Journal*, *4*(3), 183–194. https://doi.org/10.1080/15730620701464141.

Hallegatte, S., Ranger, N., Bhattacharya, S., Bachu, M., Priya, S., Dhore, K., et al. (2010). *Flood risks, climate change impacts and adaptation benefits in Mumbai: an initial assessment of socio-economic consequences of present and climate change induced flood risks and of possible adaptation options*. OECD Publishing.

Hong, C. C., Hsu, H. H., Lin, N. H., & Chiu, H. (2011). Roles of European blocking and tropical-extratropical interaction in the 2010 Pakistan flooding. *Geophysical Research Letters*, *38*(13). https://doi.org/10.1029/2011GL047583.

Houze, R. A., Rasmussen, K. L., Medina, S., Brodzik, S. R., & Romatschke, U. (2011). Anomalous atmospheric events leading to the summer 2010 floods in Pakistan. *Bulletin of the American Meteorological Society*, *92*(3), 291–298. https://doi.org/10.1175/2010BAMS3173.1.

Hu, S., & Fedorov, A. V. (2017). The extreme El Niño of 2015–2016 and the end of global warming hiatus. *Geophysical Research Letters*, *44*(8), 3816–3824. https://doi.org/10.1002/2017GL072908.

Huffman, G. J., Bolvin, D. T., Nelkin, E. J., Wolff, D. B., Adler, R. F., Gu, G., et al. (2007). The TRMM multisatellite precipitation analysis (TMPA): Quasi-global, multiyear, combined-sensor precipitation estimates at fine scales. *Journal of Hydrometeorology*, *8*, 38–55.

Hunt, K. M. R., & Menon, A. (2020). The 2018 Kerala floods: A climate change perspective. *Climate Dynamics*, *54*(3–4), 2433–2446. https://doi.org/10.1007/s00382-020-05123-7.

Hunt, K. M. R., Turner, A. G., Inness, P. M., Parker, D. E., & Levine, R. C. (2016). On the structure and dynamics of Indian monsoon depressions. *Monthly Weather Review*, *144*(9), 3391–3416. https://doi.org/10.1175/MWR-D-15-0138.1.

Indian Express. (2018). *483 dead in Kerala floods and landslides, losses more than annual plan outlay: Pinarayi Vijayan*. https://indianexpress.com/article/india/483-dead-in-kerala-floods-and-landslides-losses-more-than-annual-plan-outlay-pinarayi-vijayan-5332306.

IPCC. (2018). *Global warming of 1.5°C: An IPCC special report on the impacts of global warming of 1.5°C above pre-industrial levels and related global greenhouse gas emission pathways, in the context of strengthening the global response to the threat of climate change, sustainable development, and efforts to eradicate poverty*. World Meteorological Organization. https://www.ipcc.ch/sr15/.

Jamshadali, V. H., Reji, M. J. K., Varikoden, H., & Vishnu, R. (2021). Spatial variability of south Asian summer monsoon extreme rainfall events and their association with global climate indices. *Journal of Atmospheric and Solar-Terrestrial Physics*, *221*. https://doi.org/10.1016/j.jastp.2021.105708, 105708.

Jayasankar, C. B., Rajendran, K., & Surendran, S. (2018). Monsoon climate change projection for the orographic West Coast of India using high-resolution nested dynamical downscaling model. *Journal of Geophysical Research-Atmospheres*, *123*(15), 7821–7838. https://doi.org/10.1029/2018JD028677.

Jayasankar, C. B., Surendran, S., & Rajendran, K. (2015). Robust signals of future projections of Indian summer monsoon rainfall by IPCC AR5 climate models: Role of seasonal cycle and interannual variability. *Geophysical Research Letters*, *42*(9), 3513–3520. https://doi.org/10.1002/2015GL063659.

Jenamani, R. K., Bhan, S. C., & Kalsi, S. R. (2006). Observational/forecasting aspects of the meteorological event that caused a record highest rainfall in Mumbai. *Current Science*, *90*(10), 1344–1362. http://www.ias.ac.in/currsci/may252006/1344.pdf.

Joseph, S., Sahai, A. K., Sharmila, S., Abhilash, S., Borah, N., Chattopadhyay, R., et al. (2015). North Indian heavy rainfall event during June 2013: Diagnostics and extended range prediction. *Climate Dynamics*, *44*(7–8), 2049–2065. https://doi.org/10.1007/s00382-014-2291-5.

Joshi, U. R., & Rajeevan, M. (2006). *Trends in precipitation extremes over India. National Climate Centre*. India Meteorological Department.

Kale, V. S. (2003). *Geomorphic effects of monsoon floods on Indian Rivers. Vol. 28* (pp. 65–84). Springer Science and Business Media LLC. https://doi.org/10.1007/978-94-017-0137-2_3. Issue 1.

Kamaljit, R., Kannan, B. A. M., Stella, S., Sen, B., Sharma, P., & Thampi, S. B. (2016). Heavy rains over Chennai and surrounding areas as captured by Doppler weather radar during northeast monsoon 2015: A case study. In *Vol. 9876. Proceedings of SPIE—The International Society for optical engineering* SPIE. https://doi.org/10.1117/12.2239563.

Karmakar, N., Chakraborty, A., & Nanjundiah, R. S. (2017). Increased sporadic extremes decrease the intraseasonal variability in the Indian summer monsoon rainfall. *Scientific Reports*, *7*(1). https://doi.org/10.1038/s41598-017-07529-6.

Karuna Sagar, S., Rajeevan, M., & Vijaya Bhaskara Rao, S. (2017). On increasing monsoon rainstorms over India. *Natural Hazards*, *85*(3), 1743–1757. https://doi.org/10.1007/s11069-016-2662-9.

Kerala Post-Disaster Needs Assessment. (2018). https://www.undp.org/publications/post-disaster-needs-assessment-kerala#modal-publication-download.

Kim, I. W., Oh, J., Woo, S., & Kripalani, R. H. (2019). Evaluation of precipitation extremes over the Asian domain: Observation and modelling studies. *Climate Dynamics*, *52*(3–4), 1317–1342. https://doi.org/10.1007/s00382-018-4193-4.

Kotal, S. D., Roy, S. S., & Roy Bhowmik, S. K. (2014). Catastrophic heavy rainfall episode over Uttarakhand during 16-18 June 2013 - observational aspects. *Current Science*, *107*(2), 234–245. http://www.currentscience.ac.in/Volumes/107/02/0234.pdf.

Krishnakumar, K. N., Prasada Rao, G. S. L. H. V., & Gopakumar, C. S. (2009). Rainfall trends in twentieth century over Kerala, India. *Atmospheric Environment*, *43*(11), 1940–1944. https://doi.org/10.1016/j.atmosenv.2008.12.053.

Krishnamurthy, V., & Ajayamohan, R. S. (2010). Composite structure of monsoon low pressure systems and its relation to Indian rainfall. *Journal of Climate*, *23*(16), 4285–4305. https://doi.org/10.1175/2010JCLI2953.1.

Krishnan, R., Sabin, T. P., Vellore, R., Mujumdar, M., Sanjay, J., Goswami, B. N., et al. (2016). Deciphering the desiccation trend of the South Asian monsoon hydroclimate in a warming world. *Climate Dynamics*, *47*(3–4), 1007–1027. https://doi.org/10.1007/s00382-015-2886-5.

Krishnaswamy, J., Vaidyanathan, S., Rajagopalan, B., Bonell, M., Sankaran, M., Bhalla, R. S., et al. (2015). Non-stationary and non-linear influence of ENSO and Indian Ocean Dipole on the variability of Indian monsoon rainfall and extreme rain events. *Climate Dynamics*, *45*(1–2), 175–184. https://doi.org/10.1007/s00382-014-2288-0.

Kshirsagar, N. A., Shinde, R. R., & Mehta, S. (2006). Floods in Mumbai: Impact of public health service by hospital staff and medical students. *Journal of Postgraduate Medicine*, *52*(4), 312–314.

Kulkarni, B. D., Nandargi, S., & Mulye, S. S. (2010). Estimation zonale des hauteurs des précipitations maximales probables dans le bassin versant du Fleuve Krishna en Inde péninsulaire. *Hydrological Sciences Journal*, *55*(1), 93–103. https://doi.org/10.1080/02626660903529015.

Kumar, K. R., Pant, G. B., Parthasarathy, B., & Sontakke, N. A. (1992). Spatial and subseasonal patterns of the long-term trends of Indian summer monsoon rainfall. *International Journal of Climatology*, *12*(3), 257–268. https://doi.org/10.1002/joc.3370120303.

Kuttippurath, J., Murasingh, S., Stott, P. A., Sarojini, B. B., Jha, M. K., Kumar, P., et al. (2021). Observed rainfall changes in the past century (1901–2019) over the wettest place on earth. *Environmental Research Letters*, *16*(2), 024018. https://doi.org/10.1088/1748-9326/abcf78.

Lau, W. K. M., & Kim, K. M. (2012). The 2010 Pakistan flood and Russian heat wave: Teleconnection of hydrometeorological extremes. *Journal of Hydrometeorology*, *13*(1), 392–403. https://doi.org/10.1175/JHM-D-11-016.1.

Lavers, D. A., Villarini, G., Allan, R. P., Wood, E. F., & Wade, A. J. (2012). The detection of atmospheric rivers in atmospheric reanalyses and their links to British winter floods and the large-scale climatic circulation. *Journal of Geophysical Research-Atmospheres*, *117*(20). https://doi.org/10.1029/2012JD018027.

Lawrence, D. M., & Webster, P. J. (2001). Interannual variations of the intraseasonal oscillation in the South Asian summer monsoon region. *Journal of Climate*, *14*(13), 2910–2922. https://doi.org/10.1175/1520-0442(2001)014<2910:IVOTIO>2.0.CO;2.

Lenderink, G., & Fowler, H. J. (2017). Understanding rainfall extremes. *Nature Climate Change*, *7*(6), 391–393. https://doi.org/10.1038/nclimate3305.

Martha, T. R., Roy, P., Govindharaj, K. B., Kumar, K. V., Diwakar, P. G., & Dadhwal, V. K. (2015). Landslides triggered by the June 2013 extreme rainfall event in parts of Uttarakhand state, India. *Landslides*, *12*(1), 135–146. https://doi.org/10.1007/s10346-014-0540-7.

Martha, T. R., Roy, P., Khanna, K., Mrinalni, K., & Vinod Kumar, K. (2019). Landslides mapped using satellite data in the Western Ghats of India after excess rainfall during August 2018. *Current Science*, *117*(5), 804–812. https://doi.org/10.18520/cs/v117/i5/804-812.

Martius, O., Sodemann, H., Joos, H., Pfahl, S., Winschall, A., Croci-Maspoli, M., et al. (2013). The role of upper-level dynamics and surface processes for the Pakistan flood of July 2010. *Quarterly Journal of the Royal Meteorological Society*, *139*(676), 1780–1797. https://doi.org/10.1002/qj.2082.

Min, S. K., Zhang, X., Zwiers, F. W., & Hegerl, G. C. (2011). Human contribution to more-intense precipitation extremes. *Nature*, *470*(7334), 378–381. https://doi.org/10.1038/nature09763.

Mishra, A., & Srinivasan, J. (2013). Did a cloud burst occur in Kedarnath during 16 and 17 June 2013? *Current Science*, *105*(10), 1351–1352. http://www.currentscience.ac.in/Volumes/105/10/1351.pdf.

Mishra, A. K. (2016). Monitoring Tamil Nadu flood of 2015 using satellite remote sensing. *Natural Hazards*, *82*(2), 1431–1434. https://doi.org/10.1007/s11069-016-2249-5.

Mishra, V., Aaadhar, S., Shah, H., Kumar, R., Pattanaik, D. R., & Tiwari, A. D. (2018). The Kerala flood of 2018: Combined impact of extreme rainfall and reservoir storage. *Hydrology and Earth System Sciences Discussions*. https://doi.org/10.5194/hess-2018-480.

Mishra, V., & Shah, H. L. (2018). Hydroclimatological perspective of the Kerala flood of 2018. *Journal of the Geological Society of India*, *92*(5), 645–650. https://doi.org/10.1007/s12594-018-1079-3.

Mizuta, R., Oouchi, K., Yoshimura, H., Noda, A., Katayama, K., Yukimoto, S., et al. (2006). 20-km-mesh global climate simulations using JMA-GSM model—Mean climate states. *Journal of the Meteorological Society of Japan*, *84*(1), 165–185. https://doi.org/10.2151/jmsj.84.165.

Mohandas, S., Francis, T., Singh, V., Jayakumar, A., George, J. P., Sandeep, A., et al. (2020). NWP perspective of the extreme precipitation and flood event in Kerala (India) during August 2018. *Dynamics of Atmospheres and Oceans*, *91*. https://doi.org/10.1016/j.dynatmoce.2020.101158.

Mohapatra, M., Kumar, N., & Bandyopadhyay, B. K. (2005). Role of mesoscale low and urbanization on exceptionally heavy rainfall event of 26th July 2005 over Mumbai: Some observational evidences. *Mausam*, *60*, 317–324.

Mujumdar, M., Preethi, B., Sabin, T. P., Ashok, K., Saeed, S., Pai, D. S., et al. (2012). Increase in extreme precipitation events under anthropogenic warming in India. Weather and climate extremes. *Meteorological Applications*, *19*(2), 45–53. https://doi.org/10.1002/met.1301Mukherjee.

Mukherjee, S., Aadhar, S., Stone, D., & Mishra, V. (2018). Increase in extreme precipitation events under anthropogenic warming in India. *Weather and Climate Extremes*, *20*, 45–53. https://doi.org/10.1016/j.wace.2018.03.005.

Nandargi, S., & Dhar, O. N. (2011). Evénements de précipitations extrêmes dans l'Himalaya, entre 1871 et 2007. *Hydrological Sciences Journal*, *56*(6), 930–945. https://doi.org/10.1080/02626667.2011.595373.

Narasimhan, B., Bhallamudi, S. M., Mondal, A., Ghosh, & Mujumdar, P. (2016). *Chennai floods 2015: A rapid assessment*. Bangalore: Interdisciplinary Centre for water research, Indian Institute of Science.

Nikumbh, A. C., Chakraborty, A., & Bhat, G. S. (2019). Recent spatial aggregation tendency of rainfall extremes over India. *Scientific Reports*, *9*(1). https://doi.org/10.1038/s41598-019-46719-2.

Nikumbh, A. C., Chakraborty, A., Bhat, G. S., & Frierson, D. M. W. (2020). Large-scale extreme rainfall-producing synoptic systems of the Indian Summer Monsoon. *Geophysical Research Letters*, *47*(11). https://doi.org/10.1029/2020GL088403.

O'Gorman, P. A., & Schneider, T. (2009). The physical basis for increases in precipitation extremes in simulations of 21st-century climate change. *Proceedings of the National Academy of Sciences*, *106*(35), 14773–14777. https://doi.org/10.1073/pnas.0907610106.

Oza, M., & Kishtawal, C. M. (2014). Trends in rainfall and temperature patterns over north East India. *Earth Science India*. https://doi.org/10.31870/esi.07.4.2014.8.

Ozone, A. (1985). *Assessment of our understanding of the processes controlling its present distribution and change; global ozone research and monitoring project 16. Vol. 2* (pp. 410–411). World Meteorological Organization.

Pai, D. S., Sridhar, L., Badwaik, M. R., & Rajeevan, M. (2015). Analysis of the daily rainfall events over India using a new long period (1901–2010) high resolution ($0.25° \times 0.25°$) gridded rainfall data set. *Climate Dynamics*, *45*(3–4), 755–776. https://doi.org/10.1007/s00382-014-2307-1.

Parthasarathy, B., & Yang, S. (1995). Relationships between regional Indian summer monsoon rainfall and Eurasian snow cover. *Advances in Atmospheric Sciences*, *12*(2), 143–150. https://doi.org/10.1007/BF02656828.

Pattanaik, D. R., & Rajeevan, M. (2010). Variability of extreme rainfall events over India during southwest monsoon season. *Meteorological Applications*, *17*(1), 88–104. https://doi.org/10.1002/met.164.

Patwardhan, S., Sooraj, K. P., Varikodenw, H., Vishnu, S., Koteswararao, K., & Ramarao, M. V. S. (2020). Synoptic scale systems. In *Assessment of climate change over the Indian region: A report of the Ministry of Earth Sciences (MoES), Government of India* (pp. 143–154). Singapore: Springer. https://doi.org/10.1007/978-981-15-4327-2_7.

Pendergrass, A. G. (2018). What precipitation is extreme? *Science*, *360*(6393), 1072–1073. https://doi.org/10.1126/science.aat1871.

Pendergrass, A. G., & Hartmann, D. L. (2014). Changes in the distribution of rain frequency and intensity in response to global warming. *Journal of Climate*, *27*(22), 8372–8383. https://doi.org/10.1175/JCLI-D-14-00183.1.

Poloczanska, E., Mintenbeck, K., Portner, H. O., Roberts, D., & Levin, L. A. (2018). The IPCC special report on the ocean and cryosphere in a changing climate. In *Ocean sciences meeting*AGU.

Prajeesh, A. G., Ashok, K., & Rao, D. V. B. (2013). Falling monsoon depression frequency: A gray-Sikka conditions perspective. *Scientific Reports*, *3*. https://doi.org/10.1038/srep02989.

Prasad, A. K., Vinay Kumar, K., Singh, S., & Singh, R. P. (2006). Potentiality of multi-sensor satellite data in mapping flood Hazard. *Journal of the Indian Society of Remote Sensing, 34*(3), 219–231. https://doi.org/10.1007/BF02990651.

Prasanna, V. (2014). Impact of monsoon rainfall on the total foodgrain yield over India. *Journal of Earth System Science, 123*(5), 1129–1145. https://doi.org/10.1007/s12040-014-0444-x.

Preethi, B., & Revadekar, J. V. (2013). Kharif foodgrain yield and daily summer monsoon precipitation over India. *International Journal of Climatology, 33*(8), 1978–1986. https://doi.org/10.1002/joc.3565.

Priya, P., Krishnan, R., Mujumdar, M., & Houze, R. A. (2017). Changing monsoon and midlatitude circulation interactions over the Western Himalayas and possible links to occurrences of extreme precipitation. *Climate Dynamics, 49*(7–8), 2351–2364. https://doi.org/10.1007/s00382-016-3458-z.

Rai, P. K., Singh, G. P., & Dash, S. K. (2020). Projected changes in extreme precipitation events over various subdivisions of India using RegCM4. *Climate Dynamics, 54*(1–2), 247–272. https://doi.org/10.1007/s00382-019-04997-6.

Rajeevan, M., Bhate, J., & Jaswal, A. K. (2008). Analysis of variability and trends of extreme rainfall events over India using 104 years of gridded daily rainfall data. *Geophysical Research Letters, 35*(18). https://doi.org/10.1029/2008GL035143.

Rajendran, K., & Kitoh, A. (2008). Indian summer monsoon in future climate projection by a super high-resolution global model. *Current Science, 95*(11), 1560–1569. http://www.ias.ac.in/currsci/dec102008/1560.pdf.

Rajendran, K., Kitoh, A., Srinivasan, J., Mizuta, R., & Krishnan, R. (2012). Monsoon circulation interaction with Western Ghats orography under changing climate. *Theoretical and Applied Climatology, 110*(4), 555–571. https://doi.org/10.1007/s00704-012-0690-2.

Rajendran, K., Sajani, S., Jayasankar, C. B., & Kitoh, A. (2013). How dependent is climate change projection of Indian summer monsoon rainfall and extreme events on model resolution? *Current Science, 104*(10), 1409–1418. http://www.currentscience.ac.in/Volumes/104/10/1409.pdf.

Rakhecha, P. R., & Clark, C. (2000). Point and areal PMP estimates for durations of two and three days in India. *Meteorological Applications, 7*(1), 19–26. https://doi.org/10.1017/S1350482700001389.

Rakhecha, P. R., & Soman, M. K. (1994). Trends in the annual extreme rainfall events of 1 to 3 days duration over India. *Theoretical and Applied Climatology, 48*(4), 227–237. https://doi.org/10.1007/BF00867053.

Ramasamy, S. M., Gunasekaran, S., Rajagopal, N., Saravanavel, J., & Kumanan, C. J. (2019). Flood 2018 and the status of reservoir-induced seismicity in Kerala, India. *Natural Hazards, 99*(1), 307–319. https://doi.org/10.1007/s11069-019-03741-x.

Ramu, D. A., Chowdary, J. S., Ramakrishna, S. S. V. S., & Kumar, O. S. R. U. B. (2018). Diversity in the representation of large-scale circulation associated with ENSO-Indian summer monsoon teleconnections in CMIP5 models. *Theoretical and Applied Climatology, 132*(1–2), 465–478. https://doi.org/10.1007/s00704-017-2092-y.

Ranger, N., Hallegatte, S., Bhattacharya, S., Bachu, M., Priya, S., Dhore, K., et al. (2011). An assessment of the potential impact of climate change on flood risk in Mumbai. *Climatic Change, 104*(1), 139–167. https://doi.org/10.1007/s10584-010-9979-2.

Rao, K. P. R., & Jagannathan, P. (1953). A study of the northeast monsoon rainfall of Tamilnadu. *Indian Journal of Meteorology and Geophysics, 4*, 22–43.

Rasmusson, E. M., & Carpenter, T. H. (1983). The relationship between eastern equatorial Pacific sea surface temperatures and rainfall over India and Sri Lanka. *Monthly Weather Review, 111*(3), 517–528. https://doi.org/10.1175/1520-0493(1983)111<0517:TRBEEP>2.0.CO;2.

Revadekar, J. V., Varikoden, H., Preethi, B., & Mujumdar, M. (2016). Precipitation extremes during Indian summer monsoon: Role of cyclonic disturbances. *Natural Hazards, 81*(3), 1611–1625. https://doi.org/10.1007/s11069-016-2148-9.

Roxy, M. K., Ghosh, S., Pathak, A., Athulya, R., Mujumdar, M., Murtugudde, R., et al. (2017). A threefold rise in widespread extreme rain events over Central India. *Nature Communications, 8*(1). https://doi.org/10.1038/s41467-017-00744-9.

Roxy, M. K., Ritika, K., Terray, P., Murtugudde, R., Ashok, K., & Goswami, B. N. (2015). Drying of Indian subcontinent by rapid Indian ocean warming and a weakening land-sea thermal gradient. *Nature Communications, 6*. https://doi.org/10.1038/ncomms8423.

Roy, S. S., & Balling, R. C. (2004). Trends in extreme daily precipitation indices in India. *International Journal of Climatology, 24*(4), 457–466. https://doi.org/10.1002/joc.995.

Saji, N. H., Goswami, B. N., Vinayachandran, P. N., & Yamagata, T. (1999). A dipole mode in the tropical Indian ocean. *Nature, 401*(6751), 360–363. https://doi.org/10.1038/43854.

Sandeep, S., & Ajayamohan, R. S. (2015). Poleward shift in Indian summer monsoon low level jetstream under global warming. *Climate Dynamics, 45*(1–2), 337–351. https://doi.org/10.1007/s00382-014-2261-y.

Sandeep, S., Ajayamohan, R. S., Boos, W. R., Sabin, T. P., & Praveen, V. (2018). Decline and poleward shift in Indian summer monsoon synoptic activity in a warming climate. *Proceedings of the National Academy of Sciences, 115*(11), 2681–2686. https://doi.org/10.1073/pnas.1709031115.

Sarkar, S., & Maity, R. (2020). Increase in probable maximum precipitation in a changing climate over India. *Journal of Hydrology, 585*. https://doi.org/10.1016/j.jhydrol.2020.124806, 124806.

Sati, S. P., & Gahalaut, V. K. (2013). The fury of the floods in the north-west Himalayan region: The Kedarnath tragedy. *Geomatics, Natural Hazards and Risk, 4*(3), 193–201. https://doi.org/10.1080/19475705.2013.827135.

Selvaraj, K., Pandiyan, J., Yoganandan, V., & Agoramoorthy, G. (2016). India contemplates climate change concerns after floods ravaged the coastal city of Chennai. *Ocean and Coastal Management, 129*, 10–14. https://doi.org/10.1016/j.ocecoaman.2016.04.017.

Sen Gupta, A., McNeil, B., Henderson-Sellers, A., & McGuffie, K. (2012). *Variability and change in the ocean* (pp. 141–165). Elsevier. https://doi.org/10.1016/B978-0-12-386917-3.00006-3 (chapter 6).

Sengupta, A., & Rajeevan, M. (2013). Uncertainty quantification and reliability analysis of CMIP5 projections for the Indian summer monsoon. *Current Science, 105*(12), 1692–1703. http://www.currentscience.ac.in/Volumes/105/12/1692.pdf.

Shivaprasad Sharma, S. V., Roy, P. S., Chakravarthi, V., & Srinivasa Rao, G. (2018). Flood risk assessment using multi-criteria analysis: A case study from Kopili river basin, Assam, India. *Geomatics, Natural Hazards and Risk, 9*(1), 79–93. https://doi.org/10.1080/19475705.2017.1408705.

Shukla, J., & Mooley, D. A. (1987). Empirical prediction of the summer monsoon rainfall over India. *Monthly Weather Review, 115*(3), 695–704. https://doi.org/10.1175/1520-0493(1987)115<0695:EPOTSM>2.0.CO;2.

Shyamala, B., & Bhadram, C. V. V. (2006). Impact of mesoscale-synoptic scale interactions on the Mumbai historical rain event during 26-27 July 2005. *Current Science, 91*(12), 1649–1654. http://www.ias.ac.in/currsci/dec252006/1649.pdf.

Siderius, C., Biemans, H., Wiltshire, A., Rao, S., Franssen, W. H. P., Kumar, P., et al. (2013). Snowmelt contributions to discharge of the Ganges. *Science of the Total Environment, 468–469*, S93–S101. https://doi.org/10.1016/j.scitotenv.2013.05.084.

Sikka, D. R. (1980). Some aspects of the large scale fluctuations of summer monsoon rainfall over India in relation to fluctuations in the planetary and regional scale circulation parameters. *Proceedings of the Indian Academy of Sciences - Earth and Planetary Sciences*, *89*(2), 179–195. https://doi.org/10.1007/BF02913749.

Simmons, A. J., Willett, K. M., Jones, P. D., Thorne, P. W., & Dee, D. P. (2010). Low-frequency variations in surface atmospheric humidity, temperature, and precipitation: Inferences from reanalyses and monthly gridded observational data sets. *Journal of Geophysical Research-Atmospheres*, *115*(1). https://doi.org/10.1029/2009JD012442.

Singh, D., Horton, D. E., Tsiang, M., Haugen, M., Ashfaq, M., Mei, R., et al. (2014). Severe precipitation in Northern India in June 2013: Causes, historical context, and changes in probability. *Bulletin of the American Meteorological Society*, *95*(9), S58–S61.

Singh, D., Tsiang, M., Rajaratnam, B., & Diffenbaugh, N. S. (2014). Observed changes in extreme wet and dry spells during the south Asian summer monsoon season. *Nature Climate Change*, *4*(6), 456–461. https://doi.org/10.1038/nclimate2208.

Singh, N., Kumar, K. K., & Soman, M. K. (1989). Some features of the periods contributing specified percentages of rainfall to annual total in Kerala, India. *Theoretical and Applied Climatology*, *39*(3), 160–170. https://doi.org/10.1007/BF00868311.

Singh, O., & Kumar, M. (2013). Flood events, fatalities and damages in India from 1978 to 2006. *Natural Hazards*, *69*(3), 1815–1834. https://doi.org/10.1007/s11069-013-0781-0.

Sinha Ray, K. C., & Srivastava, A. K. (2000). Is there any change in extreme events like heavy rainfall? *Current Science*, *79*(2), 155–158.

Soman, M. K., Krishna, K., & Singh, N. (1988). Decreasing trend in the rainfall of Kerala. *Current Science*, *97*(1), 7–12.

Sudheer, K. P., Murty Bhallamudi, S., Narasimhan, B., Thomas, J., Bindhu, V. M., Vema, V., et al. (2019). Role of dams on the floods of August 2018 in Periyar River Basin, Kerala. *Current Science*, *116*(5), 780–794. https://doi.org/10.18520/cs/v116/i5/780-794.

Sun, Y., Solomon, S., Dai, A., & Portmann, R. W. (2006). How often does it rain? *Journal of Climate*, *19*(6), 916–934. https://doi.org/10.1175/JCLI3672.1.

Suthinkumar, P. S., Babu, C. A., & Varikoden, H. (2019). Spatial distribution of extreme rainfall events during 2017 southwest monsoon over Indian subcontinent. *Pure and Applied Geophysics*, *176*(12), 5431–5443. https://doi.org/10.1007/s00024-019-02282-5.

Toreti, A., Naveau, P., Zampieri, M., Schindler, A., Scoccimarro, E., Xoplaki, E., et al. (2013). Projections of global changes in precipitation extremes from coupled model intercomparison project phase 5 models. *Geophysical Research Letters*, *40*(18), 4887–4892. https://doi.org/10.1002/grl.50940.

Trenberth, K. E., Dai, A., Rasmussen, R. M., & Parsons, D. B. (2003). The changing character of precipitation. *Bulletin of the American Meteorological Society*, *84*(9), 1161–1205. https://doi.org/10.1175/BAMS-84-9-1205.

Trenberth, K. E., Fasullo, J. T., & Shepherd, T. G. (2015). Attribution of climate extreme events. *Nature Climate Change*, *5*(8), 725–730. https://doi.org/10.1038/nclimate2657.

Turner, A. G., & Annamalai, H. (2012). Climate change and the South Asian summer monsoon. *Nature Climate Change*, *2*(8), 587–595. https://doi.org/10.1038/nclimate1495.

Ullah, K., & Shouting, G. (2013). A diagnostic study of convective environment leading to heavy rainfall during the summer monsoon 2010 over Pakistan. *Atmospheric Research*, *120–121*, 226–239. https://doi.org/10.1016/j.atmosres.2012.08.021.

Vaidya, S. S., & Kulkarni, J. R. (2007). Simulation of heavy precipitation over Santacruz, Mumbai on 26 July 2005, using mesoscale model. *Meteorology and Atmospheric Physics*, *98*(1–2), 55–66. https://doi.org/10.1007/s00703-006-0233-4.

Varikoden, H., & Babu, C. A. (2015). Indian summer monsoon rainfall and its relation with SST in the equatorial Atlantic and Pacific oceans. *International Journal of Climatology*, *35*(6), 1192–1200. https://doi.org/10.1002/joc.4056.

Varikoden, H., Mujumdar, M., Revadekar, J. V., Sooraj, K. P., Ramarao, M. V. S., Sanjay, J., et al. (2018). Assessment of regional downscaling simulations for long term mean, excess and deficit Indian summer monsoons. *Global and Planetary Change*, *162*, 28–38. https://doi.org/10.1016/j.gloplacha.2017.12.002.

Varikoden, H., & Preethi, B. (2013). Wet and dry years of Indian summer monsoon and its relation with Indo-Pacific Sea surface temperatures. *International Journal of Climatology*, *33*(7), 1761–1771. https://doi.org/10.1002/joc.3547.

Varikoden, H., Preethi, B., & Revadekar, J. V. (2012). Diurnal and spatial variation of Indian summer monsoon rainfall using tropical rainfall measuring mission rain rate. *Journal of Hydrology*, *475*, 248–258. https://doi.org/10.1016/j.jhydrol.2012.09.056.

Varikoden, H., & Revadekar, J. V. (2019). On the extreme rainfall events during the southwest monsoon season in northeast regions of the Indian subcontinent. *Meteorological Applications*, *27*(1). https://doi.org/10.1002/met.1822.

Varikoden, H., Revadekar, J. V., Kuttippurath, J., & Babu, C. A. (2019). Contrasting trends in southwest monsoon rainfall over the Western Ghats region of India. *Climate Dynamics*, *52*(7–8), 4557–4566. https://doi.org/10.1007/s00382-018-4397-7.

Vellore, R. K., Kaplan, M. L., Krishnan, R., Lewis, J. M., Sabade, S., Deshpande, N., et al. (2016). Monsoon-extratropical circulation interactions in Himalayan extreme rainfall. *Climate Dynamics*, *46*(11–12), 3517–3546. https://doi.org/10.1007/s00382-015-2784-x.

Venkateswarlu, B. (2011). Rainfed agriculture in India: Issues in technology development and transfer. *Model training course on "impact of climate change in rainfed agriculture and adaptation strategies"* (pp. 22–29). Hyderabad, India: CRIDA.

Vishnu, S., Francis, P. A., Shenoi, S. S. C., & Ramakrishna, S. S. V. S. (2016). On the decreasing trend of the number of monsoon depressions in the Bay of Bengal. *Environmental Research Letters*, *11*(1), 014011. https://doi.org/10.1088/1748-9326/11/1/014011.

Viswanadhapalli, Y., Srinivas, C. V., Basha, G., Dasari, H. P., Langodan, S., Venkat Ratnam, M., et al. (2019). A diagnostic study of extreme precipitation over Kerala during August 2018. *Atmospheric Science Letters*, *20*(12). https://doi.org/10.1002/asl.941.

Wake, B. (2013). Flooding costs. *Nature Climate Change*, *3*(9), 778. https://doi.org/10.1038/nclimate1997.

Wang, S. Y., Davies, R. E., Huang, W. R., & Gillies, R. R. (2011). Pakistan's two-stage monsoon and links with the recent climate change. *Journal of Geophysical Research-Atmospheres*, *116*(16). https://doi.org/10.1029/2011JD015760.

Wang, S. Y. S., Zhao, L., Yoon, J. H., Klotzbach, P., & Gillies, R. R. (2018). Quantitative attribution of climate effects on Hurricane Harvey's extreme rainfall in Texas. *Environmental Research Letters*, *13*(5). https://doi.org/10.1088/1748-9326/aabb85.

Woo, S., Singh, G. P., Oh, J. H., & Lee, K. M. (2019). Projection of seasonal summer precipitation over Indian sub-continent with a high-resolution AGCM based on the RCP scenarios. *Meteorology and Atmospheric Physics*, *131*(4), 897–916. https://doi.org/10.1007/s00703-018-0612-7.

World Meteorological Organization. (2009). *Manual on the estimation of probable maximum precipitation*. World Meteorological Organization.

CHAPTER

6

Historical changes in hydroclimatic extreme events over Iran

Vahid Nourani and Hessam Najafi

Center of Excellence in Hydroinformatics and Faculty of Civil Engineering, University of Tabriz, Tabriz, Iran

1 Introduction

The concentration of greenhouse gases (GHGs) in the atmosphere has increased since the preindustrial era, driven largely by economic and population growth, and has reached the highest level than ever. The observed increases in well-mixed GHG concentrations since around 1750 are unequivocally caused by human activities. Land and ocean have taken up a near-constant proportion of carbon dioxide (CO_2) emissions from human activities over the past six decades, with regional differences. This has led to atmospheric concentrations of GHGs: CO_2, methane, and nitrous oxide, that are unprecedented in at least the last 800,000 years (IPCC, 2021). The effects of GHGs, together with those of other anthropogenic drivers, have been detected throughout the climate system and are the main cause of the observed overall warming across the globe since the mid-20th century. Each of the last four decades has been successively warmer than any decade that preceded it since 1850. According to the Intergovernmental Panel on Climate Change (IPCC) report in 2021, the global surface temperature in the first two decades of the 21st century (2001–20) was 0.99°C higher than 1850–1900. The global surface temperature was 1.09°C higher in 2011–20 than 1850–1900, with larger increases over land (1.59°C) than over the ocean (0.88°C) (IPCC, 2021). On annual basis, datasets show that the warmest 6 years on record have all been since 2015, with 2016, 2019 and 2020 being the top three.

Human influence has been detected in warming of the atmosphere and the ocean, in changes in the global water cycle, in reductions in snow and ice, and in global mean sea level rise since the mid-20th century. In recent decades, changes in climate have caused impacts on natural and human systems globally. The impacts are due to observed climate change, irrespective of its cause, indicating the sensitivity of natural and human systems to the changing climate (IPCC, 2014). Climate change can relate to a specific region or the entire globe. Weather patterns may become less predictable as a result of climate change, making it difficult to maintain and grow crops in regions that mainly rely on rainwater for farming (Ongoma, 2013). Climate change has also been associated with other damaging extreme weather and climate events that are becoming more frequent and intense. Such events include hurricanes, floods, downpours, winter storms, and droughts. Evidence of observed climate change impacts is strongest and most comprehensive for natural systems (IPCC, 2014). Some impacts on human systems have also been attributed to climate change, with a major or minor contribution of climate change distinguishable from other influences. Impacts on human systems are often geographically heterogeneous because they depend not only on changes in climate variables but also on social and economic factors. Hence, the changes are more easily observed at local levels, while attribution can remain challenging (IPCC, 2014, 2021).

Not all climate change impacts occur in a systematic way: some happen abruptly while others do slowly and persistently, and there can be multiple impacts that occur concurrently and in different combinations. The interactions between these impacts therefore, will have spatial and temporal implications, cutting across different settings and contexts. This will require new methods of accounting for potential damages and to inform the design of appropriate responses. Although the scope and scale of their effects are not yet well understood, empirical examples are beginning to suggest that climate change impacts and implications will propagate as cascades across physical and human systems (Hilly et al., 2018; Rocha, Peterson, Bodin, & Levin, 2018). For example, the combined effect of increased

https://doi.org/10.1016/B978-0-323-88456-3.00001-0

frequency of high-intensity storms, sea-level rise impacts, and more frequent storm tides, will have compounding impacts on the capacity of individuals, governments, and the private sector to adapt in time before loss and damages occur (Lawrence, Blackett, & Cradock-Henry, 2020). Consequently, various sectors such as water resources, health, energy, tourism, security, transport, and ecosystem have been affected by climate change.

Global climate change and its impact on water resources have received increasing attention from societies across the world due to the associated substantial environmental and economic implications (Jasper, Calanca, Gyalistras, & Fuhrer, 2004; Luo et al., 2019; Zhang, Xu, Zhang, Ren, & Chen, 2008). Consequently, water resource management and allocation, which is already creating conflicts due to water scarcity and is highly unevenly distributed in space is likely to be more challenging in the near future (Chen, Brissette, Poulin, & Leconte, 2011; Chen, Hill, Ohlemüller, Roy, & Thomas, 2011; Fröhlich, 2012; Powell, Kløcker Larsen, De Bruin, Powell, & Elrick-Barr, 2017). Climate change has negative effects on ozone (O_3) chemistry and possibly particulate matter (Doherty, Heal, & O'Connor, 2017). Hence, stronger emission controls will be needed in the future to avoid higher health risks associated with climate change-induced worsening of air quality especially in industrialized and populated regions (Fang, Mauzerall, Liu, Fiore, & Horowitz, 2013). The combined rate and magnitude of climate change are already resulting in a global-scale biological response. Marine, freshwater, and terrestrial organisms are altering distributions to stay within their preferred environmental conditions (Chen, Brissette, et al., 2011), and species are likely changing distributions more rapidly than they have in the past (Lawing & Polly, 2011; Pecl et al., 2017). From the aspects of energy demand and production, with changes in temperature over the past decade, cooling and heating have become one of the main drivers of residential electricity consumption growth, for example, by rapid growth in the adoption of air conditioners that are reversible as heat pumps (Li, Pizer, & Wu, 2019). Also, without placing climate-water impacts on individual plants in a power systems context, vulnerability assessments that aim to support adaptation and resilience strategies misgauge the extent to which regional energy systems are vulnerable (Miara et al., 2017). More recently, there has been an increased focus on climate change as a potential risk to regional security. Climate change has strong implications for national and regional security; as such, regional security should not be constrained to threats from external invasion or domestic insurgency (Seiyefa, 2019). For example, the security of states is increasingly jeopardized by climate change, examples include Hurricane Katrina in the United States, which led to civil disorder, destruction of state infrastructure and loss of lives and destruction of properties (Lynn, Healy, & Druyan, 2009; Rohland, 2018; Seiyefa, 2019). Furthermore, climate change impacts such as flash floods and drought could pose spillover effect to neighboring states and the international community (Seiyefa, 2019). Everyone and all sectors of the economy are increasingly under threat of climate change (Dube & Nhamo, 2019). To this end, it is difficult to ignore the risk of climate change given the significance of the tourism economy to many countries. Given the fact that tourism is dependent on geographic space that is greatly influenced by weather and climate variations, research into this topical issue is important especially in countries such Zimbabwe and Kenya in Africa that are prone to climate extremes, to ensure the industry's sustainability (Boko et al., 2007; Hall, 2008). As a result, given the widespread impact of climate change, bilateral/international and interdisciplinary extensive planning and cooperation are key to address climate change challenges (IPCC, 2014, 2021).

The IPCC reports show that climate change is evident by high frequency in climate extreme events including flooding, drought, and heat waves (IPCC, 2014, 2021). Studies have shown the changes in the occurrence and severity of climate extreme events, along with the variability of weather patterns, cause substantial impacts on human and natural systems (Esayas, Simane, Teferi, Ongoma, & Tefera, 2018; IPCC, 2014, 2021). It is widely projected that climate variability will increase with increase in global warming and climate change (Screen, Deser, & Sun, 2015; Sillmann, Kharin, Zwiers, Zhang, & Bronaugh, 2013; Thornton, Ericksen, Herrero, & Challinor, 2014). According to IPCC (2007), climate change is the change in the state of the climate that can be identified (e.g., using statistical tests) by changes in the mean and/or the variability of its properties, and that persists for an extended period, typically decades or longer while climate variability is a periodic variation in these statistics. Climate variability on the other hand is the variation in the mean state and other statistics (such as standard deviations, the occurrence of extremes, etc.) of the climate on all spatial and temporal scales beyond that of individual weather events. The variability may be due to natural internal processes within the climate system (internal variability), or to variations in natural or anthropogenic external forcing (external variability). In essence, the persistence of anomalous conditions is the main difference between climate variability and change. Natural variability of the climate system, in particular on seasonal and longer time scales, predominantly occurs with preferred spatial patterns and time scales, through the dynamical characteristics of the atmospheric circulation and through interactions with the land and ocean surfaces. The climate variability driven by natural cycles is associated with sea level pressure or sea surface temperature (SST) in specific parts of oceans. Such patterns are often called regimes, modes, or teleconnections. Examples are the North Atlantic Oscillation (NAO), the Pacific-North American pattern (PNA), and the El Niño-Southern Oscillation (ENSO), among others. Climate change has major implications for climate variability and the severity and frequency of hydroclimatic extreme events such as floods and droughts have changed (Tabari, 2020, 2021; Tegegne & Melesse, 2020; Van Loon et al., 2016). Changes in many extreme weather and climate events have been observed since about mid of 20th century. Some of these changes have been linked to human influences, including a decrease in cold temperature extremes, an increase in warm temperature extremes, an increase in extreme high sea levels and an increase in the number of heavy precipitation events in a number of regions. Thus, investigating the changes in the hydroclimatic extreme events is important to deal with the anthropogenic impacts (Giorgi et al., 2014).

Oceanic-atmospheric teleconnection patterns could affect hydroclimatic extreme events over large distances across the world (Mehr, Nourani, Hrnjica, & Molajou, 2017; Najafi, Nourani, Sharghi, Roushangar, & Dąbrowska, 2022; Nourani, Najafi, Sharghi, & Roushangar, 2021). Climatic observations are often correlated across long spatial distances, and extreme events, such as heat waves or floods, are typically assumed to be related to such teleconnections (Boers et al., 2019; Hong, Hsu, Lin, & Chiu, 2011; Lau & Kim, 2012). Revealing atmospheric teleconnection patterns and understanding their underlying mechanisms is of great importance for weather forecasting in general and extreme-event prediction in particular (Hoskins, 2013; Webster, Toma, & Kim, 2011), especially considering that the characteristics of extreme events have been suggested to change under ongoing anthropogenic climate change (Cho, Li, Wang, Yoon, & Gillies, 2016; Shepherd, 2014; Trenberth & Fasullo, 2012; Trenberth, Fasullo, & Shepherd, 2015).

Climatic oceanic indices such as Pacific Decadal Oscillation (PDO), Atlantic Multidecadal Oscillation (AMO), NAO, Indian Ocean Dipole (IOD) and ENSO provide important predictive information about hydrologic variability in regions around the world (Soukup, Aziz, Tootle, Piechota, & Wulff, 2009) where SST is the main factor almost in all of such indices. The SST is a fundamental variable for understanding, monitoring and predicting fluxes of heat, momentum and gases at a variety of scales that determine complex interactions between the atmosphere and ocean (O'Carroll et al., 2019; Minnett et al., 2019). The SST is critical to the exchanges of heat, moisture, momentum, and gases since it is at the interface of the ocean and the atmosphere (Bentamy et al., 2017; Wanninkhof, Asher, Ho, Sweeney, & McGillis, 2009). The SST at the ocean-atmosphere interface has a significant societal impact, through, e.g., large ocean gyres and atmospheric circulation cells influencing weather and climate, weather systems, and severe storms and local-scale phenomena, such as the generation of sea breezes and convective clouds (O'Carroll et al., 2019). The patterns of SST reveal subsurface dynamics, at least those with a surface thermal expression such as fronts and eddies (Tandeo, Chapron, Ba, Autret, & Fablet, 2013), and the modulation of the surface momentum exchanges across the temperature gradients modify the atmospheric boundary layer on the mesoscale (O'Neill, Esbensen, Thum, Samelson, & Chelton, 2010; Perlin et al., 2014) and larger (McPhaden, Zhang, Hendon, & Wheeler, 2006; Minobe, Kuwano-Yoshida, Komori, Xie, & Small, 2008). The PDO is based on the monthly SST variability of the North Pacific Ocean on the decadal-scale (Mantua & Hare, 2002). The continuing sequences of long-duration changes in the sea surface temperature of the North Atlantic Ocean are termed AMO, having cool and warm phases that can last from 20 to 40 years at a time (Enfield, Mestas-Nuñez, & Trimble, 2001).

El Niño is one of the largest oscillations of the climate system and is defined as warmer than the normal conditions of Pacific Ocean surface temperature in eastern parts. Commonly used ENSO index include regional SST values (e.g., Nino-1+2, Nino-3, Nino-4, and Nino-3.4) (Meidani & Araghinejad, 2014). The SOI is an atmospheric pressure-based index and is a standardized anomaly of the mean sea level pressure difference (Tadesse, Wilhite, Harms, Hayes, & Goddard, 2004). La Niña dominates during a positive SOI while a negative SOI indicates El Niño conditions. The NAO index is the winter climate variability mode in North Atlantic Ocean and is defined as the difference in normalized mean winter (December-March) sea level pressure anomalies between the island of Iceland and Portugal (Hurrell, 1995). The importance of oceanic-atmospheric teleconnection patterns in dealing with extreme events has been investigated by several researchers. For example, a significant relationship exists between south of the African Sahara rainfall and Atlantic SST (Camberlin, Janicot, & Poccard, 2001). Rucong, Minghua, Yongqiang, and Yimin (2001) indicated a significant relationship between SST of the Pacific Ocean and summer monsoon in Mid-Eastern China (Rucong et al., 2001). The SST of the Mediterranean Sea could affect the African Sahel rainfall (Rowell, 2003). The continuous positive periods of NAO could affect drought (Stefan, Ghioca, Rimbu, & Boroneant, 2004). The ENSO could impact India's winter extreme rainfall events and consequently, the ENSO index can be used for the estimation of extreme rainfall events in the region, 4 to 6 months in advance (Revadekar & Kulkarni, 2008). The ENSO and the IOD are the major parameters affecting extreme floods in the Omo-Gibe River, Ethiopia (Degefu & Bewket, 2017). Also, a close link exists between local SST anomalies and tropical rainfall (Ying, Huang, & Lian, 2019).

Although there is a lot literature on the influence of oceanic-atmospheric circulations on the hydroclimatic conditions of Iran, about the hydroclimatic systems of western Iran, more attention has been paid to parameters such as NAO, SOI and SST. For example, changes in SST of the Mediterranean, Black and Red Seas NAO and SOI could affect the climate conditions of western Iran (Ghasemi & Khalili, 2008; Meidani & Araghinejad, 2014). According to Meidani and Araghinejad (2014), a significant relationship was found between the stream flow of southwest Iran and the SST of the Mediterranean Sea. The results of their study showed that the application of SST (as predictors of stream flow in southwest Iran) led to better performance than when using NAO, PDO, and AMO indices. In the research by Ghasemi and Khalili (2008), the wet conditions in Iran were found to be characterized by a negative SST anomaly in the Mediterranean and the Black Sea, while dry conditions were found to be characterized by a positive SST anomaly in the Mediterranean and the Black Sea. The results of this study also revealed that the winter precipitation in Iran is mainly related to the Black and Mediterranean and Red Seas. Nazemosadat, Samani, Barry, and Molaii Niko (2006) investigated the effects of ENSO on the 49-year precipitation distribution in Iran. The results revealed that around the mid-1970s, the precipitation has significantly enhanced in the southwest and north of Iran as a result of a consistent increase in the frequency and intensity of El Niño events. Nazemosadat and Ghaedamini (2010) demonstrated a strong linkage between Madden-Julian oscillation (MJO) and daily, monthly, and seasonal precipitation over Southern Iran and Arabian Peninsula. Sabziparvar, Mirmasoudi, Tabari, Nazemosadat, and Maryanaji (2011) studied the impacts of different ENSO phases on reference evapotranspiration variability and demonstrated significant correlations between ENSO events and seasonal evapotranspiration variations in 54% of their study sites. The variations of the NAO and SOI could affect drought events in southwestern parts of Iran (Dezfuli, Karamouz, & Araghinejad, 2010). The anomaly in SSTs of the Persian Gulf and the Red Sea could affect the monthly precipitation in western Iran (Rahimikhoob, 2010). Winter and autumn precipitations in the west of Iran are also related to previous season anomalies of Mediterranean SST and the cooler condition of the sea in autumn leads to wetter winter (Rezaebanafsheh, Jahanbakhsh, Bayati, & Zeynali, 2011). Oceanic-atmospheric parameters such as ENSO, NAO, and SOI, along with SSTs could be applied to predict hydroclimatic events, but in areas surrounded by seas, using SSTs could take precedence (Nourani, Najafi, Sharghi, et al., 2021). Hydroclimatic indices such as PDO, NAO, AMO and SOI along with SSTs could be used as predictors of the long-term precipitation, but in regions surrounding by seas, priority could be on using the SSTs of the surrounding seas. Previous studies have mostly focused on the correlation of ENSO and NAO with Iran's hydrological cycle, while Iran is surrounded by seas, and the use of their SSTs could improve the forecasting results (Meidani & Araghinejad, 2014; Nazemosadat et al., 2006; Nazemosadat & Ghaedamini, 2010; Sabziparvar et al., 2011). This may be due to the fact that the Mediterranean, Black and Red Seas as sources of moisture of Iran western are placed on west of Iran, affecting various areas around. In addition, practically these seas as main sources of Iran precipitations are affected by ocean-atmospheric indices such as ENSO and NAO. Thus, the SSTs of the Mediterranean, Black and Red seas could be used as predictors for the prediction of Tabriz and Kermanshah maximum monthly precipitation (MMP) events located at Iran northwest. According to the literature, it could be concluded that the surrounding seas of studied stations (Tabriz and Kermanshah) along with NAO and SOI indices could affect the MMP events of studied stations, and could be used as predictors to predict MMP events.

In dealing with complex and uncertain systems such as hydroclimatic extreme events that are affected by remote factors, data mining may be a useful alternative to conventional approaches such as regression-based methods that are widely used for the detection of possible teleconnection and point prediction (Nourani, Najafi, Sharghi, et al., 2021). Data mining is a technique that allows decision-makers to extract hidden predictive information from big databases, allowing them to make knowledge-based judgments. The goal of data mining is to improve ways of detecting relevant, significant, and unexpected relationships in huge datasets that cannot be detected using traditional methods or manually. It is possible to discover causal linkages, describe which factors have a substantial influence on the problems of interest, and develop models that anticipate the future using data mining approaches (Dadaser-Celik, Celik, & Dokuz, 2013). (i) Anomaly detection, (ii) classification, (iii) clustering, (iv) time series forecasting, and (v) association rules are the five primary branches of data mining (Han, Kamber, & Pei, 2011). Outlier and aberrant data are identified using the anomaly detection approach. Data may be grouped into groups or clusters using classification and clustering. Due to the complexity and uncertainty of the long-distance teleconnection processes and weak correlation between predictors and predictands, the use of conventional methods and point prediction will have poor performance. Association rule mining is a technique for identifying relevant and unforeseen patterns in large datasets that cannot be detected using traditional methods or manually. By extracting patterns that demonstrate the cause-effect connection between different inputs, association rule mining can be used to investigate processes with high complexity and uncertainty such as hydroclimatic extreme events (Changpetch & Lin, 2013). Although association rules are widely employed in many disciplines of research, their application to hydroenvironmental issues is quite restricted. For example, teleconnection patterns exist among several oceanic-atmospheric parameters and drought events of Nebraska in United States (Tadesse et al., 2004). For India, precipitation time series were investigated to extracts association rules between large-scale ocean-atmospheric parameters and floods and droughts events (Dhanya & Nagesh Kumar, 2009). For Kzlrmak River Basin in Turkey, by utilizing association rules, the connections between stream flow and meteorological factors were investigated (Dadaser-Celik et al., 2013). Extracted patterns using the binary genetic programming and association rules demonstrated the teleconnection between SSTs of Red, Black and Mediterranean seas and monthly rainfall over the northwest of Iran (Mehr et al., 2017). Also, extracted patterns by the association rule mining were used to construct if-then rules (to model using a new fuzzy logic tool named Z-numbers) for predicting drought events in the northwest of Iran (Nourani, Najafi, Sharghi, et al., 2021).

Changes in climate variability and extreme events have significant consequences for sectors with strong connections to the environment, such as water, agriculture, and health. As the novelty of this research, in this study, association rule mining by extracting teleconnection patterns was used to investigate the changes in climate variability and extreme events. This chapter investigates the historical changes in maximum monthly precipitation (MMP) events, four data division strategies with different thresholds of 25%, 50%, and 75% of total samples were considered for extracting the teleconnection patterns. To find teleconnection patterns among MMP events and SSTs of Red, Black and Mediterranean seas, NAO, and SOI indices, association mining was applied. According to previous studies, the SSTs of the Persian Gulf and the Caspian Sea have no effect on precipitation over northwestern Iran, perhaps due to the obstruction of the Alborz and Zagros mountains (Nourani, Najafi, Sharghi, et al., 2021) and as a result, their SST was not used in the present research. In this regard, after extracting the patterns between the three parameters (i.e., SSTs, NAO, and SOI) and MMP events of Tabriz and Kermanshah stations (in northwestern Iran), the historical changes in the MMP events were investigated by comparing the extracted teleconnection patterns for each data division strategy.

2 Materials and methods

2.1 Study area

Monthly precipitation observations from two stations in the northwestern (Tabriz city, 38.05°N, 46.17°E and Kermanshah city, 34.21°N, 47.90°E) of Iran and SSTs of surrounding seas (Black, Mediterranean and Red Seas), NAO and SOI parameters have been considered to the association mining (Figs. 6.1 and 6.2, and Table 6.1). Both stations are located in the northwest of the country with an elevation about 1350 m above sea level. The selection of these gauges is based on their locations and length of available measurements that provide an appropriate situation for both temporal and spatial assessments of the results. The two stations are approximately 422 km apart and have different climatic conditions. The elevations of these cities are 1361 and 1318.6 m above sea level, respectively. The mean monthly precipitation at the Kermanshah synoptic station is about 12 mm greater than the mean monthly precipitation at the Tabriz synoptic station and also the mean monthly temperatures at the Kermanshah synoptic station is about 2.5°C

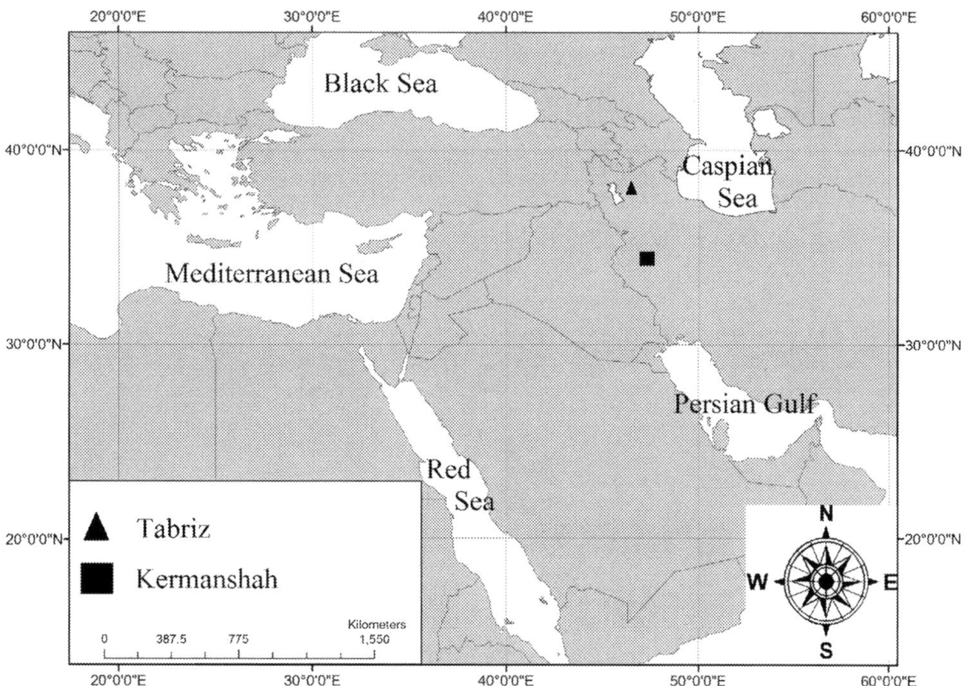

FIG. 6.1 Location of the Tabriz and Kermanshah synoptic stations and adjacent seas.

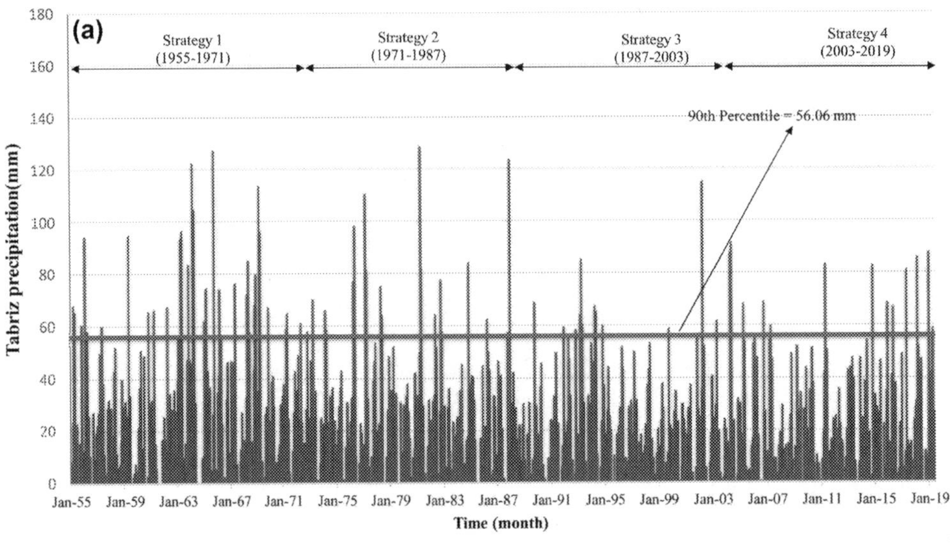

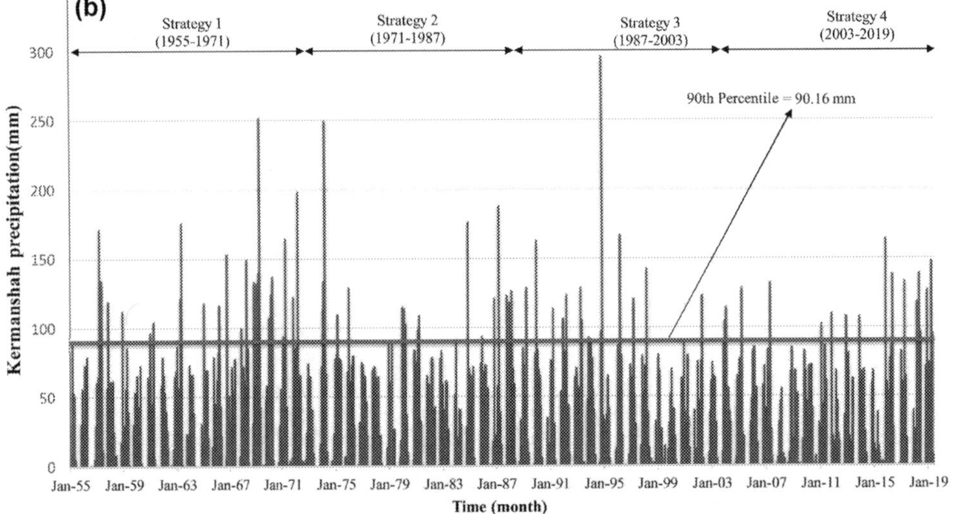

FIG. 6.2 Monthly precipitation time series and threshold limit for synoptic stations of (A) Tabriz and (B) Kermanshah.

TABLE 6.1 Statistics of observed monthly precipitation and used SSTs and indices.

Parameter	Location	Mean	Maximum	Minimum	Std.
SST (°C)	Black Sea	11.88	26.11	−1.59	7.54
	Mediterranean Sea	20.08	28.03	14.18	4.08
	Red Sea	25.42	31.76	16.76	4.14
MP (mm)	Tabriz	23.56	128.41	0	23.36
	Kermanshah	36.87	295.41	0	42.8
NAO	–	0.003	3.04	−3.18	1.029
SOI	–	0.132	2.9	−3.6	0.949

warmer than the Tabriz. Tabriz has a semiarid climate with regular seasons but the Kermanshah climate is heavily influenced by the proximity of the Zagros mountains, classified as a hot dry summer Mediterranean climate.

The Zagros and Alborz Mountains as the two large mountain chains of Iran are situated in Iran's northwest, west, and north. Fig. 6.3 depicts the spatial distribution of annual precipitation in western Iran. Precipitation fluctuations of different regions of Iran are not significant and such fluctuations can be found only in western Iran (Fig. 6.3). It is only

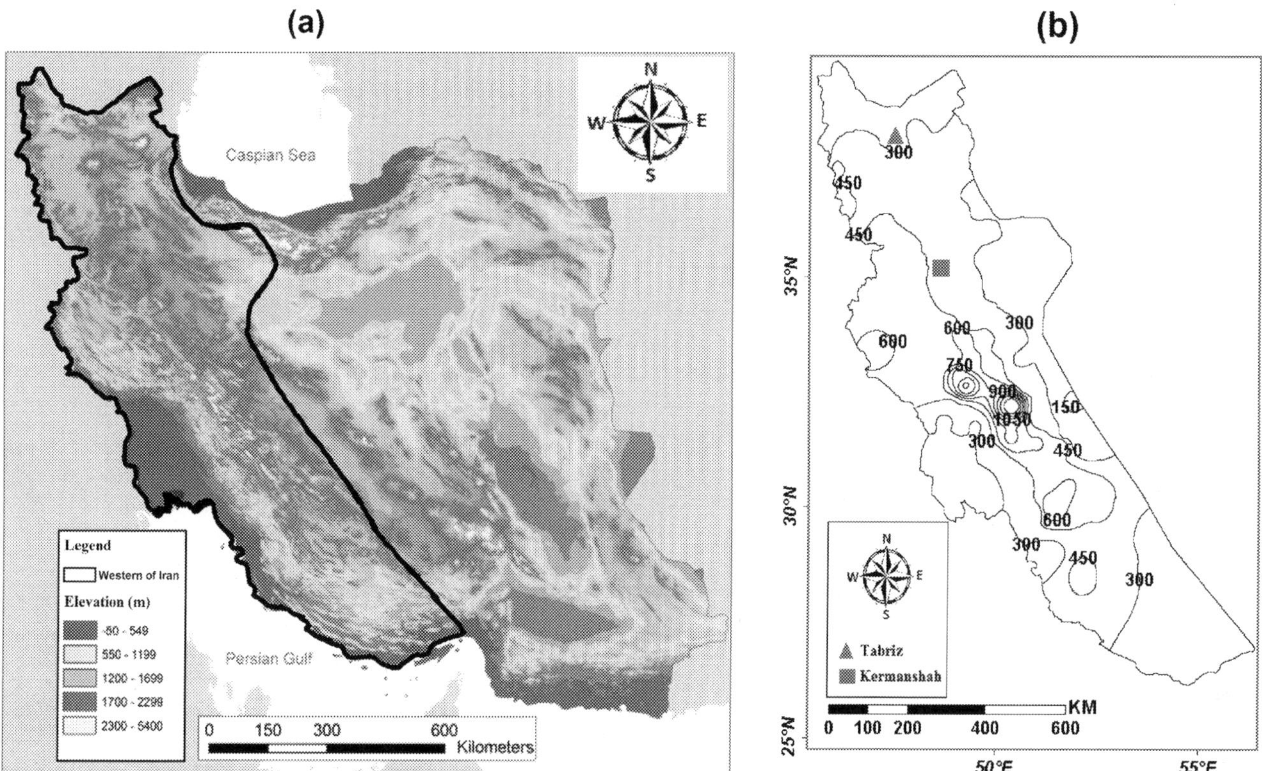

FIG. 6.3 (A) The map of Iran representing the topographic height, and (B) spatial distribution of the annual mean precipitation (mm).

the west of Iran that is affected by SSTs of the Red, Mediterranean and Black Seas. Thus, the data of Tabriz station (in northwestern Iran) and Kermanshah station (close to the center of western Iran) which have data with appropriate quality and quantity were used as the representative of western Iran and could be used to investigate the historical changes in hydroclimatic extreme events. It is notable that, in the proposed methodology, different combinations of inputs with different lag times were tested to get the best possible combination of inputs. This study did not claim that exactly the same inputs (SSTs, NAO and SOI) or even the same combination of antecedents are suitable for all over IRAN stations. But claimed that, the association rule mining could be applied to identify the teleconnection factors that affecting the precipitation of each station over the considered time periods.

2.2 Data

The statistics of monthly time series of precipitations of Tabriz and Kermanshah stations, SSTs of the Black Sea, Mediterranean Sea and Red Sea, SOI and NAO parameters for the study period are presented in Table 6.1. Monthly precipitation data running from 1955 to 2019 used in this research was sourced from the Iran Meteorological Organization (http://www.irimo.ir/) (Specialized products and services weather, 2021).

The SSTs were sourced from the National Oceanic and Atmospheric Administration (NOAA, http://www.esrl.noaa.gov/psd/cgi-bin/data/timeseries/timeseries1.pl). The data has a resolution of 1° grid squares of the seas. In this research, the mean value of the selected grid points (monthly SSTs) was used as SST time series (Kalnay et al., 1996). The overall gridded data region consists of 14, 10 and 101 cells for Red, Black and Mediterranean Sea, respectively. The average monthly values of SSTs for the selected grid points were used in this analysis. Furthermore, the SOI and NAO monthly time series were retrieved from NOAA (https://www.ncdc.noaa.gov/teleconnections/) (Teleconnections, 2021).

According to the climatic regions of western Iran which is predominantly arid to semiarid, precipitation above the threshold of the 90th could be considered as an extreme event. Fig. 6.2 shows the threshold of the 90th percentile applied for classifying the monthly precipitation data into two categories of high (H) and low (L) values. The threshold precipitation of 90th percentile that represents 56.06 mm for Tabriz and 90th percentile that is equivalent to 90.16 mm for Kermanshah synoptic stations were determined and applied to the data. Ordinarily, there is no standard threshold for the determination of extreme events. For example, Rahimikhoob (2010) considered $T=25\%$ (or percentile 75) as extreme events while Mehr et al. (2017) examined different thresholds (15%, 25% and 35%) as extreme events for the precipitation monitoring. In this regard, due to the arid to the semiarid condition of western Iran, the threshold precipitation of $T=10\%$ (or percentile 90th) was determined and applied to the data (but other threshold values may be also tried within the suggested methodology). The data was divided into four equal parts according to thresholds of 25%, 50% and 75% of total samples (four strategies) to investigate the historical changes in MMP events (Fig. 6.2).

2.3 Methodology

The data analysis process involved four distinct phases (preprocessing, data dividing, association rule mining, and finally analyzing and assessing the obtained results). In this research utilizing two indices (i.e., NAO and SOI) and SST values, the historical changes in MMP events were investigated. The methodology employed is schematically shown in Fig. 6.4. In the continuation of this subsection, each of the four phases is described in detail.

Firstly, the monthly SSTs and NAO data were discretized in groups called very high (VH), H, medium (M), L and very low (VL) (Table 6.2). The monthly SOI data were discretized in groups called extremely high (EH), VH, H, M, L, VL and extremely low (EL). Also, the monthly precipitation time series of the stations were discretized into H (MMP events) and L groups by defining the 90th percentile as the threshold of maximum precipitation in this study (but definitely other threshold values may be also tried within the proposed methodology). This discretization was performed using scientific judgment (as well as trial and error) and earlier researches, so the discretized group numbers

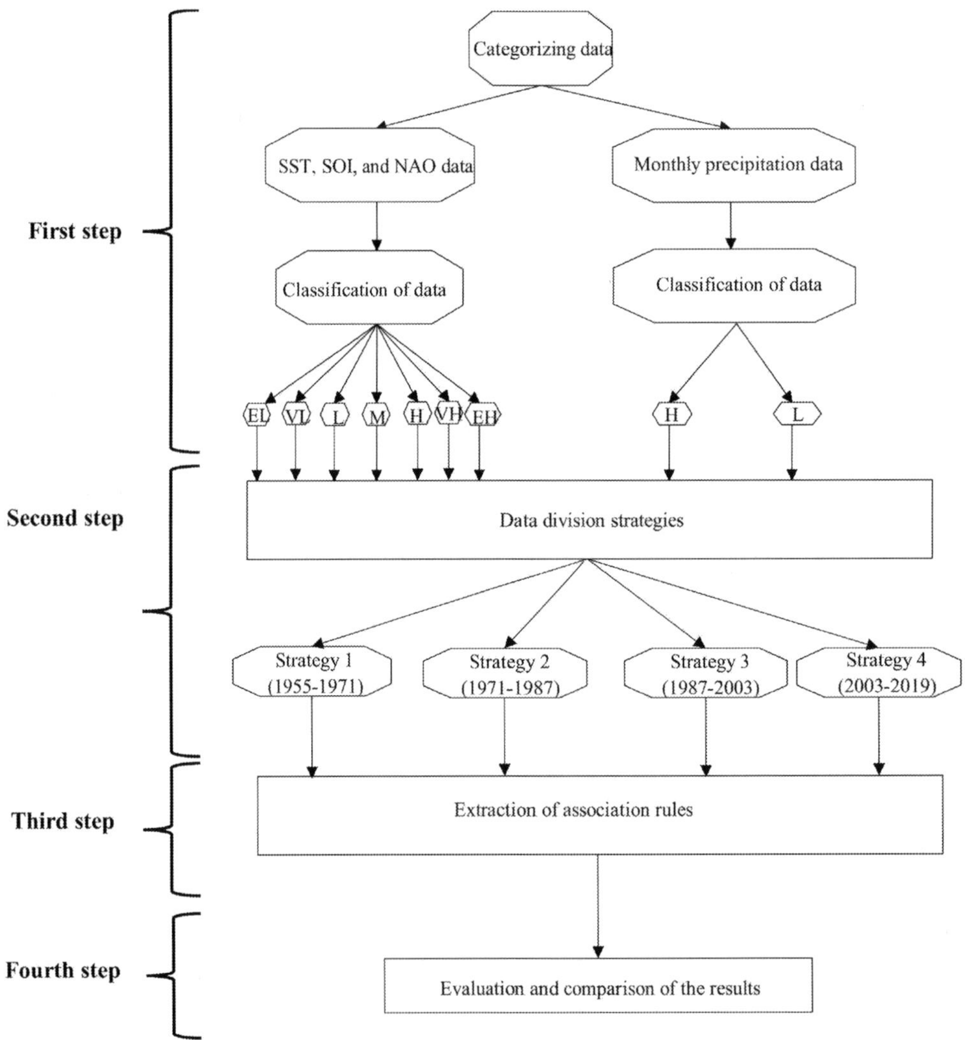

FIG. 6.4 Schematic of the proposed methodology.

TABLE 6.2 The groups of used monthly parameters (i.e., SSTs, NAO, and SOI).

	EL	VL	L	M	H	VH	EH
Black Sea SST (°C)	–	<0.57	(0.57)–(4.34)	(4.34)–(19.41)	(19.41)–(23.20)	>23.20	–
Mediterranean Sea SST (°C)	–	<13.97	(13.97)–(16.00)	(16.00)–(24.15)	(24.15)–(26.19)	>26.19	–
Red Sea SST (°C)	–	<19.23	(19.23)–(21.29)	(21.29)–(29.55)	(29.55)–(31.62)	>31.62	–
NAO	–	<−3.00	(−3.00)–(−2.00)	(−2.00)–(2.00)	(2.00)–(3.00)	>3.00	–
SOI	<−1.31	(−1.31)–(−0.87)	(−0.87)–(−0.40)	(−0.40)–(0.55)	(0.55)–(1.00)	(1.00)–(1.47)	>1.47

in the antecedents and consequent may vary. To this end, according to confidence and support criteria, the possibility of extracting appropriate patterns by association mining was investigated. In this way, for each trial with specified numbers of discretized groups for each variable, if the extracted patterns by association mining were acceptable in terms of confidence and support criteria (e.g., existing some patterns with confidence >0.6), the assumed number of groups for each variable was confirmed.

The data was then divided into four parts with defining different thresholds of 25%, 50% and 75% of the total time series for extracting the teleconnection patterns (four strategies). Due to the considered threshold for extreme events (90th percentile), the occurrence of MMP extreme events is rare. So to the extraction of teleconnection patterns with association mining, sufficient data for each strategy must be considered (in this research about 16 years of monthly data for each strategy was considered). To this end, datasets were divided into four parts to be appropriate both in terms of the length of data (for each strategy) and in terms of the number of parts to investigate changes in the MMP events.

Thirdly, the patterns between the three parameters (i.e., SSTs, NAO, and SOI) and MMP were discovered using the association rule mining technique. Fourthly, the results of all strategies were evaluated and compared with each other.

Association rule mining can be used to discover the common patterns in large datasets. An association rule $C \Rightarrow D$ states that if C happens as an antecedent, D occurs as a consequent with a certain possibility. Confidence and support criteria are used to assess the reliability of extracted association rules. The ratio of simultaneous C and D events to all events is seen in the support of the rules (Eq. 6.1). The ratio of simultaneous C and D events to total C events is seen in the confidence of rules (Eq. 6.2; Han et al., 2011).

$$\text{Support}\,(C \to D) = P(C \cup D) \tag{6.1}$$

$$\text{Confidence}\,(C \to D) = P(C|\,D) = \frac{\text{Support}\,(C \cup D)}{\text{Support}\,(C)} \tag{6.2}$$

In general, each interestingness measure is associated with a threshold, which may be controlled by the user. For example, rules that do not satisfy a confidence threshold of, for instance, 60% can be considered uninteresting. Rules below the threshold likely reflect noise, exceptions, or minority cases and are probably of less value. Rules that satisfy both a minimum support threshold and a minimum confidence threshold are called strong.

In this work, the a priori algorithm was applied to generate association rules. Among the best-known algorithms for association rule induction is the a priori algorithm (Agrawal, Imieliński, & Swami, 1993; Agrawal, Mannila, Srikant, Toivonen, & Verkamo, 1996; Borgelt & Kruse, 2002). This algorithm works in two steps: first, the frequent itemsets are determined. These are sets of items that have at least the given minimum support. In the second step, association rules are generated from the frequent itemsets found in the first step. Usually, the first step is the more important because it accounts for the greater part of the processing time. In order to make it efficient, the a priori algorithm exploits the simple observation that no superset of an infrequent itemset (i.e., an itemset not having minimum support) can be frequent (can have enough support).

3 Results and discussion

The association rule mining is used in this work to investigate teleconnections between parameters of SSTs, NAO and SOI and MMP events. According to the IPCC reports (IPCC, 2014, 2021), each of the last four decades has been successively warmer than any decade, and there has been a positive trend in global temperatures. In this regard, four data division strategies with different thresholds of 25%, 50% and 75% of total time series (according to the quality and quantity of available data) were considered for extracting the teleconnection patterns. Following the categorization of the hydroclimatic parameters (to interval sets such as VL, VH, etc.), the association mining was applied to find the teleconnection patterns between the three parameters (i.e., SSTs, NAO, and SOI) and MMP events for each data division strategy. Finally, the historical changes in the occurrence of MMP events were investigated by comparing the extracted teleconnection patterns for each data division strategy.

For association rule mining, the data should be first classified into some classes (Dadaser-Celik et al., 2013; Mehr et al., 2017; Dhanya & Nagesh Kumar, 2009; Nourani, Najafi, Sharghi, et al., 2021; Tadesse et al., 2004). To this end, as indicated in Table 6.2, the considered parameters as antecedents were discretized into several groups within bounds of $\mu \pm j\sigma$ (μ as mean and σ as std.). Different scenarios had to be tested to identify the best thresholds and class numbers for categorizing input parameters (SSTs, NAO, and SOI) and extracting rules through association mining. According to the literature (Mehr et al., 2017; Nourani, Najafi, Sharghi, et al., 2021; Tadesse et al., 2004), various values for j were checked, including $j=1,1.5$; $j=0.5,1.5$; $j=0.5,1,1.5$; $j=2,3$. However, the approach that uses $j=1, 1.5$ for SSTs, $j=0.5, 1, 1.5$ for SOI and $j=2, 3$ for the NAO provided the appropriate patterns with high confidence and support. Monthly precipitation was also discretized into two groups of L and H (MMP events). These discretized groups influence the derived patterns from association mining. Therefore, with improper discretization, the extracted patterns may be weak (according to the confidence and support of extracted patterns).

The extracted teleconnection patterns between the SSTs, NAO and SOI parameters and MMP events with confidence >0.6 and support >0.01 (at least 2-time occurrence) were investigated (see Tables 6.3 and 6.4). Detection of effective time lags of antecedents is an important step in extracting suitable patterns (Tabari & Willems, 2018). The comparative analysis indicated that using lag 1 for all inputs as antecedents could lead to the extraction of better patterns according to confidence and support metrics. So, the teleconnection patterns were extracted among the MMP(t) and SST($t-1$), NAO($t-1$) and SOI($t-1$) as:

$$\mathrm{MMP}_t = f\left(\mathrm{SST}_{t-1}^{k}, \mathrm{NAO}_{t-1}, \mathrm{SOI}_{t-1}\right) \tag{6.3}$$

where k (1, 2, and 3) denotes the SST of the related sea.

It is notable that according to the binary assumption of precipitation in this study (L or H), the MMP events (i.e., H events) were rare compared to the L events, consequently, the extracted association rules for L precipitation events were much more than MMP events (e.g., Tables 6.3 and 6.4 that in each considered time periods, only a few association rules were extracted for MMP events). Tables 6.3 and 6.4 indicate the extracted patterns for MMP events of Tabriz and Kermanshah stations. According to Tables 6.3 and 6.4, for both stations, although similar patterns are found between strategies 1 to 3 (e.g., for Tabriz station, pattern 2 for strategy 1 and pattern 1 for strategy 2), there are no similar patterns in strategies 4. Due to climate change and global warming, extracted teleconnection patterns have changed over time (strategy 1 to strategy 4) and practically different patterns were obtained for different considered strategies. For

TABLE 6.3 The extracted association rules for Tabriz station.

		Antecedent (if…occur)					Consequent (then…occur)	
		Black Sea	Mediterranean Sea	Red Sea	NAO	SOI	Tabriz monthly precipitation	
	Num.	($t-1$)	($t-1$)	($t-1$)	($t-1$)	($t-1$)	(t)	Confidence
Strategy 1 (1955–1971)	1	M	–	M	L	M	H	1
	2	–	L	M	M	VH		0.75
	3	–	–	M	L	M		0.67
	4	M	–	M	M	EL		0.67
	5	L	L	M	M	VH		0.67
Strategy 2 (1971–1987)	1	–	L	M	M	EH		0.75
	2	L	L	–	M	EH		0.67
	3	M	L	M	M	EH		0.67
Strategy 3 (1987–2003)	1	–	L	M	M	M		0.67
Strategy 4 (2003–2019)	1	H	–	M	M	L		1

TABLE 6.4 The extracted association rules for Kermanshah station.

		Antecedent (if...occur)					Consequent (then...occur)	
		Black Sea	Mediterranean Sea	Red Sea	NAO	SOI	Kermanshah monthly precipitation	
	Num.	$(t-1)$	$(t-1)$	$(t-1)$	$(t-1)$	$(t-1)$	(t)	Confidence
Strategy 1 (1955–1971)	1	L	M	–	M	M	H	1
	2	M	L	L	M	M		1
	3	M	L	L	M	–		0.75
	4	–	M	VL	M	M		0.67
Strategy 2 (1971–1987)	1	L	L	L	M	M		0.75
	2	–	–	VL	M	EH		0.67
Strategy 3 (1987–2003)	1	L	L	L	M	M		0.67
	2	L	L	–	M	M		0.6
Strategy 4 (2003–2019)	1	L	L	M	M	EL		0.67

both stations, a combination of SSTs of all considered seas, NAO and SOI parameters is the most repeating pattern. However, the discrete classes of these antecedents are different for different strategies. Also, the number of extracted patterns has decreased over time which for strategy 4 only one pattern was extracted. This can be due to the climate change and rapid changes during the strategy 3 and 4 (last 33 years) and also reduction of extreme events with the approach of the present time (see Tables 6.5 and 6.6) which makes it difficult to predict the MMP extreme events in the future. For both stations, according to the extracted patterns for lag 1, MMP events happened when SSTs from the M, L, and VL classes were combined, but no SSTs from the H or VH classes were included in the derived rules except for strategy 4 of Tabriz station (the period of 2003 to 2019) which antecedent contains the H class of Black Sea. This might also be due to climate change and global warming.

TABLE 6.5 The number of MMP events for Tabriz station.

	Winter			Spring			Summer			Autumn			
	Jan.	Feb.	Mar.	Apr.	May	Jun.	Jul.	Aug.	Sep.	Oct.	Nov.	Dec.	Sum.
Strategy 1 (1955–1971)	1	1	4	9	6	1	0	0	0	1	4	2	29
Strategy 2 (1971–1987)	0	0	2	7	6	1	0	0	0	1	3	0	20
Strategy 3 (1987–2003)	0	0	1	5	2	0	0	0	0	3	2	2	15
Strategy 4 (2003–2019)	0	1	0	7	1	0	0	0	0	2	1	1	13

TABLE 6.6 The number of MMP events for Kermanshah station.

	Winter			Spring			Summer			Autumn			
	Jan.	Feb.	Mar.	Apr.	May	Jun.	Jul.	Aug.	Sep.	Oct.	Nov.	Dec.	Sum.
Strategy 1 (1955–1971)	4	1	5	6	2	0	0	0	0	1	3	1	23
Strategy 2 (1971–1987)	1	4	5	1	0	0	0	0	0	0	3	4	18
Strategy 3 (1987–2003)	2	1	6	2	1	0	0	0	0	2	2	3	19
Strategy 4 (2003–2019)	2	1	3	4	1	0	0	0	0	2	3	1	17

The frequencies of MMP events were 13% for winter, 58% for spring, 0% for summer and 29% for autumn for Tabriz station (Tables 6.5 and 6.6). For Kermanshah station, the frequencies of MMP events were 45.5% for winter, 22% for spring, 0% for summer and 32.5% for autumn. So, for both stations, the frequencies of MMP events (with the threshold of 90th percentile) were 0% for summer. By summer occurrence, SSTs of H and VH, the MMP events were uncommon and virtually no patterns with H or VH SST group were found (except for strategy 4 of Tabriz station). This confirms a previous study (Ghasemi & Khalili, 2008) that the wet conditions in Iran are characterized by a negative SST anomaly in the Mediterranean and the Black Sea. Also for both stations, in addition to the decrease in extreme events over time, due to climate changes, the rate of extreme events in winter has decreased and has been added to spring and autumn.

Previous studies (Dezfuli et al., 2010; Hosseinzadeh Talaee, Tabari, & Sobhan Ardakani, 2014; Nazemosadat et al., 2006; Raziei, Saghafian, Paulo, Pereira, & Bordi, 2009) mostly focused on the correlation of ENSO and NAO and Iran's regional climate. However, Iran is surrounded by seas and using their SSTs might enhance the quality of the findings. The results of this study confirmed this hypothesis and indicated that SSTs along with NAO and SOI indices are appropriate variables for the investigation of historical changes of MMP events in western Iran. This may be due to the fact that approximately 70% of Iran precipitations are originated either in the Black Sea or the Mediterranean Sea. The other 30% originates in North Africa and the Red Sea and comes to Iran via Saudi Arabia and the Persian Gulf ken (Ghasemi & Khalili, 2008; Kendrew, 1922).

In this work, two classes of output (binary) were considered in the modeling (i.e., high or low monthly precipitation); however, the output could be classified into more classes if needed. The monthly precipitation events of only two stations were investigated but in order to further investigate the historical changes of MMP events, it is recommended to apply it to multiple precipitations series, with diverse characteristics over the whole country. Although only the SSTs, NAO and SOI are applied herein as predictors (inputs), other indices such as AMO, and PDO could be considered to extract teleconnection patterns between hydroclimatic parameters.

Indeed, the hydroclimatic extreme events are complex systems because of the highly nonlinear and interacting components. Consequently, the application of association rule mining might be an appropriate manner for investigating such complex processes. When compared to other techniques, especially the conventional correlation approaches, data mining has four major advantages:

(i) In time series analysis, data mining can offer a lot of precision when it comes to finding relationships between different parameters that exist at the same time or with time lags.
(ii) The rules and models provided by data mining are simple to comprehend.
(iii) Using data mining methods, a researcher would be able to get a broad view of the relationships between the various parameters by identifying all possible patterns in datasets.
(iv) When vast volumes of data with several variables are processed, data mining can be used as a reliable method for tracking extreme events like MMP.

4 Conclusions

Climate variability is expected to increase as climate change continues following globe warming. Changes in the climate variability and in the frequency, severity, spatial extent and timing of extreme events are unavoidable consequences of climate change (Field, Barros, Stocker, & Dahe, 2012; Thornton et al., 2014). Changes in extreme events will have major implications for both human and natural systems. To cope with such anthropogenic impacts, it is important to analyze changes in hydroclimatic extreme events. Large-scale oceanic-atmospheric teleconnection patterns could be used to investigate hydroclimatic extreme events including MMP events. In this study, to investigate the historical changes in MMP events, four data division strategies with different thresholds of 25%, 50%, and 75% of total samples were considered for extracting the teleconnection patterns. The association mining was applied to find the rules between the SST, NAO, and SOI parameters and MMP events for each data division strategy. Finally, by analyzing and comparing the extracted teleconnection patterns for each data division strategy, the historical changes in MMP events were investigated.

It is evident that the large-scale parameters such as the SSTs of surrounding seas SOI and NAO could be applied as predictors to predict MMP events over northwestern Iran (with over 67% confidence in both stations). For both Tabriz and Kermanshah stations, the number of extracted patterns from about 5 patterns for strategy 1 (1955–71) has decreased to only one pattern for strategy 4 (2003–19). This can be due to the climate change and rapid changes during strategies 3 and 4 (last 33 years) and reduction of extreme events with the approach of the present time (strategy 4) which makes it difficult to predict the MMP extreme events in the future. Also, the extracted teleconnection patterns for strategy 4 were different from other strategies (strategy 1 to 3). Practically, this proves the changes in MMP extreme events during the last 65 years (1955–2019) and the need to define new patterns for the prediction of MMP events in the future.

Reliable outcome of the data mining for investigating the teleconnection patterns between MMP events and SSTs, NAO, and SOI, recommends its employment to investigate historical changes of other hydroclimatic extreme events such as drought and flood.

References

Agrawal, R., Imieliński, T., & Swami, A. (1993). Mining association rules between sets of items in large databases. In *Proceedings of the 1993 ACM SIGMOD international conference on management of data* (pp. 207–216).

Agrawal, R., Mannila, H., Srikant, R., Toivonen, H., & Verkamo, A. I. (1996). Fast discovery of association rules. *Advances in Knowledge Discovery and Data Mining*, *12*(1), 307–328.

Bentamy, A., Piolle, J.-F., Grouazel, A., Danielson, R., Gulev, S., Paul, F., et al. (2017). Review and assessment of latent and sensible heat flux accuracy over the global oceans. *Remote Sensing of Environment*, *201*, 196–218.

Boers, N., Goswami, B., Rheinwalt, A., Bookhagen, B., Hoskins, B., & Kurths, J. (2019). Complex networks reveal global pattern of extreme-rainfall teleconnections. *Nature*, *566*(7744), 373–377.

Boko, M., Niang, I., Nyong, A., Vogel, A., Githeko, A., Medany, M., et al. (2007). *Africa climate change 2007: Impacts, adaptation and vulnerability: Contribution of working group II to the fourth assessment report of the intergovernmental panel on climate change*. Cambridge University Press.

Borgelt, C., & Kruse, R. (2002). Induction of association rules: A priori implementation. In *Compstat* (pp. 395–400). Physica. https://doi.org/10.1007/978-3-642-57489-4_59.

Camberlin, P., Janicot, S., & Poccard, I. (2001). Seasonality and atmospheric dynamics of the teleconnection between African rainfall and tropical sea-surface temperature: Atlantic vs. ENSO. *International Journal of Climatology*, *21*(8), 973–1005.

Changpetch, P., & Lin, D. K. J. (2013). Model selection for logistic regression via association rules analysis. *Journal of Statistical Computation and Simulation*, *83*(8), 1415–1428.

Chen, I.-C., Hill, J. K., Ohlemüller, R., Roy, D. B., & Thomas, C. D. (2011). Rapid range shifts of species associated with high levels of climate warming. *Science*, *333*(6045), 1024–1026.

Chen, J., Brissette, F. P., Poulin, A., & Leconte, R. (2011). Overall uncertainty study of the hydrological impacts of climate change for a Canadian watershed. *Water Resources Research*, *47*(12), W12509. https://doi.org/10.1029/2011WR010602.

Cho, C., Li, R., Wang, S.-Y., Yoon, J.-H., & Gillies, R. R. (2016). Anthropogenic footprint of climate change in the June 2013 northern India flood. *Climate Dynamics*, *46*(3–4), 797–805.

Dadaser-Celik, F., Celik, M., & Dokuz, A. S. (2013). Associations between stream flow and climatic variables at Kizilirmak river basin in Turkey. *Global NEST Journal*, *14*(3), 354–361. https://doi.org/10.30955/gnj.000881.

Degefu, M. A., & Bewket, W. (2017). Variability, trends, and teleconnections of stream flows with large-scale climate signals in the Omo-Ghibe River Basin, Ethiopia. *Environmental Monitoring and Assessment*, *189*(4). https://doi.org/10.1007/s10661-017-5862-1.

Dezfuli, A. K., Karamouz, M., & Araghinejad, S. (2010). On the relationship of regional meteorological drought with SOI and NAO over Southwest Iran. *Theoretical and Applied Climatology*, *100*(1), 57–66. https://doi.org/10.1007/s00704-009-0157-2.

Dhanya, C. T., & Nagesh Kumar, D. (2009). Data mining for evolving fuzzy association rules for predicting monsoon rainfall of India. *Journal of Intelligent Systems*, *18*(3), 193–210.

Doherty, R. M., Heal, M. R., & O'Connor, F. M. (2017). Climate change impacts on human health over Europe through its effect on air quality. *Environmental Health*, *16*(1), 33–44.

Dube, K., & Nhamo, G. (2019). Climate change and potential impacts on tourism: Evidence from the Zimbabwean side of the Victoria Falls. *Environment, Development and Sustainability*, *21*(4), 2025–2041.

Enfield, D. B., Mestas-Nuñez, A. M., & Trimble, P. J. (2001). The Atlantic multidecadal oscillation and its relation to rainfall and river flows in the continental US. *Geophysical Research Letters*, *28*(10), 2077–2080.

Esayas, B., Simane, B., Teferi, E., Ongoma, V., & Tefera, N. (2018). Trends in extreme climate events over three agroecological zones of southern Ethiopia. *Advances in Meteorology*, *2018*. https://doi.org/10.1155/2018/7354157, 7354157.

Fang, Y., Mauzerall, D. L., Liu, J., Fiore, A. M., & Horowitz, L. W. (2013). Impacts of 21st century climate change on global air pollution-related premature mortality. *Climatic Change*, *121*(2), 239–253.

Field, C. B., Barros, V., Stocker, T. F., & Dahe, Q. (2012). *Managing the risks of extreme events and disasters to advance climate change adaptation: Special report of the intergovernmental panel on climate change*. Cambridge University Press.

Fröhlich, C. J. (2012). Water: Reason for conflict or catalyst for peace? The case of the Middle East. *L'Europe En Formation*, *3*, 139–161.

Ghasemi, A. R., & Khalili, D. (2008). The association between regional and global atmospheric patterns and winter precipitation in Iran. *Atmospheric Research*, *88*(2), 116–133. https://doi.org/10.1016/j.atmosres.2007.10.009.

Giorgi, F., Coppola, E., Raffaele, F., Diro, G. T., Fuentes-Franco, R., Giuliani, G., et al. (2014). Changes in extremes and hydroclimatic regimes in the CREMA ensemble projections. *Climatic Change*, *125*(1), 39–51. https://doi.org/10.1007/s10584-014-1117-0.

Hall, C. M. (2008). Tourism and climate change: Knowledge gaps and issues. *Tourism Recreation Research*, *33*(3), 339–350.

Han, J., Kamber, M., & Pei, J. (2011). *Data mining: Concepts and techniques* (3rd ed.). Morgan Kaufmann.

Hilly, G., Vojinovic, Z., Weesakul, S., Sanchez, A., Hoang, D. N., Djordjevic, S., et al. (2018). Methodological framework for analysing cascading effects from flood events: The case of Sukhumvit area, Bangkok, Thailand. *Water*, *10*(1), 81.

Hong, C., Hsu, H., Lin, N., & Chiu, H. (2011). Roles of European blocking and tropical-extratropical interaction in the 2010 Pakistan flooding. *Geophysical Research Letters*, *38*(13), L13806.

Hoskins, B. (2013). The potential for skill across the range of the seamless weather-climate prediction problem: A stimulus for our science. *Quarterly Journal of the Royal Meteorological Society*, *139*(672), 573–584.

Hosseinzadeh Talaee, P., Tabari, H., & Sobhan Ardakani, S. (2014). Hydrological drought in the west of Iran and possible association with large-scale atmospheric circulation patterns. *Hydrological Processes*, *28*(3), 764–773.

Hurrell, J. W. (1995). Decadal trends in the North Atlantic oscillation: Regional temperatures and precipitation. *Science*, *269*(5224), 676–679.

IPCC. (2007). Climate change 2007: Synthesis report. In Core Writing Team, R. K. Pachauri, & A. Reisinger (Eds.), *Contribution of working groups I, II and III to the fourth assessment report of the intergovernmental panel on climate change*IPCC.

IPCC. (2014). Climate change 2014: Synthesis report. In Core Writing Team, R. K. Pachauri, & L. A. Meyer (Eds.), *Contribution of working groups I, II and III to the fifth assessment report of the intergovernmental panel on climate change*IPCC.

IPCC. (2021). Climate change 2021: The physical science basis. In V. Masson-Delmotte, P. Zhai, A. Pirani, S. L. Connors, C. Péan, S. Berger, et al. (Eds.), *Contribution of working group I to the sixth assessment report of the intergovernmental panel on climate change.* Cambridge, UK/New York, NY: Cambridge University Press. https://doi.org/10.1017/9781009157896 (in press) https://www.ipcc.ch/report/ar6/wg1/about/how-to-cite-this-report/.

Jasper, K., Calanca, P., Gyalistras, D., & Fuhrer, J. (2004). Differential impacts of climate change on the hydrology of two alpine river basins. *Climate Research*, *26*(2), 113–129.

Kalnay, E., Kanamitsu, M., Kistler, R., Collins, W., Deaven, D., Gandin, L., et al. (1996). The NCEP/NCAR reanalysis 40-year project. *Bulletin of the American Meteorological Society*, *77*(3), 437–471.

Kendrew, W. G. (1922). *The climates of the continents.* Claredon Press.

Lau, W. K., & Kim, K.-M. (2012). The 2010 Pakistan flood and Russian heat wave: Teleconnection of hydrometeorological extremes. *Journal of Hydrometeorology*, *13*(1), 392–403.

Lawing, A. M., & Polly, P. D. (2011). Pleistocene climate, phylogeny, and climate envelope models: An integrative approach to better understand species' response to climate change. *PLoS One*, *6*(12), e28554.

Lawrence, J., Blackett, P., & Cradock-Henry, N. A. (2020). Cascading climate change impacts and implications. *Climate Risk Management*, *29*, 100234.

Li, Y., Pizer, W. A., & Wu, L. (2019). Climate change and residential electricity consumption in the Yangtze River Delta, China. *Proceedings of the National Academy of Sciences*, *116*(2), 472–477.

Luo, M., Liu, T., Meng, F., Duan, Y., Bao, A., Xing, W., et al. (2019). Identifying climate change impacts on water resources in Xinjiang, China. *Science of the Total Environment*, *676*, 613–626.

Lynn, B. H., Healy, R., & Druyan, L. M. (2009). Investigation of Hurricane Katrina characteristics for future, warmer climates. *Climate Research*, *39*(1), 75–86.

Mantua, N. J., & Hare, S. R. (2002). The Pacific decadal oscillation. *Journal of Oceanography*, *58*(1), 35–44.

McPhaden, M. J., Zhang, X., Hendon, H. H., & Wheeler, M. C. (2006). Large scale dynamics and MJO forcing of ENSO variability. *Geophysical Research Letters*, *33*(16).

Mehr, A. D., Nourani, V., Hrnjica, B., & Molajou, A. (2017). A binary genetic programing model for teleconnection identification between global sea surface temperature and local maximum monthly rainfall events. *Journal of Hydrology*, *555*, 397–406.

Meidani, E., & Araghinejad, S. (2014). Long-lead streamflow forecasting in the southwest of Iran by sea surface temperature of the Mediterranean sea. *Journal of Hydrologic Engineering*, *19*(8). https://doi.org/10.1061/(ASCE)HE.1943-5584.0000965.

Miara, A., Macknick, J. E., Vörösmarty, C. J., Tidwell, V. C., Newmark, R., & Fekete, B. (2017). Climate and water resource change impacts and adaptation potential for US power supply. *Nature Climate Change*, *7*(11), 793–798.

Minnett, P., Alvera-Azcárate, A., Chin, T., Corlett, G., Gentemann, C., Karagali, I., et al. (2019). Half a century of satellite remote sensing of sea-surface temperature. *Remote Sensing of Environment*, *233*, 111366.

Minobe, S., Kuwano-Yoshida, A., Komori, N., Xie, S.-P., & Small, R. J. (2008). Influence of the Gulf Stream on the troposphere. *Nature*, *452*(7184), 206–209.

Najafi, H., Nourani, V., Sharghi, E., Roushangar, K., & Dąbrowska, D. (2022). Application of Z-numbers to teleconnection modeling between monthly precipitation and large scale sea surface temperature. *Hydrology Research*, *53*(1), 1–13.

Nazemosadat, M., & Ghaedamini, H. (2010). On the relationships between the Madden–Julian oscillation and precipitation variability in southern Iran and the Arabian Peninsula: Atmospheric circulation analysis. *Journal of Climate*, *23*(4), 887–904.

Nazemosadat, M., Samani, N., Barry, D., & Molaii Niko, M. (2006). ENSO forcing on climate change in Iran: Precipitation analysis. *Iranian Journal of Science and Technology, Transaction B: Engineering*, *30*, 555–565.

Nourani, V., Najafi, H., Sharghi, E., & Roushangar, K. (2021). Application of Z-numbers to monitor drought using large-scale oceanic-atmospheric parameters. *Journal of Hydrology*, *598*, 126198. https://doi.org/10.1016/j.jhydrol.2021.126198.

O'Carroll, A. G., Armstrong, E. M., Beggs, H. M., Bouali, M., Casey, K. S., Corlett, G. K., et al. (2019). Observational needs of sea surface temperature. *Frontiers in Marine Science*, *6*, 420.

O'Neill, L. W., Esbensen, S. K., Thum, N., Samelson, R. M., & Chelton, D. B. (2010). Dynamical analysis of the boundary layer and surface wind responses to mesoscale SST perturbations. *Journal of Climate*, *23*(3), 559–581.

Ongoma, V. (2013). A review of the effects of climate change on occurrence of aflatoxin and its impacts on food security in semi-arid areas of Kenya. *International Journal of Agricultural Science Research*, *2*(11), 307–311.

Pecl, G. T., Araújo, M. B., Bell, J. D., Blanchard, J., Bonebrake, T. C., Chen, I.-C., et al. (2017). Biodiversity redistribution under climate change: Impacts on ecosystems and human well-being. *Science*, *355*(6332), eaai9214.

Perlin, N., De Szoeke, S. P., Chelton, D. B., Samelson, R. M., Skyllingstad, E. D., & O'Neill, L. W. (2014). Modeling the atmospheric boundary layer wind response to mesoscale sea surface temperature perturbations. *Monthly Weather Review*, *142*(11), 4284–4307.

Powell, N., Kløcker Larsen, R., De Bruin, A., Powell, S., & Elrick-Barr, C. (2017). Water security in times of climate change and intractability: Reconciling conflict by transforming security concerns into equity concerns. *Water*, *9*(12), 934.

Rahimikhoob, A. (2010). Forecasting of maximum monthly precipitation of Ilam using data mining techniques. *Iranian Journal of Soil and Water Research*, *42*(1), 1–7 (in Persian).

Raziei, T., Saghafian, B., Paulo, A. A., Pereira, L. S., & Bordi, I. (2009). Spatial patterns and temporal variability of drought in western Iran. *Water Resources Management*, *23*(3), 439–455.

Revadekar, J. V., & Kulkarni, A. (2008). The El Nino-southern oscillation and winter precipitation extremes over India. *International Journal of Climatology, 28*(11), 1445–1452. https://doi.org/10.1002/joc.1639.

Rezaebanafsheh, M., Jahanbakhsh, S., Bayati, M., & Zeynali, B. (2011). Forecasting autumn and winter precipitation of west of Iran applying Mediterranean SSTs in summer and autumn. *Physical Geography Research Quarterly, 74*, 47–62.

Rocha, J. C., Peterson, G., Bodin, Ö., & Levin, S. (2018). Cascading regime shifts within and across scales. *Science, 362*(6421), 1379–1383.

Rohland, E. (2018). Adapting to hurricanes. A historical perspective on New Orleans from its foundation to Hurricane Katrina, 1718–2005. *Wiley Interdisciplinary Reviews: Climate Change, 9*(1), e488.

Rowell, D. P. (2003). The impact of Mediterranean SSTs on the Sahelian rainfall season. *Journal of Climate, 16*(5), 849–862.

Rucong, Y., Minghua, Z., Yongqiang, Y., & Yimin, L. (2001). Summer monsoon rainfalls over Mid-Eastern China lagged correlated with global SSTs. *Advances in Atmospheric Sciences, 18*(2), 179–196.

Sabziparvar, A., Mirmasoudi, S., Tabari, H., Nazemosadat, M., & Maryanaji, Z. (2011). ENSO teleconnection impacts on reference evapotranspiration variability in some warm climates of Iran. *International Journal of Climatology, 31*(11), 1710–1723.

Screen, J. A., Deser, C., & Sun, L. (2015). Projected changes in regional climate extremes arising from Arctic Sea ice loss. *Environmental Research Letters, 10*(8), 084006.

Seiyefa, E. (2019). How climate change impacts on regional security in West Africa: Exploring the link to organised crime. *African Security Review, 28*(3–4), 159–171.

Shepherd, T. G. (2014). Atmospheric circulation as a source of uncertainty in climate change projections. *Nature Geoscience, 7*(10), 703–708.

Sillmann, J., Kharin, V. V., Zwiers, F., Zhang, X., & Bronaugh, D. (2013). Climate extremes indices in the CMIP5 multimodel ensemble: Part 2. Future climate projections. *Journal of Geophysical Research-Atmospheres, 118*(6), 2473–2493.

Soukup, T. L., Aziz, O. A., Tootle, G. A., Piechota, T. C., & Wulff, S. S. (2009). Long lead-time streamflow forecasting of the North Platte River incorporating oceanic–atmospheric climate variability. *Journal of Hydrology, 368*(1–4), 131–142.

Specialized products and services weather. (2021). https://www.irimo.ir/eng/wd/720-Products-Services.html.

Stefan, S., Ghioca, M., Rimbu, N., & Boroneant, C. (2004). Study of meteorological and hydrological drought in southern Romania from observational data. *International Journal of Climatology, 24*(7), 871–881.

Tabari, H. (2020). Climate change impact on flood and extreme precipitation increases with water availability. *Scientific Reports, 10*(1), 13768. https://doi.org/10.1038/s41598-020-70816-2.

Tabari, H. (2021). Extreme value analysis dilemma for climate change impact assessment on global flood and extreme precipitation. *Journal of Hydrology, 593*, 125932. https://doi.org/10.1016/j.jhydrol.2020.125932.

Tabari, H., & Willems, P. (2018). Lagged influence of Atlantic and Pacific climate patterns on European extreme precipitation. *Scientific Reports, 8*(1), 1–10.

Tadesse, T., Wilhite, D. A., Harms, S. K., Hayes, M. J., & Goddard, S. (2004). Drought monitoring using data mining techniques: A case study for Nebraska, USA. *Natural Hazards, 33*(1), 137–159. https://doi.org/10.1023/B:NHAZ.0000035020.76733.0b.

Tandeo, P., Chapron, B., Ba, S., Autret, E., & Fablet, R. (2013). Segmentation of mesoscale ocean surface dynamics using satellite SST and SSH observations. *IEEE Transactions on Geoscience and Remote Sensing, 52*(7), 4227–4235.

Tegegne, G., & Melesse, A. M. (2020). Multimodel ensemble projection of hydro-climatic extremes for climate change impact assessment on water resources. *Water Resources Management, 34*(9), 3019–3035. https://doi.org/10.1007/s11269-020-02601-9.

Teleconnections. (2021). https://www.ncdc.noaa.gov/teleconnections/.

Thornton, P. K., Ericksen, P. J., Herrero, M., & Challinor, A. J. (2014). Climate variability and vulnerability to climate change: A review. *Global Change Biology, 20*(11), 3313–3328.

Trenberth, K. E., & Fasullo, J. T. (2012). Climate extremes and climate change: The Russian heat wave and other climate extremes of 2010. *Journal of Geophysical Research-Atmospheres, 117*, D17103.

Trenberth, K. E., Fasullo, J. T., & Shepherd, T. G. (2015). Attribution of climate extreme events. *Nature Climate Change, 5*(8), 725–730.

Van Loon, A. F., Stahl, K., Di Baldassarre, G., Clark, J., Rangecroft, S., Van Lanen, H. A., et al. (2016). Drought in a human-modified world: Reframing drought definitions, understanding, and analysis approaches. *Hydrology and Earth System Sciences, 20*(9), 3631–3650.

Wanninkhof, R., Asher, W. E., Ho, D. T., Sweeney, C., & McGillis, W. R. (2009). Advances in quantifying air-sea gas exchange and environmental forcing. *Annual Review of Marine Science, 1*, 213–244.

Webster, P., Toma, V. E., & Kim, H. (2011). Were the 2010 Pakistan floods predictable? *Geophysical Research Letters, 38*(4), L04806.

Ying, J., Huang, P., & Lian, T. (2019). Changes in the sensitivity of tropical rainfall response to local sea surface temperature anomalies under global warming. *International Journal of Climatology, 39*(15), 5801–5814.

Zhang, Q., Xu, C.-Y., Zhang, Z., Ren, G., & Chen, Y. (2008). Climate change or variability? The case of Yellow river as indicated by extreme maximum and minimum air temperature during 1960–2004. *Theoretical and Applied Climatology, 93*(1), 35–43.

Further reading

Lenoir, J., & Svenning, J. (2015). Climate-related range shifts—A global multidimensional synthesis and new research directions. *Ecography, 38*(1), 15–28.

Poloczanska, E. S., Brown, C. J., Sydeman, W. J., Kiessling, W., Schoeman, D. S., Moore, P. J., et al. (2013). Global imprint of climate change on marine life. *Nature Climate Change, 3*(10), 919–925.

Sharghi, E., Nourani, V., Najafi, H., & Gokcekus, H. (2019). Conjunction of a newly proposed emotional ANN (EANN) and wavelet transform for suspended sediment load modeling. *Water Supply, 19*(6), 1726–1734.

CHAPTER

7

Intensification of precipitation extremes in the United States under global warming

Akintomide Afolayan Akinsanola[a,b] *and Gabriel J. Kooperman*[a]

[a]Department of Geography, University of Georgia, Athens, GA, United States

[b]Environmental Science Division, Argonne National Laboratory, Lemont, IL, United States

1 Introduction

Precipitation-related extremes are among the most impact-relevant consequences of a warmer climate (Akinsanola, Kooperman, Reed, Pendergrass, & Hannah, 2020; Zhang, Zhou, Zou, et al., 2018). Globally, extreme precipitation events have significantly increased in frequency and magnitude under climate change, causing more water-related disasters such as droughts, floods, and even landslides and debris flows in mountainous areas (Amarnath et al., 2017; Dai, 2013; Milly, Wetherald, Dunne, & Delworth, 2002; Trenberth et al., 2014). Many studies based on both observations and global climate models (GCMs) have demonstrated a general increase in extreme precipitation events (e.g., Fischer & Knutti, 2016; Kharin, Zwiers, Zhang, & Wehner, 2013; Lehmann, Coumou, & Frieler, 2015; Li et al., 2020; Noor, Ismail, Shahid, & Nashwan, 2020; Scoccimarro, Gualdi, Bellucci, Zampieri, & Navarra, 2013; Uranchimeg, Kwon, Kim, & Kim, 2020; Westra, Alexander, & Zwiers, 2013). In recent decades, these climate change-induced natural disasters, such as the severe flood in northern India in 2013, heavy rainfall over northern Pakistan in July 2010, severe debris flow in northwest China in 2010, the winter floods in southern England in 2014, and the persistent drought in California in 2014 (Agha Kouchak, Cheng, Mazdiyasni, & Farahmand, 2014; Cho, Li, Wang, Yoon, & Gillies, 2016; Lau & Kim, 2012; Schaller et al., 2016; Wang, Wang, & Hong, 2016) have caused enormous loss of lives and destruction of property, and have therefore become a major threat to human security and sustainable socioeconomic development (Jongman et al., 2014; Jonkman, 2005; Smith & Katz, 2013). Considering the continued global reliance on fossil fuels and the associated emissions of greenhouse gases (GHGs), the world's population is likely to be exposed to increased hazards that could have adverse impacts on human systems and the economy (Mora et al., 2018). Thus, assessments of changes in precipitation extremes in response to global warming are critical for mitigation and adaptation management.

The United States is vulnerable to precipitation-related extremes from a variety of storm types ranging from organized thunderstorms (e.g., Haberlie & Ashley, 2019) to hurricanes (e.g., Kunkel & Champion, 2019; Liu & Smith, 2016; Reed, Wehner, Stansfield, & Zarzycki, 2021) to atmospheric rivers (e.g., Lavers & Villarini, 2013; Slinskey, Loikith, Waliser, Guan, & Martin, 2020), and associated disasters have resulted in significant loss of lives and properties. Many studies have investigated future changes in climate extremes anticipated for the United States under different scenarios of GHG emissions (e.g., Kirchmeier-Young & Zhang, 2020; Lopez-Cantu, Prein, & Samaras, 2020; Prein et al., 2017; Sillmann, Kharin, Zwiers, Zhang, & Bronaugh, 2013; Wuebbles et al., 2014) and most recent of these results are based mainly on the GCMs simulations available from the Coupled Model Intercomparison Project phases 5 & 6 (CMIP5 & CMIP6). For example, Janssen, Sriver, Wuebbles, and Kunkel (2016) and Janssen, Wuebbles, Kunkel, Olsen, and Goodman (2014) reported that the observed increase in heavy precipitation events will continue in the future in CMIP5 scenarios with moderate and high GHG increases. Similarly, Akinsanola, Kooperman, et al. (2020) documented a robust intensification of winter precipitation extremes under the CMIP6 high emissions scenario, particularly over the northern parts of the United States, despite pronounced uncertainty that dominates the summer season for most of the country (with the exception of a consistent increase in consecutive dry days in the Central United States). While

Climate Impacts on Extreme Weather
https://doi.org/10.1016/B978-0-323-88456-3.00010-1

previous studies have detailed the potential changes in future precipitation extremes at the end of the 21st century in CMIP5 simulations, an updated synthesis based on CMIP6 of the impacts of climate change on annual precipitation extremes over the United States at the middle and end of the century is needed.

In view of this, this chapter aims to document the future changes in annual precipitation extremes in the United States by employing multimodel projections from the CMIP6 models. In this chapter, dry and wet precipitation extremes are assessed using two indices defined by the Expert Team on Climate Change Detection and Indices (ETCCDI, Zhang et al., 2011); maximum consecutive dry days (CDD) and maximum 5-day precipitation (RX5day), respectively. These two indices encapsulate the most consistent seasonal changes previously identified over the United States in CMIP6 simulations (Akinsanola, Kooperman, et al., 2020) and severe to illustrate changes on both sides of precipitation extremes. The CDD is derived as the maximum number of consecutive days with daily precipitation less than 1 mm and is often used as one of the meteorological drought or dryness indicators; while the RX5day is a frequently used extreme precipitation index in flood risk assessments (Seneviratne et al., 2012; Zhang et al., 2011).

2 Data and methodology

Daily precipitation data from 19 CMIP6 model simulations (Eyring et al., 2016; details in Table 7.1) from the first realization ("r1i1p1f1") for both the historical and the projected fossil-fuel-based economic (i.e., high greenhouse gas emissions) scenario (Shared Socioeconomic Pathways; SSP5-8.5; O'Neill et al., 2016) experiments were used for the analysis presented in this chapter. Both experiments included fully coupled simulations with interactive atmosphere, land, ocean, and sea-ice components. The SSP5-8.5 scenario is closely related to the CMIP5 RCP8.5 scenario, which produces a radiative forcing of about 8.5 Wm^{-2} by the end of the 21st century. The National Oceanic and Atmosphere Administration Climate Prediction Center Unified Gauge-Based Analysis (CPC) daily precipitation data at $0.25° \times 0.25°$ resolution was used for model validation (Chen and Knutson, 2008). The present-day and projected simulations cover the period of 1985–2014 and 2015–99, respectively. The model outputs and observations were regridded to a common grid of $2.81° \times 2.81°$ using a distance-weighted interpolation method in the Climate Data Operators (https://code.zmaw.de/projects/cdo) to produce multimodel summary statistics based on the lowest model resolution. This approach uses the first-order conservative regridding method described in Jones (1999) and reflects the viewpoint that precipitation output from climate models represents an area-average over the grid cell (Chen & Knutson, 2008). In order to reduce uncertainty, we used the multimodel ensemble mean of all the CMIP6 simulations, referred to herein as "EnsMean," but include several metrics to assess the consistency and spread across models.

The CDD and RX5day indices defined by the ETCCDI (Klein Tank, Zwiers, & Zhang, 2009; Zhang et al., 2011) were used as indicators for dry and wet extreme precipitation, respectively. These are nonparametric indices that describe moderate extremes with a recurrence time of at most a year and are calculated from daily precipitation. These indices were calculated for the individual models on the common grid and averaged to form the EnsMean.

Previous studies have shown that the spatial characteristics of precipitation extremes can differ significantly across seasons (Akinsanola et al., 2020; Srivastava, Richard, & Paul, 2020). Therefore, to have a complete explanation of the projected changes in annual and seasonal precipitation extremes over the United States, all the analyses and calculations presented in this study for annual (maps) and seasonal (time series) timescales are defined herein as: ANN (January-December), DJF (December-January-February), MAM (March-April-May), JJA (June-July-August), and SON (September-October-November), which were integrated over the United States and further assessed over the seven subregions defined in the National Climate Assessment Report (https://nca2018.globalchange.gov/; see Fig. 7.1).

The ability of the CMIP6 models to reproduce the annual dry and wet precipitation extremes is first evaluated by comparing the present-day simulations with CPC observations (Section 3.1). Broad spatial assessment and several summary statistics (pattern correlation coefficient, PCC, and normalized root mean square error, NRMSE) were used to evaluate the models. The projected changes were computed by comparing two 30-year time slices from the projections, 2040–69 and 2070–99, to the present-day period of 1985–2014. Furthermore, to assess the intermodel agreement and robustness of the results, we assessed grid points where at least 70% of the ensemble members agree on the sign of change in the ensemble mean. We further assessed grid points where the changes are statistical significant based on student *t*-test. Changes in annual mean precipitation, maximum consecutive dry days, and maximum 5-day precipitation are presented in Sections 3.2, 3.3 and 3.4, respectively; followed by a summary of the major conclusions and implications in Section 4.

TABLE 7.1 Information of the 19 CMIP6 climate models used in this study.

S/N	Model	Institute	Resolution (°lon × °lat)	References
1	ACCESS-CM2	Commonwealth Scientific and Industrial Research Organisation	1.88 × 1.25	Dix, Bi, Dobrohotoff, Fiedler, et al. (2019a, 2019b)
2	ACCESS-ESM1-5	Commonwealth Scientific and Industrial Research Organisation	1.88 × 1.24	Ziehn, Chamberlain, Lenton, et al. (2019a, 2019b)
3	CanESM5	Canadian Earth System Model	2.81 × 2.81	Swart et al. (2019a, 2019b)
4	CESM2-WACCM	National Center for Atmospheric Research	1.25 × 0.94	Danabasoglu (2019a, 2019b)
5	CMCC-CM2-SR5	Euro-Mediterranean Centre on Climate Change coupled climate model	1.25 × 0.94	Lovato and Peano (2020a, 2020b)
6	CMCC-ESM2	Euro-Mediterranean Centre on Climate Change Earth System Model	1.25 × 0.94	Lovato, Peano, and Butenschön (2021a, 2021b)
7	EC-Earth3-CC	EC-EARTH consortium	0.70 × 0.70	EC-Earth Consortium (EC-Earth) (2021a, 2021b)
8	EC-Earth3	EC-EARTH consortium	0.70 × 0.70	EC-Earth Consortium (EC-Earth) (2019a, 2019b)
9	EC-Earth3-Veg-LR	EC-EARTH consortium	1.13 × 1.13	EC-Earth Consortium (EC-Earth) (2020a, 2020b)
10	EC-Earth3-Veg	EC-EARTH consortium	0.70 × 0.70	EC-Earth Consortium (EC-Earth) (2019c, 2019d)
11	INM-CM4-8	Institute of Numerical Mathematics	2.00 × 1.50	Volodin, Mortikov, Gritsun, Lykossov, et al. (2019a, 2019b)
12	INM-CM5-0	Institute of Numerical Mathematics	2.00 × 1.50	Volodin, Mortikov, Gritsun, Lykossov, et al. (2019c, 2019d)
13	IPSL-CM6A-LR	Institute Pierre-Simon Laplace (IPSL)	2.50 × 1.26	Boucher, Denvil, Caubel, and Foujols (2018, 2019)
14	MIROC6	Japanese Modeling Community	1.41 × 1.41	Tatebe and Watanabe (2018) and Shiogama, Abe, and Tatebe (2019)
15	MPI-ESM1-2-HR	Max Planck Institute	0.94 × 0.94	Jungclaus, Bittner, Wieners, Wachsmann, et al. (2019) and Schupfner, Wieners, Wachsmann, Steger, et al. (2019)
16	MPI-ESM1-2-LR	Max Planck Institute	1.88 × 1.88	Wieners, Giorgetta, Jungclaus, Reick, et al. (2019a, 2019b)
17	MRI-ESM2-0	Meteorological Research Institute (MRI)	1.13 × 1.13	Yukimoto, Koshiro, Kawai, et al. (2019a, 2019b)
18	NESM3	Nanjing University of Information Science and Technology Earth System Model	1.88 × 1.88	Cao and Wang (2019) and Cao (2019)
19	TaiESM1	Taiwan Earth System Model	1.25 × 0.94	Lee and Liang (2020a, 2020b)

3 Results and discussion

3.1 Evaluation of present-day annual dry and wet extremes

The spatial distributions of annual CDD and RX5day are presented in Fig. 7.2 for the CPC observation and the CMIP6 EnsMean. The observed CDD averaged over the period of 1985–2014 shown in Fig. 7.2A is dominated by a zonal dipole, with the maximum (minimum) values located west (east) of the Rocky Mountains. This spatial pattern is reasonably reproduced by the CMIP6 EnsMean (Fig. 7.2B), although the magnitude of the CDD is considerably

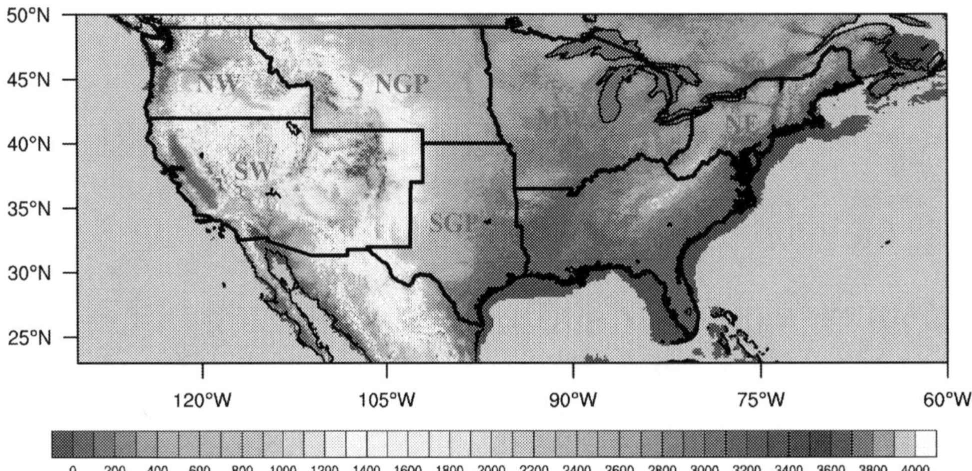

FIG. 7.1 Study domain showing the seven CONUS subregions (*NW*, Northwest; *SW*, Southwest; *NGP*, Northern Great Plains; *SGP*, Southern Great Plains; *MW*, Midwest; *SE*, Southeast; *NE*, Northeast; *US*, all of United States). The color scale shows topography in meters.

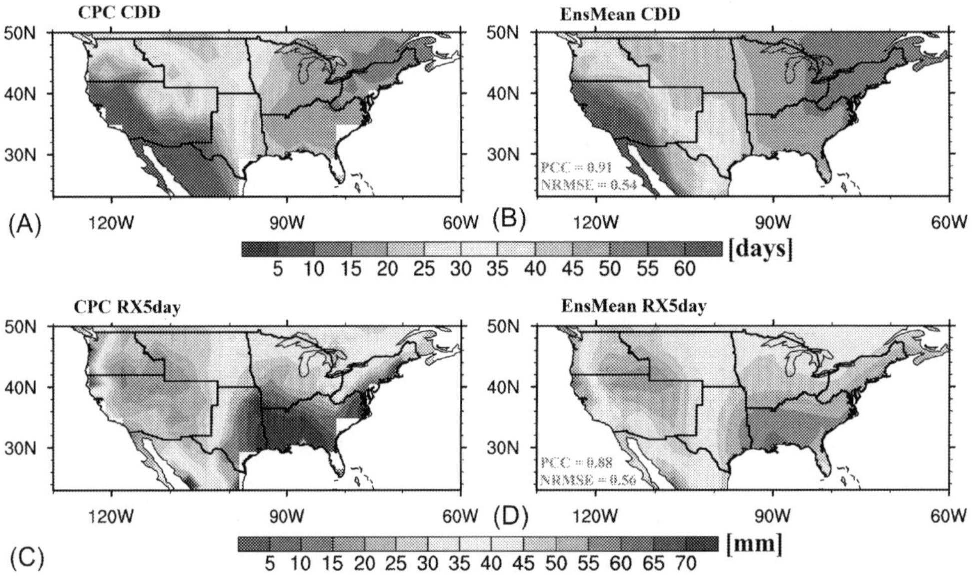

FIG. 7.2 CPC and EnsMean annual mean extreme precipitation indices for (A,B) consecutive dry days (CDD), (C,D) maximum 5-day precipitation (RX5day). Bottom left values in (B,D) are the pattern correlation coefficient (PCC) and normalized root mean square error (NRMSE) of EnsMean relative to CPC.

underestimated over the northwest, southwest, northern great plains, and southern great plains subregions. Relative to CPC observation, the CMIP6 EnsMean has a pattern correlation coefficient (PCC) of 0.91 and Normalized root mean square error (NRMSE) of 0.54; with a span from 0.78 to 0.88 and 0.51 to 1.08 across ensemble members for PCC and NRMSE, respectively (Table 7.2).

We further show the observed and simulated spatial patterns of the RX5day index over the United States. The observed RX5day increases from the western to eastern United States (Fig. 7.2C), with a maximum center in southeast (Gulf coast) subregion (higher than 65mm). The locations of the observed dry/wet centers are well captured by the CMIP6 EnsMean, but the CMIP6 EnsMean underestimates the magnitude of RX5day, particularly over the southeast subregion (Fig. 7.2D). Relative to CPC observation, the CMIP6 EnsMean exhibits considerable skill with PCC of 0.88 and NRMSE of 0.56. Also, RX5day characteristic is reasonably reproduced by the individual ensemble members, with PCC and NRMSE spanning from 0.79 to 0.85 and 0.56 to 0.85, respectively (Table 7.2). Our results are consistent with previously detailed evaluation studies over the United States who have separately reported similar biases in CDD and

TABLE 7.2 Descriptive statistics values of CMIP6 models relative to CPC observations.

	CDD		RX5day	
Model	PCC	NRMSE	PCC	NRMSE
ACCESS-CM2	0.86	0.51	0.85	0.56
ACCESS-ESM1-5	0.83	0.81	0.84	0.62
CanESM5	0.83	0.65	0.8	0.63
CESM2-WACCM	0.85	0.67	0.84	0.63
CMCC-CM2-SR5	0.82	0.59	0.82	0.74
CMCC-ESM2	0.83	0.57	0.82	0.79
EC-Earth3-CC	0.87	0.56	0.85	0.64
EC-Earth3	0.87	0.61	0.84	0.66
EC-Earth3-Veg-LR	0.86	0.57	0.85	0.73
EC-Earth3-Veg	0.88	0.57	0.84	0.66
INM-CM4-8	0.83	1.02	0.79	0.85
INM-CM5-0	0.84	1.08	0.81	0.73
IPSL-CM6A-LR	0.83	0.71	0.81	0.64
MIROC6	0.84	0.69	0.84	0.57
MPI-ESM1-2-HR	0.84	0.71	0.83	0.62
MPI-ESM1-2-LR	0.8	0.73	0.83	0.78
MRI-ESM2-0	0.83	0.77	0.82	0.64
NESM3	0.78	0.65	0.85	0.62
TaiESM1	0.84	0.58	0.83	0.73

RX5day (Akinsanola et al., 2020; Srivastava et al., 2020), and a long-standing tendency of global models to simulate precipitation that is too frequent and weak compared to observations (Sillmann, Kharin, Zhang, Zwiers, & Bronaugh, 2013). Based on the descriptive statistics presented in Table 7.2 and consistent with previous studies, we found that the CMIP6 EnsMean outperforms most individual models at capturing the two extreme precipitation indices, particularly in comparison to CPC observations, and thus will be the main focus of subsequent sections.

3.2 Projected changes in annual mean precipitation over the United States

Before investigating the potential future changes in dry and wet precipitation extremes over the United States, it is important to investigate how the mean annual precipitation would change under global warming. The spatial distribution of projected changes in mean annual precipitation over the United States is shown in Fig. 7.3 for the period 2040–69 and 2070–99 relative to the present-day period of 1985–2014. Consistent with previous assessments based on CMIP5 RCP8.5 simulations (e.g., USGCRP, 2014, 2017), CMIP6 SSP5-8.5 projected increases of about 20% (10%) are evident in annual mean precipitation over most northern parts of the United States in the 2070–99 (2040–69) period. This increase is robust as at least 70% of the ensemble members agree on the sign change in the EnsMean and changes in most grid point is statistically significant at 95% level.

There is also a slight reduction (around −8%) over the southern great plains and southern parts of the southwest subregion, but this is only robust in the 2070–99 period. The spatial pattern is generally consistent between the two time periods, but the magnitude of the projected changes is higher at the end of the 21st century (2070–99) than at the mid-century (2040–69). This expected future increase in annual precipitation at higher latitudes across the United States should have great impacts on the regional agriculture and water resources of the United States.

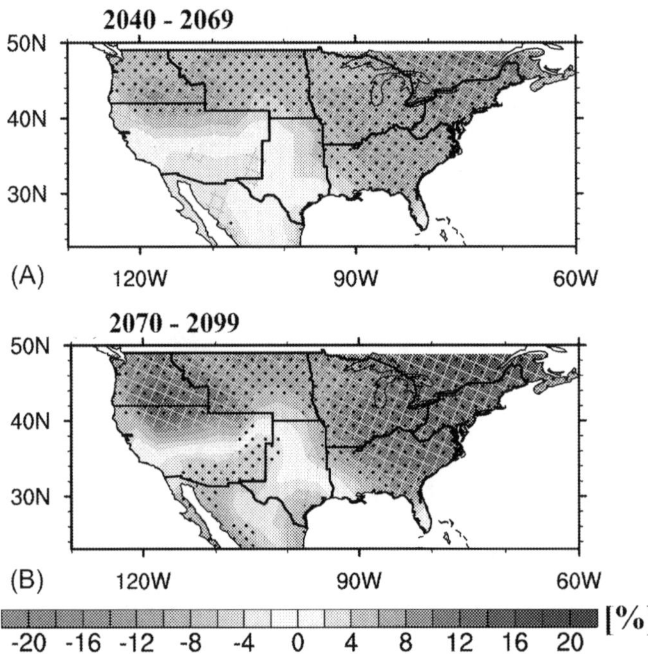

FIG. 7.3 Projected changes in annual precipitation over the period of (A) 2040–69, (B) 2070–99 relative to the reference period 1985–2014. Hatching indicates grid points where at least 70% of the GCMs agree on the sign of the change in EnsMean, and areas with statistically significance differences at 95% level are marked with black stippling.

3.3 Projected changes in maximum consecutive dry days over the Unites States

The projected changes in annual CDD under SSP5-8.5 for the mid and end of the 21st century are presented in Fig. 7.4A–D. Spatial changes of the index are shown for the EnsMean (Fig. 7.4A and B), and area-weighted changes for the entire United States and its subregions are shown for each individual model and EnsMean (Fig. 7.4C and D).

The CDD index is designed to assess the maximum length of dry periods and is generally inversely related to other precipitation indices (Akinsanola & Zhou, 2019). Robust projected increases in CDD are evident over many parts of the United States at the end of the 21st century, while the mid-century changes are smaller in magnitude and less statistically significant. Also, the projected changes in 2040–69 exhibit high inter-model spread as fewer models agree on the sign of change in the EnsMean. In the 2070–99 period, the NW, SGP, and NE subregions have the largest projected increase in CDD (similar to CMIP5 RCP8.5; USGCRP, 2014), between 8% and 18% for EnsMean and up to 58% for individual ensemble members. There is also strong intermodel agreement among the ensemble members in these regions, with at least 70% of the models agreeing on the sign of statistically significant changes for many grid points. In particular, all but three models agree on the overall sign of the change in the NW, all but four agree in most subregions and the entire United States (Fig. 7.4D).

Furthermore, in order to quantify the relative contributions of seasonal changes in CDD to the overall annual changes, the area-weighted temporal evolution of the projected changes in seasonal and annual CDD is assessed for all the subregions and the entire United States (Fig. 7.5).

Across the northern United States, the JJA and SON (DJF and MAM) seasons exhibit a future increase (decrease) in CDD alongside the annual time-scale over the MW, NE, NGP, and NW subregions (Fig. 7.5A–D). The increases are most pronounced after 2050, with JJA (SON) having a higher increase in the MW and NGP (NE), while the annual and JJA evolution is similar over NW subregion. In the southern subregions and United States as a whole, the MAM season shows the largest increases by the end of the century, which is most significant in the SW. All seasons alongside the annual have pronounced increases in the SGP subregion, while the SE subregion changes are less distinct and the SW exhibits a decrease in JJA (Fig. 7.5E–G). Across the entire country, except SGP, there is a small decrease or no change in DJF. Overall, the robust projected annual increase in CDD over the MW, NE, NGP, and NW subregions, especially after 2050, is largely contributed by increases in the JJA and SON seasons.

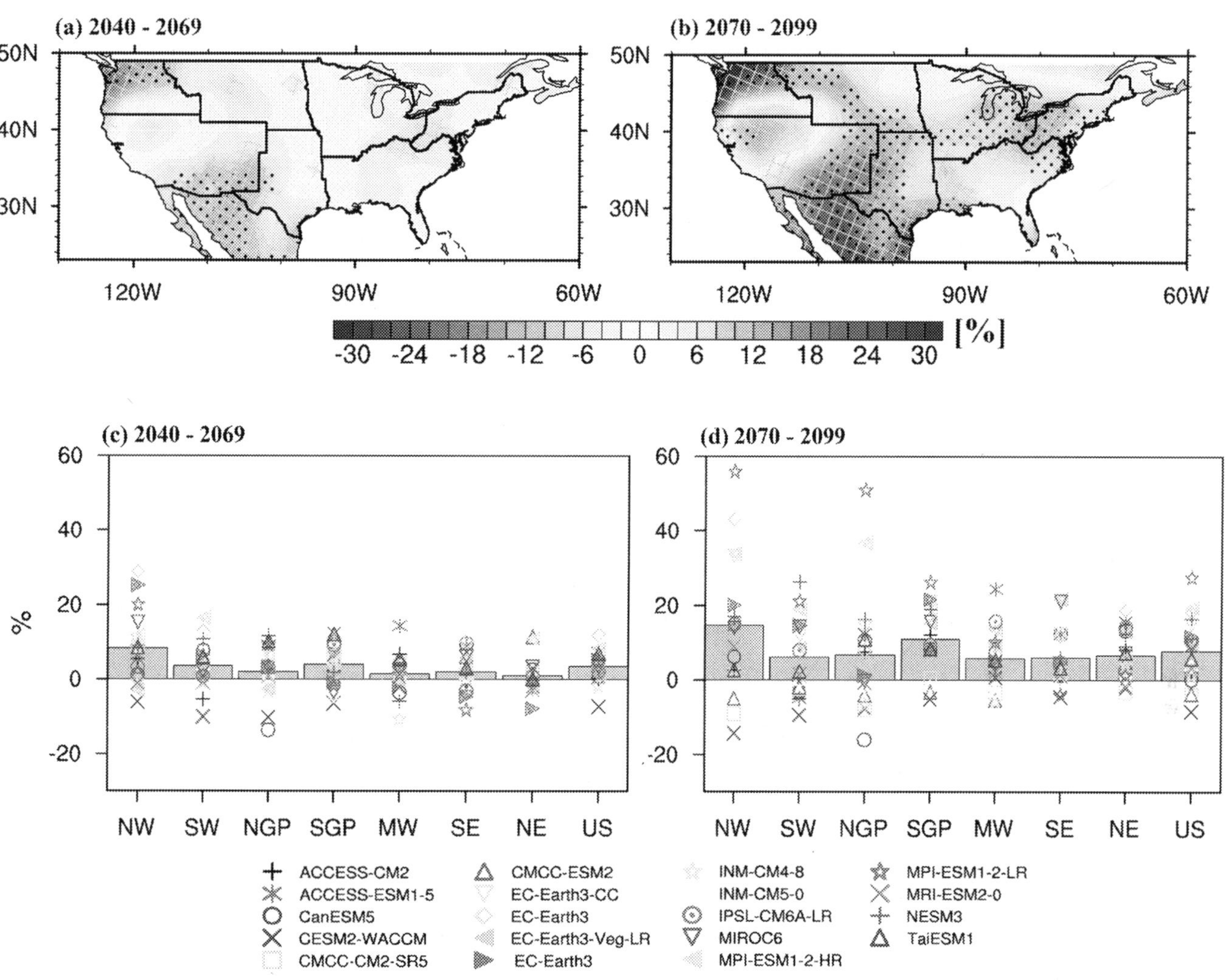

FIG. 7.4 Projected changes in annual maximum consecutive dry days (CDD) over the period of (A,C) 2040–69, (B,D) 2070–99 relative to the reference period 1985–2014 for (A,B) spatial changes, (C,D) area-weighted subregional changes (*NW*, Northwest; *SW*, Southwest; *NGP*, Northern Great Plains; *SGP*, Southern Great Plains; *MW*, Midwest; *SE*, Southeast; *NE*, Northeast; *US*, all of United States). Hatching in (A,B) indicates grid points where at least 70% of the GCMs agree on the sign of the change in EnsMean, and areas with statistically significance differences at 95% level are marked with black stippling. *Gray bars* in (C,D) is the EnsMean values. Figures (C,D) are computed as % change in regional weighted-average.

3.4 Projected changes in maximum 5-day precipitation over the Unites States

The spatial and subregional distributions of the projected changes in annual RX5day are shown in Fig. 7.6. Robust intensification (about 30%) of RX5day is projected over the entire United States. This is consistent with earlier CMIP5 RCP8.5 results (USGCRP, 2014), regardless of the CDD changes. There is a high intermodel agreement in the projected changes as at least 70% of the ensemble members agreed on the sign of change in the EnsMean and the changes at most grid points are statistically significant at the 95% level. The projected increase in RX5day is evident in both the mid and end of 21st century, although the magnitude is higher in the latter (Fig. 7.6A and B). The projected increase by the EnsMean (individual models) range from 8% to 12% (−3% to +28%) for 2040–69 and 15% to 30% (−1% to +47%) for 2070–99 over the United States as a whole and its subregions (Fig. 7.6C and D).

Interestingly, all of the seasons contributed positively to the increase in the annual RX5day across all the subregions of the United States, except for JJA in NGP and NW where the temporal evolution is almost zero (Fig. 7.7). In fact, the JJA RX5day changes are the smallest and most uncertain across all the US subregions, while the projected increase in the other seasons is consistently high in most subregions (except the SW during MAM) and the US-average. This result is consistent with Akinsanola, Kooperman, et al. (2020) who reported that the summer season wet extreme precipitation projections in the United States is dominated by pronounced uncertainty, but the winter has a significant amplification.

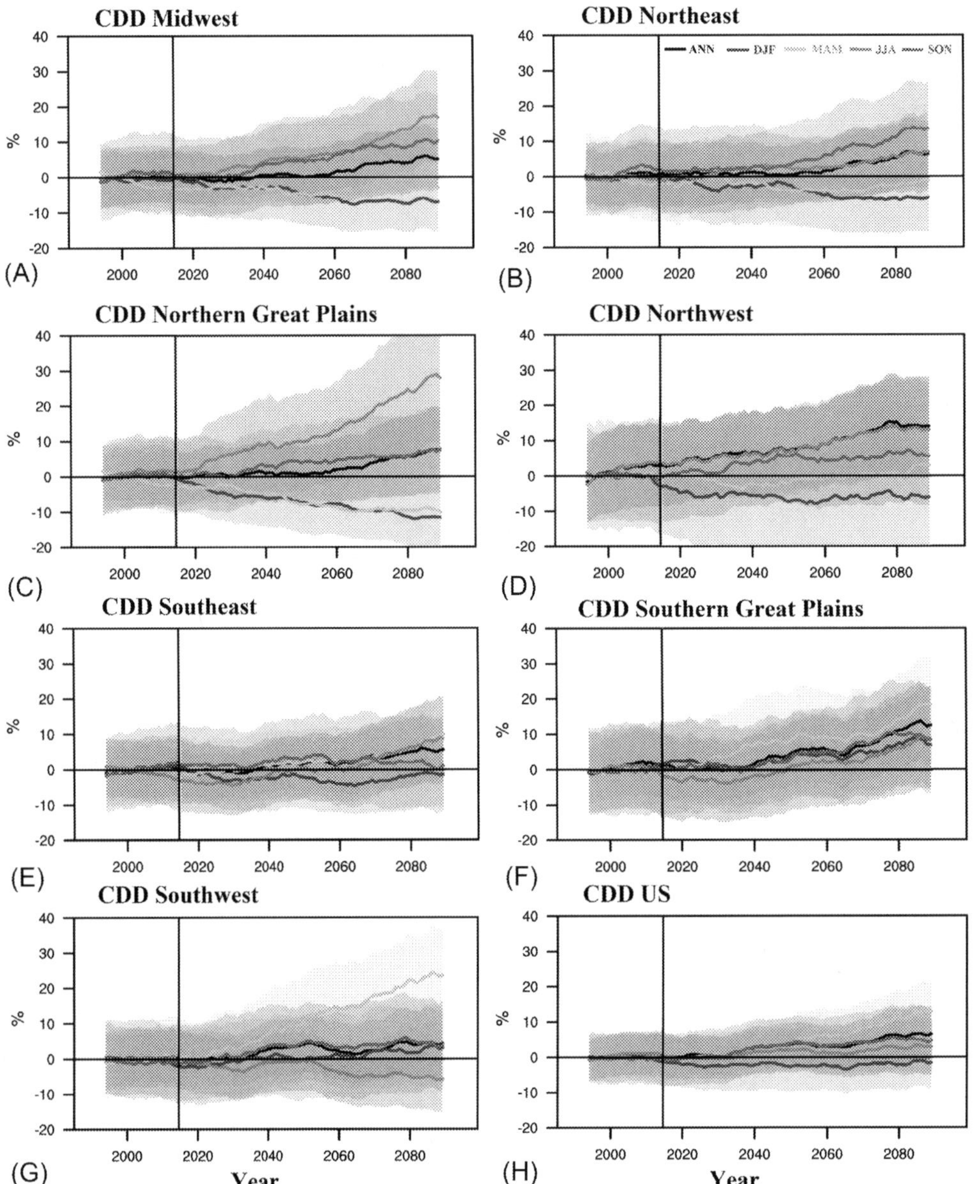

FIG. 7.5 Annual and seasonal maximum consecutive dry days (CDD) changes (%) relative to the reference period (1985–2014) mean for the 21st century under the SSP5-8.5 scenario averaged over the entire United States and its subregions. *Solid lines* indicate the EnsMean and the shadings show 95% confidence in the CMIP6 ensemble members. Time series are smoothed with a 20-year running mean filter. The figures are computed as % change in regional weighted-average (*y*-axis).

4 Summary and conclusion

In this chapter, we examined the projected changes in two ETCCDI extreme precipitation indices over the United States, which are based on daily precipitation data as simulated by the CMIP6 multimodel ensemble under the SSP5-8.5 forcing scenario. The present-day representations of precipitation statistics from CMIP6 models were first validated against CPC observations using both spatial distribution assessment and two descriptive statistics (PCC and NRMSE), and results suggested that the CMIP6 EnsMean can reasonably reproduce both the CDD and RX5day extreme precipitation indices, although noticeable biases exist, consistent with previous extreme precipitation evaluation studies over the United States (e.g., Akinsanola et al., 2020; Srivastava et al., 2020). The CMIP6 EnsMean outperforms most individual models across the two extreme precipitation indices, which is in line with previous multimodel studies

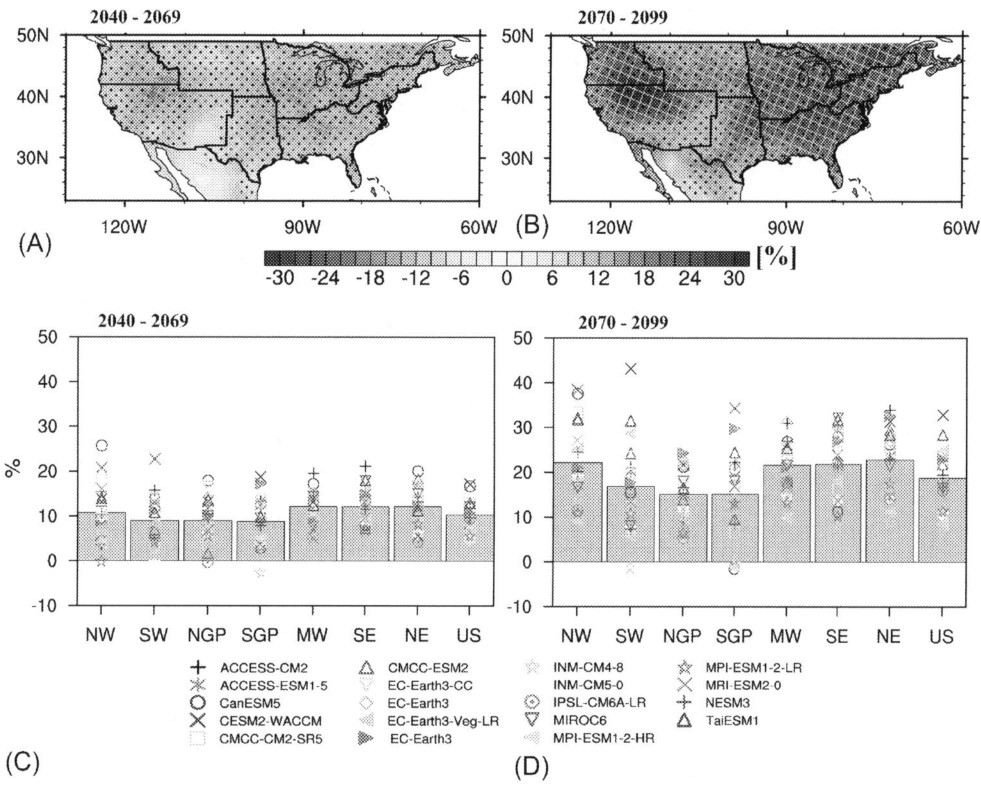

FIG. 7.6 (A,B) Same as Fig. 7.4 but for maximum 5-day precipitation (RX5day). (C,D) Same as Fig. 7.4 but for maximum 5-day precipitation (RX5day).

(Akinsanola & Zhou, 2019; Sillmann, Kharin, Zhang, et al., 2013; Zhou, Wen, Xu, & Song, 2014) and is the primary focus of the analysis in this chapter. Before assessing the projected changes in the precipitation extreme indices, we documented the future changes in mean annual precipitation, and the results show a projected increase of about 20% (10%) over most northern and southeastern parts of the United States in the 2070–99 (2040–69) period. In the Southwest and Southern Great Plains subregions, there is a decrease evident, which becomes robust and statistically significant by the end of the century (2070–99 period).

Furthermore, a projected intensification of both dry and wet annual precipitation extremes is evident over much of the United States; and the patterns and magnitude of the changes remain consistent with earlier projections based on CMIP5 models (USGCRP, 2014, 2017). The magnitude of the projected changes is higher at the end (2070–99) than middle (2040–69) of the 21st century. While robust future intensification dominates the RX5day changes across the mid and end of 21st century, only the 2070–99 period is robust for CDD changes, as the magnitude of the 2040–69 projection is smaller and less significant. Compared to the earlier CMIP6-based projection studies, this chapter documented the annual changes in future precipitation extremes in the United States and its subregions, and also quantified the relative contribution of different seasons to the annual change. Across all of the United States and its subregions, all seasons except summer significantly contributed to the annual projected increase in RX5day, with the winter season having major dominance in most subregions. The seasonal contributions to the CDD annual projection can vary across US subregions, but tends to occur in summer for the northern United States. Uncertainties in the projection are evident in CDD and are generally larger in the mid-century projection.

The projected increase in both dry and wet extremes, especially at the end of 21st century implies a higher probability of combined precipitation-related natural disasters and may threaten the available water resources for the country. This also has implications for the regional availability of the food supply, sustainability of natural ecosystems, maintenance of infrastructure, and flood-related damages and loss of life. Therefore, these projections provide valuable information for informing adaptation strategies and mitigation measures to combat the potential impact of these climate extremes.

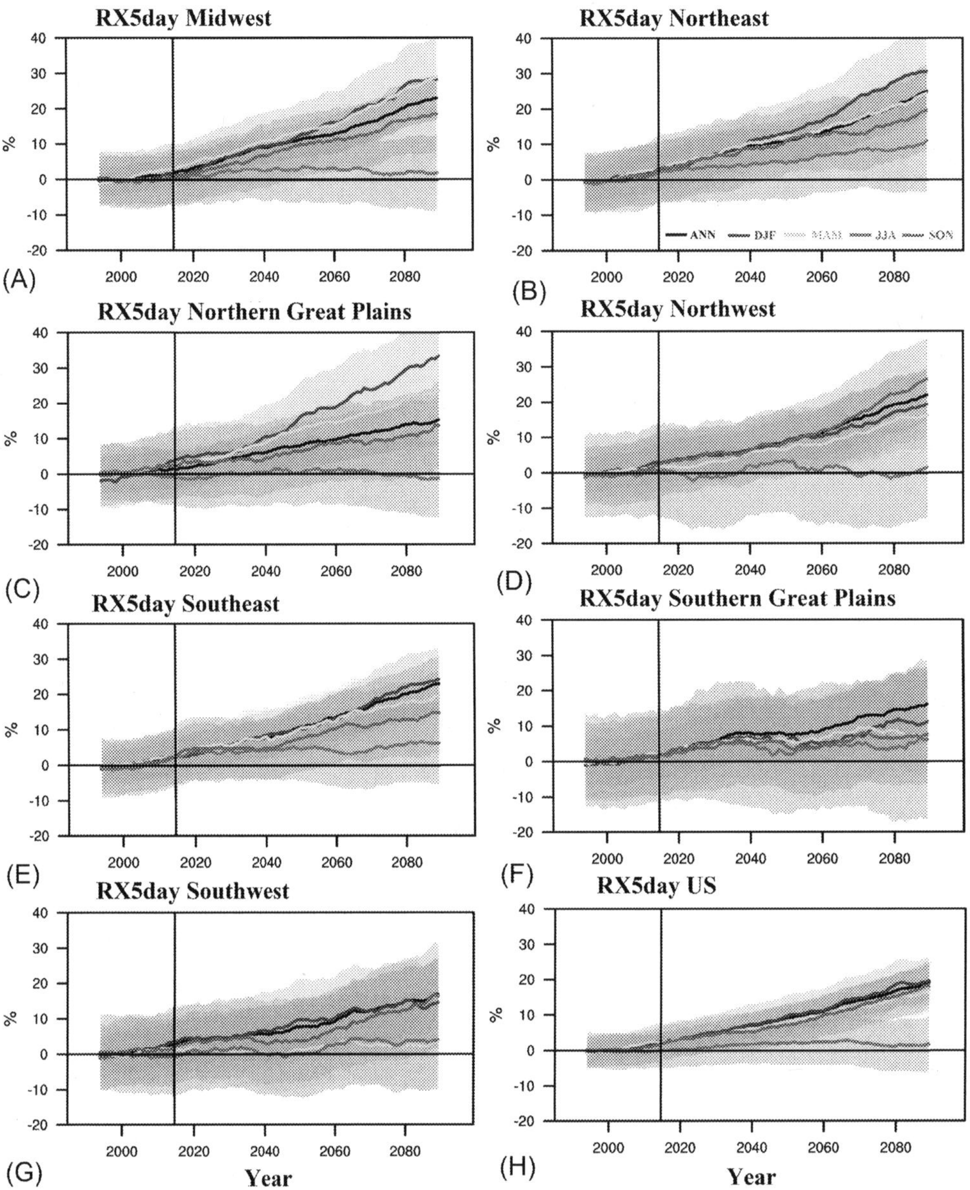

FIG. 7.7 Same as Fig. 7.5 but for maximum 5-day precipitation (RX5day).

Acknowledgments

The authors acknowledge the support of the US Department of Energy (DOE) Regional and Global Model Analysis (RGMA) Program (DE-SC0019459 and DE-SC0021209). We acknowledge the World Climate Research Program's Working Group on Coupled Modeling, which is responsible for coordinating CMIP6. Special appreciation goes to the climate modeling groups listed in Table 7.1 for producing and making available their model outputs and ESGF for archiving the model outputs and providing access. We are also grateful to the services that have operated the CPC dataset.

Data availability statement

The data that support the findings of this study are openly available at the following URL/DOI: https://esgf-node.ipsl.upmc.fr/projects/cmip6-ipsl/ and https://psl.noaa.gov/data/gridded/data.unified.daily.conus.html.

References

Agha Kouchak, A., Cheng, L., Mazdiyasni, O., & Farahmand, A. (2014). Global warming and changes in risk of concurrent climate extremes: Insights from the 2014 California drought. *Geophysical Research Letters, 41*, 8847–8852. https://doi.org/10.1002/2014GL062308.

Akinsanola, A. A., Kooperman, G. J., Reed, K. A., Pendergrass, A. G., & Hannah, W. M. (2020). Projected changes in seasonal precipitation extremes over the United States in CMIP6 simulations. *Environmental Research Letters, 15*, 104078. https://doi.org/10.1088/17489326/abb397.

Akinsanola, A. A., & Zhou, W. (2019). Projections of West African summer monsoon rainfall extremes from two CORDEX models. *Climate Dynamics, 52*, 2017. https://doi.org/10.1007/s0038 2-018-4238-8.

Akinsanola, A. A., et al. (2020). Seasonal representation of extreme precipitation indices over the United States in CMIP6 present-day simulations. *Environmental Research Letters, 15*, 094003. https://doi.org/10.1088/17489326/ab92c1.

Amarnath, G., Yoshimoto, S., Goto, O., Fujihara, M., Smakhtin, V., Aggarwal, P. K., et al. (2017). *Global trends in water-related disasters using publicly available database for hazard and risk assessment.* https://cgspace.cgiar.org/bitstream/handle/10568/93032/H048407.pdf. (Accessed August 2019).

Boucher, O., Denvil, S., Caubel, A., & Foujols, M. A. (2018). *IPSL IPSL-CM6A-LR model output prepared for CMIP6 CMIP historical.* https://doi.org/10.22033/ESGF/CMIP6.5195.

Boucher, O., Denvil, S., Caubel, A., & Foujols, M. A. (2019). *IPSL IPSL-CM6A-LR model output prepared for CMIP6 ScenarioMIP ssp585.* Earth System Grid Federation. https://doi.org/10.22033/ESGF/CMIP6.5271.

Cao, J. (2019). *NUIST NESMv3 model output prepared for CMIP6 ScenarioMIP ssp585.* Earth System Grid Federation. https://doi.org/10.22033/ESGF/CMIP6.8790.

Cao, J., & Wang, B. (2019). *NUIST NESMv3 model output prepared for CMIP6 CMIP historical.* Earth System Grid Federation. https://doi.org/10.22033/ESGF/CMIP6.8769.

Chen, C., & Knutson, T. (2008). On the verification and comparison of extreme rainfall indices from climate models. *Journal of Climate, 21*(7), 1605–1621.

Cho, C., Li, R., Wang, S. Y., Yoon, J. H., & Gillies, R. (2016). Anthropogenic footprint of climate change in the June 2013 northern Indian flood. *Climate Dynamics, 46*, 797–805.

Dai, A. (2013). Increasing drought under global warming in observations and models. *Nature Climate Change, 3*, 52–58.

Danabasoglu, G. (2019a). *NCAR CESM2-WACCM model output prepared for CMIP6 CMIP historical.* Earth System Grid Federation. https://doi.org/10.22033/ESGF/CMIP6.10071.

Danabasoglu, G. (2019b). *NCAR CESM2-WACCM model output prepared for CMIP6 ScenarioMIP ssp585.* Earth System Grid Federation. https://doi.org/10.22033/ESGF/CMIP6.10115.

Dix, M., Bi, D., Dobrohotoff, P., Fiedler, R., et al. (2019a). *CSIRO-ARCCSS ACCESS-CM2 model output prepared for CMIP6 CMIP historical.* Earth System Grid Federation. https://doi.org/10.22033/ESGF/CMIP6.4271.

Dix, M., Bi, D., Dobrohotoff, P., Fiedler, R., et al. (2019b). *CSIRO-ARCCSS ACCESS-CM2 model output prepared for CMIP6 ScenarioMIP ssp585.* Earth System Grid Federation. https://doi.org/10.22033/ESGF/CMIP6.4332.

EC-Earth Consortium (EC-Earth). (2019a). *EC-Earth-Consortium EC-Earth3 model output prepared for CMIP6 CMIP historical.* Earth System Grid Federation. https://doi.org/10.22033/ESGF/CMIP6.4700.

EC-Earth Consortium (EC-Earth). (2019b). *EC-Earth-Consortium EC-Earth3 model output prepared for CMIP6 ScenarioMIP ssp585.* Earth System Grid Federation. https://doi.org/10.22033/ESGF/CMIP6.4912.

EC-Earth Consortium (EC-Earth). (2019c). *EC-Earth-Consortium EC-Earth3-Veg model output prepared for CMIP6 CMIP historical.* Earth System Grid Federation. https://doi.org/10.22033/ESGF/CMIP6.4706.

EC-Earth Consortium (EC-Earth). (2019d). *EC-Earth-Consortium EC-Earth3-Veg model output prepared for CMIP6 ScenarioMIP ssp585.* Earth System Grid Federation. https://doi.org/10.22033/ESGF/CMIP6.4914.

EC-Earth Consortium (EC-Earth). (2020a). *EC-Earth-Consortium EC-Earth3-Veg-LR model output prepared for CMIP6 CMIP historical.* Earth System Grid Federation. https://doi.org/10.22033/ESGF/CMIP6.4707.

EC-Earth Consortium (EC-Earth). (2020b). *EC-Earth-Consortium EC-Earth3-Veg-LR model output prepared for CMIP6 ScenarioMIP ssp585.* Earth System Grid Federation. https://doi.org/10.22033/ESGF/CMIP6.4915.

EC-Earth Consortium (EC-Earth). (2021a). *EC-Earth-Consortium EC-Earth-3-CC model output prepared for CMIP6 CMIP historical.* Earth System Grid Federation. https://doi.org/10.22033/ESGF/CMIP6.4702.

EC-Earth Consortium (EC-Earth). (2021b). *EC-Earth-Consortium EC-Earth3-CC model output prepared for CMIP6 ScenarioMIP ssp585.* Earth System Grid Federation. https://doi.org/10.22033/ESGF/CMIP6.15636.

Eyring, V., Bony, S., Meehl, G. A., Senior, C. A., Stevens, B., Stouffer, R. J., & Taylor, K. E. (2016). Overview of the coupled model intercomparison project phase 6 (CMIP6) experimental design and organization. *Geoscientific Model Development, 9*, 1937–1958.

Fischer, E. M., & Knutti, R. (2016). Observed heavy precipitation increase confirms theory and early models. *Nature Climate Change, 6*(11), 986.

Haberlie, A. M., & Ashley, W. S. (2019). A radar-based climatology of mesoscale convective systems in the United States. *Journal of Climate, 32*(5), 1591–1606.

Janssen, E., Sriver, R. L., Wuebbles, D. J., & Kunkel, K. E. (2016). Seasonal and regional variations in extreme precipitation event frequency using CMIP5. *Geophysical Research Letters, 43*, 5385–5393. https://doi.org/10.1002/2016GL069151.

Janssen, E., Wuebbles, D. J., Kunkel, K. E., Olsen, S. C., & Goodman, A. (2014). Observational- and model-based trends and projections of extreme precipitation over the contiguous United States. *Earth's Future, 2*, 99–113. https://doi.org/10.1002/2013EF000185.

Jones, P. W. (1999). First- and second-order conservative remapping schemes for grids in spherical coordinates. *Monthly Weather Review, 127*, 2204–2210.

Jongman, B., Hochrainer-Stigler, S., Feyen, L., Aerts, J. C., Mechler, R., Botzen, W. W., et al. (2014). Increasing stress on disaster-risk finance due to large floods. *Nature Climate Change, 4*, 264–268.

Jonkman, S. N. (2005). Global perspectives on loss of human life caused by floods. *Natural Hazards, 34*, 151–175.

Jungclaus, J., Bittner, M., Wieners, K.-H., Wachsmann, F., et al. (2019). *MPI-M MPI-ESM1.2-HR model output prepared for CMIP6 CMIP historical*. Earth System Grid Federation. https://doi.org/10.22033/ESGF/CMIP6.6594.

Kharin, V. V., Zwiers, F. W., Zhang, X., & Wehner, M. (2013). Changes in temperature and precipitation extremes in the CMIP5 ensemble. *Climatic Change, 119*(2), 345–357.

Kirchmeier-Young, M. C., & Zhang, X. (2020). Human influence has intensified extreme precipitation in North America. *Proceedings of the National Academy of Sciences, 117*(24), 13308–13313. https://doi.org/10.1073/pnas.1921628117.

Klein Tank, A. M. G., Zwiers, F. W., & Zhang, X. (2009). *Guidelines on analysis of extremes in a changing climate in support of informed decision for adaptation*. World Meteorological Organization. WMO/TD-No. 1500; WCDMP-No. 72.

Kunkel, K. E., & Champion, S. M. (2019). An assessment of rainfall from Hurricanes Harvey and florence relative to other extremely wet storms in the United States. *Geophysical Research Letters, 46*, 13500–13506. https://doi.org/10.1029/2019GL085034.

Lau, W. K. M., & Kim, K. (2012). The 2010 Pakistan flood and Russian heat wave: Teleconnection of hydrometeorological extremes. *Journal of Hydrometeorology, 13*, 392–403.

Lavers, D. A., & Villarini, G. (2013). Atmospheric Rivers and flooding over the Central United States. *Journal of Climate, 26*(20), 7829–7836.

Lee, W.-L., & Liang, H.-C. (2020a). *AS-RCEC TaiESM1.0 model output prepared for CMIP6 CMIP historical*. Earth System Grid Federation. https://doi.org/10.22033/ESGF/CMIP6.9755.

Lee, W.-L., & Liang, H.-C. (2020b). *AS-RCEC TaiESM1.0 model output prepared for CMIP6 ScenarioMIP ssp585*. Earth System Grid Federation. https://doi.org/10.22033/ESGF/CMIP6.9823.

Lehmann, J., Coumou, D., & Frieler, K. (2015). Increased record-breaking precipitation events under global warming. *Climatic Change, 132*(4), 501–515.

Li, L. C., Zou, Y. F., Li, Y., Lin, H. X., Liu, D. L., Wang, B., … Song, S. B. (2020). Trends, change points and spatial variability in extreme precipitation events from 1961 to 2017 in China. *Hydrology Research, 51*(3), 484–504.

Liu, M., & Smith, J. A. (2016). Extreme rainfall from landfalling tropical cyclones in the Eastern United States: Hurricane Irene (2011). *Journal of Hydrometeorology, 17*(11), 2883–2904.

Lopez-Cantu, T., Prein, A. F., & Samaras, C. (2020). Uncertainties in future U.S. extreme precipitation from downscaled climate projections. *Geophysical Research Letters, 47*. https://doi.org/10.1029/2019GL086797. e2019GL086797.

Lovato, T., & Peano, D. (2020a). *CMCC CMCC-CM2-SR5 model output prepared for CMIP6 CMIP historical*. Earth System Grid Federation. https://doi.org/10.22033/ESGF/CMIP6.3825.

Lovato, T., & Peano, D. (2020b). *CMCC CMCC-CM2-SR5 model output prepared for CMIP6 ScenarioMIP ssp585*. Earth System Grid Federation. https://doi.org/10.22033/ESGF/CMIP6.3896.

Lovato, T., Peano, D., & Butenschön, M. (2021a). *CMCC CMCC-ESM2 model output prepared for CMIP6 CMIP historical*. Earth System Grid Federation. https://doi.org/10.22033/ESGF/CMIP6.13195.

Lovato, T., Peano, D., & Butenschön, M. (2021b). *CMCC CMCC-ESM2 model output prepared for CMIP6 ScenarioMIP ssp585*. Earth System Grid Federation. https://doi.org/10.22033/ESGF/CMIP6.13259.

Milly, P., Wetherald, R., Dunne, K. A., & Delworth, T. L. (2002). Increasing risk of great floods in a changing climate. *Nature, 415*, 514–517.

Mora, C., Spirandelli, D., Franklin, E. C., Lynham, J., Kantar, M. B., Miles, W., Smith, C. Z., et al. (2018). Broad threat to humanity from cumulative climate hazards intensified by greenhouse gas emissions. *Nature Climate Change, 8*, 1062–1071. https://doi.org/10.1038/s41558-018-0315-6.

Noor, M., Ismail, T., Shahid, S., & Nashwan, M. S. (2020). Development of multi-model ensemble for projection of extreme rainfall events in peninsular Malaysia. *Hydrology Research, 50*(6), 1772–1788.

O'Neill, B. C., et al. (2016). The scenario model intercomparison project (ScenarioMIP) for CMIP6. *Geoscientific Model Development, 9*, 3461–3482.

Prein, A. F., Rasmussen, R. M., Ikeda, K., Liu, C., Clark, M. P., & Holland, G. J. (2017). The future intensification of hourly precipitation extremes. *Nature Climate Change, 7*(1), 48–52. https://doi.org/10.1038/nclimate3168.

Reed, K., Wehner, M. F., Stansfield, A. M., & Zarzycki, C. M. (2021). Anthropogenic influence on Hurricane Dorian's extreme rainfall. *Bulletin of the American Meteorological Society, 102*(1), S9–S15.

Schaller, N., Kay, A. L., Lamb, R., Massey, N. R., van Oldenborgh, G. J., Otto, F. E. L., Sparrow, S. N., et al. (2016). Human influence on climate in the 2014 southern England winter floods and their impacts. *Nature Climate Change, 6*, 627–634. https://doi.org/10.1038/nclimate2927.

Schupfner, M., Wieners, K.-H., Wachsmann, F., Steger, C., et al. (2019). *DKRZ MPI-ESM1.2-HR model output prepared for CMIP6 ScenarioMIP ssp585*. Earth System Grid Federation. https://doi.org/10.22033/ESGF/CMIP6.4403.

Scoccimarro, E., Gualdi, S., Bellucci, A., Zampieri, M., & Navarra, A. (2013). Heavy precipitation events in a warmer climate: Results from CMIP5 models. *Journal of Climate, 26*(20), 7902–7911.

Seneviratne, S. I., et al. (2012). Changes in climate extremes and their impacts on the natural physical environment. In *Managing the risks of extreme events and disasters to advance climate change adaptation, a special report of working groups I and II of the intergovernmental panel on climate change (IPCC), rep.* (pp. 109–230). Cambridge, UK, and New York: Cambridge University Press.

Shiogama, H., Abe, M., & Tatebe, H. (2019). *MIROC MIROC6 model output prepared for CMIP6 ScenarioMIP ssp585*. Earth System Grid Federation. https://doi.org/10.22033/ESGF/CMIP6.5771.

Sillmann, J., Kharin, V., Zwiers, F., Zhang, X., & Bronaugh, D. (2013). Climate extremes indices in the CMIP5 multimodel ensemble: Part 2. Future climate projections. *Journal of Geophysical Research: Atmospheres, 118*, 2473–2493. https://doi.org/10.1002/jgrd.50188.

Sillmann, J., Kharin, V. V., Zhang, X., Zwiers, F. W., & Bronaugh, D. (2013). Climate extremes indices in the CMIP5 multimodel ensemble: Part 1. Model evaluation in the present climate. *Journal of Geophysical Research, 118*, 1716–1733.

Slinskey, E. A., Loikith, P. C., Waliser, D. E., Guan, B., & Martin, A. (2020). A climatology of atmospheric rivers and associated precipitation for the seven U.S. national climate assessment regions. *Journal of Hydrometeorology, 21*(11), 2439–2456.

Smith, A. B., & Katz, R. W. (2013). US billion-dollar weather and climate disasters: Data sources, trends, accuracy and biases. *Natural Hazards, 67*, 387–410.

Srivastava, A. K., Richard, G., & Paul, A. U. (2020). Evaluation of historical CMIP6 model simulations of extreme precipitation over contiguous US regions. *Weather and Climate Extremes*, 100268. https://doi.org/10.1016/j.wace.2020.100268.

Swart, N. C., Cole, J. N. S., Kharin, V. V., Lazare, M., Scinocca, J. F., Gillett, N. P., et al. (2019a). *CCCma CanESM5 model output prepared for CMIP6 CMIP historical*. Earth System Grid Federation. https://doi.org/10.22033/ESGF/CMIP6.3610.

Swart, N. C., Cole, J. N. S., Kharin, V. V., Lazare, M., Scinocca, J. F., Gillett, N. P., et al. (2019b). *CCCma CanESM5 model output prepared for CMIP6 ScenarioMIP ssp585*. Earth System Grid Federation. https://doi.org/10.22033/ESGF/CMIP6.3696.

Tatebe, H., & Watanabe, M. (2018). *MIROC MIROC6 model output prepared for CMIP6 CMIP historical*. Earth System Grid Federation. https://doi.org/10.22033/ESGF/CMIP6.5603.

Trenberth, K. E., Dai, A., van der Schrier, G., Jones, P. D., Barichivich, J., Briffa, K. R., et al. (2014). Global warming and changes in drought. *Nature Climate Change*, *4*, 17–22.

Uranchimeg, S., Kwon, H.-H., Kim, B., & Kim, T.-W. (2020). Changes in extreme rainfall and its implications for design rainfall using a Bayesian quantile regression approach. *Hydrology Research*, *51*(4), 699–719.

USGCRP. (2014). In J. M. Melillo, T. C. Richmond, & G. W. Yohe (Eds.), *Climate change impacts in the United States: The third National Climate Assessment*. Washington, DC, USA: U.S. Global Change Research Program (USGCRP). http://nca2014.globalchange.gov.

USGCRP. (2017). In D. J. Wuebbles, D. W. Fahey, K. A. Hibbard, D. J. Dokken, B. C. Stewart, & T. K. Maycock (Eds.), *Vol. I. Climate science special report: Fourth National Climate Assessment*. Washington, DC, USA: U.S. Global Change Research Program. https://doi.org/10.7930/J0J964J6. 470 pp.

Volodin, E., Mortikov, E., Gritsun, A., Lykossov, V., et al. (2019a). *INM INM-CM4-8 model output prepared for CMIP6 CMIP historical*. Earth System Grid Federation. https://doi.org/10.22033/ESGF/CMIP6.5069.

Volodin, E., Mortikov, E., Gritsun, A., Lykossov, V., et al. (2019b). *INM INM-CM4-8 model output prepared for CMIP6 ScenarioMIP ssp585*. Earth System Grid Federation. https://doi.org/10.22033/ESGF/CMIP6.12337.

Volodin, E., Mortikov, E., Gritsun, A., Lykossov, V., et al. (2019c). *INM INM-CM5-0 model output prepared for CMIP6 CMIP historical*. Earth System Grid Federation. https://doi.org/10.22033/ESGF/CMIP6.5070.

Volodin, E., Mortikov, E., Gritsun, A., Lykossov, V., et al. (2019d). *INM INM-CM5-0 model output prepared for CMIP6 ScenarioMIP ssp585*. Earth System Grid Federation. https://doi.org/10.22033/ESGF/CMIP6.12338.

Wang, J., Wang, H. J., & Hong, Y. (2016). Comparison of satellite-estimated and model-forecasted rainfall data during a deadly debris-flow event in Zhouqu, Northwest China. *Atmospheric and Oceanic Science Letters*, *9*, 139–145.

Westra, S., Alexander, L. V., & Zwiers, F. W. (2013). Global increasing trends in annual maximum daily precipitation. *Journal of Climate*, *26*(11), 3904–3918.

Wieners, K.-H., Giorgetta, M., Jungclaus, J., Reick, C., et al. (2019a). *MPI-M MPI-ESM1.2-LR model output prepared for CMIP6 CMIP historical*. Earth System Grid Federation. https://doi.org/10.22033/ESGF/CMIP6.6595.

Wieners, K.-H., Giorgetta, M., Jungclaus, J., Reick, C., et al. (2019b). *MPI-M MPI-ESM1.2-LR model output prepared for CMIP6 ScenarioMIP ssp585*. Earth System Grid Federation. https://doi.org/10.22033/ESGF/CMIP6.6705.

Wuebbles, D., et al. (2014). CMIP5 climate model analyses: Climate extremes in the United States. *Bulletin of the American Meteorological Society*, *95*, 571–583. https://doi.org/10.1175/BAMS-D-12-00172.1.

Yukimoto, S., Koshiro, T., Kawai, H., et al. (2019a). *MRI MRI-ESM2.0 model output prepared for CMIP6 CMIP historical*. https://doi.org/10.22033/ESGF/CMIP6.6842.

Yukimoto, S., Koshiro, T., Kawai, H., et al. (2019b). *MRI MRI-ESM2.0 model output prepared for CMIP6 ScenarioMIP ssp585*. Earth System Grid Federation. https://doi.org/10.22033/ESGF/CMIP6.6929.

Zhang, W., Zhou, T., Zou, L., et al. (2018). Reduced exposure to extreme precipitation from 0.5 °C less warming in global land monsoon regions. *Nature Communications*, *9*, 3153. https://doi.org/10.1038/s41467-018-05633-3.

Zhang, X. B., Alexander, L., Hegerl, G. C., Jones, P., Tank, A. K., Peterson, T. C., … Zwiers, F. W. (2011). Indices for monitoring changes in extremes based on daily temperature and precipitation data wires. *Climatic Change*, *2*, 851–870. https://doi.org/10.1002/wcc.147.

Zhou, B. T., Wen, H. Q. Z., Xu, Y., & Song, C. L. (2014). Projected changes in temperature and precipitation extremes in China by the CMIP5 multi-model ensembles. *Journal of Climate*, *27*, 659–6611. https://doi.org/10.1175/JCLI-D-13-00761.1.

Ziehn, T., Chamberlain, M., Lenton, A., et al. (2019a). *CSIRO ACCESS-ESM1.5 model output prepared for CMIP6 CMIP historical*. Earth System Grid Federation. https://doi.org/10.22033/ESGF/CMIP6.4272.

Ziehn, T., Chamberlain, M., Lenton, A., et al. (2019b). *CSIRO ACCESS-ESM1.5 model output prepared for CMIP6 ScenarioMIP ssp585*. Earth System Grid Federation. https://doi.org/10.22033/ESGF/CMIP6.4333.

CHAPTER

8

A review on observed historical changes in hydroclimatic extreme events over Europe

Kristian Förster and Larissa Nora van der Laan

Institute for Hydrology and Water Resources Management, Leibniz University Hannover, Hannover, Germany

1 Climate variability in the past millennium

The past millennium has seen two remarkable changes in climatic conditions, suggesting a subdivision into three phases (Fig. 8.1 gives an overview of most relevant quantities and their evolvement over time, while Table 8.1 compiles the data sources):

(i) The Medieval Climate Anomaly (MCA, until ~1250CE): In this period, temperatures were above average (in comparison to the entire millennium 1000CE to 2000CE), suggesting that viniculture was even possible in England and Norway (Sirocko, 2009). In this period, total solar irradiance was higher than average (Lean, 2018). Summers were warm and dry as a consequence of higher persistence of high pressure zones through blocking conditions (Mann et al., 2009). Consequently, MCA was subject to dryer conditions when compared to today's climate, which has been reconstructed from tree ring analyses (dendroclimatic reconstruction) and is characterized by the coinciding dryness in Europe and regions affected by El Niño Southern Oscillation (ENSO). Especially the coupled atmosphere–ocean system of ENSO reflected much smaller variability than today with no El Niño related flooding in Peru (Helama, Meriläinen, & Tuomenvirta, 2009; Rein, Luckge, & Sirocko, 2004). At the same time, the Multidecadal Atlantic Oscillation was subject to positive anomalies in the MCA, which explains the high persistence of high pressure zones and this coincidence of dryness in the Pacific region and Europe is related to teleconnections between ENSO and AMO. Another important driver for positive temperature anomalies in Europe is that northward monsoonal transport was highest, which is in line with increased meridional transport of energy (Sirocko, 2009).

(ii) The Little Ice Age (LIA) followed the MCA. As the name of this period suggests, cold temperatures prevailed throughout this period, which extends from ~1350CE to ~1850CE. Major drivers of change are a decrease in sun activity and volcanic activity (Fig. 8.1A). The Wolf minimum of solar activity (reconstructed from Carbon and Beryllium isotopes) is one of the major minima of solar activity in the past millennium. As a consequence, temperature decreased throughout the 13th century and the increase in extreme coincides with the decrease in solar activity and volcanic eruption (Fig. 8.1C). Around 1400CE, temperatures increased again but did not reach the level of the MCA. Instead a second major minimum of solar irradiance was observed, the Spörer minimum. Even though first half of the 16th century was subject to only small negative temperature anomalies, the second cold phase of the LIA started around 1560, which was likely sustained through the occurrence of the third minimum of solar activity (Maunder minimum). Lower irradiance triggered a southward shift of precipitation-bearing westerlies during winter (Ait Brahim et al., 2018). As a consequence, winters were cold and as discussed later, the intensified accumulation of snow led to more intensive snowmelt floods in spring. At the same time glaciers expanded (Fig. 8.1E). The last decade of the 20th century reached temperatures peaks that have only been reconstructed only 14 times before, in the last millennium (Guiot et al., 2005), suggesting that climate conditions markedly changed after the end of the LIA.

Climate Impacts on Extreme Weather
https://doi.org/10.1016/B978-0-323-88456-3.00015-0

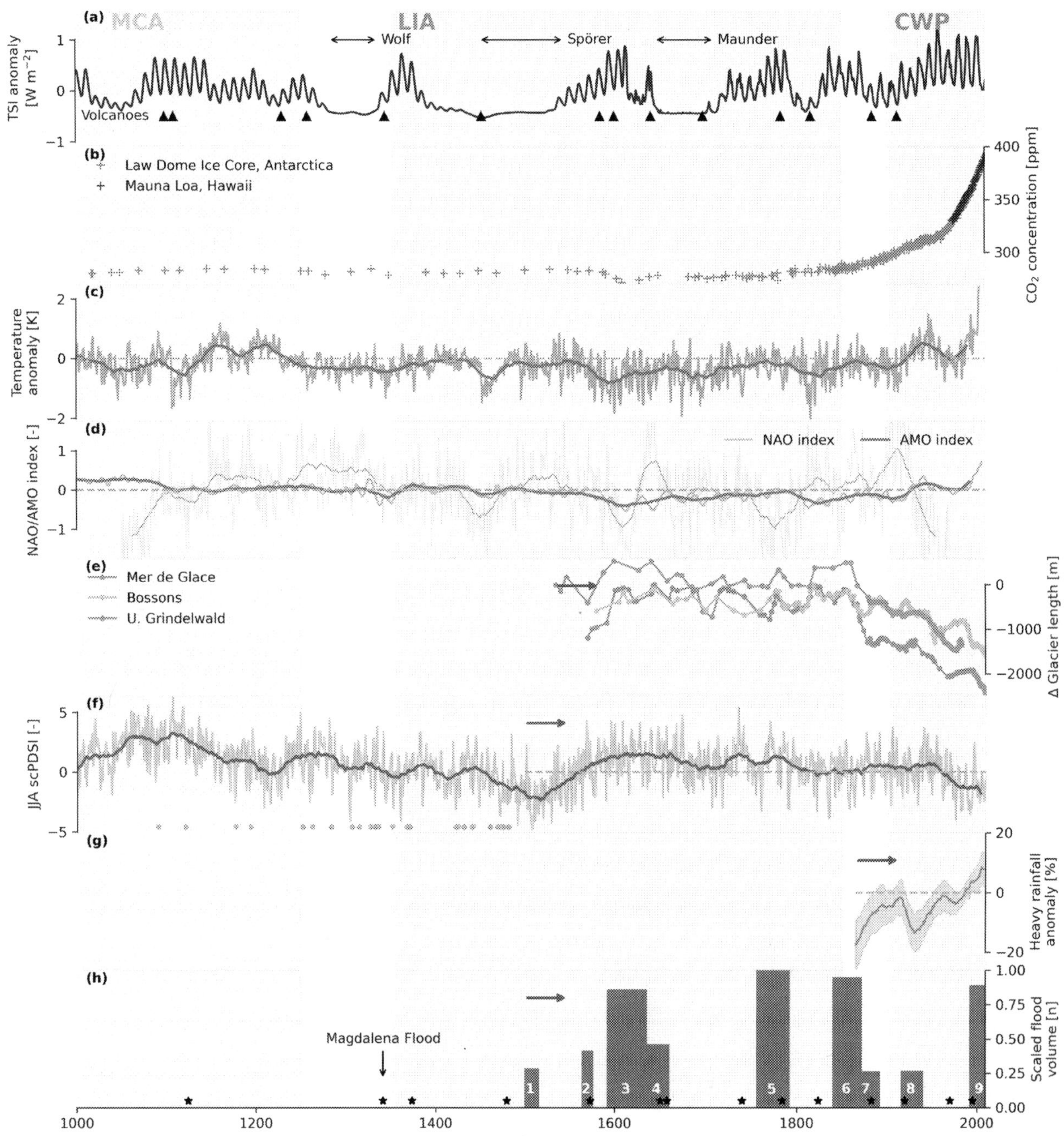

FIG. 8.1 **The past millennium and hydroclimatic extremes.** Changes in hydroclimatic extremes in the past millennium, including (from upside-down): (A) Total Solar Irradiance (TSI), (B) CO2 concentration, (C) air temperature, (D) Atlantic Multidecadal Oscillation (AMO) index and North Atlantic Oscillation (NAO) index, (E) changes in glacier length, (F) Self-calibrated JJA Palmer Drought Severity Index (scPDSI), (G) changes in 10-min heavy precipitation percentiles, and (H) scaled volume of flood-rich periods. Please refer to Table 8.1 for details on data sources. *No permission required.*

(iii) The Current Warm Period (CWP, since 1850) represents the most recent period and extends until now. It is mostly characterized by a temperature increase that is associated to the rise in CO_2 concentration (Fig. 8.1B) and hence to anthropogenic forcing. While the first part of this period is mostly linked to a temperature increase related to increased solar activity, anthropogenic forcing is the major driver that determines the unprecedented rise in air temperature in the most recent decade. The radiative forcing effect on air temperature associated to greenhouse gas emissions amounts to ~0.6 in the period 1951–2010 (Jones, Stott, & Christidis, 2013). For this period our knowledge about hydroclimatic extremes is best, since it coincides with the instrumental period, for which direct

TABLE 8.1 Sources of data.

Hydroclimatic quantity	Source
(a) Total Solar Irradiance ($W m^{-2}$), Volcanoes	Lean (2018), Schneider, Smerdon, Pretis, Hartl-Meier, and Esper (2017)
(b) CO2 concentration (ppm)	Rubino et al. (2019) and Thoning, Crotwell, and Mund (2021)
(c) Air temperature (°C)	PAGES 2k Consortium (2013)
(d) Atlantic Multidecadal Oscillation (AMO) index and North Atlantic Oscillation (NAO) index	Mann et al. (2009), Ortega et al. (2015)
(e) Changes in glacier length (m)	Leclercq et al. (2014)
(f) Palmer Drought Severity Index (PDSI) (–), droughts in Germany and Czech Republic	Büntgen et al. (2021), Glaser and Kahle (2020), Brázdil et al. (2013)
(g) Anomalies in high percentiles of heavy precipitation (%)	Förster and Thiele (2020)
(g) Scaled flood volume in flood-rich periods (–)	Blöschl et al. (2020)

measurements of numerous hydroclimatic quantities exist, like, for instance, rainfall intensity, water levels and discharge as well as glacier mass balances. The temperature increase in the 20th century relates to 98% of the globe and this magnitude of spatially coherent warming has not been observed in the past two millennia (Neukom, Steiger, Gómez-Navarro, Wang, & Werner, 2019).

2 Precipitation extremes

Short and intensive rainfall events can cause severe flooding (pluvial floods) affecting, e.g., small creeks which might turn into torrents or urban areas whose drainage structure is not capable of draining rainfall events that exceed the intensity of typical design rainfall events. Precipitation extremes are expected to change either as a response to changes in circulation patterns or to changes in temperature. A lot of research has been carried out to unravel the relationship between temperature and rainfall which follows the Clausius–Clapeyron (CC) equation suggesting that rainfall extremes increase by 7% per °C of warming (Berg et al., 2009; Lenderink & Van Meijgaard, 2008; Zeder & Fischer, 2020) or even exceeding this rate (Berg, Moseley, & Haerter, 2013; Lenderink, Barbero, Loriaux, & Fowler, 2017; Westra et al., 2014). The latter is called super CC scaling which can be explained by an amplified atmospheric moisture convergence holding responsible for this increase in scaling (Tabari, 2020). Some authors also found a decrease in rainfall rates above 24°C, which might be associated to limited moisture availability or arid surface conditions not considered by the CC and super CC scaling, respectively (Lenderink, Mok, Lee, & Van Oldenborgh, 2011; Westra et al., 2014). Other researchers argue that this decrease could be explained by undersampling, i.e., a small sample size of events (Boessenkool, Bürger, & Heistermann, 2017).

In contrast to other past quantities, reconstructing past precipitation is challenging, especially when considering short temporal scales. While rainfall events on larger temporal and spatial scales (e.g., over weeks and spatial extent at the catchment scale) can be reconstructed from river bed or lake sediments, like, e.g., the Magdalena flood in 1342 (Herget et al., 2015), less information are available for small-scale events, like thunderstorms. While it is virtually impossible to reconstruct quantitative rainfall estimates from different sources of historic documentary, first recording rain gages emerged in the late 19th century (Kurtyka & Madow, 1952; Strangeways, 2010), which allowed a continuous of rainfall intensities with small temporal increments (e.g., subdaily recordings). Fig. 8.2 shows an example plot of a recording rain gage, which plots mass curves on paper. Even early devices were capable to reconstruct rainfall events on subdaily increments of time, like, e.g., the rainfall station at Uccle, Belgium, which provides subdaily rainfall recordings from 1898 onward (Demarée, 2003). However, this station is a remarkable exception, which is why most studies focusing on past changes in heavy precipitation frequency and magnitude consider daily recordings. These historical daily recordings of precipitation are more readily available, since rain gages haven been emptied and read each day, but they mostly also cover the second half of the CWP. Therefore, instrumental evidence on historical changes in extreme precipitation is limited to the most recent decades in the CWP and only some combined observational and modeling studies allow for extending the study period further into the past.

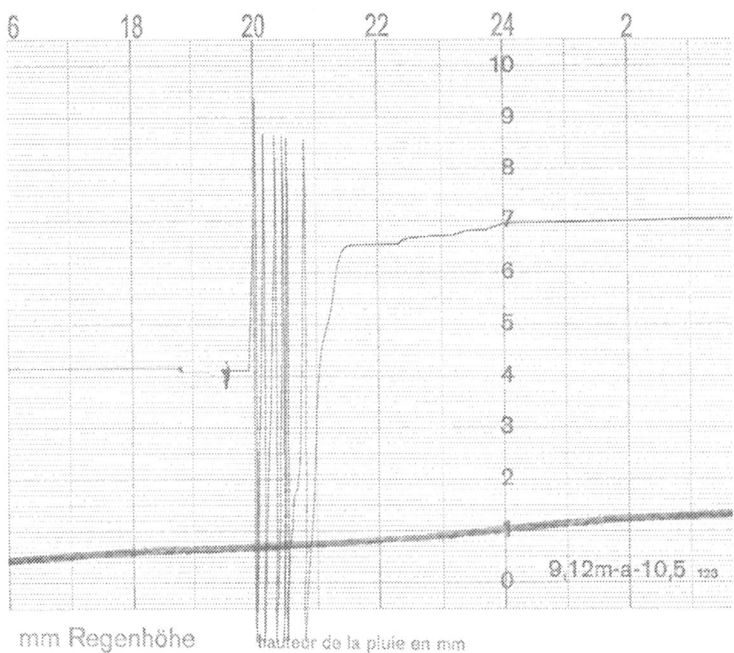

FIG. 8.2 Sample plot of mechanically recording rain gage. Sample plot of registering rain gage. These types of recordings were common types of recordings before digital recordings techniques emerged. Source: Own measurement at 51.00 N, 9.15E on May 30, 2008. The maximum hourly precipitation intensity amounts to 57.4 mm/h, which corresponds to a 20 years return period, and the precipitation total is 64.4 mm. *No permission required.*

For instance, Łupikasza (2017) analyses daily precipitation recordings from 1950 onward across Europe and in all seasons positive significant trends are identified in all seasons, expect for summer when positive and negative trends have been reported similarly (Łupikasza, 2017; Zolina, 2012). The study moreover concludes what has been already identified as most challenging in analyzing changes in extreme over time, more specifically that "significant extreme precipitation trends [are] rare" (Łupikasza, 2017). This can be related to relatively short record lengths in comparison to temperature, which is obvious from the brief summary on trends in heavy precipitation based on observational data. Moreover, upward trends are detectable for all seasons, except for summer when negative trends are found to the same extent (Łupikasza, 2017; Zolina et al., 2014). Similar analyses considering subdaily data exist, especially those which focus on short-term rainfall that might cause erosion. These studies confirm an overall increasing trend in frequency and magnitude over Europe also for subdaily time series in Czechia and Germany, respectively (Hanel, Pavlásková, & Kyselý, 2016; Mueller & Pfister, 2011). Mueller and Pfister report an increase in 0.5 events per year increase for the highest intensity class. In contrast to the trend analyses that focus on rainfall extremes of daily recordings, which also reveal negative trends in summer (Łupikasza, 2017; Zolina et al., 2014), the authors found that the most pronounced increase of heavy precipitation events, relevant for erosion (i.e., subdaily scale), occurred in summer (Mueller & Pfister, 2011).

Other work also acknowledges the fact that natural climate variability affects the detection trends (Martel, Mailhot, Brissette, & Caya, 2018). Ntegeka and Willems (2008) investigate the trends and oscillations in a 107-year time series of 10-min rainfall data from Uccle station in Belgium (Ntegeka & Willems, 2008). They consider different aggregation levels of precipitation intensities up to the monthly scale, and utilize different moving windows from 5 to 15 years for analyzing the trend and oscillations. Results show considerable changes in rainfall quantiles persisting for periods of 10 to 15 years. They also analyze the changes for different seasons and the results imply that for summer and winter, high extremes are prevailing in the 1910s–20s, 1960s, and 1990s. Willems (2013) uses the same method as in Ntegeka and Willems (2008) to further investigate the mentioned oscillations and to check the consistency of these findings using daily long rainfall records at 724 stations in Europe and the Middle East. Here, rainfall peaks over threshold are considered as extremes and therefore used in calculating the oscillations. Results show that multidecadal oscillation in the precipitation extremes exist over the Europe. The driving forces have been addressed in a more recent study, which demonstrates a correlation between NAO and ENSO winter anomalies and extreme precipitation anomalies in winter, and even a delayed effect on the following seasons is detected (Tabari & Willems, 2018).

Likewise, Lenderink et al. (2011) investigate the relationship between subdaily precipitation extremes and near surface humidity using observations from two regions with completely different climates namely Hong Kong and the Netherlands (Lenderink et al., 2011). They analyze the dependency of hourly precipitation extremes (considering the 95th, 99th, and 99.5th percentiles) on the dew point temperature. Results show that with increasing dew point temperature, an increase in hourly precipitation extremes by 14% per degree is observed, which is twice as large as the CC relation (super CC scaling). However, this dependency is only valid for dew point temperatures up to 23°C, eventually highlighting the relevance of undersampling (Boessenkool et al., 2017). Surprisingly, the results obtained for both study areas, are very similar in spite of different climates but show different phases in terms of multidecadal oscillations.

In order to extend changes in rainfall intensities back in time beyond the instrumental period of subdaily precipitation recordings, modeling studies might help to understand the variability of heavy precipitation events. For instance, Regional Climate Models (RCM) can be used to downscale atmospheric reanalysis data (Förster & Thiele, 2020). In this study, subdaily precipitation intensities in Central Europe have been obtained this way and changes in high percentiles have been derived through adopting the methods from Willems (2013) and Lenderink et al. (2011) (Fig. 8.1G). This study underlines that high percentiles in heavy precipitation (e.g., 10min duration) have increased in the past decades from the end of the LIA onward, whereby the signal is subject to multidecadal oscillations, thus confirming earlier work (Lenderink et al., 2011; Willems, 2013). The modeling results are confirmed by observational evidence. However, the modeling study allows to extend the analysis beyond the instrumental period for a few stations to larger spatial scales (Förster & Thiele, 2020). In essence, it could be shown that in the last two decades heavy precipitation intensities have been higher than ever before in the last 150years in Central Europe.

Similarly, a study unraveling millennial scale signal from General Circulation Models (GCM) reveals that the increase in heavy rainfall occurred in the CWP, similar to the findings of the previous study, while precipitation extremes reflect quasistationarity in earlier periods of the past millennium, suggesting that precipitation extremes have not changed throughout the MCA and the LIA. The increase in the CWP is therefore unprecedented in the past millennium and assumed to be linked to anthropogenic forcing (Zhang, Villarini, Scoccimarro, & Vecchi, 2017). The impact of anthropogenic effects on extreme precipitation amounts to up to 26% in Central and Northern Europe between 50° and 60° (Tabari, Madani, & Willems, 2020), while the anthropogenic influence is found to be lower in Southern Europe (8%), where extreme precipitation is mostly convective and uncertainties in modeling those events is expected to be related to higher uncertainties (Tabari et al., 2020; Tabari, Hosseinzadehtalaei, Aghakouchak, & Willems, 2019). The anthropogenic contribution changes in extreme precipitation amount to 26%, 26%, 22%, and 24% in winter, spring, summer, and autumn, respectively (Tabari et al., 2020). The lowest value occurs in summer due to the more convective events, for which the signal is masked by internal variability to a greater extent (Martel et al., 2018; Tabari et al., 2020).

However, the majority of modeling studies that aim to predict future changes in heavy precipitation provided by climate changes projections agree on the fact that heavy precipitation increases in terms of frequency and magnitude as a response of global warming and the associated (super) CC scaling effect (Ban, Schmidli, & Schär, 2015; Berg et al., 2019; Martel, Mailhot, & Brissette, 2020; Prein et al., 2017). Myhre et al. (2019) conclude that the "precipitation from these intense events almost doubles per degree of warming, mainly due to changes in frequency, while the intensity changes are relatively weak" (Myhre et al., 2019). However, especially rare events are expected to intensify more than less rare events (Hosseinzadehtalaei, Tabari, & Willems, 2020). Regarding the spatial distribution of change signals, projected changes in extreme precipitation show a bipolar pattern, characterized by an increase in Northern Europe and decrease in Southern Europe, related to thermodynamic effects (i.e., scaling) in the North and dynamic effects (i.e., changes in circulation patterns) in the South (Hosseinzadehtalaei, Tabari, & Willems, 2019; Tabari et al., 2020).

3 Glacier mass balances

Glaciers are often regarded one of the most linear recorders of climatic change; their mass changes considered a direct consequence of atmospheric variables (Rupper & Roe, 2008). This mass change, summarized in the term glacier surface mass balance (SMB), refers to the difference between glacier ablation and accumulation, and is an essential characteristic of glacier health (Gardelle, Berthier, & Arnaud, 2012; Harrison, Cox, Hock, March, & Pettit, 2009). Concurrently, it is an important parameter in the analysis of past and present climatic change. As the climate has evolved over the last millennium, so have glaciers and their seasonal cycles, mainly through the length of the accumulation and ablation periods. This seasonal cycle is essential in the global hydrological cycle, being the driver of water supply in a large part of the world: seasonal snow and glacier melt comprises the main water resource for up to 22% of the global population (Immerzeel et al., 2020; Kaser, Großhauser, & Marzeion, 2010). In the past, inhabitants of glacier-dominated regions have been spared when drought was imminent through lack of precipitation (Pritchard, 2019), such as in the 1540 drought in Central Europe (Brázdil et al., 2013; Pfister, 1999). Beyond these indirect impacts, SMB change can affect and cause other hydroclimatic events, including extremes such as droughts and floods. Single, negative, cumulative glacier SMB values have been observed periodically over the last century, with significant overall retreat since the mid-nineteenth century, in line with global temperature trends (Intergovernmental Panel on Climate Change IPCC, 2020; Roe, Baker, & Herla, 2017). In order to understand the relationship between climate, glacier surface mass balance, and hydroclimatic extremes over the last millennium, glacier and climate reconstructions are essential.

Historical glacier reconstructions are among the most ubiquitous records of climatic change in many parts of the world (Clapperton, Sugden, Birnie, & Wilson, 1989; Porter & Orombelli, 1985). These are based on historical material, ranging from pictorial documents to verbal descriptions, as well as early measurements of glacier lengths and terminus location, paleogeomorphological information (Le Quesne, Acuña, Boninsegna, Rivera, & Barichivich, 2009; Leclercq & Oerlemans, 2012; Matthews et al., 2005; Nussbaumer & Zumbühl, 2012). From these reconstructions, climatic conditions at the time are inferred, based on the assumption that glaciers respond sensitively and relatively linearly to changes in climate. This simplifies the complex relationship between glacier SMB and climate, but is found to hold largely true on the regional to global scale and in the case of temperate glaciers (Beniston, 2012; Leclercq et al., 2014). On temperate glaciers, this is because accumulation is almost solely the consequence of winter precipitation and ablation is largely driven by melt, with sublimation and condensation being negligible (Braithwaite, 1981; Sicart, Hock, & Six, 2008). The seasonal melt variability is governed by short-wave radiation, which correlates well with air temperature (Pellicciotti et al., 2008). Consequently, the changes in the dynamical system of a glacier can, to an extent, be directly related to changes in regional climate (Oerlemans & Hoogendoorn, 1989; Thibert, Eckert, & Vincent, 2013).

Prime examples of these types of reconstructions are the work by LeClercq and Oerlemans (2012) and Nussbaumer and Zumbühl (2012). The former operate on a global scale, whereas Nussbaumer and Zumbühl focus on the LIA history of the Glacier de Bossons (Mont Blanc massif, France) through glacier length reconstruction. The Glacier des Bossons is one of the most well-documented glaciers globally, with the reconstructed length record now dating back to the late 16th century. As Nussbaumer and Zumbühl also note, the Mont Blanc area includes another very well documented glacier: the Mer de Glace. The centennial length record of these two glaciers, as well as the U. Grindelwald Glacier in the Swiss Alps, are depicted in Fig. 8.1E (Leclercq et al., 2014). For the Glacier des Bossons, the team evaluates 250 historical documents, yielding a length record from 1570 to 2005, clearly depicting the LIA maximum in 1818 and various other maxima. The results for all three reconstructed glaciers are remarkably similar and in line with available temperature records, such as seen in Fig. 8.1C. LeClercq and Oerlemans (2012) use these glacier length reconstructions as part of a larger dataset of 308 glaciers, to reconstruct temperature over the last two centuries. They produce a decadal temperature proxy, entirely independent of other temperature time series. The method is based on the assumption discussed above, resulting in the use of a linear response equation and analytical glacier model. Their findings of global cumulative warming of 0.94 ± 0.31 K over the period 1830–2000 and a cumulative warming of 0.84 ± 0.35 K over the period 1600–2000 are in good agreement with observed temperature data and other, multiproxy temperature reconstructions, thus confirming the usefulness of glacier fluctuation in informing us of past climatic change. The same was concluded on a longer time-scale, focusing on the past millennium, in a comprehensive study by Goosse et al. (2018), using global climate models to drive the Open Global Glacier Model (OGGM) (Goosse et al., 2018; Maussion et al., 2019).

Rather than inferring climate from glacier reconstruction, there is a multitude of modeling efforts to reconstruct glacier mass balance from climatic records too, with methods ranging from climate data reanalysis, glacier models to complex deep learning applications (Bolibar, Rabatel, Gouttevin, & Galiez, 2020; Hanna et al., 2011; Malles & Marzeion, 2020). Regular glacier surface mass balance (SMB) observations have only become common in the 20th century, but models can be successfully validated using the few longer observational records available, as well as including previously mentioned data such as length records (Eis, van der Laan, Maussion, & Marzeion, 2021). These SMB reconstructions help us link climate minima and maxima to hydrological extremes, e.g., in the case of flooding through the bursting of proglacial lakes, in years of extremely high melt (Pfister, 1999). The research by Wilhelm et al. (2012) reconstructs a flood calendar of the past 270 years, which, through the comparison of climatic records, glacier records and paleolimnological information, suggests the relationship between flood hazard and melt peaks through higher temperatures (Wilhelm et al., 2012). The other side of the coin relates glacier mass balance to drought, which becomes a hazard after the peak water of a glacier has been reached, referring to a tipping point in a glacier with continual negative cumulative SMB, after which runoff steadily declines (Fig. 8.3) (Huss & Hock, 2018). This can cause severe drought in regions dependent on glacier runoff, as has been observed in the past (Beniston, 2012), and is expected to increase as climate change progresses under anthropogenic forcing (Marzeion, Cogley, Richter, & Parkes, 2014).

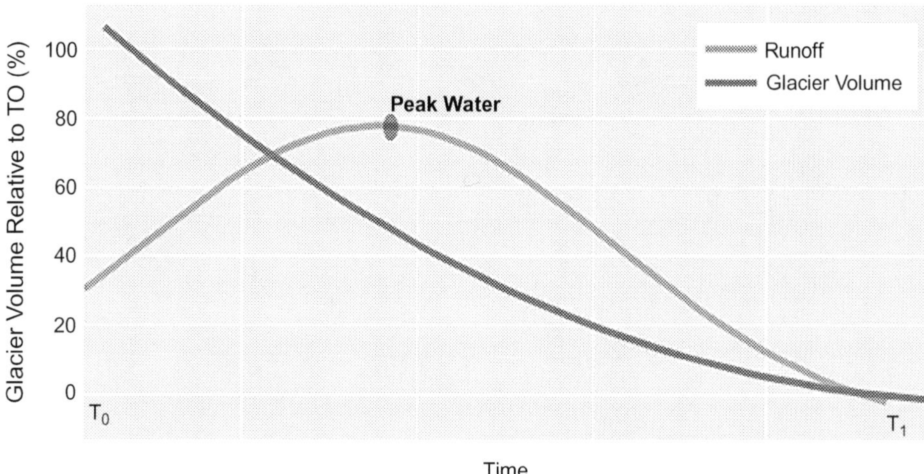

FIG. 8.3 Peak water. The volume and consequent runoff from meltwater of a hypothetical glacier over time. As the glacier loses mass (negative SMB), runoff increases until the maximum is reached (peak water), after which it steadily declines. *No permission required.*

4 Floods

Reconstructing floods in the past millennium mainly covers three different periods for which different methodological approaches exist (Brázdil, Kundzewicz, & Benito, 2006): (i) The paleo-flood hydrology is based on analyzing sediments from which a flood stratigraphy is derived. This type of data is used as proxy in order to derive floods prior to the increasing availability of documentary reports. For instance, Schulte et al. (2015) provide a 2600 years history of floods in the Bernese Alps (Switzerland) (Schulte et al., 2015). (ii) The emergence of documentary reports about floods represent the transition to historical hydrology in the 13th century (Brázdil et al., 2006). Among others, information about flooding is available in annals, chronicles, memory books and memoirs, weather diaries, correspondence (letters), special prints, official economic and administrative records as well as newspapers and journals (Brázdil et al., 2012). However, since these types of documentary are subject to possible biases due to contemporary perception, a careful review of events has been identified as crucial step toward a systematic classification of flood events, which is required to estimate their severity (Brázdil et al., 2006). For instance, the flooding perception level decreased over time due to progressive human settlement toward rivers (Brázdil et al., 2006, 2006), thus representing an increase in flood risk over time.

In medieval times, first occasional water level records complement our picture of historic floods and these observations can be related to the period of (iii) instrumental hydrology, which has seen major advance in observational methods, leading to continuous time series of hydrometric data from the 19th century onward. The different types of reconstructing historic floods highlight that especially the documentary reports requires expert knowledge to relate them to classifications in a homogeneous way (i.e., to make them comparable to instrumental in terms of magnitude and frequency). Therefore, studies that systematically unravel flooding in the past centuries rely on expert judgment regarding individual events from preinstrumental periods (Blöschl et al., 2020).

As an example of flooding in Europe in the past millennium, is the study from Schulte et al. (2015). They analyzed highly resolved delta plain sediments in a floodplain in the northern slope Swiss Alps in order to study flooding in the past 2600 years. For the past millennium, they identified several flood-rich periods: 1250–1350, 1420–90, 1550–1620, 1650–1720, and 1811–67 (Schulte et al., 2015). These flood-rich periods mostly coincide with at least one of the following climate anomalies in the LIA: cool summers, lower solar irradiance, volcanic eruptions (see, Fig. 8.1), and periods with dryer spring temperatures (Schulte et al., 2015). The authors argue that especially colder climate suggests a greater fraction of winter precipitation that is accumulated in the snowpack, which might suggest higher floods and spring higher base flow that extends until summer.

Blöschl et al. (2020) analyzed spatiotemporal clusters of flood events across Europe in order to identify flood-rich periods from 1500 onward. In contrast to Schulte et al. (2015) their study relies on documentary data of floods (historical hydrology) and more than 100 individual flooding time series in Europe. The flood-rich periods they identified

TABLE 8.2 Flood-rich periods in Europe.

Period #	Temporal range	Spatial range	Temperature anomaly (relative to 1000–2010 CE)
1	1500–20	Western and Central Europe	0.1°C
2	1560–80	Western and Central Europe	−0.1°C
3	1590–1640	Iberian Peninsula and Southern France	−0.5°C
4	1630–60	Western Europe and western part of Central Europe and Northern Italy	−0.2°C
5	1750–1800	Western, Central and Southern Europe, Scandinavia	0.0°C
6	1840–80	Western and Southern Europe	−0.0°C
7	1860–1900	Eastern Central Europe	−0.0°C
8	1910–40	Scandinavia	0.4°C
9	1990 to present	Western and Central Europe, Italy	1.1°C

Adopted from Blöschl, G., Kiss, A., Viglione, A., Barriendos, M., Böhm, O., Brázdil, R., … Wetter, O. (2020). Current European flood-rich period exceptional compared with past 500 years. Nature, *583(7817), 560–566. https://doi.org/10.1038/s41586-020-2478-3.*

are displayed in Fig. 8.1H and their characteristics are compiled in Table 8.2 including temporal and spatial coverage as well as the temperature anomaly (in contrast to Blöschl et al. (2020) anomalies relative to 1000 to 2010 are computed in this chapter). In this chapter the scaled volume is visualized in Fig. 8.1H together with major flood events along the Rhine River from the Undine platform. The latter also includes flood events prior to 1500, thus, also covering the Magdalena flood (1342), which is viewed as the "millennium" event in Central Europe, as it exceeded in damage any other reported event in historic and recent times (Herget et al., 2015). This flood event coincides with the transition from the MCA to the LIA, which was accompanied with a decrease in annual average temperatures and an increase in the occurrence of weather extremes (Sirocko, 2009). Events of similar magnitude have not been observed in Europe, which is why it is viewed as unprecedented flood event or even the "millennium" event. The fact that the Magdalena flood affected a region subject to below average precipitation totals in Central Europe highlights the relevance of studying past extremes.

However, a more coherent picture of flooding in Europe is only available from 1500 onward if one is interested in historic changes in floods on the continental scale. For this very reason Blöschl et al. (2020) compared the past 500 years with the instrumental period in terms of both frequency and magnitude of floods. They identified nine flood-rich periods in total (see Fig. 8.1H and Table 8.2) including the past three decades, which constitutes the last flood rich period in their record. In contrast to former flood-rich periods the most recent periods (i.e., 1910–1940 in Scandinavia, and 1990 to present in Central and Western Europe as well as in Italy) coincide with positive temperature anomalies. In contrast, the flood-rich-periods of the LIA also occurred during negative temperature anomalies and these periods were on average 0.3 lower than corresponding interflood periods (Blöschl et al., 2020). However, this suggests that the flood mechanisms differ between LIA floods and those observed in the CWP.

In the most recent flood-rich period, which is characterized by anthropogenic climate change, increasing temperatures led to both an increase and a decrease of floods, since a decrease in snowmelt resulted in a decrease in flooding in Eastern Europe (Blöschl et al., 2019). Likewise, increases in autumn and winter rainfall led to increasing trends in flooding in Northwestern Europe, while the temperature-driven evapotranspiration increase in Southern Europe resulted in a decrease in flooding (Blöschl et al., 2019). Therefore, seasonality of floods is also sensitive to changes in the climate: In the same period, a shift in timing of floods can be observed, which is linked to earlier snowmelt and lower glacier melt after peak water, increased flooding activity in winter due to polar warming and the higher number of winter storms (Blöschl et al., 2017). Compared to earlier flood-rich periods in the past millennium, the current flood-rich period is characterized by a change in flood frequency toward the dominant flood season (Blöschl et al., 2020), suggesting a higher frequency of summer floods in Central Europe, winter floods in Western Europe, and autumn floods in Southern Europe. These changes are linked to climate change induced shifts in processes (Blöschl et al., 2017): earlier snowmelt and fewer ice-jam floods in Central Europe, more frequent winter storms in Western Europe, and enhanced convection in autumn in Southern Europe (Blöschl et al., 2020).

Blöschl et al. (2020) conclude that "hydroclimatic conditions over Europe have shifted to their millennial boundaries with a dry anomaly in southern Europe and a wet anomaly in central and northern Europe," which might be linked to more frequent low pressure systems with higher persistence, which is also expressed by the positive AMO anomaly (Fig. 8.1D). This bipolar pattern was also found for extreme rainfall (Hosseinzadehtalaei et al., 2019; Tabari et al., 2020), suggesting that spatial patterns in floods follow those of extreme precipitations. However, Blöschl et al. (2020) also emphasize that the "exact mix" of atmospheric drivers that cause flood-rich periods on the millennial scale still remains unknown. However, it is concluded that—based on evidence from GCM simulations—future intensification might be more linked to thermodynamic effects (i.e., the (super) CC effect) than to changes in circulation patterns (Blöschl et al., 2020). This effect is most pronounced in regions where water availability increases (Tabari, 2020), which confirms the bipolar pattern of recent and projected changes across Europe, with a more pronounced intensification in the North. Future projections suggest an increase in magnitude and intensity of flood peaks in Western and Central Europe, while a reduction is detected for Spain and Scandinavia as a result of higher evapotranspiration and less snow, respectively (Alfieri, Burek, Feyen, & Forzieri, 2015). Floods above the 100-year return period are projected to rise in terms of frequency in most parts of Europe, which even holds for regions where the overall number of events decreases (Alfieri et al., 2015).

5 Droughts

Historical droughts are usually reconstructed from proxy data (such as tree rings, which are sensitive to drought stress and temperature) and different sources of documentary (Brázdil, Kiss, Luterbacher, Nash, & Řezníčková, 2018). When considering the past millennium and the occurrence of droughts in Europe, particular dry conditions were prevailing in the MCA with dry summer especially between 1010 and 1165 (Büntgen et al., 2010; Cook et al., 2015), while the LIA between 1430 and 1720 was subject to overall drier conditions (e.g., 15th and 16th centuries) but less pronounced compared to the MCA (Büntgen et al., 2010). Severe droughts occurred in 1540, 1590, 1616, and 1718 including 1719 (Brázdil et al., 2013). Even though their analyses focus on the Czech Republic, the authors of this study assume that this region is representative for surrounding areas in Central Europe, since they form a homogeneous region, which acknowledges the fact that droughts are large-scale events in terms of both spatial (up to continental scale) and temporal dimension (several months and beyond). For numerous historic drought events, the exact timing cannot be reconstructed accurately in terms of the affected period within in the year from the available sources, even though exceptions exist (Brázdil et al., 2013). The 1540 drought, for instance, was especially severe in terms of its length, since it covered a dry period from March 1540 until October 1540. In agreement with the previous study, 1540 was also severe in Switzerland with only 6 days of rainfall between mid of March until end of September (Pfister, 1999), which is why it is sometimes called a "black swan" event. A black swan has a low probability and it is "a high-impact event of an unimaginable magnitude" (Pfister, 2018). Though droughts exhibit homogeneity over large scales, drought patterns and characteristics are not uniform over Europe, as depicted in (Stagge, Kingston, Tallaksen, & Hannah, 2017), who show a divergence in drought likelihood between Northern and Southern Europe in a changing climate.

Different concepts exist for quantifying droughts from data that is more readily available, such as temperature reconstructions or proxy data. A very common approach is representing interannual drought variability through employing the Palmer drought severity index (PDSI) and a modification, called the self-calibrated PDSI (scPDSI), which replaces empirical constants by dynamically calculated parameters in order to make different locations better comparable (Wells, Goddard, & Hayes, 2004). The scPDSI represents a moisture estimation based on temperature and in this way it was the first comprehensive method to quantify moisture storage (Mishra & Singh, 2010). It is a standardized index, generally spanning from -10 to 10, while the former is related to extreme dry conditions and the latter to extreme wet conditions.

Besides the PDSI and scPDSI, droughts can be quantified through different metrics, in particular precipitation, evapotranspiration and runoff. From precipitation data, droughts are commonly quantified using the Standardized Precipitation Index (SPI), describing the onset of drought through precipitation deficit on different time scales and in different climatic regions (Cancelliere, Mauro, Bonaccorso, & Rossi, 2007; Guttman, 1998; McKee, Doesken, & Kleist, 1993). The 1998 study by Guttmann, comparing historical time series of the PDSI and the SPI, shows that the former varies from site to site, whereas the latter does not. It also shows that the complexity and long memory of the PDSI, while the SPI is an easier to interpret moving average. Similar to the SPI in its calculation is the Standardized Precipitation Evapotranspiration Index (SPEI), developed and suggested for use in 2010 (Vicente-Serrano, Beguería, & López-Moreno, 2010). The difference is that the SPEI uses climatic water balance—precipitation minus reference evapotranspiration—as input, rather than precipitation (Beguería, Vicente-Serrano, Reig, & Latorre, 2014). Finally, also based on the SPI is the Standardized Runoff Index (SRI), used to derive hydrological drought from runoff (Shukla & Wood, 2008). These indexes are often used together, to provide a more comprehensive picture of droughts in the framework of different climatic zones or time scales, e.g., in Li, Zhou, and Hu (2016) and Touma, Ashfaq, Nayak, Kao, and Diffenbaugh (2015).

A dataset which is based upon scPDSI data is the Old World Drought Atlas (OWDA) (Cook et al., 2015), which provides spatiotemporal distributions of the scPDSI (i.e., a map of Europe for each year). In this chapter, we visualize the temporal evolution of the June–July–August (JJA) scPDSI in Europe for the past millennium (Fig. 8.1F), based on data from (Büntgen et al., 2021). In contrast to earlier reconstructions, this dataset suggests a significant multimillennial drying trend in Europe, exhibiting a decrease in scPDSI of 0.12 per century. Individual events superimpose this evolvement over time, suggesting that the more recent droughts reflect lower initial values of the scPDSI than earlier events.

In addition to the scPDSI time series for Europe, severe drought events from documentary sources in the Czech Republic and Germany are plotted in Fig. 8.1F (Brázdil et al., 2013; Glaser & Kahle, 2020). The reason which might hold responsible for the low number of events in the MCA might be related to an underestimation of frequency due to the small number of documentary sources (Brázdil et al., 2019). Therefore, deriving frequencies from this type of data is challenging. However, a visual inspection reveals that droughts have been reconstructed in all centuries in the past millennium (subject to lower evidence in the MCA from documentary sources). This also includes wet periods with positive values of scPDSI. However, the density of points is higher in the 15th and 16th centuries, which is in line with the identification of a "megadrought" period from 1437 to 1473 (Cook et al., 2015), which falls into the transition period between MCA and LIA—period lower solar irradiance, a drop in temperature and more weather extremes. In contrast, a drought poor period coincides with the LIA pluvial (wet period between ~1600 and 1800). As possible driver of this drought rich periods might be the cold state of the Atlantic Ocean with winter atmospheric blocking (Ionita, Dima, Nagavciuc, Scholz, & Lohmann, 2021), which is expressed by a lower AMO index (see, Fig. 8.1D). As will be shown in the next section, more flood-rich periods occurred in this period.

Studies in other continents likewise exist. Similar to Europe, Sinha et al. (2011) identified monsoon droughts in China in the 14th and 15th centuries with duration up to decades, which coincide with the onset of the LIA (Sinha et al., 2011). These results are confirmed by Hao et al. (2020) who analyzed harvests. They also identified drought-rich periods between 1481 and 1530 and 1581–1650. Compared to the MCA, droughts in the LIA were less severe, which is why the authors conclude that a warmer climate could weaken the impact of droughts on harvests. Changes in climate, such as exhibited here, have always had an impact on drought occurrence and severity. Recent advances in paleoclimatology and the use of standardized indices have made attribution studies more prevalent, ranging from global studies to regional climatic change discussion. Examples of these studies of climatic changes and their impact on drought and drought effects are the works by (Polemio & Casarano, 2008), looking at a historical record for Italy, and (Büntgen et al., 2010; Cook et al., 2015), who look at different climatic drivers of drought over time, including precipitation decline and increased evapotranspiration linked to higher temperatures. A comprehensive overview on droughts worldwide is given by Brázdil et al. (2018).

The most recent drought events in 2003, 2015, and 2018 occurred within a relatively short period of time and these droughts have been compared to the historical droughts previously discusses. Büntgen et al. (2021) conclude that especially the 2015 event is unprecedented in the past 2110 years in terms of severity due to the multimillennial drying trend and its impact on the overall lower moisture situation prior to the events. Another study highlights that most recent severe drought events are not unprecedented compared to the historical droughts in the LIA, since they are within the range of natural climate variability (Ionita et al., 2021). However, the latter study focuses on the meteorological aspects rather than on the trends in scPDSI but emphasizes that the recent droughts have been warmer. Hanel et al. (2018) also highlight that they are less uncommon in terms of precipitation deficit when reviewed in a historical context considering the past 250 years. However, under consideration of the hydrological relevance of soil moisture for the emergence of droughts and the intensified increase of temperature suggest that the impact of droughts might change in the future (Hanel et al., 2018). This is of great relevance given that projected future changes suggest an increase in droughts for all of Europe, with an intensification, especially in spring and summer, and a decrease in Northern European winters (Spinoni, Vogt, Naumann, Barbosa, & Dosio, 2018).

6 Summary on past changes in hydroclimatic extremes

In this chapter we compiled reconstructions of several hydroclimatic quantities and how their extremes evolved over time in the past millennium in Europe. For these extremes, direct measurements—as they are available to us for current events nowadays—are virtually unavailable, which is why historic changes in hydroclimatic extremes have been compiled from proxy data, such as tree ring analyses, which reflect the temperature and rainfall conditions in each year in the past, or from different written sources, such as books or other documentary sources in libraries. Apart from documentary sources and proxy data, models help to gain insight into the question how hydroclimatic extremes have evolved over time (Brönnimann, 2017).

Our brief review reveals that concurrent (or at least more recent extremes) are not generally unprecedented but that event frequencies of a certain magnitude might suggest a shift toward more extreme weather conditions, associated with anthropogenic climate change. This holds especially true for heavy precipitation, floods and droughts. Studies suggest an increase in drought severity and occurrence, amplified by anthropogenic climate change, causing glacier mass loss and evapotranspiration increase through rising temperatures, as well as past and recent decreases in precipitation rates (Hanel et al., 2018). For floods and especially heavy precipitation, the (super) CC scaling suggests an intensification of both types of hydroclimatic extremes (Blöschl et al., 2020; Lenderink & Van Meijgaard, 2008). A look into the past is therefore also helpful in case of hydroclimatic extremes, since it complements our picture of the possible magnitude of extreme events that might undergo a change in terms of frequency in the future.

References

Ait Brahim, Y., Wassenburg, J. A., Cruz, F. W., Sifeddine, A., Scholz, D., Bouchaou, L., … Cheng, H. (2018). Multi-decadal to centennial hydro-climate variability and linkage to solar forcing in the Western Mediterranean during the last 1000 years. *Scientific Reports, 8*(1), 17446. https://doi.org/10.1038/s41598-018-35498-x.

Alfieri, L., Burek, P., Feyen, L., & Forzieri, G. (2015). Global warming increases the frequency of river floods in Europe. *Hydrology and Earth System Sciences, 19*(5), 2247–2260. https://doi.org/10.5194/hess-19-2247-2015.

Ban, N., Schmidli, J., & Schär, C. (2015). Heavy precipitation in a changing climate: Does short-term summer precipitation increase faster? *Geophysical Research Letters, 42*(4), 1165–1172. https://doi.org/10.1002/2014GL062588.

Beguería, S., Vicente-Serrano, S. M., Reig, F., & Latorre, B. (2014). Standardized precipitation evapotranspiration index (SPEI) revisited: Parameter fitting, evapotranspiration models, tools, datasets and drought monitoring. *International Journal of Climatology, 34*(10), 3001–3023. https://doi.org/10.1002/joc.3887.

Beniston, M. (2012). Impacts of climatic change on water and associated economic activities in the Swiss Alps. *Journal of Hydrology, 412–413*, 291–296. https://doi.org/10.1016/j.jhydrol.2010.06.046.

Berg, P., Christensen, O. B., Klehmet, K., Lenderink, G., Olsson, J., Teichmann, C., & Yang, W. (2019). Summertime precipitation extremes in a EURO-CORDEX 0.11° ensemble at an hourly resolution. *Natural Hazards and Earth System Sciences, 19*(4), 957–971. https://doi.org/10.5194/nhess-19-957-2019.

Berg, P., Haerter, J. O., Thejll, P., Piani, C., Hagemann, S., & Christensen, J. H. (2009). Seasonal characteristics of the relationship between daily precipitation intensity and surface temperature. *Journal of Geophysical Research, 114*(D18), D18102. https://doi.org/10.1029/2009JD012008.

Berg, P., Moseley, C., & Haerter, J. O. (2013). Strong increase in convective precipitation in response to higher temperatures. *Nature Geoscience, 6*(3), 181–185. https://doi.org/10.1038/ngeo1731.

Blöschl, G., Hall, J., Parajka, J., Perdigão, R. A. P., Merz, B., Arheimer, B., Aronica, G. T., Bilibashi, A., Bonacci, O., Borga, M., Čanjevac, I., Castellarin, A., Chirico, G. B., Claps, P., Fiala, K., Frolova, N., Gorbachova, L., Gül, A., Hannaford, J., … Živković, N. (2017). Changing climate shifts timing of European floods. *Science, 357*(6351), 588–590. https://doi.org/10.1126/science.aan2506.

Blöschl, G., Hall, J., Viglione, A., Perdigão, R. A. P., Parajka, J., Merz, B., Lun, D., Arheimer, B., Aronica, G. T., Bilibashi, A., Boháč, M., Bonacci, O., Borga, M., Čanjevac, I., Castellarin, A., Chirico, G. B., Claps, P., Frolova, N., Ganora, D., … Živković, N. (2019). Changing climate both increases and decreases European river floods. *Nature, 573*(7772), 108–111. https://doi.org/10.1038/s41586-019-1495-6.

Blöschl, G., Kiss, A., Viglione, A., Barriendos, M., Böhm, O., Brázdil, R., Coeur, D., Demarée, G., Llasat, M. C., Macdonald, N., Retsö, D., Roald, L., Schmocker-Fackel, P., Amorim, I., Bělínová, M., Benito, G., Bertolin, C., Camuffo, D., Cornel, D., … Wetter, O. (2020). Current European flood-rich period exceptional compared with past 500 years. *Nature, 583*(7817), 560–566. https://doi.org/10.1038/s41586-020-2478-3.

Boessenkool, B., Bürger, G., & Heistermann, M. (2017). Effects of sample size on estimation of rainfall extremes at high temperatures. *Natural Hazards and Earth System Sciences, 17*(9), 1623–1629. https://doi.org/10.5194/nhess-17-1623-2017.

Bolibar, J., Rabatel, A., Gouttevin, I., & Galiez, C. (2020). A deep learning reconstruction of mass balance series for all glaciers in the French Alps: 1967–2015. *Earth System Science Data, 12*(3), 1973–1983. https://doi.org/10.5194/essd-12-1973-2020.

Braithwaite, R. J. (1981). On glacier energy balance, ablation, and air temperature. *Journal of Glaciology, 27*(97), 381–391. https://doi.org/10.1017/S0022143000011424.

Brázdil, R., Dobrovolný, P., Trnka, M., Kotyza, O., Řezníčková, L., Valášek, H., … Štěpánek, P. (2013). Droughts in the Czech Lands, 1090–2012 AD. *Climate of the Past, 9*(4), 1985–2002. https://doi.org/10.5194/cp-9-1985-2013.

Brázdil, R., Dobrovolný, P., Trnka, M., Řezníčková, L., Dolák, L., & Kotyza, O. (2019). Extreme droughts and human responses to them: The Czech Lands in the pre-instrumental period. *Climate of the Past, 15*(1), 1–24. https://doi.org/10.5194/cp-15-1-2019.

Brázdil, R., Kiss, A., Luterbacher, J., Nash, D. J., & Řezníčková, L. (2018). Documentary data and the study of past droughts: A global state of the art. *Climate of the Past, 14*(12), 1915–1960. https://doi.org/10.5194/cp-14-1915-2018.

Brázdil, R., Kundzewicz, Z. W., & Benito, G. (2006). Historical hydrology for studying flood risk in Europe. *Hydrological Sciences Journal, 51*(5), 739–764. https://doi.org/10.1623/hysj.51.5.739.

Brázdil, R., Kundzewicz, Z. W., Benito, G., Demarée, G., Macdonald, N., & Roald, L. A. (2012). Historical floods in Europe in the Past Millennium. In Z. W. Kundzewicz (Ed.), *Changes in flood risk in Europe* (p. 46). CRC Press.

Brönnimann, S. (2017). Weather extremes in an ensemble of historical reanalyses. *Geographica Bernensia, G92*, 7–22. https://doi.org/10.4480/GB2017.G92.01.

Büntgen, U., Trouet, V., Frank, D., Leuschner, H. H., Friedrichs, D., Luterbacher, J., & Esper, J. (2010). Tree-ring indicators of German summer drought over the last millennium. *Quaternary Science Reviews, 29*, 1005–1016.

Büntgen, U., Urban, O., Krusic, P. J., Rybníček, M., Kolář, T., Kyncl, T., … Trnka, M. (2021). Recent European drought extremes beyond common era background variability. *Nature Geoscience*, *14*(4), 190–196. https://doi.org/10.1038/s41561-021-00698-0.

Cancelliere, A., Mauro, G. D., Bonaccorso, B., & Rossi, G. (2007). Drought forecasting using the standardized precipitation index. *Water Resources Management*, *21*(5), 801–819. https://doi.org/10.1007/s11269-006-9062-y.

Clapperton, C. M., Sugden, D. E., Birnie, J., & Wilson, M. J. (1989). Late-glacial and holocene glacier fluctuations and environmental change on South Georgia, Southern Ocean. *Quaternary Research*, *31*(2), 210–228. https://doi.org/10.1016/0033-5894(89)90006-9.

Cook, E. R., Seager, R., Kushnir, Y., Briffa, K. R., Büntgen, U., Frank, D., Krusic, P. J., Tegel, W., van der Schrier, G., Andreu-Hayles, L., Baillie, M., Baittinger, C., Bleicher, N., Bonde, N., Brown, D., Carrer, M., Cooper, R., Čufar, K., Dittmar, C., … Zang, C. (2015). Old World megadroughts and pluvials during the common era. *Science Advances*, *1*(10). https://doi.org/10.1126/sciadv.1500561, e1500561.

Demarée, G. R. (2003). Le pluviographe centenaire du plateau d'Uccle: son histoire, ses données et ses applications. *La Houille Blanche*, *4*, 1–8.

Eis, J., van der Laan, L., Maussion, F., & Marzeion, B. (2021). Reconstruction of past glacier changes with an ice-flow glacier model: Proof of concept and validation. *Frontiers in Earth Science*, *9*. https://doi.org/10.3389/feart.2021.595755.

Förster, K., & Thiele, L. B. (2020). Variations in sub-daily precipitation at centennial scale. *npj Climate and Atmospheric Science*, *3*(1). https://doi.org/10.1038/s41612-020-0117-1.

Gardelle, J., Berthier, E., & Arnaud, Y. (2012). Slight mass gain of Karakoram glaciers in the early twenty-first century. *Nature Geoscience*, *5*(5), 322–325. https://doi.org/10.1038/ngeo1450.

Glaser, R., & Kahle, M. (2020). Reconstructions of droughts in Germany since 1500—Combining hermeneutic information and instrumental records in historical and modern perspectives. *Climate of the Past*, *16*(4), 1207–1222. https://doi.org/10.5194/cp-16-1207-2020.

Goosse, H., Barriat, P. Y., Dalaiden, Q., Klein, F., Marzeion, B., Maussion, F., … Vlug, A. (2018). Testing the consistency between changes in simulated climate and Alpine glacier length over the past millennium. *Climate of the Past*, *14*(8), 1119–1133. https://doi.org/10.5194/cp-14-1119-2018.

Guiot, J., Nicault, A., Rathgeber, C., Edouard, J. L., Guibal, F., Pichard, G., & Till, C. (2005). Last-millennium summer-temperature variations in Western Europe based on proxy data. *The Holocene*, *15*(4), 489–500. https://doi.org/10.1191/0959683605hl819rp.

Guttman, N. B. (1998). Comparing the palmer drought index and the standardized precipitation index. *Journal of the American Water Resources Association*, *34*(1), 113–121. https://doi.org/10.1111/j.1752-1688.1998.tb05964.x.

Hanel, M., Pavlásková, A., & Kyselý, J. (2016). Trends in characteristics of sub-daily heavy precipitation and rainfall erosivity in the Czech Republic. *International Journal of Climatology*, *36*(4), 1833–1845. https://doi.org/10.1002/joc.4463.

Hanel, M., Rakovec, O., Markonis, Y., Máca, P., Samaniego, L., Kyselý, J., & Kumar, R. (2018). Revisiting the recent European droughts from a long-term perspective. *Scientific Reports*, *8*(1), 9499. https://doi.org/10.1038/s41598-018-27464-4.

Hanna, E., Huybrechts, P., Cappelen, J., Steffen, K., Bales, R. C., Burgess, E., … Savas, D. (2011). Greenland ice sheet surface mass balance 1870 to 2010 based on twentieth century reanalysis, and links with global climate forcing. *Journal of Geophysical Research-Atmospheres*, *116*(24). https://doi.org/10.1029/2011JD016387.

Hao, Z., Wu, M., Zheng, J., Chen, J., Zhang, X., & Luo, S. (2020). Patterns in data of extreme droughts/floods and harvest grades derived from historical documents in eastern China during 801–1910. *Climate of the Past*, *16*(1), 101–116.

Harrison, W. D., Cox, L. H., Hock, R., March, R. S., & Pettit, E. C. (2009). Implications for the dynamic health of a glacier from comparison of conventional and reference-surface balances. *Annals of Glaciology*, *50*(50), 25–30. https://doi.org/10.3189/172756409787769654.

Helama, S., Meriläinen, J., & Tuomenvirta, H. (2009). Multicentennial megadrought in northern Europe coincided with a global El Niño–southern oscillation drought pattern during the medieval climate anomaly. *Geology*, *37*(2), 175–178. https://doi.org/10.1130/G25329A.1.

Herget, J., Kapala, A., Krell, M., Rustemeier, E., Simmer, C., & Wyss, A. (2015). The millennium flood of July 1342 revisited. *Catena*, *130*, 82–94.

Hosseinzadehtalaei, P., Tabari, H., & Willems, P. (2019). Regionalization of anthropogenically forced changes in 3 hourly extreme precipitation over Europe. *Environmental Research Letters*, *14*(12). https://doi.org/10.1088/1748-9326/ab5638.

Hosseinzadehtalaei, P., Tabari, H., & Willems, P. (2020). Climate change impact on short-duration extreme precipitation and intensity–duration–frequency curves over Europe. *Journal of Hydrology*, *590*. https://doi.org/10.1016/j.jhydrol.2020.125249, 125249.

Huss, M., & Hock, R. (2018). Global-scale hydrological response to future glacier mass loss. *Nature Climate Change*, *8*(2), 135–140. https://doi.org/10.1038/s41558-017-0049-x.

Immerzeel, W. W., Lutz, A. F., Andrade, M., Bahl, A., Biemans, H., Bolch, T., Hyde, S., Brumby, S., Davies, B. J., Elmore, A. C., Emmer, A., Feng, M., Fernández, A., Haritashya, U., Kargel, J. S., Koppes, M., Kraaijenbrink, P. D. A., Kulkarni, A. V., Mayewski, P. A., … Baillie, J. E. M. (2020). Importance and vulnerability of the world's water towers. *Nature*, *577*(7790), 364–369. https://doi.org/10.1038/s41586-019-1822-y.

Intergovernmental Panel on Climate Change (IPCC). (2020). *IPCC fifth assessment report (AR5) observed climate change impacts database, version 2.01*. NASA Socioeconomic Data and Applications Center (SEDAC).

Ionita, M., Dima, M., Nagavciuc, V., Scholz, P., & Lohmann, G. (2021). Past megadroughts in central Europe were longer, more severe and less warm than modern droughts. *Communications Earth & Environment*, *2*(1), 61. https://doi.org/10.1038/s43247-021-00130-w.

Jones, G. S., Stott, P. A., & Christidis, N. (2013). Attribution of observed historical near-surface temperature variations to anthropogenic and natural causes using CMIP5 simulations. *Journal of Geophysical Research-Atmospheres*, *118*(10), 4001–4024. https://doi.org/10.1002/jgrd.50239.

Kaser, G., Großhauser, M., & Marzeion, B. (2010). Contribution potential of glaciers to water availability in different climate regimes. *Proceedings of the National Academy of Sciences of the United States of America*, *107*(47), 20223–20227. https://doi.org/10.1073/pnas.1008162107.

Kurtyka, J. C., & Madow, L. (1952). *Precipitation measurements study*. Illinois Univ at Urbana-Chamapaign.

Le Quesne, C., Acuña, C., Boninsegna, J. A., Rivera, A., & Barichivich, J. (2009). Long-term glacier variations in the Central Andes of Argentina and Chile, inferred from historical records and tree-ring reconstructed precipitation. *Palaeogeography, Palaeoclimatology, Palaeoecology*, *281*(3–4), 334–344. https://doi.org/10.1016/j.palaeo.2008.01.039.

Lean, J. L. (2018). Estimating solar irradiance since 850 CE. *Earth and Space Science*, *5*(4), 133–149. https://doi.org/10.1002/2017EA000357.

Leclercq, P. W., & Oerlemans, J. (2012). Global and hemispheric temperature reconstruction from glacier length fluctuations. *Climate Dynamics*, *38*(5–6), 1065–1079. https://doi.org/10.1007/s00382-011-1145-7.

Leclercq, P. W., Oerlemans, J., Basagic, H. J., Bushueva, I., Cook, A. J., & Le Bris, R. (2014). A data set of worldwide glacier length fluctuations. *The Cryosphere*, *8*(2), 659–672. https://doi.org/10.5194/tc-8-659-2014.

Lenderink, G., Barbero, R., Loriaux, J. M., & Fowler, H. J. (2017). Super-Clausius–Clapeyron scaling of extreme hourly convective precipitation and its relation to large-scale atmospheric conditions. *Journal of Climate*, *30*(15), 6037–6052. https://doi.org/10.1175/JCLI-D-16-0808.1.

Lenderink, G., Mok, H. Y., Lee, T. C., & Van Oldenborgh, G. J. (2011). Scaling and trends of hourly precipitation extremes in two different climate zones—Hong Kong and the Netherlands. *Hydrology and Earth System Sciences*, *15*(9), 3033–3041. https://doi.org/10.5194/hess-15-3033-2011.

Lenderink, G., & Van Meijgaard, E. (2008). Increase in hourly precipitation extremes beyond expectations from temperature changes. *Nature Geoscience*, *1*(8), 511–514. https://doi.org/10.1038/ngeo262.

Li, J., Zhou, S., & Hu, R. (2016). Hydrological drought class transition using SPI and SRI time series by loglinear regression. *Water Resources Management*, *30*(2), 669–684. https://doi.org/10.1007/s11269-015-1184-7.

Łupikasza, E. B. (2017). Seasonal patterns and consistency of extreme precipitation trends in Europe, December 1950 to February 2008. *Climate Research*, *72*(3), 217–237. https://doi.org/10.3354/cr01467.

Malles, J. H., & Marzeion, B. (2020). 20th century global glacier mass change: An ensemble-based model reconstruction. *The Cryosphere Discussions*, 1–30.

Mann, M. E., Zhang, Z., Rutherford, S., Bradley, R. S., Hughes, M. K., Shindell, D., … Ni, F. (2009). Global signatures and dynamical origins of the little ice age and medieval climate anomaly. *Science*, *326*(5957), 1256–1260. https://doi.org/10.1126/science.1177303.

Martel, J. L., Mailhot, A., & Brissette, F. (2020). Global and regional projected changes in 100-yr subdaily, daily, and multiday precipitation extremes estimated from three large ensembles of climate simulations. *Journal of Climate*, *33*(3), 1089–1103. https://doi.org/10.1175/JCLI-D-18-0764.1.

Martel, J. L., Mailhot, A., Brissette, F., & Caya, D. (2018). Role of natural climate variability in the detection of anthropogenic climate change signal for mean and extreme precipitation at local and regional scales. *Journal of Climate*, *31*(11), 4241–4263. https://doi.org/10.1175/JCLI-D-17-0282.1.

Marzeion, B., Cogley, J. G., Richter, K., & Parkes, D. (2014). Attribution of global glacier mass loss to anthropogenic and natural causes. *Science*, 919–921. https://doi.org/10.1126/science.1254702.

Matthews, J. A., Berrisford, M. S., Quentin Dresser, P., Nesje, A., Olaf Dahl, S., Elisabeth Bjune, A., … Barnett, C. (2005). Holocene glacier history of Bjørnbreen and climatic reconstruction in central Jotunheimen, Norway, based on proximal glaciofluvial stream-bank mires. *Quaternary Science Reviews*, *24*(1–2), 67–90. https://doi.org/10.1016/j.quascirev.2004.07.003.

Maussion, F., Butenko, A., Champollion, N., Dusch, M., Eis, J., Fourteau, K., … Marzeion, B. (2019). The open global glacier model (OGGM) v1.1. *Geoscientific Model Development*, *12*(3), 909–931. https://doi.org/10.5194/gmd-12-909-2019.

McKee, T., Doesken, & Kleist, J. (1993). The relationship of drought frequency and duration to time scales. In *Paper presented at 8th conference on applied climatology*.

Mishra, A. K., & Singh, V. P. (2010). A review of drought concepts. *Journal of Hydrology*, *391*(1–2), 202–216. https://doi.org/10.1016/j.jhydrol.2010.07.012.

Mueller, E. N., & Pfister, A. (2011). Increasing occurrence of high-intensity rainstorm events relevant for the generation of soil erosion in a temperate lowland region in Central Europe. *Journal of Hydrology*, *411*(3–4), 266–278. https://doi.org/10.1016/j.jhydrol.2011.10.005.

Myhre, G., Alterskjær, K., Stjern, C. W., Hodnebrog, Marelle, L., Samset, B. H., … Stohl, A. (2019). Frequency of extreme precipitation increases extensively with event rareness under global warming. *Scientific Reports*, *9*(1). https://doi.org/10.1038/s41598-019-52277-4.

Neukom, R., Steiger, N., Gómez-Navarro, J. J., Wang, J., & Werner, J. P. (2019). No evidence for globally coherent warm and cold periods over the preindustrial Common Era. *Nature*, *571*(7766), 550–554. https://doi.org/10.1038/s41586-019-1401-2.

Ntegeka, V., & Willems, P. (2008). Trends and multidecadal oscillations in rainfall extremes, based on a more than 100-year time series of 10 min rainfall intensities at Uccle, Belgium. *Water Resources Research*, *44*(7). https://doi.org/10.1029/2007WR006471.

Nussbaumer, S. U., & Zumbühl, H. J. (2012). The little ice age history of the Glacier des Bossons (Mont Blanc massif, France): A new high-resolution glacier length curve based on historical documents. *Climatic Change*, *111*(2), 301–334. https://doi.org/10.1007/s10584-011-0130-9.

Oerlemans, J., & Hoogendoorn, N. C. (1989). Mass-balance gradients and climatic change. *Journal of Glaciology*, *35*(121), 399–405. https://doi.org/10.1017/S0022143000009333.

Ortega, P., Lehner, F., Swingedouw, D., Masson-Delmotte, V., Raible, C. C., Casado, M., & Yiou, P. (2015). A model-tested North Atlantic Oscillation reconstruction for the past millennium. *Nature*, *523*(7558), 71–74. https://doi.org/10.1038/nature14518.

PAGES 2k Consortium. (2013). Continental-scale temperature variability during the past two millennia. *Nature Geoscience*, *6*(5), 339–346. https://doi.org/10.1038/ngeo1797.

Pellicciotti, F., Helbing, J., Rivera, A., Favier, V., Corripio, J., Araos, J., … Carenzo, M. (2008). A study of the energy balance and melt regime on Juncal Norte Glacier, semi-arid Andes of central Chile, using melt models of different complexity. *Hydrological Processes*, *22*(19), 3980–3997. https://doi.org/10.1002/hyp.7085.

Pfister, C. (1999). *Wetternachhersage—500 Jahre Klimavariationen und Naturkatastrophen*. Haupt Verlag.

Pfister, C. (2018). The "Black Swan" of 1540: Aspects of a European Megadrought. In C. Leggewie, & F. Mauelshagen (Eds.), *Climate change and cultural transition in Europe* (pp. 156–194). Brill.

Polemio, M., & Casarano, D. (2008). Climate change, drought and groundwater availability in southern Italy. *Geological Society, London, Special Publications*, *288*(1), 39–51. https://doi.org/10.1144/sp288.4.

Porter, S. C., & Orombelli, G. (1985). Glacier contraction during the middle Holocene in the western Italian Alps: Evidence and implications. *Geology*, *13*(4), 296–298. https://doi.org/10.1130/0091-7613(1985)13<296:GCDTMH>2.0.CO;2.

Prein, A. F., Rasmussen, R. M., Ikeda, K., Liu, C., Clark, M. P., & Holland, G. J. (2017). The future intensification of hourly precipitation extremes. *Nature Climate Change*, *7*(1), 48–52. https://doi.org/10.1038/nclimate3168.

Pritchard, H. D. (2019). Asia's shrinking glaciers protect large populations from drought stress. *Nature*, *569*(7758), 649–654. https://doi.org/10.1038/s41586-019-1240-1.

Rein, B., Luckge, A., & Sirocko, F. (2004). A major Holocene ENSO anomaly during the Medieval period. *Geophysical Research Letters*, *31*(L17211), 4. https://doi.org/10.1029/2004GL020161.

Roe, G. H., Baker, M. B., & Herla, F. (2017). Centennial glacier retreat as categorical evidence of regional climate change. *Nature Geoscience*, *10*(2), 95–99. https://doi.org/10.1038/ngeo2863.

Rubino, M., Etheridge, D. M., Thornton, D. P., Howden, R., Allison, C. E., Francey, R. J., … Smith, A. M. (2019). Revised records of atmospheric trace gases CO_2, CH_4, N_2, and $\delta^{13}C$-CO_2 over the last 2000 years from Law Dome, Antarctica. *Earth System Science Data*, *11*(2), 473–492. https://doi.org/10.5194/essd-11-473-2019.

Rupper, S., & Roe, G. (2008). Glacier changes and regional climate: A mass and energy balance approach. *Journal of Climate*, *21*(20), 5384–5401. https://doi.org/10.1175/2008JCLI2219.1.

Schneider, L., Smerdon, J. E., Pretis, F., Hartl-Meier, C., & Esper, J. (2017). A new archive of large volcanic events over the past millennium derived from reconstructed summer temperatures. *Environmental Research Letters*, *12*(9). https://doi.org/10.1088/1748-9326/aa7a1b, 094005.

Schulte, L., Peña, J. C., Carvalho, F., Schmidt, T., Julià, R., Llorca, J., & Veit, H. (2015). A 2600-year history of floods in the Bernese Alps, Switzerland: Frequencies, mechanisms and climate forcing. *Hydrology and Earth System Sciences*, *19*, 3047–3071. https://doi.org/10.5194/hess-19-3047-2015.

Shukla, S., & Wood, A. W. (2008). Use of a standardized runoff index for characterizing hydrologic drought. *Geophysical Research Letters*, *35*(2). https://doi.org/10.1029/2007GL032487.

Sicart, J. E., Hock, R., & Six, D. (2008). Glacier melt, air temperature, and energy balance in different climates: The Bolivian Tropics, the French Alps, and northern Sweden. *Journal of Geophysical Research-Atmospheres*, *113*(24). https://doi.org/10.1029/2008JD010406.

Sinha, A., Stott, L., Berkelhammer, M., Cheng, H., Edwards, R. L., Buckley, B., … Mudelsee, M. (2011). A global context for megadroughts in monsoon Asia during the past millennium. *Quaternary Science Reviews*, *30*(1–2), 47–62. https://doi.org/10.1016/j.quascirev.2010.10.005.

Sirocko, F. (2009). *Wetter, Klima, Menschheitsentwicklung: von der Eiszeit bis ins 21. Jahrhundert*. Theiss.

Spinoni, J., Vogt, J. V., Naumann, G., Barbosa, P., & Dosio, A. (2018). Will drought events become more frequent and severe in Europe? *International Journal of Climatology*, *38*(4), 1718–1736. https://doi.org/10.1002/joc.5291.

Stagge, J. H., Kingston, D. G., Tallaksen, L. M., & Hannah, D. M. (2017). Observed drought indices show increasing divergence across Europe. *Scientific Reports*, *7*(1). https://doi.org/10.1038/s41598-017-14283-2.

Strangeways, I. (2010). A history of rain gauges. *Weather*, *65*(5), 133–138. https://doi.org/10.1002/wea.548.

Tabari, H. (2020). Climate change impact on flood and extreme precipitation increases with water availability. *Scientific Reports*, *10*(1). https://doi.org/10.1038/s41598-020-70816-2.

Tabari, H., Hosseinzadehtalaei, P., Aghakouchak, A., & Willems, P. (2019). Latitudinal heterogeneity and hotspots of uncertainty in projected extreme precipitation. *Environmental Research Letters*, *14*(12). https://doi.org/10.1088/1748-9326/ab55fd.

Tabari, H., Madani, K., & Willems, P. (2020). The contribution of anthropogenic influence to more anomalous extreme precipitation in Europe. *Environmental Research Letters*, *15*(10). https://doi.org/10.1088/1748-9326/abb268.

Tabari, H., & Willems, P. (2018). Lagged influence of Atlantic and Pacific climate patterns on European extreme precipitation. *Scientific Reports*, *8*(1). https://doi.org/10.1038/s41598-018-24069-9.

Thibert, E., Eckert, N., & Vincent, C. (2013). Climatic drivers of seasonal glacier mass balances: An analysis of 6 decades at Glacier de Sarennes (French Alps). *The Cryosphere*, *7*(1), 47–66. https://doi.org/10.5194/tc-7-47-2013.

Thoning, K. W., Crotwell, A. M., & Mund, J. W. (2021). *Atmospheric carbon dioxide dry air mole fractions from continuous measurements at Mauna Loa, Hawaii, Barrow, Alaska, American Samoa and South Pole. 1973-2020, version 2021-08-09*. Boulder, CO: National Oceanic and Atmospheric Administration (NOAA), Global Monitoring Laboratory (GML). https://doi.org/10.15138/yaf1-bk21.

Touma, D., Ashfaq, M., Nayak, M. A., Kao, S. C., & Diffenbaugh, N. S. (2015). A multi-model and multi-index evaluation of drought characteristics in the 21st century. *Journal of Hydrology*, *526*, 196–207. https://doi.org/10.1016/j.jhydrol.2014.12.011.

Vicente-Serrano, S. M., Beguería, S., & López-Moreno, J. I. (2010). A multiscalar drought index sensitive to global warming: The standardized precipitation evapotranspiration index. *Journal of Climate*, *23*(7), 1696–1718. https://doi.org/10.1175/2009JCLI2909.1.

Wells, N., Goddard, S., & Hayes, M. J. (2004). A self-calibrating palmer drought severity index. *Journal of Climate*, 17.

Westra, S., Fowler, H. J., Evans, J. P., Alexander, L. V., Berg, P., Johnson, F., … Roberts, N. M. (2014). Future changes to the intensity and frequency of short-duration extreme rainfall. *Reviews of Geophysics*, *52*(3), 522–555. https://doi.org/10.1002/2014RG000464.

Wilhelm, B., Arnaud, F., Enters, D., Allignol, F., Legaz, A., Magand, O., … Malet, E. (2012). Does global warming favour the occurrence of extreme floods in European Alps? First evidences from a NW Alps proglacial lake sediment record. *Climatic Change*, *113*(3–4), 563–581. https://doi.org/10.1007/s10584-011-0376-2.

Willems, P. (2013). Multidecadal oscillatory behaviour of rainfall extremes in Europe. *Climatic Change*, *120*(4), 931–944. https://doi.org/10.1007/s10584-013-0837-x.

Zeder, J., & Fischer, E. M. (2020). Observed extreme precipitation trends and scaling in Central Europe. *Weather and Climate Extremes*, *29*. https://doi.org/10.1016/j.wace.2020.100266, 100266.

Zhang, W., Villarini, G., Scoccimarro, E., & Vecchi, G. A. (2017). Stronger influences of increased CO2 on subdaily precipitation extremes than at the daily scale. *Geophysical Research Letters*, *44*(14), 7464–7471. https://doi.org/10.1002/2017GL074024.

Zolina, O. (2012). Changes in intense precipitation in Europe. In Z. W. Kundzewicz (Ed.), *Changes in flood risk in Europe* (pp. 97–119). CRC Press.

Zolina, O., Simmer, C., Kapala, A., Shabanov, P., Becker, P., MäcHel, H., … Groisman, P. (2014). Precipitation variability and extremes in Central Europe: New view from STAMMEX results. *Bulletin of the American Meteorological Society*, *95*(7), 995–1002. https://doi.org/10.1175/BAMS-D-12-00134.1.

CHAPTER

9

Meteorological droughts in semi-arid Eastern Kenya

Charles W. Recha[a], *Grace W. Kibue*[b], *and A.P. Dimri*[c]

[a]Department of Geography, Environment and Development Studies, Bomet University College, Bomet, Kenya

[b]Department of Natural Resources, Egerton University, Egerton, Kenya

[c]School of Environmental Sciences, Jawaharlal Nehru University, New Delhi, India

1 Introduction

Drought is a natural hazard that relates to prolonged lack of rainfall that leads to a temporary deficit in natural water availability (Spinoni, Naumann, Carrao, Barbosa, & Vogt, 2014). The temporary period of rainfall deficit is relative to the statistical multiyear mean of a region and may range from a season, a year, or several years. Drought constitutes lack of water for needs in economic sectors (such as agriculture, livestock, and industry) and human population. According to the World Meteorological Department (WMO, 2006), droughts are often characterized by three main aspects: intensity, duration, and spatial coverage. To better monitor and quantify drought, various drought indicators and indices have been developed. Indicators are variables (such as rainfall, temperature, stream flow, and soil moisture) used to describe drought conditions, while indices are computed numerical representations of drought severity (WMO and GWP, 2016).

There are different types of droughts—meteorological, agricultural, hydrological, and socioeconomic (Dai, 2011; Mishra & Singh, 2010). Meteorological drought is a period of months or years with below normal rainfall. Agricultural drought refers to a period with declining soil moisture and consequent crop failure without any reference to surface water resources. Hydrological drought occurs when river stream flow and water storages in aquifers, lakes, or reservoirs fall below long-term mean levels. Socioeconomic drought is associated with failure of water resources systems to meet water demand for some commodity or economic goods such as water, livestock forage and hydroelectric power generation. There is a fifth category of drought—ground water drought (Mishra and Singh, 2010), although there is lack of clarity on how different it is from hydrological drought. From these definitions, it is observed that unlike meteorological drought, agricultural, hydrological and socioeconomic droughts place emphasis on human and biophysical aspects of drought. According to WMO (2006), all other types of drought originate from and occur less frequently than meteorological drought.

Droughts in Sub-Saharan Africa are fairly well documented; in West Africa (Padgham et al., 2016), Southern Africa (Ujeneza & Abiodun, 2015) and the Great Horn of Africa (Lyon, 2014; Nicholson, 2014). These droughts pose a threat to the population through loss of livelihoods and lives (FAO, 2017; Gbetibouo, Ringler, & Hassan, 2010; Nikoloski, Christiaensen, & Hill, 2018). In addition, droughts act as catalysts for malnutrition (Davenport, Grace, Funk, & Shukla, 2017), environmental degradation (Gizaw & Gan, 2017), and, in some cases, civil conflicts (Uexkull, 2014). These challenges have prompted actions that range from humanitarian assistance (USAID, 2018) to the use of seasonal weather forecasts in decision-making (Poterie et al., 2018) and promotion of climate risk management practices at household level (Padgham et al., 2016). Thus, the importance of droughts in Sub-Saharan Africa needs not to be overemphasized. In the Great Horn of Africa (GHA) region (which constitutes, among others, Kenya, Ethiopia, Somalia, Eritrea, and Djibouti), droughts have increased in severity and frequency, impacting on eco- and livelihood systems within and across national borders. According to Bartel and Muller (2007), more than 60% of the area coverage of each of the GHA countries is susceptible to drought—way ahead of other hazards such as floods and locust infestation.

Climate Impacts on Extreme Weather
https://doi.org/10.1016/B978-0-323-88456-3.00006-X

A number of studies have been undertaken on drought in the GHA. The role of sea surface temperatures and El Niño Southern Oscillation (ENSO) in the prediction of droughts in Africa has been captured by Lyon (2014) (GHA region), Gizaw and Gan (2017), Yuan et al. (2013) (Africa), Ujeneza and Abiodun (2015) (Southern Africa), and Uhe et al. (2018) (Kenya). In addition, fluctuations in rainfall amounts in Sub-Saharan Africa are influenced by monsoons, the Intertropical Convergence Zone (ITCZ), subtropical anticyclones and African jet streams, among others (Onyango, 2014). These studies, whereas they contribute to improvement of the skill of drought predictions, fail to examine spatial coverage, frequency and magnitude of past drought events. Other drought related studies have focused on recent droughts (Nicholson, 2014) (1998–2012) and Klisch and Atzberger (2016) (2001–2012). While studies by Asfaw, Simane, Hassen, and Bantider (2018), Elkollaly, Khadr, and Zeidan (2018), Geng et al. (2016), and Al-Qinna, Hammori, Obeidat, and Ahmed (2011) focused on the occurrence of droughts at regional and subregional levels with reliance on satellite and gridded data. Use of satellite and gridded data in drought studies is often attributed to limited record length of in situ data in most developing countries (Kaluba, Verbist, Cornelis, & Ranst, 2017; Washington et al., 2006). Most of the satellite data, although available, does not perform well in Africa and does not take into account the complex terrains that characterize the continent (Nicholson, 2014). In Kenya, Nathan et al. (2020) compared satellite and observed rain gauge data in the tropical central highlands. This study, however, focused on onset, cessation, rainfall trends and length of growing seasons. Ochieng, Recha, and Bebe (2017), Karanja, Ondimu, and Recha (2017), and Onyango (2014) had varied findings on the magnitude (severity) of droughts in the North Rift Valley and North Eastern regions. Okal, Ungetich, and Okeyo (2020) used selected indices to establish the occurrence of droughts in the Upper Tana River watershed of Kenya. In these studies, findings have delved into annual drought events and not rainfall seasons. Yet, crop farming and livestock activities are influenced by seasonal rainfall. Barret et al. (2020) and Klisch and Atzberger (2016) explored forecasting vegetation indices for drought monitoring in Kenya. It is important to note that accuracy of drought prediction is largely dependent on the quality/type of data used.

This work sought to establish the frequency and magnitude of droughts in Eastern Kenya, which is dominantly arid and semiarid land (ASAL). More than 80% of Kenya's land surface is ASAL and 71% of land area has a 25% to 50% drought probability (Bartel & Muller, 2007). For the period 1998 and 2011, the Kenyan government and international humanitarian agencies spent a combined total of US$ 1278.9 on drought mitigation (USAID, 2018). To reverse this situation, the Kenyan government has put in place four components dubbed the Kenya Drought Management System that targets the arid and semiarid counties. These are drought early warning systems, county level contingency plans, National Drought Emergency Fund (NDEF) and drought coordination and response structures (USAID, 2018). At a regional level, the government of Kenya will need stronger cross-border approaches to solving the impacts of droughts. This should include regional monitoring and coordination (FAO, 2017). To actualize these plans, details of local level characteristics of drought are required. Eastern Kenya, a semiarid region, is prone to drought. Against this background, the study used both in situ and gridded rainfall data to analyze the characteristics of meteorological droughts for March-May (MAM) and October-December (OND) rainfall seasons in Eastern Kenya. Specifically, the study sought to provide answers to the questions; (i) what is the spatial and temporal spread of meteorological droughts in Eastern Kenya?, and (ii) what is the difference between gridded and in situ data in revealing meteorological droughts in Eastern Kenya?

Analysis of drought, depending on the type, can be based on different time-scales as exemplified by Fabeku and Okogbue (2014), World Bank (2013), and (Elkollaly et al., 2018). This study focused on meteorological drought with the rainfall amount as the indicator of analysis. The period of analysis based on two main rainfall seasons: MAM and OND.

2 Materials and methods

2.1 Study area

The study area is Eastern Kenya, an ASAL that falls within the former Eastern Province of Kenya. Fig. 9.1 shows a map of the study area—counties, altitude, and in situ rainfall stations. Moyale and Isiolo, part of the former Eastern province, were excluded because they are predominantly under pastoralism (Cecchi et al., 2010). According to the 2019 national census (Republic of Kenya, 2019), the six counties of study (Makueni, Machakos, Embu, Kitui, Tharaka-Nithi, and Meru) have a total population of 6,093,262, accounting for 13% of the country's population. The study area has a varied elevation that ranges from about 600 m to 1600 m a.s.l.

Eastern Kenya falls within the tropics, a factor that explains the bi-modal rainfall season that characterizes the area. The two main growing seasons in Kenya are MAM and OND. Although both seasons are used for crop farming, OND

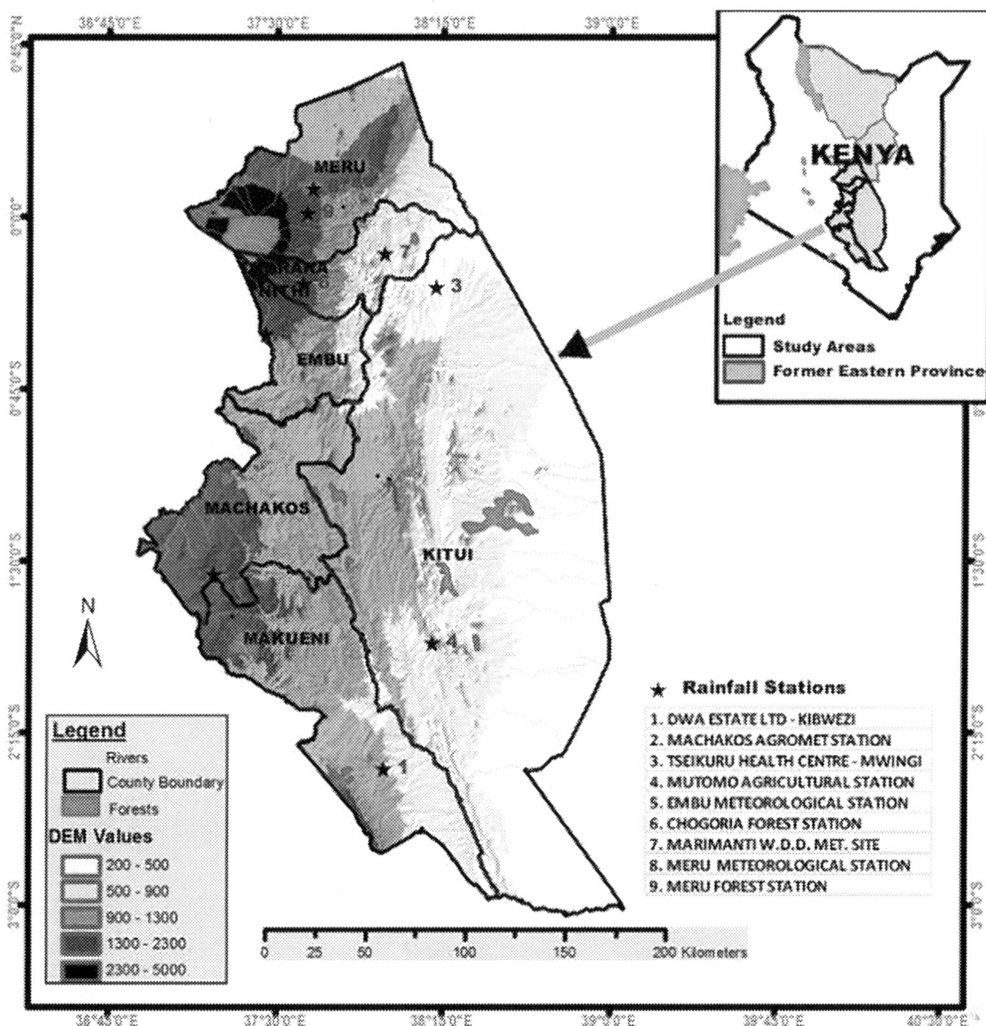

FIG. 9.1 A map showing the study area and distribution of in situ rainfall stations.

rainfall has been found to be more reliable and dependent on by the agro-pastoralists in the region (Hansen & Indeje, 2004). Both the timing and length of growing seasons vary from year to year due to the high interannual variability of rainfall that characterizes the area (Ngugi, Rao, Oyoo, & Kwena, 2014). A variety of cereals (maize, sorghum, and millet), industrial crops (tea and coffee), and legumes (beans, cowpeas, green grams and pigeon peas) are cultivated in the region. The choice of the study area was informed by the following reasons. First, Eastern Kenya, in terms of area coverage, has the largest agro-pastoral system in Kenya (Cecchi et al., 2010). Thus, study findings on drought occurrence will offer grounds for understanding the effect of drought on the region's economic mainstay—agro-pastoralism. Second, the diversity of agro-ecological zones—from the upper midland in the highlands (north) to the inner lowlands (southern and eastern) (Jaetzold, Schmidt, Hornetz, & Shisanya, 2006)—would suggest variations in the magnitude of droughts. Identification of the spatial spread and frequency of droughts gives a platform for county-level drought intervention strategies.

2.2 Data collection

First, the choice of counties in this study was purposive, guided by the region's national level contribution to dryland crop farming. Second was the selection of in situ rainfall stations. The choice of in situ rainfall stations was guided by representation of the agro-ecological zones (Jaetzold et al., 2006) in Eastern Kenya (Table 9.1). The selection of agro-ecological zones was guided by areal coverage (size), availability of in situ rainfall data exceeding 20 years (WMO and GWP, 2016) and representation of each county. Based on these, a total of 22 rainfall stations were identified and a request was made to the Kenya Meteorological Department (KMD). Due to gaps (missing data) and cost for data,

TABLE 9.1 Rainfall stations and their respective altitude and agro-ecological zones.

County	Sr. no.	Rainfall station	Agro-ecological zone	Altitude in m a.s.l	Rainfall data length		% of missing data (in situ)
					Period	No. of years	
Makueni	1	Kibwezi	Lower Midland 5	914	1985–2014	30	10
Machakos	2	Machakos Agro-met	Upper Midland 4	1600	1970–2016	47	0
Kitui	3	Tseikuru	Inner Lowlands 6	594	1970–2016	47	23
	4	Mutomo	Lower Midland 5	914	1974–2011	38	22
Embu	5	Embu Met	Upper Midland 3	1494	1976–2016	41	0
Tharaka-Nithi	6	Chogoria Forest	Upper Midland 1	1372	1970–2013	44	8
	7	Marimanti	Inner Lowlands 5	587	1969–1997	28	14
Meru	8	Meru Met	Upper Midlands 2	1524	1970–2015	46	1
	9	Meru Forest	Lower Midlands 3	1585	1970–2013	44	13

rainfall stations were reduced to nine as shown in Fig. 9.1 and Table 9.1. Significantly, all counties and major agro-ecological zones in Eastern Kenya were represented.

Two types of rainfall data sets were used in this study—gridded and in situ. The gridded monthly data was downloaded from the University of Delaware at $0.5° \times 0.5°$ horizontal resolution version 4.01 while in situ data was acquired from the KMD. In selecting the University of Delaware gridded data version 4.01, it was found to have a reasonably high resolution, longer temporal records with no missing observation at no cost (Matsuura & Wilmott, 2015). In addition, gridded data mitigated the uneven spread and incomplete in situ data that characterize Eastern Kenya. Informed by the findings of Manzanas, Amekudzi, Preko, Herrera, and Gutiérrez (2014), in situ and gridded rainfall data sets of nine stations for the period of 40 years, 1973–2013 were compared.

The period of analysis was considered because a majority of the study sites had in situ data falling within this period (Table 9.1). According to WMO and GWP (2016), analysis of drought using SPI should ideally have a minimum of 30 years of data. Thus, although gridded data was available for the period 1900–2014, its use in this study was confined to the period 1974–2013.

2.3 Data analysis

The first step in data analysis was to check the quality of data and fill in missing values. There was no missing data for the gridded data. In situ data were, however, found to have gaps as shown in Table 9.1. Although WMO sets the threshold for filling in missing data at 10%, this study set the threshold at 15% for purposes of increasing representation in the study area. Thus, because in situ data for Tseikuru and Mutomo stations exceeded 15% (Table 9.1), they were excluded. Multiple imputation methods were used to fill in missing data because it does not suffer from the problem of underestimating the sample error as it appropriately adjusts the standard error of the missing data (Enders, 2010). A paired *t*-test was applied to determine a significant difference between gridded and in situ data for OND and MAM seasons.

The Standard Precipitation Index (SPI) was used to estimate the occurrence of drought for the 3 months of the main rainfall seasons in Eastern Kenya; MAM and OND seasons. Analysis SPI based on 3 months—for MAM and OND—was to capture the impact of droughts on agriculture as the study area has a bimodal rainfall. The choice of SPI was informed by its computational flexibility for multiple timescales. Second, SPI's probabilistic nature (which involves fitting a Gamma probability density function) gives it a historical context. Thirdly, SPI uses rainfall data only as the input parameter. For scarce data regions such as Eastern Kenya, SPI is the most common method to quantify rainfall deficit. Overall, the SPI ranks top among drought indicators, and is robust and reliable (Spinoni et al., 2014). However, SPI has limitations as well. According to Mishra and Singh (2010) and WMO (2012), similar and consistent results are observed when the SPI values are computed from different lengths of record. This is because SPI applies a similar Gamma distribution over the same period. To address this limitation, 1974–2013 was the set period of analysis.

McKee, Doesken, and Kleist (1993) came up with values to measure the severity (or magnitude) of droughts as illustrated in WMO (2012) and Tigkas, Vangelis, and Tsakiris (2015). Accordingly, a drought event occurs when the SPI is continuously negative and reaches an intensity of -1.0 or less. Thus, in this study, the severity of droughts should be measured and classified as moderate (-1.0 to -1.49), severe (-1.5 to -1.99) and extremely dry (-2 and less). This should exclude the mild drought which occurs at SPI values of between -0.0 and -0.99. Drought Indices Calculator (*DrinC*) software version 1.5.73 was used to calculate the SPI. *DrinC* software was selected because it was freely available, user friendly and has an in-built distribution function (Gamma and log-normal). By using *DrinC*, monthly rainfall data was arranged in the *Ms Excel* spreadsheet in the format from October of that year *x* to September of the subsequent year. This was to conform to the *DrinC* requirements. Data was then imported into *DrinC* (from *Ms Excel*). For each station, data was run into *DrinC* for calculation of SPI values for MAM and OND seasons for both gridded and in situ data. Occurrence rate of drought severity (moderate, severe and extreme) for MAM and OND between 1973 and 2013 was calculated. For example, the occurrence rate of moderate drought was calculated as using Eq. (9.1)

$$MD = (n_{mds} \div n_{ds}) \times 100\% \tag{9.1}$$

where MD is Moderate Drought, n_{mds} is the total number of moderate drought months in a given season for the period on record, and n_{ds} is the total number of months for the period on record. The same formula should apply for severe and extreme droughts, symbolized as SD and ED respectively. Drought occurrence rate for MAM and OND was calculated.

A *t*-test was used to test the null hypothesis "there is no difference between gridded and in situ data to cause variations in the occurrence of meteorological drought." As such, the means of seasonal rainfall—MAM and OND for gridded and in situ data were compared. To carry out a *t*-test, mean, standard deviation, and standard error of mean were calculated for each of the datasets—gridded and in situ. The hypothesis was tested at 95% confidence level.

As noted in Table 9.1, there were variations in data availability for in situ data. Thus, four stations (Machakos Agro-Met, Chogoria, Meru Met, and Meru Forest) which had complete data for 1973 to 2013 were used for comparison between gridded and in situ data.

3 Results and discussions

3.1 Spatiotemporal occurrence of meteorological droughts

Results in Fig. 9.2A show that extreme drought was recorded at all stations during the 2000 MAM season. Extreme drought events during the MAM season were also recorded in 1984 (Marimanti) and 2007 (Kibwezi). Severe droughts were experienced in 1984 (on 5/9 stations) and 2009 (7/9 stations). Although moderate droughts were the most common, they highly varied in space and time. In terms of spatial spread, Tseikuru recorded the highest number of MAM droughts—seven (4—moderate, 2—severe, and 1—extreme drought). Stations in the north (Embu, Chogoria, Marimanti, Meru Met and Meru Forest) experienced six droughts during the MAM season. Kibwezi and Machakos Agro-Met stations had five drought events each for the period of study during the MAM season.

Extreme droughts were not experienced during the OND in Eastern Kenya (Fig. 9.2B). However, severe and moderate droughts occurred at various time scales. For instance, 1987 (6/9 stations) and 2005 (7/9 stations) were years of severe drought in Eastern Kenya during the OND season. Moderate droughts were recorded in most stations in 1973, 1975, 1985, 1998 and 2000 during the OND season. There are noticeable spatial distinctions in the occurrence of droughts. An example is 1987 where stations in the south (Kibwezi, Machakos Agro-Met and Mutomo) recorded moderate drought; while stations in the north (Tseikuru, Embu, Chogoria, Marimanti, Meru Met and Meru Forest) experienced severe drought. The spatial variation of drought is further reflected in 1980 (moderate drought at stations in the south), 1985 and 1998 (moderate drought at stations in the north) and 2010 (moderate drought at stations in the south). Overall, stations in Kitui County, a lowland, had the highest number of OND droughts on record (Mutomo—9; Tseikuru—8). Machakos Agro-Met station had the least number of OND droughts—six. All the other stations had seven drought events each for the period on record.

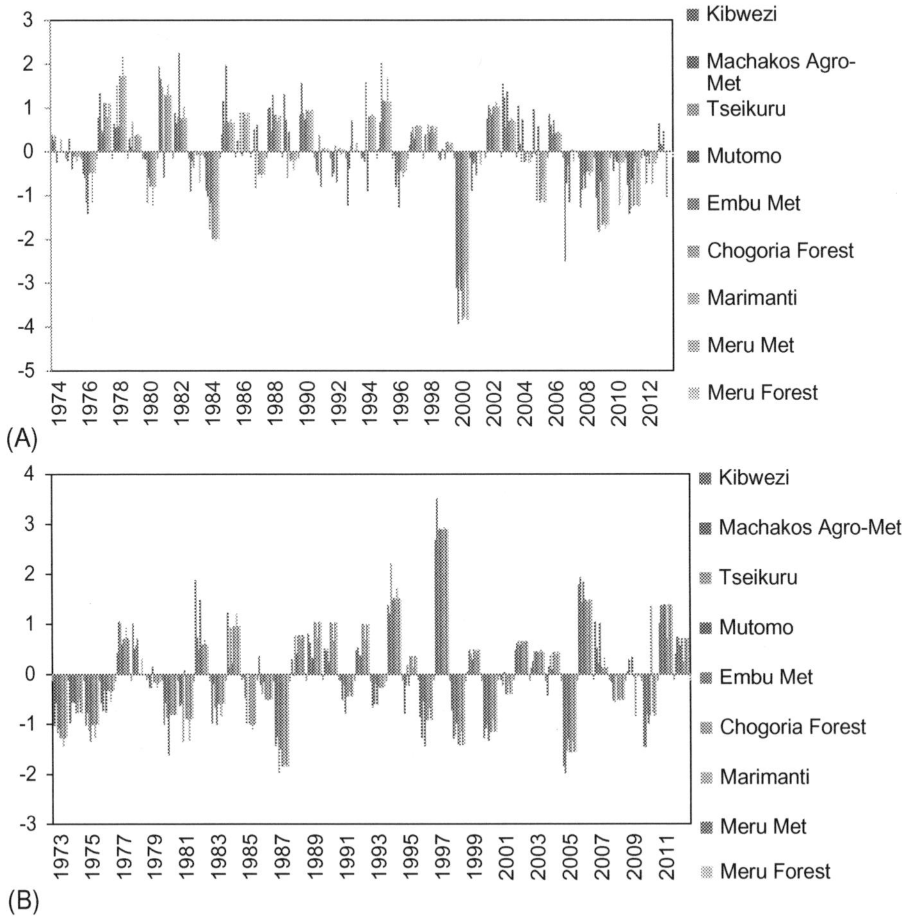

FIG. 9.2 (A) SPI results of gridded data for MAM rainfall season. (B) SPI results of gridded data for OND rainfall season.

In Eastern Kenya, there were instances when the two rain seasons failed in the same year or followed each other (Table 9.2). The years 2000 and 2005 experienced droughts during MAM and OND seasons. It was further observed that whereas the 2000 drought was spread in the whole region, the drought of 2005 was limited to stations in the north. In 1975–1976 season, Tseikuru, Mutomo and Marimanti experienced moderate droughts during OND and MAM seasons. These three stations fall in the inner lowlands (IL) 5 and 6 agro-ecological zones—the driest in Eastern Kenya.

TABLE 9.2 Subsequent drought incidences in Eastern Kenya.

Year	Station (s)	Sequence of seasons	Magnitude of drought
1975–1976	Tseikuru, Mutomo, Marimanti	OND	
		MAM	
1983–1984	Mutomo	OND	
		MAM	
1996	Mutomo	MAM	
		OND	
2000	Kibwezi, Machakos Agro-Met, Tseikuru, Mutomo, Embu Met, Chogoria, Meru Met, Meru Forest	MAM	
		OND	
2005	Tseikuru, Embu, Chogoria, Meru Met, Meru Forest	MAM	
		OND	
2007	Kibwezi	MAM	
2008			
2009			
2010–2011	Machakos Agro-Met	OND	
		MAM	

Key: Moderate drought; Severe drought; Extreme drought

Consecutive droughts aggravate the impact of droughts on livelihoods and natural resources upon which people depend on.

Temporarily, 1984, 2000, 2005, 2009, 2011 (MAM), 1973, 1975, 1985, 1987, 1998, 2000 and 2005 (OND) were years of moderate, severe and extreme droughts in Eastern Kenya. The findings in part, corroborate Okal et al. (2020) who established drought years in 1984, 1996, 2000 and 2008. Ayugi et al. (2020) established extreme drought incidences in 1984, 1987, 2000, 2006, and 2009 using a 3-month Standardized Precipitation-Evaporation Index (SPEI). In this study however, extreme drought was only observed in 2000 in all stations during MAM.

The spatial and temporal variation of droughts in Eastern Kenya is an indication of the contrasting processes involved in the development of drought. For instance, Lott, Christidis, and Stott (2013) attributed failure of the 2010 OND rains to La Nina, and the 2011 MAM rains to human influence. Similarly, Benitez and Domecq (2014) established spatial variations and the influence of La Nina events in the occurence of droughts in Paraguay. Marthews et al. (2019) determined that human-induced climate change may have resulted in Lake Victoria and Northern Kenya regions receive less rainfall during the MAM rains of 2014. These findings suggest that human activities such as deforestation, overgrazing and cultivation of land in semiarid environments (which opens up the soil to evaporation of soil moisture) are potential determinants of occurrence of droughts. This is in addition to the teleconnections associated with the oceans, atmosphere, and land. This calls for improvement of climate forecast skills where land surface characteristics are included in the modeling at regional or local levels.

The distinct variation in the occurrence of droughts between the lowlands (Kibwezi, Mutomo, and Tseikuru) and highland areas on the leeward side of Mt. Kenya (Chogoria, Embu, Meru Forest, and Meru Met) underscore the place of relief in influencing droughts. There is a likelihood that proximity to Mt. Kenya is influencing the magnitude of drought in Eastern Kenya. These findings are supported by Buttaffuoco, Caloiero, Ricca, and Guagliardi (2018) who established a strong relationship between orography and Reconnaissance Drought Index (RDI) in southern Italy. Similarly, Ramkar and Yadav (2018) and Mbiriri, Mukwanda, and Manatsa (2018) found that highland areas were more prone to droughts in the Upper Girna Basin (India) and Free State Province (South Africa), respectively. But the exact contribution of altitude and relief needs further investigation since there are instances when lowlands experience droughts of high magnitude than highland areas. An example is 1980 and 2010 OND seasons when the lowlands (Kibwezi and Mutomo) experienced moderate drought while the highland areas (Embu, Chogoria, Meru Met and Meru Forest) experienced mild drought. The findings corroborate Okal et al. (2020) who found the lowlands of the Upper Tana River recorded the worst drought in 2014 and 2013 when compared to highland areas.

There could also be a possibility that distance from the Indian Ocean is influencing magnitude of drought in Eastern Kenya. This would explain why stations in the south (Kibwezi, Mutomo, and Machakos Agro-Met) experienced moderate or mild droughts in 1984, 2009 (MAM), and 1987 (OND) when stations in the north experienced a higher severe or extreme droughts. Spatial variation in the occurrence of drought also highlights the centrality of agro-ecology in increasing climate resilience as articulated by Leippert, Darmaun, Bermoux, and Mpheshea (2020). Noting distinctions in occurrence of droughts between the highlands and lowlands of Eastern Kenya, climate change adaptation—especially when planning for drought events, must be take cognizance of agro-ecological zones.

Moderate droughts have the highest occurrence rate in Eastern Kenya during MAM and OND rainfall seasons. During MAM, the highest occurrence rate of moderate drought was about 10% at Mutomo, Meru Met, Embu and Chogoria (Fig. 9.3A). While during OND season, the highest occurrence rate of moderate drought was at Meru Forest (18%), Embu (18%), and Chogoria (17%) (Fig. 9.3B). All stations had an occurrence rate of less than 10% for severe and extreme

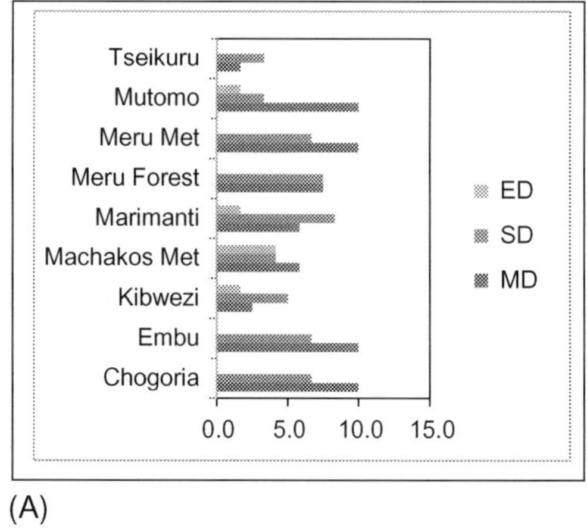

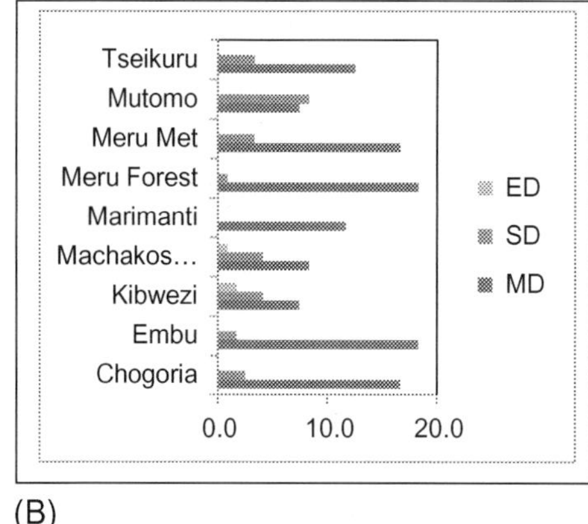

FIG. 9.3 Occurrence rate of drought events in Eastern Kenya. (A) MAM and (B) OND. Key: *ED*, extreme drought; *SD*, severe drought; *MD*, moderate drought.

droughts for both MAM and OND season. The likelihood of severe drought was highest during the MAM season than OND season in the study area. Drier areas of Inner Lowlands 5 (Marimanti—8%) and Lower Midlands 5 (Mutomo—8%) had the highest occurrence rate of severe droughts during MAM and OND seasons respectively. In a related study, Benitez and Domecq (2014) established the occurence of severe and extreme drought events in Paraguay. A study by Haile et al. (2020) showed that the East Africa region is likely to experience an increase in area affected by extreme drought, more than severe and moderate droughts. This view is reinforced by Naumann et al. (2018) who established that the magnitude of droughts in Eastern Africa (alongside Australia, Central Argentina and North-Eastern Brazil) would double within the 21st century. This would call for incorporation of extreme drought events in climate change adaptation planning especially at county levels in Kenya.

The vulnerability of OND season to drought when it's the most dependent (Recha, Makokha, Shisanya, & Mukopi, 2017) on by the agro-pastoralists of Eastern Kenya is of concern. The relatively high occurrence rate of severe droughts in the lowlands of Kitui County (Tseikuru), Tharaka-Nithi (Marimanti), Makueni (Kibwezi) (MAM) and Mutomo (OND) underscores their vulnerability to a variable climate. These findings corroborate Mutua and Runguma (2020) who found a deterioration of the MAM rains. The likelihood of severe droughts during MAM season in Eastern Kenya could be a pointer or a justification of why farmers invest less in the season when compared to OND season. Recha et al. (2017) established that households in Tharaka-Nithi County commit more land area to crop farming in the OND season than MAM season. Although it is important to note that the probability of occurrence of severe and extreme droughts in most parts of the world lower than that of moderate droughts (Geng et al., 2016). Thus, the low occurrence rate of moderate droughts during MAM season should be used to encourage farmers to invest more in the season. The occurrence rate of more than 15% during OND season when compared to Geng et al. (2016) should be of concern as this poses a risk to the agro-pastoral enterprise that characterize Eastern Kenya. The high occurrence of moderate droughts in Eastern Kenya is in agreement with the findings of Ayugi et al. (2020) who established a high frequency of moderate droughts across Kenya.

3.2 Gridded verses in situ data: A comparison of meteorological drought events

The study sought to compare occurrence and magnitude of meteorological drought events based on gridded and in situ datasets. *t*-test results show that there is a difference between gridded and in situ data in all the four stations that were compared as shown in Table 9.3A and B. Indeed, gridded data had much lower values for rainfall data when compared to in situ data—a factor that contributed to the lower mean values of rainfall. Previous studies conducted in Eastern Kenya such as Mutua and Runguma (2020), Shisanya, Recha, and Anyamba (2011), and Recha, Shisanya, Anyamba, and Okolla (2014) show that MAM and OND seasonal rainfall are in the range of 200 to 500 mm and 300 to 900 mm, respectively. For these studies, in situ (or gauged data) was used. The similarity in gridded values between Meru Forest and Meru Met can be attributed to the two stations being within a range of less than 55 km apart.

TABLE 9.3 *t*-Test results of comparison between gridded and station data sets for MAM (A) and OND (B) rainfall seasons.

(A)	MAM—Gridded			MAM—In situ				
Station	Mean (mm)	SD	SEM	Mean (mm)	SD	SEM	*t*-Value	*p*-Value
Chogoria	28.7	11.9	1.8	909.8	385.3	60.2	15.0	0.0001
Machakos Agro-Met	27.8	9.3	1.5	281.4	123.9	19.4	13.7	0.0001
Meru Forest	29.1	11.9	1.9	489.2	154.4	24.1	20.3	0.0001
Meru Met	29.1	11.9	1.9	450.8	178.6	27.9	16.0	0.0001
(B)	**OND—Gridded**			**OND—In situ**				
Station	**Mean (mm)**	**SD**	**SEM**	**Mean (mm)**	**SD**	**SEM**	***t*-Value**	***p*-Value**
Chogoria	35.2	18.6	2.9	897.3	304.5	47.6	19.0	0.0001
Machakos Agro-Met	25.7	14.1	2.2	280.8	126.3	19.7	14.0	0.0001
Meru Forest	35.2	18.6	2.9	702.7	311.2	48.6	14.5	0.0001
Meru Met	35.2	18.6	2.9	698.5	298.9	43.6	16.1	0.0001

95% confidence interval, df = 40. Key: *SD*, standard deviation; *SEM*, standard error of mean.

MAM

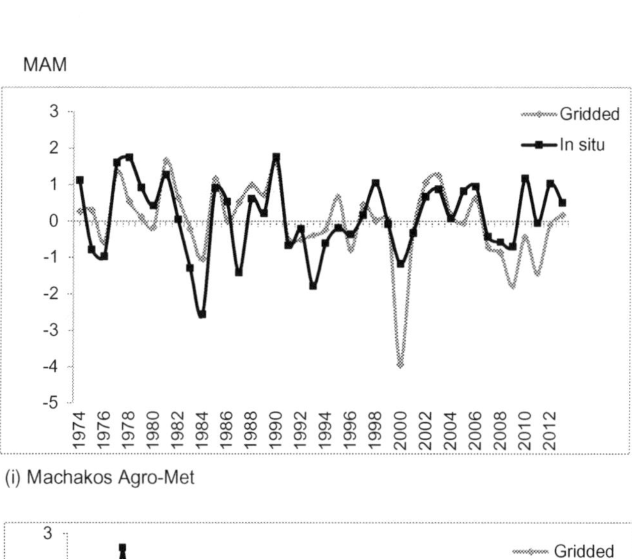

(i) Machakos Agro-Met

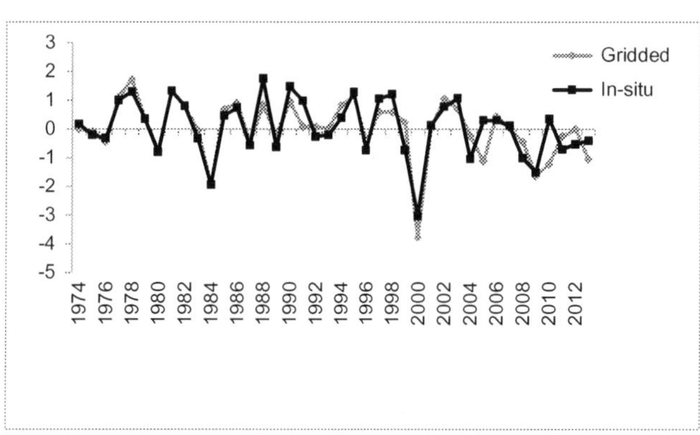

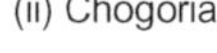
(ii) Chogoria

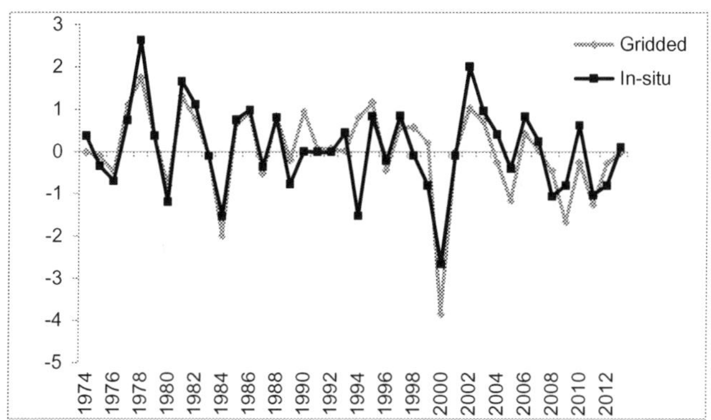

(iii) Meru Met

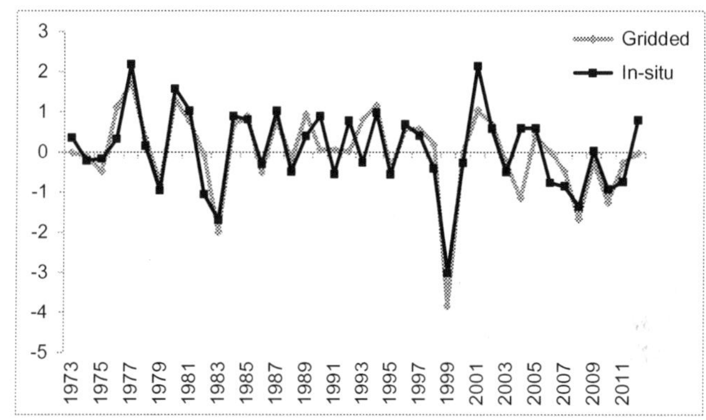

(iv) Meru Forest

(A)

OND

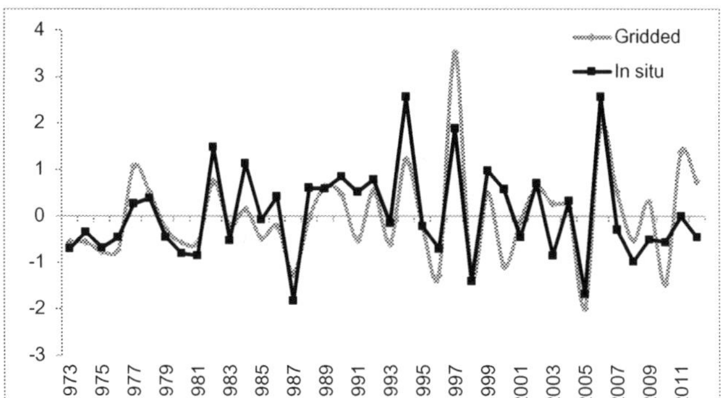

(i) Machakos Agro-Met

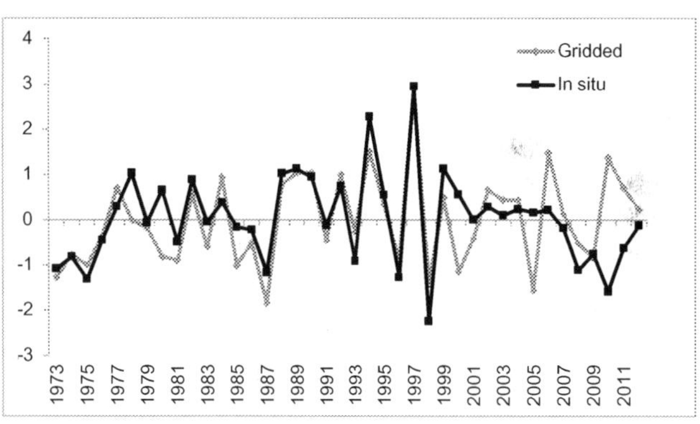

(ii) Chogoria

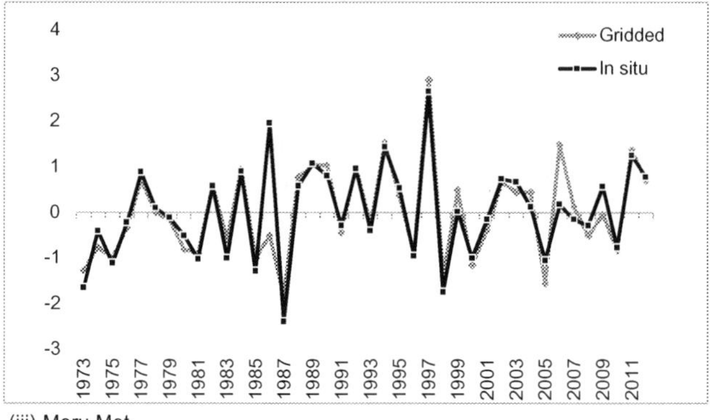

(iii) Meru Met

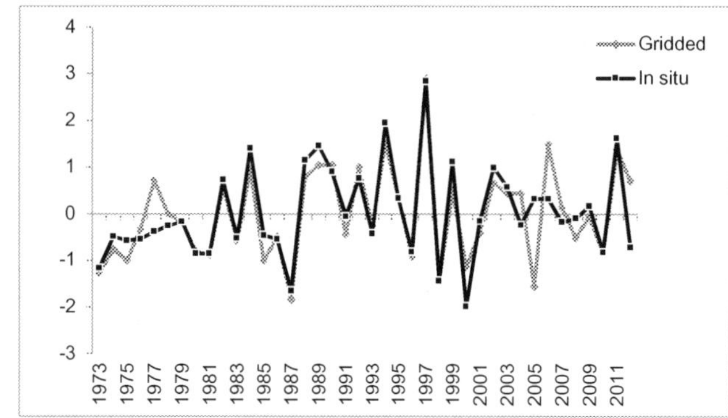

(iv) Meru Forest

(B)

FIG. 9.4 (A) A comparison of drought events in the March-May season for selected stations under gridded and in situ data. (B) A comparison of drought events in the October-December season for selected stations under gridded and in situ data.

The difference in the two data sets is further reflected in the number of drought events and when the droughts occurred. Fig. 9.4A and B shows drought events and their magnitude for both MAM and OND seasons. It is evident that the number of drought events and when they occurred varied between gridded and in situ data sets. For instance, at Machakos Agro-Met, there were a total of four (1984, 2000, 2009, and 2011) and five (1983, 1984, 1987, 1993, and 2000) drought events under gridded and in situ data sets, respectively, during MAM season. Similarly, Chogoria, Meru Met and Meru Forest experienced each six droughts during MAM season under gridded data. But under in situ data, Chogoria, Meru Met, and Meru Forest stations, respectively, experienced five, six, and four droughts in the period 1973–2013. During the OND season, gridded data show that all the four stations experienced 6 to 7 droughts (Machakos Agro-Met-6; Chogoria, Meru Met, and Meru Forest—7 droughts each). But under in situ data, Machakos Agro-Met had the lowest number of droughts (3), while Meru Met had the highest number of droughts (9). Significantly, these drought events occurred at different times (years) and varied in magnitude. For instance, at Chogoria, 2005 and 2010 experienced moderate droughts under gridded data and normal rainfall under in situ data. Similarly, Meru Forest experienced seven droughts (1973, 1975, 1985, 1987, 1998, 2000, and 2006) under gridded data and four droughts (1973, 1987, 1998, and 2000) under in situ data during the OND season. In 2009 MAM season, results of gridded data show severe drought in all stations while in situ data presented varied magnitudes of drought in the four stations. In 1983, a severe drought was recorded at Machakos Agro-Met for in situ data during the MAM season; but all other stations (for both in situ and gridded) recorded mild drought. Gridded and in situ data show that the 1985 OND season experienced moderate and mild droughts respectively. In 1998, the drought of the OND season was captured by both data-sets but with varying magnitudes again. In the 1999 MAM season, gridded and in situ data had +ve and −ve SPI values, respectively—but all fell within a near normal range (−0.99 to 0.99).

There were also instances of concurrence between gridded and in situ data. This is reflected in the extreme drought of 2000 and the severe drought of 1984 during the MAM season. It is also observed that for most drought years, such as 1983, 2009, 2011 (MAM), 1973, 1987, and 2000 (OND), both data sets indicated below normal rainfall (but with varying magnitude).

Rarieya and Fortun (2010), Mateche (2011), and Huho, Ngaira, and Ogindo (2010) identified droughts in the period 1970s to the 2000s. Their findings are in agreement with analysis of gridded data for OND. For in situ data, the MAM seasons of 1975 and 1977 experienced mild drought (−0.00 to 0.99) and near normal (−0.99 to 0.99) rainfall respectively. The 1984 MAM drought is well documented in most studies (Huho et al., 2010; Mateche, 2011; Tilahun, 2006) and is captured by both data sets in this study. According to Rarieya and Fortun (2010), 1992–1993, 1996–1997, 1999–2000, 2005–2006, and 2008–2009 were disaster years in Kenya. A run through these individual years confirmed that at most stations the rains were below normal and in some instances, drought occurred. The findings are also in agreement with Nathan et al. (2020) who established that in the Central highlands of Kenya, both in situ (observed rain gauge) and satellite data had a similar trend, but with a varying magnitude.

Variation in magnitude of droughts for the two datasets puts into focus the criteria for declaration of national disasters and vulnerability thresholds for affected counties. In Kenya, the National Drought Management Authority (NDMA) uses gridded rainfall data (WFP-VAM, CHIRPS/MODIS) alongside household livelihood status (e.g., livestock death, milk quantity, and distance to water points) and vegetation conditions to establish thresholds for drought. This is in combination with other impact indicators such as the vegetation condition index (www.ndma.go.ke/). In India, however, in situ rainfall data (alongside other variables) is primarily used to determine droughts (Government of India, 2016).

Analysis of gridded data shows that Meru Met (10.0%) and Meru Forest (10.0%) had the highest occurrence rate of moderate droughts during the MAM season (Table 9.4A). Occurrence of severe droughts during MAM season was highest at Meru Forest (7.5%—gridded data), Chogoria (6.7%—gridded data), Machakos Agro-Met (6.7%—in situ data), and Meru Met (6.7%—gridded data). These results do not show a clear pattern in the occurrence rate of droughts for gridded and in situ data sets for the MAM season.

In the OND season, gridded data had higher values for moderate drought in all the stations (Table 9.4B). Meru Forest (18.3%), Meru Met (16.7%) and Chogoria (16.7%) had the highest occurrence rate of moderate droughts using gridded data. Occurrence rate of moderate droughts was lowest at Meru Forest (3.3%) under in situ data during the OND season. At Machakos Agro-Met, the occurrence rate of moderate droughts was the same for both in situ and gridded data. In situ data appeared to yield a high occurrence rate for severe droughts at Chogoria (4.2%), Machakos Agro-Met (7.5%) and Meru Forest (6.7%) during the OND season. Similarly, in situ data yield high values for extreme drought at Chogoria, Machakos agro-Met and Meru Forest.

Findings of high values of occurrence rate for moderate droughts during OND contradict Nathan et al. (2020) who found that satellite data underestimate observed rain gauge data. Quesada-Montano, Wetterhall, Westerberg, Hidalgo, and Halldin (2019) compared Climate Hazards Group Infra-Red Precipitation (CHIRPS), Climate Research Unit (CRU) and station data-sets and large differences between the datasets were established. In this study, large

TABLE 9.4 A comparison of gridded and in situ drought occurrence rate in percentage.

	Chogoria		Machakos Agro-Met		Meru Forest		Meru Met	
Stations/drought severity	G	I	G	I	G	I	G	I
(A) *MAM*								
Moderate drought	10.0	7.5	5.8	4.2	7.5	8.3	10.0	10.8
Severe drought	6.7	4.2	4.2	6.7	7.5	4.2	6.7	3.4
Extreme drought	0.0	5.8	4.2	1.7	0.0	3.3	0.0	2.5
(B) *OND*								
Moderate drought	16.7	5.8	8.3	8.3	18.3	3.3	16.7	15.0
Severe drought	2.5	4.2	4.2	7.5	0.8	6.7	3.3	3.3
Extreme drought	0.0	2.5	0.8	6.7	0.0	0.0	0.0	1.7

NB: *G*, gridded data; *I*, in situ data.

differences between in situ and gridded data were observed at Chogoria and Meru Forest for moderate droughts during the OND season. Although there are also instances of concurrence, such as at Machakos Agro-Met (moderate drought) and Meru Met (severe drought) for the OND season. The notable difference between in situ and gridded data makes data-set selection an important aspect in the characterization of drought. The results are a pointer to the need for more studies to compare and determine ideal sources of climate data for prediction and estimation of droughts. This should subsequently improve the quality of forecasts and management of droughts.

4 Limitations of the study

A limitation of the study was lack of adequate in situ stations in the study area, and a high cost of climate data levied by the Kenya Meteorological Department (KMD) (rainfall data is charged by station and length of duration). Although there was data for 22 stations, only 12 rainfall stations had adequate data to support a climatological analysis. Based on the rates of KMD, a fee of KES 244,200 (approx. US$2442) was required to purchase data for 12 stations. This cost was way off what the project offered for the purchase of data. To mitigate the limitation of adequacy of stations and representation of the study area, gridded data was used to compliment available in situ climate data. While missing data was filled using the approach as described in Section 2.3.

5 Conclusions

This chapter sought to; (i) analyze the spatial and temporal spread of meteorological droughts in Eastern Kenya, and, (ii) determine the difference between gridded and in situ data in analyzing meteorological drought in Eastern Kenya. *DrinC* software was used to calculate SPI values for the main rainfall seasons; March-May and October-December. The study established that Eastern Kenya has experienced more moderate droughts during the MAM and OND seasons for the period 1973–2013. Whereas stations close to Mt Kenya (Tseikuru, Meru Met, Meru Forest, Chogoria, Marimanti, and Embu) had the highest number of drought events during MAM, stations in the lowlands (in Kitui County) had the highest number of droughts during OND. In Eastern Kenya, there are times when droughts occur consecutively for the two rainy seasons of MAM and OND as was recorded in 1975, 1984, 2000 and 2005 for some stations. The occurrence rate of moderate droughts was higher during OND than MAM rainfall seasons. The occurrence rate of severe drought was highest in the driest agro-ecological zones of Inner Lowlands 5 (at Tseikuru and Marimanti) and Lower Midlands 5 (Kibwezi) during the MAM rainfall season. A comparison of gridded and in situ data shows concurrence in the occurrence of drought; there is, however, variation in magnitude. There is no discernible pattern in the occurrence of drought using both gridded and in situ during the MAM season. However, during the OND season, the occurrence of moderate droughts is highest using gridded data, while the occurrence of severe and extreme droughts was highest using in situ data. Based on these findings, it is recommended that the lower vulnerability of Eastern Kenya to MAM season droughts (compared to OND) should be explored with a view to tapping the season's full potential in supporting livelihoods. It is also of importance to give significance to land surface characteristics (including agro-ecology) in climate forecasting and climate change adaptation at subnational levels. Noting that there is a significant difference between gridded and in situ data sets, more research is required to offer insights on each, and therefore inform in the management of droughts.

Acknowledgments

This research was supported by the Federation of Indian Chambers of Commerce and Industry (FICCI) through a C.V. Raman International Fellowship for African Researchers to the first author. The first author was awarded a 3-month visiting fellowship at the Jawaharlal Nehru University, New Delhi, India.

References

Al-Qinna, M. I., Hammori, N. A., Obeidat, M. M., & Ahmed, F. Y. (2011). Drought analysis in Jordan under current and future climates. *Climate Change, 106*, 421–440. https://doi.org/10.1007/s10584-010-9954-y.

Asfaw, A., Simane, B., Hassen, A., & Bantider, A. (2018). Variability and time series trend analysis of rainfall and temperature in north central Ethiopia: A case study in the Woleka sub-basin. *Weather and Climate Extremes, 19*, 29–41. https://doi.org/10.1016/j.wace.2017.12.002.

Ayugi, B., Tan, G., Rouyun, N., Zeyao, D., Ojara, M., Mumo, L., et al. (2020). Evaluation of meteorological drought and flood scenarios over Kenya, East Africa. *Atmosphere, 11*, 307. https://doi.org/10.3390/atmos11030307.

Barret, A. B., Duivenvoorden, S., Salakpi, E. E., Muthoka, J. M., Mwangi, J., & Rowhani, P. (2020). Forecasting vegetation condition for drought early warning system in pastoral communities in Kenya. *Remote Sensing of Environment, 248*, 111886. https://doi.org/10.1016/j.rse.2020.111886.

Bartel, P., & Muller, J. (2007). *Horn of Africa natural hazard probability and risk analysis*. https://www.unisdr.org/files/3868_HOA.pdf. (Accessed 13 April 2018).

Benitez, J. B, & Domecq, R. M. (2014). Analysis of meteorological drought episodes in Paraguay. *Climate Change, 127*. https://doi.org/10.1007/s10584-014-1260-7.

Buttaffuoco, G., Caloiero, T., Ricca, N., & Guagliardi, I. (2018). Assessment of drought and its uncertainty in a southern Italy area (Calabria region). *Measurement, 113*, 205–210. https://doi.org/10.1016/j.measurement.2017.08.007.

Cecchi, G., Wint, W., Shaw, A., Marletta, A., Mattioli, R., & Robinson, T. (2010). Geographic distribution and environmental characterization of livestock production systems in eastern Africa. *Agriculture, Ecosystems and Environment, 135*, 98–110. https://doi.org/10.1016/j.agee.2009.08.011.

Dai, A. (2011). Drought under global warming: A review. WIRES. *Climate Change, 2*, 45–65. https://doi.org/10.1002/wcc.81.

Davenport, F., Grace, K., Funk, C., & Shukla, S. (2017). Child health outcomes in sub-Saharan Africa: A comparison of changes in climate and socio-economic factors. *Global Environmental Change, 46*, 72–87. https://doi.org/10.1016/j.gloenvcha.2017.04.009.

Elkollaly, M., Khadr, M., & Zeidan, B. (2018). Drought analysis in the eastern Nile basin using the standard precipitation index. *Environmental Science and Pollution Research, 25*, 30772–30786. https://doi.org/10.1007/s11356-016-8347-9.

Enders, C. K. (2010). *Applied missing data analysis*. New York: The Guilford Press, ISBN:978-1-60623-639-0. Available at: https://books.google.co.in/books?id. (Accessed 25 April 2018).

Fabeku, B. B., & Okogbue, E. C. (2014). Trends in vegetation response to drought in the Sudano-Sahelian part of northern Nigeria. *Atmospheric and Climate Sciences, 4*, 569–588. https://doi.org/10.4236/acs.2014.44052.

Food and Agriculture Organization (FAO). (2017). *Horn of Africa cross-border drought action plan for 2017*. Available at: http://www.fao.org/3/i6968e/i6968e.pdf. (Accessed 5 May 2021).

Gbetibouo, G. A., Ringler, C., & Hassan, R. (2010). Vulnerability of the south African farming sector to climate change and variability: An indicator approach. *Natural Resources Forum, 34*, 175–187.

Geng, G., Wu, J., Wang, Q., Lei, T., He, B., Li, et al. (2016). Agricultural drought hazard analysis during 1980-2008: A global perspective. *International Journal of Climatology, 36*, 389–399. https://doi.org/10.1002/joic.4356.

Gizaw, M. S., & Gan, T. W. (2017). Impacts of climate change and El Niño episodes on droughts in sub-Saharan Africa. *Climate Dynamics, 49*, 665–682. https://doi.org/10.1007/s00382-016-3366-2.

Government of India. (2016). *Manual for drought management*. New Delhi: Ministry of Agriculture and Farmers Welfare. Available at: https://agricoop.nic.in/sites/default/files/Manual%20Drought%202016.pdf. (Accessed 3 May 2018).

Haile, G. G., Tang, Q., Hosseini-Mogari, S., Liu, X., Gebremicael, T. G., Leng, G., et al. (2020). Projected impacts of climate change on drought pattern over East Africa. *Earth's Future, 8*. https://doi.org/10.1029/2020EF001502. e2020EF001502.

Hansen, J., & Indeje, M. (2004). Linking dynamic seasonal climate forecasts with crop simulation for maize yield prediction in semi-arid Kenya. *Agricultural and Forest Meteorology, 125*, 143–157. https://doi.org/10.1016/j.agrformet.2004.02.006.

Huho, J. M., Ngaira, J. K. W., & Ogindo, H. O. (2010). Drought severity and their effects on rural livelihoods in Laikipia district, Kenya. *Journal of Geography and Regional Planning*, 2070-1845. 3(3), 35–43.

Jaetzold, R., Schmidt, H., Hornetz, B., & Shisanya, C. (2006). *Farm management handbook of Kenya: Part C1, Eastern Province. Vol. II* (2nd ed.). Nairobi: Ministry of Agriculture.

Kaluba, P., Verbist, K. M. J., Cornelis, W. M., & Ranst, E. (2017). Spatial mapping of drought in Zambia using regional frequency analysis. *Hydrological Sciences Journal, 62*, 1825–1839. https://doi.org/10.1080/02626667.2017.1343475.

Karanja, A., Ondimu, K., & Recha, C. (2017). Analysis of temporal drought characteristics using SPI drought index based on rainfall data in Laikipa west sub-county, Kenya. *Open Access Library Journal, 4*, e3765. https://doi.org/10.4236/oalib.1103765.

Klisch, A., & Atzberger, C. (2016). Operational drought monitoring in Kenya using MODIS NDVI time series. *Remote Sensing, 8*, 267. https://doi.org/10.3390/rs8040267.

Leippert, F., Darmaun, M., Bermoux, M., & Mpheshea, M. (2020). *The potential of agro-ecology to build climate-resilient livelihoods and food systems*. Rome: FAO and Biovision. https://doi.org/10.4060/cb0438en.

Lott, F. C., Christidis, N., & Stott, P. A. (2013). Can the 2011 east African drought be attributed to human-induced climate change? *Geophysical Research Letters, 40*, 117–1181. https://doi.org/10.1002/grl.50235.

Lyon, B. (2014). Seasonal drought in the greater horn of Africa and its recent increase during the march-may long rains. *Journal of Climate, 27*, 7953–7975. https://doi.org/10.1175/JCLI-D-13-00459.1.

Manzanas, R., Amekudzi, L. K., Preko, K., Herrera, S., & Gutiérrez, J. M. (2014). Precipitation variability and trends in Ghana: An inter-comparison of observational and reanalysis products. *Climatic Change, 124*(4), 805–819. https://doi.org/10.1007/s10584-014-1100-9.

Marthews, T. R., Jones, R. G., Dadson, S. J., Otto, F. E. L., Mitchell, D., Guillod, B. P., et al. (2019). The impact of human-induced climate change on regional drought in the horn of Africa. *Journal of Geophysical Research. Atmospheres, 124*, 4549–4566. https://doi.org/10.1029/2018JD030085.

Mateche, D. E. (2011). *The cycle of droughts in Kenya as looming humanitarian crisis*. Institute for Security Studies. https://issafrica.org/iss-today/the-cycle-of-drought-in-kenya-a-looming-humanitarian-crisis. (Accessed 15 May 2018).

Matsuura, K., & Wilmott, C. J. (2015). *Terrestrial precipitation: 1900–2014 (version 4.01)*. University of Delaware, Department of Geography Version 4.01. http://climate.geog.udel.edu/~climate/html_pages/Global2014/README.GlobalTsP2014.html. (Accessed 27 April 2018).

Mbiriri, M., Mukwanda, G., & Manatsa, D. (2018). Influence of altitude on spatio-temporal variations of meteorological droughts in mountain regions of the Free State Province, Africa (1960-2013). *Advances in Meteorology, 18*, 5206151. https://doi.org/10.1155/2018/5206151.

McKee, T. B., Doesken, N. J., & Kleist, J. (1993). The relationship of drought frequency and duration to time scales. In *Proceedings of the 8th conference on applied climatology, 17-23 January 1993, Anaheim, CA*. Boston, MA: American Meteorological Society.

Mishra, A. K., & Singh, V. P. (2010). A review of drought concepts. *Journal of Hydrology, 391*, 202–216.

Mutua, T. M., & Runguma, S. N. (2020). Annual and seasonal rainfall variability for the Kenyan highlands from 1900-2012. *Journal of Climatology & Weather Forecasting, 8*(3), 260. https://doi.org/10.35248/2332 2594.2020.8.260.

Nathan, O. O., Ngetich, F. K., Kiboi, N. M., Muriuki, A., Adamtey, N., & Mugendi, D. N. (2020). Suitability of different data sources in rainfall pattern characterization in the tropical central highlands of Kenya. *Heliyon, 6*, e05375. https://doi.org/10.1016/j.heliyon.2020.e05375.

Naumann, G., Alfieri, L., Wyser, K., Mentaschi, L., Betts, R. A., Carrao, H., et al. (2018). Global changes in drought conditions under different levels of warming. *Geophysical Research Letters, 45*, 3285–3296. https://doi.org/10.1002/2017GL076521.

Ngugi, L. W., Rao, K. P. C., Oyoo, A., & Kwena, K. (2014). Opportunities for coping with climate change and variability through adoption of soil and water conservation technologies in semi-arid eastern Kenya. In W. L. Filho, A. O. Esilaba, K. P. C. Rao, & G. Sridha (Eds.), *Climate change management: Adapting African agriculture to climate change, transforming rural livelihoods* (pp. 149–157). New York: Springer. https://doi.org/10.1007/978-3-319-13000-2.

Nicholson, S. E. (2014). A detailed look at the recent drought situation in the Greater Horn of Africa. *Journal of Arid Environments, 103*, 71–79. https://doi.org/10.1016/j.jaridenv.2013.12.003.

Nikoloski, Z., Christiaensen, L., & Hill, R. (2018). Household shocks and coping mechanism: Evidence from sub-Saharan Africa. In Christiaensen, Luc, & Demery (Eds.), *Agriculture in Africa: Telling myth from facts. Direction in development—Agriculture and rural development* (pp. 123–134). Washington, DC: World Bank.

Ochieng, R., Recha, C. W., & Bebe, B. O. (2017). Rainfall variability and droughts in the drylands of Baringo County—Kenya. *Open Access Library Journal, 4*, e3827. https://doi.org/10.4236/oalib.1103827.

Okal, H. A., Ungetich, F. K., & Okeyo, J. M. (2020). Spatial-temporal characterization of droughts using selected indices in upper Tana River watershed, Kenya. *Scientific African, 7*, e00275. https://doi.org/10.1016/j.sciaf.2020.e00275.

Onyango, A. O. (2014). Analysis of meteorological drought in the north-Eastern Province of Kenya. *Journal of Earth Science and Climate Change, 5*(8), 219. https://doi.org/10.4172/2157-7617.1000219.

Padgham, J., Abubakari, A., Ayivor, J., Dietrich, K., Fosu-Mensoh, B., Gordon, C., et al. (2016). *Vulnerability and adaptation to climate change in semi-arid areas in West Africa*. A technical report of ASSAR https://start.org/wp-content/uploads/West-Africa-RDS.pdf. (Accessed 17 April 2021).

Poterie, A. S. T., Jjemba, W. E., Singh, R., Perez, E. C., Costella, C. V., & Arrighi, J. (2018). Understanding the use of 2015-2016 El Niño forecasts in shaping early humanitarian action in eastern and southern Africa. *International Journal of Disaster Risk Reduction, 30*, 81–94. https://doi.org/10.1016/j.ijdrr.2018.02.025.

Quesada-Montano, B., Wetterhall, F., Westerberg, I. K., Hidalgo, H. G., & Halldin, S. (2019). Characterizing droughts in Central America with uncertain hydro-meteorological data. *Theoretical and Applied Climatology, 137*, 2125–2138. https://doi.org/10.1007/s00704-018-2730-z.

Ramkar, P., & Yadav, S. M. (2018). Spatial temporal drought assessment of a semi-arid part of the middle Tapi River Basin, India. *International Journal of Disaster Risk Reduction, 28*, 414–426. https://doi.org/10.1016/j.ijdrr.2018.03.025.

Rarieya, M., & Fortun, K. (2010). Food security and seasonal climate information: Kenyan challenges. *Sustainability Science, 5*, 99–114. https://doi.org/10.1007/s11625-009-0099-8.

Recha, C. W., Makokha, G. L., Shisanya, C. A., & Mukopi, M. N. (2017). Climate variability: Attributes and indicators of adaptive capacity in semi-arid Tharaka sub-county, Kenya. *Open Access Library Journal, 4*, 1–14. https://doi.org/10.4236/oalib.1103505.

Recha, C. W., Shisanya, C. A., Anyamba, A., & Okolla, J. (2014). Determinants of seasonal rainfall and forecast skills in semi-arid Southeast Kenya. *Journal of the Geographical Association of Tanzania, 34*, 1–13.

Republic of Kenya. (2019). *2019 Kenya population and housing census. Vol. I*. Nairobi: Kenya National Bureau of Statistics. Available at: https://www.knbs.or.ke/?wpdmpro=2019-kenya-population-and-housing-census-volume-i-population-by-county-and-sub-county. (Accessed 21 February 2021).

Shisanya, C. A., Recha, C., & Anyamba, A. (2011). Rainfall variability and its impact on normalized difference vegetation index in arid and semi-arid lands of Kenya. *International Journal of Geosciences, 2*, 36–47. https://doi.org/10.4236/ijg.2011.21004.

Spinoni, J., Naumann, G., Carrao, H., Barbosa, P., & Vogt, J. (2014). World drought frequency, duration, and severity for 1951–2010. *International Journal of Climatology, 34*, 2792–2804. https://doi.org/10.1002/joc.3875.

Tigkas, D., Vangelis, H., & Tsakiris, G. (2015). DrinC: A software for drought analysis based on drought indices. *Earth Science Informatics, 8*, 697–709.

Tilahun, K. (2006). Analysis of rainfall climate and evapo-transpiration in arid and semi- arid regions of Ethiopia using data over the last half a century. *Journal of Arid Environments, 64*, 474–487. https://doi.org/10.1016/j.jaridenv.2005.06.013.

Uexkull, N. (2014). Sustained drought, vulnerability and civil conflict in sub-Saharan Africa. *Political Geography, 43*, 16–26. https://doi.org/10.1016/j.polgeo.2014.10.003.

Uhe, P., Phillip, S., Kew, S., Shah, K., Kimutai, J., Mwangi, E., et al. (2018). Attributing drivers of the 2016 Kenyan drought. *International Journal of Climatology, 38*, e554–e568. https://doi.org/10.1002/joc.5389.

Ujeneza, E. L., & Abiodun, B. J. (2015). Drought regimes in southern Africa and how well GCMs simulate them. *Climate Dynamics, 44*, 1595–1609. https://doi.org/10.1007/s00382-014-2325-z.

USAID. (2018). *Economics of resilience to drought: Kenya analysis.* https://www.usaid.gov/sites/default/files/documents/1867/Kenya_Economics_of_Resilience_Final_Jan_4_2018_-_BRANDED.pdf. (Accessed 13 April 2018).

Washington, R., Harrison, M., Conway, D., Black, E., Challinor, A., Grimes, D., et al. (2006). African climate change: Taking the shorter route. *Bulletin of the American Meteorological Society, 87*, 1355–1366. https://doi.org/10.1175/BAMS-87-10-1355.

World Bank. (2013). *The agricultural sector risk assessment in Niger: Moving from crisis response to long-term risk management.* Report No. 74322-NE.

World Meteorological Organization (WMO). (2006). *Drought monitoring and early warning: Concepts, progress and future challenges.* WMO-No. 1006, ISBN:92-63-11006-9.

World Meteorological Organization (WMO). (2012). *Standardized precipitation index: User guide.* WMO-No. 1090, ISBN:978-92-63-11091-6.

World Meteorological Organization (WMO) and Global Water Partnership (GWP). (2016). *Handbook of drought indicators and indices.* WMO-No 1173, ISBN:978-92-63-11173-9.

Yuan, X., Wood, E. F., Chaney, N. W., Sheffield, J., Kam, J., Liang, M., et al. (2013). Probabilistic seasonal forecasting of African drought by dynamical models. *Journal of Hydrometeorology, 14*, 1706–1720. https://doi.org/10.1175/JHM-D-13-054.1.

CHAPTER

10

Drought across East Africa under climate variability

Charles Onyutha[a], *Brian Ayugi*[b,c], *Hossein Tabari*[d], *Hamida Ngoma*[e], *and Victor Ongoma*[f]

[a]Department of Civil and Environmental Engineering, Kyambogo University, Kyambogo, Kampala, Uganda
[b]Department of Civil Engineering, Seoul National University of Science and Technology, Republic of Korea
[c]Organization of African Academic Doctors (OAAD), Nairobi, Kenya
[d]Hydraulics Laboratory, KU Leuven, Leuven, Belgium
[e]Department of Geosciences, University of Connecticut, Storrs, CT, United States
[f]International Water Research Institute, Mohammed VI Polytechnic University, Benguerir, Morocco

1 Introduction

Rainfall across Africa exhibits high spatiotemporal variability. The economy of most regions across the continent including East Africa (see Fig. 10.1) is heavily reliant on rain-fed agriculture. This makes the region vulnerable to climate variability given that weather and climate extremes are known to cause massive destruction of property and loss of lives. East African region and the Greater Horn of Africa (GHA) as a whole whose most of the landmass is arid and semiarid land (ASAL) is prone to unusual weather occurrences. This is known to limit socio-economic growth of the region since the region's livelihood is mainly driven from rainfall dependent sectors. The occurrence of climate extremes causes crop failure that leads to food insecurity and the related suffering.

Droughts are recurrent phenomena across Africa owing to rain failure in one season or up to one or more years (Masih, Maskey, Mussá, & Trambauer, 2014). As a result, the most common extreme events in the region are floods and droughts (Ayugi et al., 2020; Haile et al., 2020; Onyutha, 2020). Between 1960 and 2017, a total of 9, 15, 6, 10, and 6 severe drought episodes occurred in Uganda, Kenya, Burundi, Tanzania, and Rwanda, respectively, according to Emergency Events Database (EM-DAT, 2018). Considering 1964–2015, Haile et al. (2020) observed more frequent, persistent, and intense droughts events in Sudan and Tanzania. Furthermore, a number of severe droughts were experienced in Somalia, Ethiopia, and Kenya. With increasing frequency of occurrence, the magnitude of droughts tends to be low. For instance, major droughts in Kenya occur every 10 years while the minor ones are experienced every 3 to 4 years (Global Water Partnership Eastern Africa (GWPEA), 2015). In Uganda, a total of 7 minor droughts with varying severities occurred between 1991 and 2000. Over the period 1991–2007, Karamoja region in the northeastern part of Uganda experienced severe droughts that led to a depletion of pasture and acute water shortage (GWPEA, 2015). Between 1960 and 2020, various drought episodes across the entire East African region led to a huge economic loss of not less than 1.5 million USD and more than 57,000 deaths (Haile et al., 2020; Onyutha, 2020). The total number of people affected by droughts in Uganda, Kenya, and Tanzania over the periods 1967–2010, 1965–2012, and 1967–2011 was at least 4,975,000, 47,200,000, and 12,737,000, respectively (EM-DAT, 2018). In 2015–16, the GHA experienced a severe drought that left nearly 21.6 million people without food and water.

Among the documented disastrous droughts, the 2011 East African drought (from July 2011 till mid-2012) was quite outstanding in terms of its severity in many countries of East Africa (Nicholson, 2014). The drought that was as a result of failure of October to December 2010 seasonal rainfall left over 8 million people without food (OCHA, 2011). During drought, the severity of food insecurity is higher among pastoralists that mainly inhabit ASALs as compared to

Climate Impacts on Extreme Weather
https://doi.org/10.1016/B978-0-323-88456-3.00002-2

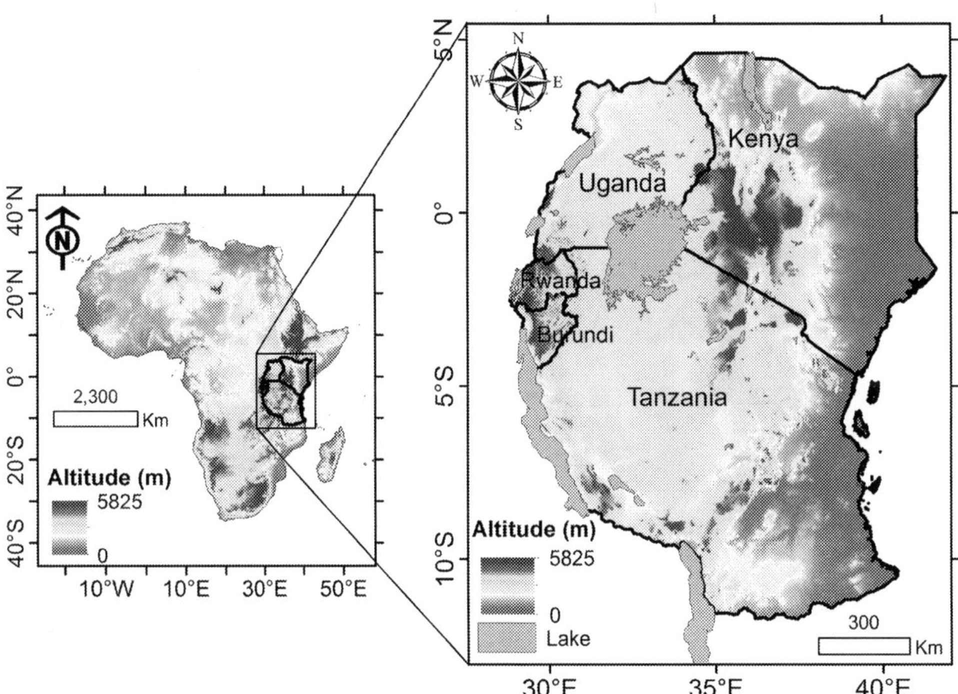

FIG. 10.1 Location of East African countries considered as the study area (enclosed in black rectangle) in Africa along longitudes 28°E–42°E and latitudes 12°S–5°N and the map on the right shows elevation (m) and physical features. The digital elevation model (DEM) datasets was obtained from shuttle radar topography mission (SRTM) 90 m spatial resolution (3 arcsec). The Lowest elevation is represented by red color in the eastern sides while highest elevation by deep blue (Mt Rwenzori in the southwest, Mt Elgon in the east, and Mt Kenya in central Kenya region).

nonpastoralists (Coughlan de Perez et al., 2019). Occurrences of drought tend to be influenced by a number of climatic variables such as precipitation amount and intensity, evaporation rates, soil moisture, and temperature. These climatic variables across East Africa and the GHA have been found to be characterized by changes due to natural variability (Liebmann et al., 2014; Lyon & Dewitt, 2012; Ngoma, Wen, Ojara, & Ayugi, 2021; Niang et al., 2014; Ongoma & Chen, 2017; Ongoma, Chen, & Gao, 2018; Onyutha, 2016, 2017). Lyon (2014) observed that the occurrence of drought over the region is partly dependent on the behavior of Sea Surface Temperature (SSTs) in the tropical Pacific and Indian Oceans. The study further reported that this may develop in response to, or independent of El Nino Southern Oscillation (ENSO). Composite analysis of the 10-driest rainfall seasons between 1950 and 2010 associated an anomalous Walker circulation cell across the Indian Ocean with the March-May (MAM) drought over GHA (Lyon, 2014). It is thus imperative to understand how global SSTs may impact drought events across East Africa during the recent decades, owing to low adaptive capacity to cope with the frequent reoccurrence of drought events witnessed over the recent years.

Unfortunately, timely detection and monitoring of drought remains a challenge not only in East Africa but in many parts of the world. The challenge is very common in most developing countries that have limited observed data (Funk et al., 2015). The datasets available in understanding drought do not only dictate the length and spatial coverage of the study but the choice of indices of climate extremes. The index of choice has an influence on the outcome of the study.

For informed planning to save lives and minimize destruction of property, a systematic framework which can adequately detect and provide information on drought onset and termination along with severity, duration, and propagation is important (Onyutha, 2020). One way to provide such information is to analyze drought under natural climate variability. This is something considered to be central to the present work since it was not undertaken before especially for the case of East Africa.

2 Materials and methods

2.1 Area of study

The study area comprises five countries: Uganda, Kenya, Tanzania, Rwanda, and Burundi (Fig. 10.1). Kenya and Tanzania have coastlines, the Indian Ocean coast. However, Uganda, Rwanda and Burundi are land-locked countries. Based on the latest data https://data.worldbank.org/indicator/AG.LND.TOTL.K2 (accessed: 10th April 2021), land areas covered by Rwanda, Burundi, Uganda, Kenya and Tanzania are 24,670, 25,680, 200,520, 569,140, and 885,800 km^2, respectively. The lowest point in Uganda is about 615 m at the border with South Sudan. The highest points in Uganda, Kenya and Tanzania are Margherita Peak (5111 m) in the Rwenzori Mountains, Batian peak (5199 m) of Mount Kenya, and Mount Kilimanjaro (5895 m), respectively. Mounts Kilimanjaro, Kenya, and Rwenzori are the first, second, and third highest mountains in the African continent, respectively. The study area has the Great Rift Valley and African Great Lakes. The African Great lakes include Lake Victoria (the second largest fresh water lake in the world in terms of area), Lake Tanganyika (the world's second largest freshwater lake in terms of volume and depth), and Lake Malawi. Examples of other lakes in East Africa are Lake Kyoga, Albert, Edward, Kyoga, Turkana, Naivasha, Baringo, Magadi, Eyasi, Natron, Manyara, Rukwa, Kivu, Muhazi, and Ihema. The study area has the East African Rift valley consisting of the western and eastern branches also called the Albertine Rift and Gregory Rift, respectively. The western Rift Valley runs along the western border of Uganda, and also the western boundary of Tanzania. In the study area, the western rift valley transects Uganda, Rwanda, Burundi, and Tanzania. The Eastern rift value transects through Kenya and Tanzania.

Rainfall over East Africa region shows high variability in both space and time. Lying in the tropics, the region is generally warm throughout the year. The months of December to February (DJF) are the warmest while June-August (JJA) season is the coldest in the region. The region's rainfall seasonality is mainly influenced by the north-south oscillation of the Intertropical Convergence Zone (ITCZ). This explains the region's two dominant rainfall seasons: March-May (MAM "long rains") and October-December (OND "short rains"). Local factors such as topography and large water bodies greatly influence the spatial variability of rainfall, ranging from relatively wet mountainous areas to dry lowlands. The ENSO, Indian Ocean Dipole (IOD), Madden-Julian Oscillation (MJO), and phase of the quasi-biennial oscillation (QBO) have a remarkable influence on the region's interannual rainfall variability (Indeje, Semazzi, & Ogallo, 2000; Lyon, 2014; Vellinga & Milton, 2018). Some studies have shown that decadal and multidecadal variability in the rainfall (a major determining factor for drought) across East Africa is linked to variation in the SST of the Atlantic Ocean. For instance, the recent (1983–2012) significant drying for the Central Equatorial Africa was highly correlated with Atlantic Multidecadal Oscillation (AMO) during boreal summer and autumn (Diem, Ryan, Hartter, & Palace, 2014). Furthermore, Lüdecke, Müller-Plath, Wallace, and Lüning (2021) showed that the coefficients of correlation between AMO and January precipitation in Uganda, Kenya, and Tanzania (the three countries considered in our study) were in the ranges −0.5 to −0.59, −0.3 to −0.39, and >−0.3, respectively. However, it is worth noting that the debate on the forcing mechanism controlling AMO or the North Atlantic SST variations has been considerable since the 1980s (Delworth & Mann, 2000; Folland, Palmer, & Parker, 1986; Mann et al., 2009; Otterå, Bentsen, Drange, & Suo, 2010; Parker et al., 2007). Based on several studies (Delworth & Mann, 2000; Folland et al., 1986; Mann et al., 2009; Parker et al., 2007), one side of the debate holds that the North Atlantic SST variations is dominantly characterized by multidecadal variability. In other words, AMO is forced by the internal ocean variability and is related to multidecadal fluctuations in the Atlantic Meridional Overturning Circulation (Knudsen, Jacobsen, Seidenkrantz, & Olsen, 2014). Eventually, AMO possesses potential impacts on regional climate (Knight, Folland, & Scaife, 2006). Another side of the said debate argues that variations of the North Atlantic SSTs especially over the past century were driven by volcanic and anthropogenic emissions of aerosols (Booth, Dunstone, Halloran, Andrews, & Bellouin, 2012). However, a follow-up study (Zhang et al., 2013) showed a considerable discrepancy between observations and model results thereby casting substantial doubt on whether aerosol forcing drives the bulk of multidecadal variability in North Atlantic SST. Nevertheless, to reconcile the opposing theories on the origin of the AMO, Knudsen et al. (2014) suggested that the linkage of North Atlantic SSTs with external forcing could be importantly explored based on the Atlantic Meridional Overturning Circulation. Notwithstanding these arguments, this work considers that AMO may potentially impact regional climate.

2.2 Data

Monitoring of drought has been hampered in most parts of Africa by limited quality observed data. The available reanalyzed, satellite, and model datasets present an opportunity to improve drought detection and monitoring. Analyses in this study were performed using different datasets which were obtained from various sources.

2.2.1 Self-calibrating Palmer Drought Severity Index (scPDSI)

Monthly scPDSI based on the Climatic Research Unit—Time Series version 4.04 (or CRU TS 4.04) (Barichivich, Osborn, Harris, Schrier, & Jones, 2020) for global land over the period 1901–2019 was obtained in gridded (0.5° latitude-longitude resolution) form via https://crudata.uea.ac.uk/cru/data/drought/ (accessed: 15th December 2020).

The scPDSI metric was introduced by Wells, Goddard, and Hayes (2004) to overcome the problem of the inconsistent behavior of the original PDSI across different climate regimes. The scPDSI automatically adjusts the empirical constants in the PDSI derivation with dynamically calculated values. Calculation of scPDSI requires monthly precipitation series, monthly potential evapotranspiration, and information on available soil water capacity of the soil layer. The details of how to compute scPDSI can be found in Wells et al. (2004).

2.2.2 *Climate indices*

Four monthly climate indices based on SST including the AMO, Niño 3 index, IOD, and QBO were obtained from various sources.

1. The AMO index is defined as the SST averaged over 25–60° N, 7–70° W minus the regression on the global mean temperature (van Oldenborgh, te Raa, Dijkstra, & Philip, 2009). The monthly AMO index (Enfield, Mestas-Nuñez, & Trimble, 2001; Rayner et al., 2003) covering the period 1985–2020 was obtained from https://psl.noaa.gov/gcos_wgsp/Timeseries/AMO/ (accessed: 20th January 2021).
2. The Niño 3 (Rayner et al., 2003) is the area-averaged SST for the Tropical Pacific region 90° W to 150° W and 5° N to 5° S. The monthly Niño 3 data covering the period 1870 to 2020 was obtained via https://psl.noaa.gov/gcos_wgsp/Timeseries/Nino3/ (accessed: 25th January 2021).
3. The IOD is the climate mode associated with the state of the SST over Western (50° E to 70° E and 10° S to 10° N) Equatorial and Southeastern (90 to 110° E and 10° S to 0° N) Indian Ocean (Saji, Goswami, Vinayachandran, & Yamagata, 1999). The monthly IOD series obtained from the Japan Agency for Marine-Earth Science and Technology (JAMEST) via the link http://www.jamstec.go.jp/frcgc/research/d1/iod (accessed: 20th January 2014).
4. The QBO is a regular variation of the winds that blow high above the Equator (Met Office, 2021). The QBO is a tropical, lower stratospheric, quasiperiodic oscillation of the equatorial zonal wind between easterlies and westerlies, with an average period of about 28 months. Monthly QBO data from 1948 to February 2019 was downloaded via https://www.daculaweather.com/4_qbo_index.php (20th February 2021).

2.3 Methods

Spatial variation in the long-term characteristics of the scPDSI across the study area was assessed in terms of the standard deviation (STDEV) and the average (or MEAN) of annual scPDSI series. The MEAN measures central tendency of a given dataset while STDEV is a measure of spread around the mean. These metrics MEAN and STDEV were computed using annual scPDSI of each grid point. Spatialization of the computed statistics was obtained through interpolation. Several methods exist for spatial interpolation including an ordinary kriging, universal kriging, and inverse distance weighted (IDW) approach. The IDW approach was adopted for this study though we deemed that the influence from the choice of a particular interpolation on the results was minimal given that the scPDSI series were obtained at equidistant grid points. Another method of spatialization of statistics from climatic data is regionalization. However, this was out of the scope of this work.

The association of annual scPDSI variations with changes in large-scale ocean-atmosphere conditions was assessed in terms of wavelet coherence. Through bivariate wavelet analysis, wavelet coherence allows estimation of the measure of correlation between the variables under consideration at all the periodicities in time. Annual scPDSI was averaged over the entire spatial domain considered in this study. Before analysis of wavelet coherence, each series was standardized. In the standardization, the mean of the data was subtracted and the next procedure was division by the standard deviation. Wavelet coherence was finally computed for the various combinations of the climatic variables including (i) scPDSI and AMO, (ii) scPDSI and IOD, (iii) scPDSI and Niño 3, (iv) scPDSI and QBO. To each pair of the climatic series, continuous wavelet transform was applied to determine the regions in time frequency where the two-time climatic series covary. Significance of the coherence was quantified using a total of 1000 Monte Carlo randomizations.

We applied empirical orthogonal function (EOF) to analyze climate variability. For EOF, let a collection of all precipitation time series comprise a set referred to as a H. Furthermore, consider that we have m precipitation time series in H. In other words, each location has a separate precipitation time series. Importantly, the average of each of the m series should be removed such that each column has a zero mean. The time series can be arranged in such a way that we obtain m columns in H. At each location, the sample size or number of data points is n. From every row of H we can obtain one map. Thus, h_i can be taken as the position vector of the ith row in m dimensional space. Any irregularities in the data would force the observations to be ordered along a particular direction; otherwise, we would have a "blob" in space in case the points are completely random. We can establish a new coordinate system for the m-dimensional space such that there can be a coordinate system going through each cluster (or organized set) of points. To do so, the method of EOF analysis determines a set of orthonormal basis vectors s_w that can maximize the projection of h_i on the basis vectors (Björnsson & Venegas, 1997). So, the mathematical problem in the EOF analysis is to maximize the sum of squared product of s_w and h_i (Eq. 10.1) following (Björnsson & Venegas, 1997) such that

$$\sum_{i=1}^{n} (s_w h_i)^2 \tag{10.1}$$

for $w = 1, \ldots, m$, subject to an orthonormality condition on s_w such that

$$s_i^t s_j = \delta_{ij}. \tag{10.2}$$

When we consider the covariance metric R such that $R = H^t H$ the expression in Eq. (10.1) leads to Eq. (10.3)

$$\sum_{i=1}^{n} (s_w h_i)^2 = s_w H^t H s_w^t = s_w R s_w^t \tag{10.3}$$

Let ∇ denote the gradient operator while λ is the Lagrange multiplier. Maximizing Eq. (10.1) subject to constraint in Eq. (10.2) yields

$$\nabla\left(\vec{y}^t R \vec{y}\right) - \lambda \nabla\left(\vec{y}^t \vec{y}\right) = 0 \tag{10.4}$$

Manipulation of Eq. (10.4) while taking advantage of the symmetry of R yields

$$R\vec{y} = \lambda \vec{y}. \tag{10.5}$$

Eq. (10.5) is another form of the eigenvalue problem such that

$$RC = C\wedge \tag{10.6}$$

where $\wedge$ is a diagonal matrix containing the eigenvalues λ_i and the c_i columns of C contain eigenvectors of R representing the eigenvalues . The size of $\wedge$ or C in Eq. (10.6) is m by m. For each selected eigenvalue, a corresponding eigenvector can be found. Each of these eigenvectors is what we regard as the EOF or a map and the eigenvectors can be ordered according to the size. The eigenvector associated with the first, second, ..., and kth biggest eigenvalue is EOF1, EOF2, ..., and EOFk, respectively. An eigenvalue indicates the measure of the fraction of the total variance in R explained by the mode. The eigenvector matrix C comprises the typical property that $C^t C = CC^t = I$ where I is an identity matrix thereby indicating the EOFs are uncorrelated over space (or the eigenvectors are orthogonal to each other) (Björnsson & Venegas, 1997).

Based on the symmetry of R (following the spectral representation theorem), the eigenvalues and the eigenvectors for the EOFs decompose R in such a way that (Björnsson & Venegas, 1997)

$$R = \lambda_1 c_1 c_1^t + \lambda_2 c_2 c_2^t + \cdots + \lambda_p c_p c_p^t \tag{10.7}$$

On the basis of Eq. (10.7)

$$EOF_i = \frac{\lambda_i}{\sum_{i=1}^{n} \lambda_i}. \tag{10.8}$$

Given the rationale behind EOF analysis (i.e., reducing the data to few modes of variability), a few of the eigenvalues can dominate others. Each EOF can be taken as a mode of variability while how it oscillates in time becomes the principal component (or expansion coefficients of the variability mode).

Localized structures expectedly tend to emerge over the spatial domain of analysis. Eventually, rotation of the eigenvectors can be considered to reduce the effect of orthogonality constraint and achieve more stable spatial patterns (Horel, 1984; Richman, 1986). Rotations allow isolation of regions with similar temporal variations. In this way, we improve the identification of regions with the maximum correlation between the variables and the components. There are several methods for rotation of eigenvectors such as, Orthomax, Varimax, Parsimax, Oblimin, Qartimax, Quartimin, Promax, and Equamax. The Varimax method preserves orthogonality, and gives more physically explainable variability patterns than other approaches (Richman, 1986). Furthermore, Varimax also exhibits less sensitivity to the number of variables in the EOF analysis (Onyutha & Willems, 2017).

To apply EOF, we divided the scPDSI data over a period of about 30 years. The time slice of at least 30 years in length is deemed suitable to characterize climate fluctuations. Eventually, EOF was applied to scPDSI over various subperiods including 1901–30, 1931–60, 1961–90, 1991–2019, 1901–60, and 1961–2019. This was to determine (i) spatial variation in EOF factor loadings, and (ii) temporal EOF factor scores. Furthermore, drought severity based on scPDSI averaged over the various subperiods was compared across the study area. Correlation between climate indices and scPDSI over the various subperiods was analyzed. The strength of the co-variation of scPDSI with climate indices was also assessed through wavelet coherence analysis.

3 Results and discussion

The spatial variation in selected descriptive statistics obtained from annual scPDSI is shown in Fig. 10.2. The mean of scPDSI was to indicate areas that were mostly dry or wet (Fig. 10.2). The various categories of the PDSI are presented in Table 10.1. Areas with negative scPDSI (close to −2 indicating moderate drought) are confined to northeastern Uganda (Karamoja region), northeastern Kenya around Lake Turkana, and southwestern Kenya near the coast. Wet conditions existed in the western part of East Africa. Standard deviation (STDEV) of scPDSI shows spatial differences in the extent of dry or wet conditions with respect to the reference (or long-term mean) (Fig. 10.2b). In other words, STDEV gives insight on the variability in the scPDSI. As a common practice, coefficient of variation (CV) tends to be used to characterize variability. However, the use of CV may be limited to variables with a positive mean such that CV values are nonnegative. In this work, it is noticeable that the mean of scPDSI yielded negative values. Therefore, the use of CV as the ratio of STDEV to mean would yield negative values of CV thereby making it inappropriate to characterize relative measure of the variability.

Fig. 10.3 shows the spatial pattern of the first EOF of annually-averaged scPDSI. The associated temporal variation of the first EOF after Varimax rotation is shown in Fig. 10.8. The first EOF explained 17.6% of the total variance in the annual scPDSI across East Africa. The second to fifth EOFs explained 13.9%, 8.6%, 5.1%, and 3.8% of the total variance, respectively. The first EOF indicates a trend in scPDSI of annual time scale, revealing areas with possible substantial drying and wetting over time (Fig. 10.3a–f). The area of a particular subregion that loaded positively varied across periods. Similarly, the areas with negative EOF loadings varied in size across subperiods. The magnitude (and sign) of the EOF loading for a particular location changed from one subperiod to another. Thus, drought statistics (for instance, severity) of a particular location or region depends on the period selected for analyses. The spatial variation in the EOF loadings across East Africa also shows regional differences in the drought variability driving forces. Regional differences in drought statistics (like a dry spell, precipitation deficiency) across East Africa can be influenced by micro-climate or features such as topography and water bodies (Onyutha & Willems, 2017).

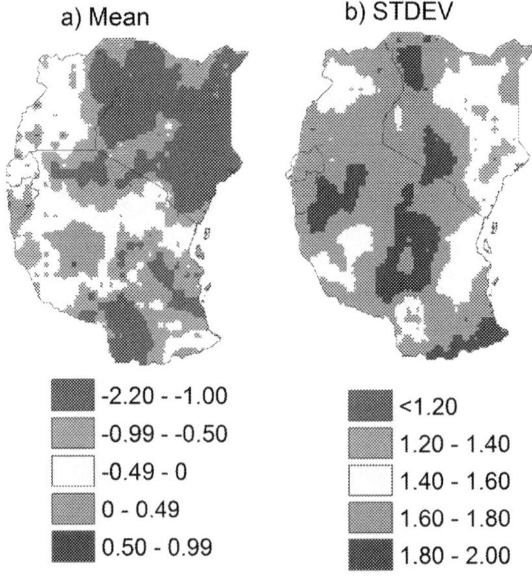

FIG. 10.2 Long-term (a) mean and (b) standard deviation (STDEV) of annual scPDSI over the period 1901–2019.

TABLE 10.1 Classification of PDSI (Wells et al., 2004).

PDSI value	PDSI category
Above 4.0	Extreme wet spell
3.00–3.99	Severe wet spell
2.00–2.99	Moderate wet spell
1.00–1.99	Mild wet spell
0.50–0.99	Incipient wet spell
0.49 to −0.49	Normal
−0.50 to −0.99	Incipient drought
−1.00 to −1.99	Mild drought
−2.00 to −2.99	Moderate drought
−3.00 to −3.99	Severe drought
Below −4.00	Extreme drought

Fig. 10.4 shows the temporal behavior of the annual scPDSI characterized by spatial variation in EOF loadings as presented in Fig. 10.3. It is vital to note that the division of EOF factor loadings was for the ease of visualizing the spatial variations. Considering loadings based on the long-term period 1901–2019, the EOF factors were mainly positive from 1920 to the mid-1960s and negative afterward (mid-1960s till 2020). Thus, the East African climate cycle is suggestively about 40 years. This, however, requires further verification using series longer than the one used herein. Nevertheless, the findings depict how a wet subperiod tends to be punctuated by dry conditions in East Africa as a semiarid environment. At a given location or subregion, rainfall can be below the long-term mean over a selected epoch. However, rainfall total goes above the reference over the subsequent epoch. This suggests variation in the strength of the major driver of the variability of wet and dry climatic conditions over time. The EOF loadings for the subperiods 1901–30,

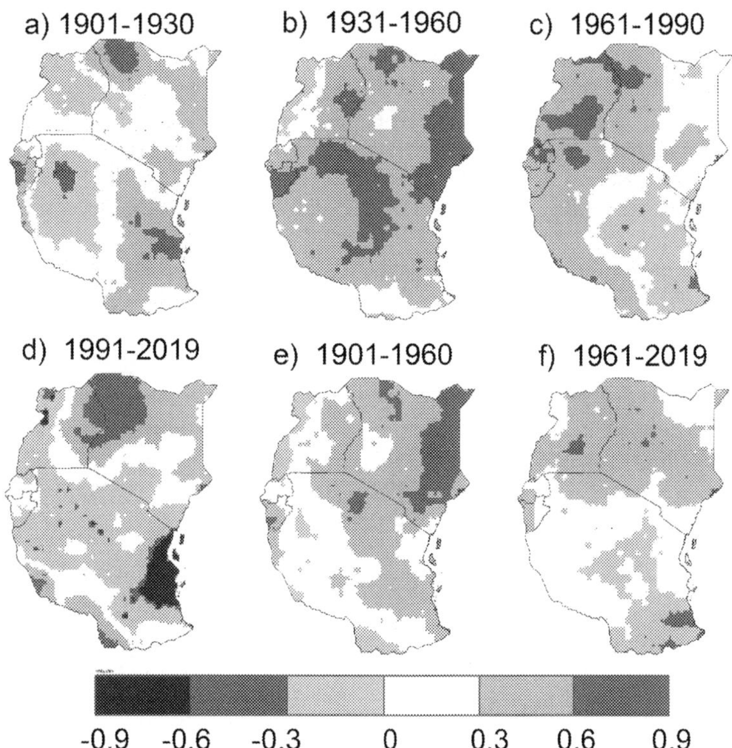

FIG. 10.3 Leading EOF factor loadings after Varimax rotation for annual scPDSI for the subperiods (a) 1901–30, (b) 1931–60, (c) 1961–90, (d) 1991–2019, (e) 1901–60, and (f) 1961–2019.

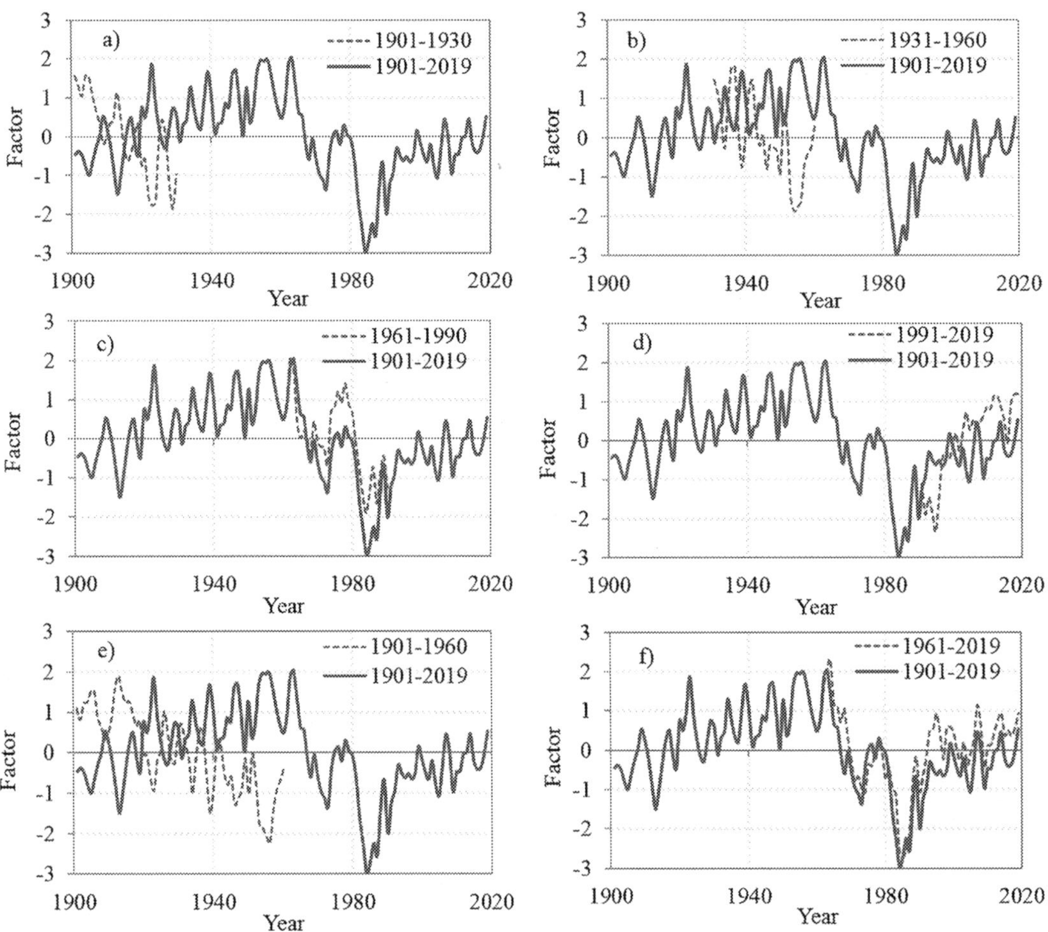

FIG. 10.4 Leading temporal EOF factor scores after Varimax rotation for annual scPDSI over the period 1901–2019 as the benchmark together with the scores for the subperiods (a) 1901–30, (b) 1931–60, (c) 1961–90, (d) 1991–2019, (e) 1901–60, and (f) 1961–2019.

1931–60, and 1901–60 (Fig. 10.4a, b, and e) are opposite to those of the corresponding epochs when the full time series 1901–2019 were used. However, the temporal loadings for the subperiods after 1960 were found to resonate quite well with the loadings obtained using the full time series (1901–2019) (Fig. 10.4c, d, and f).

Fig. 10.5 shows the amount of variability explained in annual scPDSI by the first five EOFs. Considering a 30-year period, the amount of variability explained by EOF1 was 20.5% on average. The percentage reduced to about 19% when 60-year periods were considered. For a given subperiod, the amount of explained variability reduces with increasing EOF. The cumulative amount of variability for the first two EOFs considering a 30-year period was 36%. For a 60-year period, both EOF1 and EOF2 explained up to 34% of the variability. Considering the full time period (1901–2019), both EOF1 and EOF2 cumulatively explained 31.5% of the variability. These results show that East Africa is characterized by a considerable amount of temporal variability in droughts.

Fig. 10.6 shows the space-time variation in the 2011 East African drought severity. Various categories of PDSI in Fig. 10.6 followed classifications in Table 10.1 by Wells et al. (2004) and Palmer (1965). The 2011 East African drought event occurred from July 2011 to mid-2012 and was experienced throughout the study area. From September 2011 to December 2011, a moderate wet condition was confined to the subregion comprising Rwanda, Burundi, northwestern Tanzania, and southwestern Uganda (Fig. 10.6c–f). From April 2012 to June 2012, eastern Uganda and western Kenya experienced somewhat moderate wet conditions (Fig. 10.6j–l). The rest of the study area experienced droughts of varying severities (Fig. 10.6a–l). Extremely dry conditions (with scPDSI between −4 and −5) were confined to southeastern Kenya and northeastern Tanzania.

A number of drought indices exist such as PDSI (Palmer, 1965), scPDSI (Wells et al., 2004), Standardized Precipitation Index (SPI) (McKee, Doesken, & Kleist, 1993), Standardized Groundwater Level Index (SGLI) (Bloomfield & Marchant, 2013), Standardized Precipitation Evapotranspiration Index (SPEI) (Vicente-Serrano, Beguería, & López-Moreno, 2010), and Standardized Nonparametric Indices of Precipitation and Evapotranspiration (SNIPE)

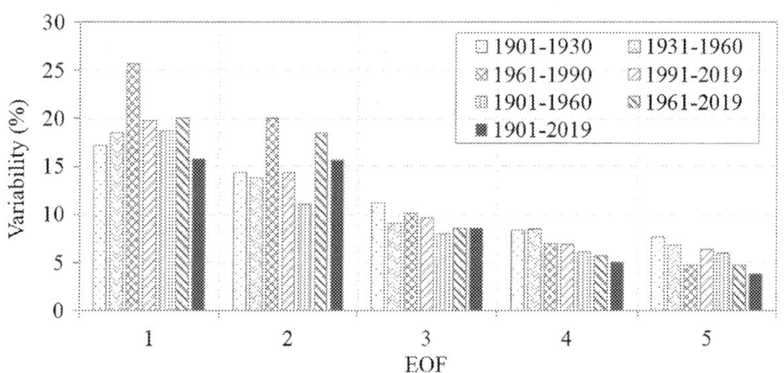

FIG. 10.5 Percentage of variability explained in annual scPDSI.

(Onyutha, 2017, 2021). Each method has its own advantages and disadvantages. For instance, multiscalar feature of these indices such as SPEI and SPI allows tailoring the drought impact for different sectors. However, most of these methods like SPI and SPEI yield skewed indices. Moreover, interpretation of results of drought analysis may be based on the method selected to derive the indicator of dry conditions. In this line, the application of each method can be based on the purpose of the study or intended risk-based water resources application. Nevertheless, drought can be caused by a number of factors such as reduced precipitation total, increased evaporation or evapotranspiration rates, and vapor pressure deficit, mostly linked to higher temperature and low relative humidity. Other features such as persistent circulation anomalies, possibly strengthened by land—atmosphere feedback contribute to drought occurrence (IPCC, 2014). Since there is no single method known to directly combine all these factors, what each drought index indicates tends to be limited in one way or the other. For instance, SPI which is based on only precipitation

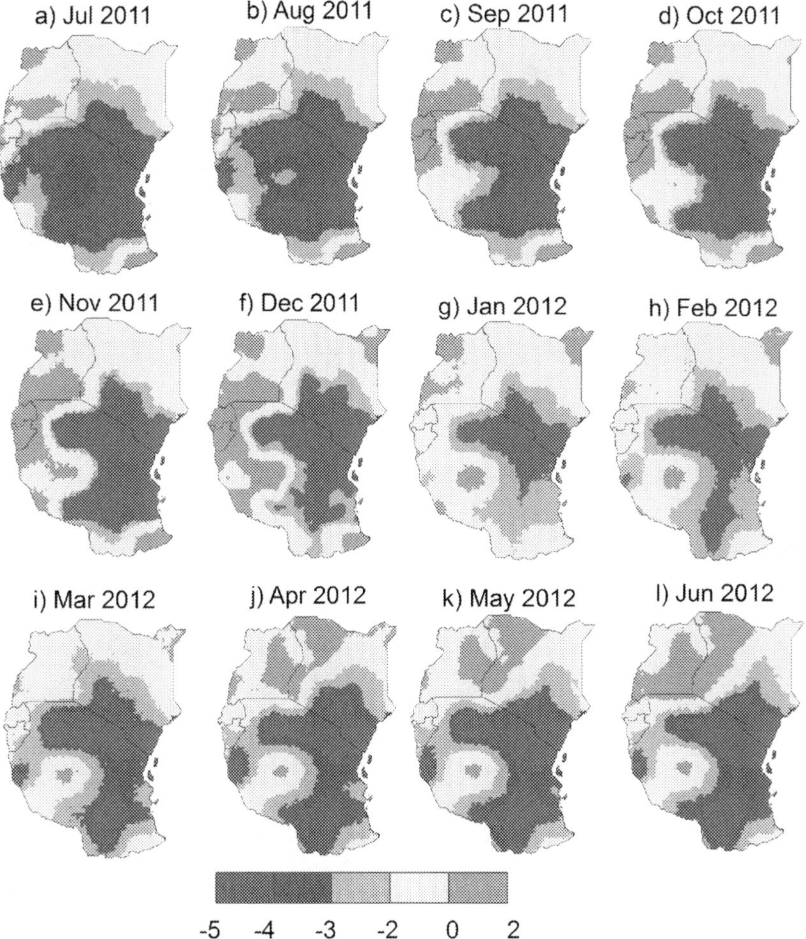

FIG. 10.6 scPDSI-based severity of the 2011 East African drought for each month (panels a–l) during the period July 2011-June 2012.

may be limited to meteorological drought. The scPDSI, however, takes into account precipitation, evapotranspiration, and soil water capacity, and better measures the atmospheric demand for moisture and thus the water balance for the region under study. We remark that to avoid the uncertainty stemming from the choice of a particular method, drought analyses and attribution (though not considered in this study for brevity) should comprise indicators of dry conditions from various approaches.

The widespread 2011 East African drought was believed to have been caused by anthropogenic climate change (Climate Home News, 2013; Hua et al., 2016; Lyon & Dewitt, 2012; Nicholson, 2014). However, no evidence was found for human influence on the 2010 short rains which would have led to the 2011 drought (Lott, Christidis, & Stott, 2013). Notably, the probability of long rains as dry as, or drier than, 2011 was found to be exacerbated by human influence (Lott et al., 2013). The periods 2010–11 and 2011–12 were strong and moderate La Niña years (Golden Gate Weather Services (GGWS), 2021). It means that the 2011 East African drought was caused by natural variability. Reducing precipitation or drying trend across East Africa was thought of in terms of the Indian SSTs which was warming faster than the Pacific SSTs (Williams & Funk, 2011). However, a follow-up study by Lott et al. (2013) established that the warming of tropical Indian Ocean and Pacific Ocean SSTs were more uniform, though with less warming in the Arabian Sea. Therefore, in this work, we conclude that the East African regional rainfall trend which leads to drying or wetting trends across East Africa is driven by the changes in SSTs due to both anthropogenic and natural origin. This is also in line with findings from other studies (Hoerling, Hurrell, Eischeid, & Phillips, 2006; Lott et al., 2013; Ngoma et al., 2021; Ogwang, Ongoma, Xing, & Ogou, 2015; Tierney, Smerdon, Anchukaitis, & Seager, 2013).

Fig. 10.7 shows measures of possible linkage of annual scPDSI to climate indices over various subperiods. It is vital to note that the division of the legend in Fig. 10.7 was to simplify visualization of variation in correlation coefficients but not for assessing significance of the correlation between scPDSI and climate indices. Otherwise, for two-tailed probabilities at $\alpha = 0.01$, the critical value for the correlation was 0.463 and 0.330 for the 30- and 60-year periods,

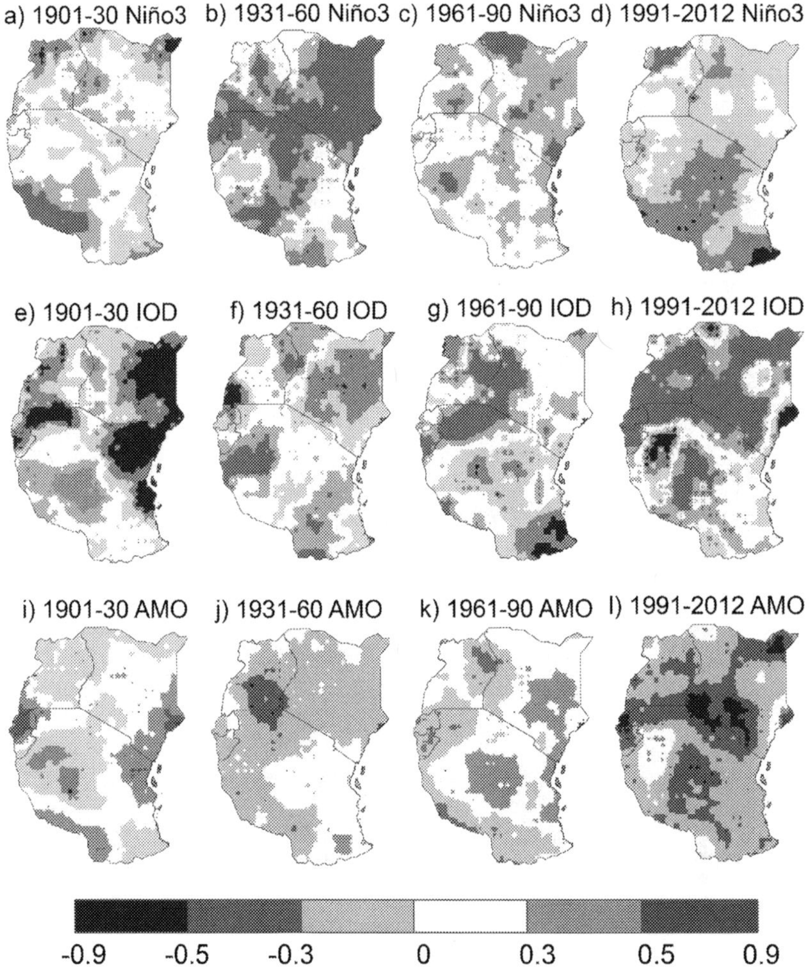

FIG. 10.7 Correlation between subseries of scPDSI and (a–d) Niño 3, (e–h) IOD, and (i–l) AMO.

respectively. It is noticeable that the spatial variation in the correlation coefficients tended to differ among the subperiods (Fig. 10.7a–l). For instance, coefficients of correlation between Niño 3 and scPDSI were spatially more positive over the period 1931–60 than those of other subperiods (Fig. 10.7a–d). In the same line, coefficients of correlation between scPDSI and IOD (AMO) were spatially more positive (negative) over the period 1991–2012 than those of other subperiods (Fig. 10.7e–l). In summary, the magnitude of the influence of a particular driver on the variation of dry or wet conditions varies from one subperiod to another. This is important in exploring mechanisms which influence variation in dry and wet conditions over a particular subperiod across. The question would be: why do the atmospheric drivers of drought vary in space across the study area? Most of the large-scale ocean-atmosphere conditions (such as El Niño Southern Oscillation) are characterized by phases or episodes. These phases tend to be described as positive or negative (for IOD), and warm or cold (in the case of ENSO and AMO). The variation of wet and dry conditions across East Africa is influenced by the positive and negative phases of IOD. Furthermore, wetter and drier conditions than the normal one are known to be brought about by the warm (El Niño) and cold (La Niña) phases, respectively. A major uptake of heat by the Pacific Ocean and the heat's storage in the Tropical western Pacific occurs during the La Niña phase (Trenberth, Fasullo, O'Dell, & Wong, 2010). Next, the ocean cools as the atmosphere responds with typical El Niño weather patterns forced from the region (Trenberth et al., 2010). This El Niño phase tends to be a typical source of influence on rainfall across various areas of the world. The atmospheric patterns following the positive phase of the AMO, include increased precipitation in the eastern tropical Atlantic, and reduced wind speeds over the tropical Atlantic; and the negative AMO leads to the opposite conditions (Alexander, Halimeda Kilbourne, & Nye, 2014). In other words, a positive AMO (or warmer than the normal Atlantic Ocean) leads to increased rainfall in certain areas of the world. A cooler than the normal Atlantic Ocean (or negative AMO) causes reduced rainfall in some parts of the world. This brief background on the phases of the different measures of the large-scale ocean-atmosphere conditions shows that a driver can be strong over one phase and weaker over the other. For instance, the Indian Walker cell can be strong during one subperiod and weak over another (Nicholson, 2017). Over corresponding periods like for the Indian Walker cell, the strength of the descending or ascending poles of the Atlantic or Pacific cell can also change (Nicholson, 2017). Over a particular period in which the negative phase of IOD is more dominant than the positive one (or when there are more La Niña than El Niño events), the East African region may experience an increased number of drought events.

When considering the long-term (1901–2012) series of scPDSI, IOD, AMO, and Niño3 as well as the entire QBO from 1948 to 2012, the correlation over the various subperiods got averaged (Fig. 10.8a–d). Again, division of the legend in Fig. 10.8 was for the ease of visualizing the variation in the correlation. Otherwise, for two-tailed probabilities at $\alpha = 0.01$, the critical correlation value was 0.236 for AMO, IOD, and Nino 3 (Fig. 10.8a–c) and 0.31 for QBO (Fig. 10.8d). It is noticeable that the variability in scPDSI was mainly linked to the variation in IOD and Niño 3 (Fig. 10.8b–c). Furthermore, results from Figs. 10.4 and 10.5 show that the predictability of the drought variability across East Africa can be based on a number of predictors (such as the negative IOD or La Niña phase of the ENSO). However, it is worth noting that the correlation between scPDSI and climate indices may merely indicate statistical measures of association. Thus, the correlation (for some factors such as intercorrelation among the climate indices, and high signal-to-noise ratio) may not adequately quantify the measure of real dynamics regarding how the variation in drought statistics (like severity) across East Africa which may be influenced by large-scale ocean-atmosphere conditions. The existence of interconnection among climate indices may inflate the correlation between extreme events and atmospheric circulations and lead to biased results (Tabari & Willems, 2018). Furthermore, other factors may need to be taken into consideration to enhance the predictability of the variation in dry conditions across East Africa. Such factors

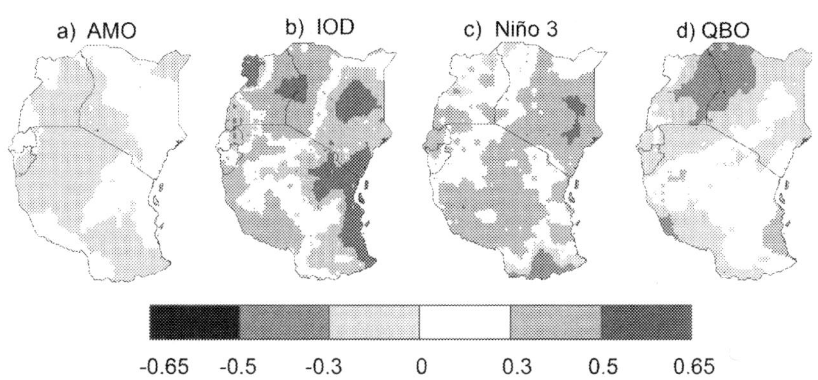

FIG. 10.8 Correlation between long-term (1901–2012) timeseries for scPDSI and (a) AMO, (b) IOD, (c) Niño 3, and (d) QBO.

include; possible nonlinearity of the relationship between drought indicators and the drivers, and time lag or precedence of variation in large-scale ocean-atmospheric conditions by the drought variability. In this regard, oceanic-atmospheric circulation patterns were found to be influential not only concurrently on European climate anomaly, but in a delayed way on the anomaly in the following seasons (Tabari & Willems, 2018). Duzenli, Tabari, Willems, and Yilmaz (2018) also found a superiority of nonlinear models to explore the relationships between precipitation extremes and large-scale atmospheric drivers.

The results of wavelet coherence analysis are shown in Fig. 10.9. The strength of the co-variation of scPDSI with climate indices was patchy through time (Fig. 10.9a–d). Coherence between scPDSI and AMO was strong at multiple scales 1–3, 6–7, 14–15, and 18–30 years. The coherence was strong over the periods 1901–15, 1960–90, and 1920–80 (Fig. 10.9a). The scPDSI cycled in phase with (or scPDSI was strongly influenced positively by) IOD and Niño 3 over the periods 1950–2000 at scales of 14–15 and 14–32 years (Fig. 10.9b and c). scPDSI cycled downward with AMO and IOD over the periods 1960–2000 and 1935–80, respectively (Fig. 10.9a and b). This means that AMO and IOD led scPDSI by $\pi/2$ and this was at scales 14–15 and 24–30 years, respectively. scPDSI cycled upward with AMO and QBO over the periods 1920–80 and 1901–65, respectively (Fig. 10.9a and d). This means that scPDSI led AMO and QBO by $\pi/2$ and this was at scales 18–30 and 16–20 years, respectively. The results show that the strength (and sometimes the sign) of the co-variation of drought indicator with the possible driver can depend on the selected subperiod as already found based on Fig. 10.7.

Drought affects various sectors in a number of ways. For instance, prolonged dry conditions can lead to (i) lack of water for livestock which brings about competition and conflicts among the nomads (like Karamojong), (ii) water scarcity which reduces hydroelectric power produced, (iii) increased forest fires, (iv) crop failure, and (v) increased pest and diseases. A few of the priority activities recommended given the strong temporal variation of droughts in the region include (Global Water Partnership Eastern Africa (GWPEA), 2015).

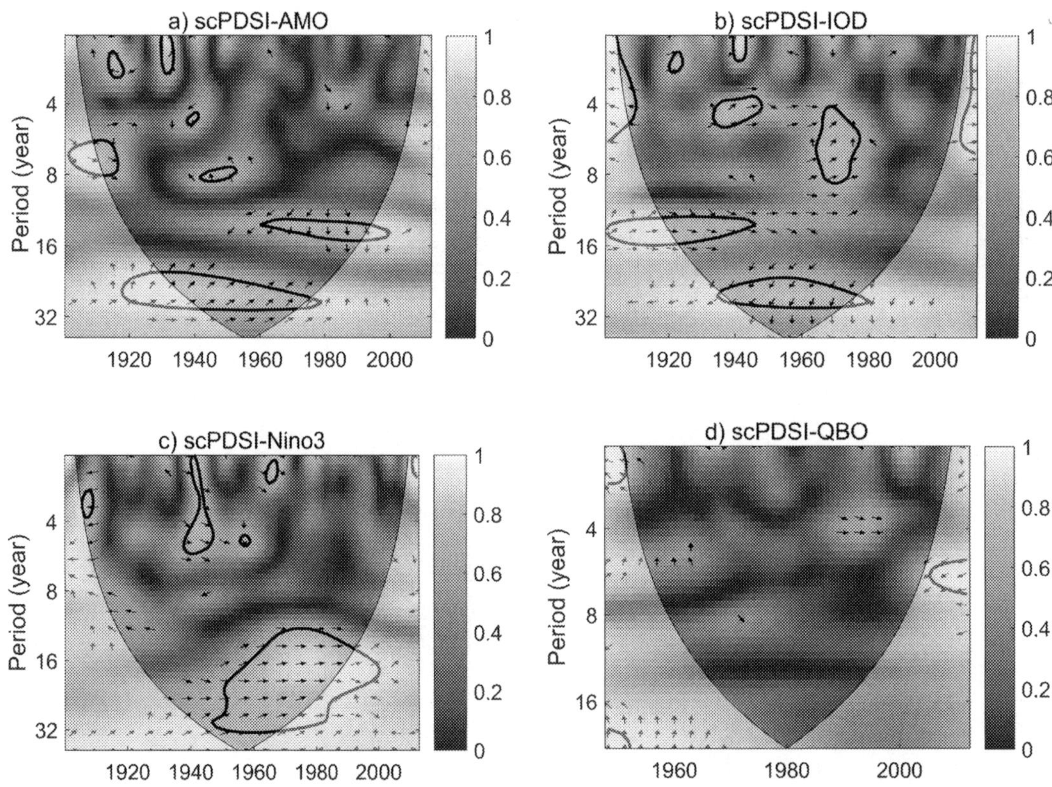

FIG. 10.9 Wavelet coherence plots for (a) scPDSI and AMO, (b) scPDSI and IOD (c) scPDSI and Niño 3, (d) scPDSI and QBO; The strength of coherence between the time series at each period through time is shown in terms of the colors. The color at the top (bottom) of the legend bar shows high (low) coherency. The reliable data are contained within the "cone of influence" away from the edge effects indicated by the colored region. Areas enclosed in *bold black lines* were derived using the Monte Carlo randomizations and indicate significant coherency. The *arrows* show phase relationships between the variables within the areas of strong coherence. Some arrows point to the right and left indicating that the variables are in phase and antiphase, respectively.

1. Planting trees (plantations, agro-forestry, woodlots, fruit orchards),
2. Supporting the provision of water for production,
3. Strengthening the early warning systems,
4. Promoting energy-efficient technologies and energy diversification,
5. Promoting drought-resistant crop varieties across various agro-ecological zones,
6. Proper management of land to reduce the rampart degradation,
7. Proper pasture and rangeland management,
8. Mainstreaming drought mitigation and adaptation strategies in various sectors.

Some of the national priority intervention areas in line with the strong temporal variation of drought events in the region include (i) developing capacity for predictive adaptation to drought, (iii) ensuring national plans comprise drought management, and (iv) embracing integrated drought management practices. The key areas which should be of focus regarding drought management include (i) adaptation to the oscillation low period of precipitation leading to severe droughts, (ii) disaster risk management, (iii) environmental rehabilitation, (iv) enhancing agricultural productivity and household poverty alleviation, and (vi) development of livestock, pasture or rangelands.

4 Conclusions

This study examined the historical variations in droughts across East Africa during the period 1901–2019. To this end, drought was characterized using the self-calibrating Palmer Drought Severity Index (scPDSI) and analyzed by empirical orthogonal function (EOF) over different subperiods. The connections between the temporal variations of droughts and large-scale oceanic-atmospheric circulation patterns were also determined, by exploring the correlations of scPDSi series with AMO, Niño 3 index, IOD, and QBO. Finally, the severity of the 2011 East African drought, as the worst drought in 60 years, was assessed by means of scPDSI. The findings revealed that the droughts in East Africa were characterized by a considerable amount of temporal variability so that both EOF1 and EOF2 cumulatively explained 31.5% of the variability for the full time period (1901–2019). Furthermore, the variability rate explained in a particular EOF was highly dependent on the subperiod for analysis. More specifically, the amount of variability explained by EOF1 was 20.5% for a 30-year data period, while it decreased to about 19% for a 60-year period. The exploration of the atmospheric drivers of the droughts in terms of the climate indices showed that the strength (and even for some cases the sign) of the co-variation of the drought indicator and possible drivers changes with the selected subperiod. Our results further indicated that the variability in dry conditions over large areas across East Africa can be mainly explained by the IOD and ENSO phenomenon and particularly by their negative and cool phases, respectively.

This study investigated the historical variations of droughts across East Africa and their connections with large-scale circulation patterns. Due to a relatively poor observation data coverage in the region, the occurrence of anomalously dry or wet spells might be underestimated, which is especially the case for the early decades of the 20th century. An assessment of nonlinear and lagged relationships between the temporal variations of droughts and large-scale patterns would also provide valuable information for a better understanding of the physical processes triggering drought conditions and for improving the predictability of dry periods in the future. Considering the documented anthropogenic hydro-climatic changes in Africa, future research is required to investigate the influence of historical and future climate change on droughts in the region.

Acknowledgment

The authors acknowledge that this study was based on scPDSI from CRU TS 4.04.

Conflict of interest

The authors declare no conflict of interest and no competing financial interests.

References

Alexander, M. A., Halimeda Kilbourne, K., & Nye, J. A. (2014). Climate variability during warm and cold phases of the Atlantic multidecadal oscillation (AMO) 1871-2008. *Journal of Marine Systems*, *133*, 14–26. https://doi.org/10.1016/j.jmarsys.2013.07.017.

Ayugi, B., Tan, G., Niu, R., Dong, Z., Ojara, M., Mumo, L., et al. (2020). Evaluation of meteorological drought and flood scenarios over Kenya, East Africa. *Atmosphere*, *11*(3), 307. https://doi.org/10.3390/atmos11030307.

Barichivich, J., Osborn, T. J., Harris, I., Schrier, & Jones, P. D. (2020). State of the climate in 2020. *Bulletin of the American Meteorological Society, 101*, S1–S475. https://doi.org/10.1175/2020BAMSStateoftheClimate.1.

Björnsson, H., & Venegas, S. (1997). *A manual for EOF and SVD analyses of climate data*. Department of Atmospheric and Oceanic Sciences and Centre for Climate and Global Change Research, McGill University. 54 pp.

Bloomfield, J. P., & Marchant, B. P. (2013). Analysis of groundwater drought building on the standardised precipitation index approach. *Hydrology and Earth System Sciences, 17*(12), 4769–4787. https://doi.org/10.5194/hess-17-4769-2013.

Booth, B. B. B., Dunstone, N. J., Halloran, P. R., Andrews, T., & Bellouin, N. (2012). Aerosols implicated as a prime driver of twentieth-century North Atlantic climate variability. *Nature, 484*(7393), 228–232. https://doi.org/10.1038/nature10946.

Climate Home News. (2013). https://www.climatechangenews.com/2013/02/21/climate-change-linked-to-2011-east-africa-drought/.

Coughlan de Perez, E., van Aalst, M., Choularton, R., van den Hurk, B., Mason, S., Nissan, H., et al. (2019). From rain to famine: Assessing the utility of rainfall observations and seasonal forecasts to anticipate food insecurity in East Africa. *Food Security, 11*(1), 57–68. https://doi.org/10.1007/s12571-018-00885-9.

Delworth, T. L., & Mann, M. E. (2000). Observed and simulated multidecadal variability in the northern hemisphere. *Climate Dynamics, 16*(9), 661–676. https://doi.org/10.1007/s003820000075.

Diem, J. E., Ryan, S. J., Hartter, J., & Palace, M. W. (2014). Satellite-based rainfall data reveal a recent drying trend in central equatorial Africa. *Climatic Change, 126*(1–2), 263–272. https://doi.org/10.1007/s10584-014-1217-x.

Duzenli, E., Tabari, H., Willems, P., & Yilmaz, M. T. (2018). Decadal variability analysis of extreme precipitation in Turkey and its relationship with teleconnection patterns. *Hydrological Processes, 32*(23), 3513–3528. https://doi.org/10.1002/hyp.13275.

EM-DAT. (2018). *The emergency events database – Université Catholique de Louvain (UCL) – CRED, D. Guha-Sapir.*.

Enfield, D. B., Mestas-Nuñez, A. M., & Trimble, P. J. (2001). The Atlantic multidecadal oscillation and its relation to rainfall and river flows in the continental U.S. *Geophysical Research Letters, 28*(10), 2077–2080. https://doi.org/10.1029/2000GL012745.

Folland, C. K., Palmer, T. N., & Parker, D. E. (1986). Sahel rainfall and worldwide sea temperatures, 1901-85. *Nature, 320*(6063), 602–607. https://doi.org/10.1038/320602a0.

Funk, C., Nicholson, S. E., Landsfeld, M., Klotter, D., Peterson, P., & Harrison, L. (2015). The centennial trends greater horn of Africa precipitation dataset. *Scientific Data, 2*. https://doi.org/10.1038/sdata.2015.50.

GGWS. (2021). *El Niño and La Niña years and intensities*. https://ggweather.com/enso/oni.htm.

Global Water Partnership Eastern Africa (GWPEA). (2015). Assessment of drought resilience frameworks in the Horn of Africa. In *Integrated drought management program in the Horn of Africa (IDMP HOA), Entebbe, Uganda*. https://www.gwp.org/en/GWP-Eastern-Africa.

Haile, G. G., Tang, Q., Hosseini-Moghari, S. M., Liu, X., Gebremicael, T. G., Leng, G., et al. (2020). Projected impacts of climate change on drought patterns over East Africa. *Earth's Future, 8*(7). https://doi.org/10.1029/2020EF001502.

Hoerling, M., Hurrell, J., Eischeid, J., & Phillips, A. (2006). Detection and attribution of twentieth-century northern and southern African rainfall change. *Journal of Climate, 19*(16), 3989–4008. https://doi.org/10.1175/JCLI3842.1.

Horel, J. D. (1984). Complex principal component analysis: Theory and examples. *Journal of Climate and Applied Meteorology, 23*(12), 1660–1673. https://doi.org/10.1175/1520-0450(1984)023<1660:CPCATA>2.0.CO;2.

Hua, W., Zhou, L., Chen, H., Nicholson, S. E., Raghavendra, A., & Jiang, Y. (2016). Possible causes of the central equatorial African long-term drought. *Environmental Research Letters, 11*(12). https://doi.org/10.1088/1748-9326/11/12/124002.

Indeje, M., Semazzi, F. H. M., & Ogallo, L. J. (2000). ENSO signals in east African rainfall seasons. *International Journal of Climatology, 20*(1), 19–46. https://doi.org/10.1002/(SICI)1097-0088(200001)20:1<19::AID-JOC449>3.0.CO;2-0.

IPCC. (2014). Climate change 2014: Synthesis report. In R. K. Pachauri, & L. A. Meyer (Eds.), *Contribution of working groups I, II and III to the fifth assessment report of the intergovernmental panel on climate change*. Geneva, Switzerland: IPCC.

Knight, J. R., Folland, C. K., & Scaife, A. A. (2006). Climate impacts of the Atlantic multidecadal oscillation. *Geophysical Research Letters, 33*(17). https://doi.org/10.1029/2006GL026242.

Knudsen, M. F., Jacobsen, B. H., Seidenkrantz, M. S., & Olsen, J. (2014). Evidence for external forcing of the Atlantic multidecadal oscillation since termination of the little ice age. *Nature Communications, 5*. https://doi.org/10.1038/ncomms4323.

Liebmann, B., Hoerling, M. P., Funk, C., Bladé, I., Dole, R. M., Allured, D., et al. (2014). Understanding recent eastern Horn of Africa rainfall variability and change. *Journal of Climate, 27*(23), 8630–8645. https://doi.org/10.1175/JCLI-D-13-00714.1.

Lott, F. C., Christidis, N., & Stott, P. A. (2013). Can the 2011 East African drought be attributed to human-induced climate change? *Geophysical Research Letters, 40*(6), 1177–1181. https://doi.org/10.1002/grl.50235.

Lüdecke, H. J., Müller-Plath, G., Wallace, M. G., & Lüning, S. (2021). Decadal and multidecadal natural variability of African rainfall. *Journal of Hydrology: Regional Studies, 34*. https://doi.org/10.1016/j.ejrh.2021.100795.

Lyon, B. (2014). Seasonal drought in the Greater Horn of Africa and its recent increase during the March-May long rains. *Journal of Climate, 27*(21), 7953–7975. https://doi.org/10.1175/JCLI-D-13-00459.1.

Lyon, B., & Dewitt, D. G. (2012). A recent and abrupt decline in the East African long rains. *Geophysical Research Letters, 39*(2). https://doi.org/10.1029/2011GL050337.

Mann, M. E., Zhang, Z., Rutherford, S., Bradley, R. S., Hughes, M. K., Shindell, D., et al. (2009). Global signatures and dynamical origins of the little ice age and medieval climate anomaly. *Science, 326*(5957), 1256–1260. https://doi.org/10.1126/science.1177303.

Masih, I., Maskey, S., Mussá, F. E. F., & Trambauer, P. (2014). A review of droughts on the African continent: A geospatial and long-term perspective. *Hydrology and Earth System Sciences, 18*(9), 3635–3649. https://doi.org/10.5194/hess-18-3635-2014.

McKee, T., Doesken, N., & Kleist, J. (1993). The relationship of drought frequency and duration to time scales. In *Proceedings of the 8th conference of applied climatology* (pp. 179–184).

Met Office. (2021). *Quasi-biennial oscillation (QBO)*. https://www.metoffice.gov.uk/weather/learn-about/weather/atmosphere/quasi-biennial-oscillation.

Ngoma, H., Wen, W., Ojara, M., & Ayugi, B. (2021). Assessing current and future spatiotemporal precipitation variability and trends over Uganda, East Africa, based on CHIRPS and regional climate model datasets. *Meteorology and Atmospheric Physics, 133*, 823–843. https://doi.org/10.1007/s00703-021-00784-3.

Niang, I., Ruppel, O. C., Abdrabo, M. A., Essel, A., Lennard, C., Padgham, J., & Urquhart, P. (2014). Africa. In V. R. Barros, C. B. Field, D. J. Dokken, M. D. Mastrandrea, K. J. Mach, T. E. Bilir, et al. (Eds.), *Climate change 2014: Impacts, adaptation, and vulnerability. Part B: Regional aspects. Contribution of working group II to the fifth assessment report of the intergovernmental panel on climate change* (pp. 1199–1265). Cambridge, United Kingdom/New York, NY: Cambridge University Press.

Nicholson, S. E. (2014). A detailed look at the recent drought situation in the Greater Horn of Africa. *Journal of Arid Environments, 103*, 71–79. https://doi.org/10.1016/j.jaridenv.2013.12.003.

Nicholson, S. E. (2017). Climate and climatic variability of rainfall over eastern Africa. *Reviews of Geophysics, 55*(3), 590–635. https://doi.org/10.1002/2016RG000544.

OCHA. (2011). *Severe drought affects millions in Eastern Africa.* https://www.unocha.org/story/severe-drought-affects-millions-eastern-africa.

Ogwang, B. A., Ongoma, V., Xing, L., & Ogou, F. K. (2015). Influence of mascarene high and Indian Ocean dipole on East African extreme weather events. *Geographica Pannonica, 19*(2), 64–72. https://doi.org/10.5937/geopan1502064o.

Ongoma, V., & Chen, H. (2017). Temporal and spatial variability of temperature and precipitation over East Africa from 1951 to 2010. *Meteorology and Atmospheric Physics, 129*(2), 131–144. https://doi.org/10.1007/s00703-016-0462-0.

Ongoma, V., Chen, H., & Gao, C. (2018). Projected changes in mean rainfall and temperature over East Africa based on CMIP5 models. *International Journal of Climatology, 38*(3), 1375–1392. https://doi.org/10.1002/joc.5252.

Onyutha, C. (2016). Geospatial trends and decadal anomalies in extreme rainfall over Uganda, East Africa. *Advances in Meteorology, 2016*, 1–15. https://doi.org/10.1155/2016/6935912.

Onyutha, C. (2017). On rigorous drought assessment using daily time scale: Non-stationary frequency analyses, revisited concepts, and a new method to yield non-parametric indices. *Hydrology, 4*(4). https://doi.org/10.3390/hydrology4040048.

Onyutha, C. (2020). Analyses of rainfall extremes in East Africa based on observations from rain gauges and climate change simulations by CORDEX RCMs. *Climate Dynamics, 54*(11–12), 4841–4864. https://doi.org/10.1007/s00382-020-05264-9.

Onyutha, C. (2021). Long-term climatic water availability trends and variability across the African continent. *Theoretical and Applied Climatology, 146*(1), 1–17. https://doi.org/10.1007/s00704-021-03669-y.

Onyutha, C., & Willems, P. (2017). Influence of spatial and temporal scales on statistical analyses of rainfall variability in the River Nile basin. *Dynamics of Atmospheres and Oceans, 77*, 26–42. https://doi.org/10.1016/j.dynatmoce.2016.10.008.

Otterå, O. H., Bentsen, M., Drange, H., & Suo, L. (2010). External forcing as a metronome for Atlantic multidecadal variability. *Nature Geoscience, 3*(10), 688–694. https://doi.org/10.1038/ngeo955.

Palmer, W. C. (1965). *Meteorological drought.* Washington DC: US Department of Commerce Weather Bureau. Research paper no 45.

Parker, D., Folland, C., Scaife, A., Knight, J., Colman, A., Baines, P., et al. (2007). Decadal to multidecadal variability and the climate change background. *Journal of Geophysical Research. Atmospheres, 112*(18). https://doi.org/10.1029/2007JD008411.

Rayner, N. A., Parker, D. E., Horton, E. B., Folland, C. K., Alexander, L. V., Rowell, D. P., et al. (2003). Global analyses of sea surface temperature, sea ice, and night marine air temperature since the late nineteenth century. *Journal of Geophysical Research. Atmospheres, 108*(14). https://doi.org/10.1029/2002jd002670.

Richman, M. B. (1986). Rotation of principal components. *Journal of Climatology, 6*(3), 293–335. https://doi.org/10.1002/joc.3370060305.

Saji, N. H., Goswami, B. N., Vinayachandran, P. N., & Yamagata, T. (1999). A dipole mode in the tropical Indian ocean. *Nature, 401*(6751), 360–363. https://doi.org/10.1038/43854.

Tabari, H., & Willems, P. (2018). Lagged influence of Atlantic and Pacific climate patterns on European extreme precipitation. *Scientific Reports, 8*(1). https://doi.org/10.1038/s41598-018-24069-9.

Tierney, J. E., Smerdon, J. E., Anchukaitis, K. J., & Seager, R. (2013). Multidecadal variability in East African hydroclimate controlled by the Indian Ocean. *Nature, 493*(7432), 389–392. https://doi.org/10.1038/nature11785.

Trenberth, K. E., Fasullo, J. T., O'Dell, C., & Wong, T. (2010). Relationships between tropical sea surface temperature and top-of-atmosphere radiation. *Geophysical Research Letters, 37*(3). https://doi.org/10.1029/2009GL042314.

van Oldenborgh, G. J., te Raa, L. A., Dijkstra, H. A., & Philip, S. Y. (2009). Frequency- or amplitude-dependent effects of the Atlantic meridional overturning on the tropical Pacific Ocean. *Ocean Science, 5*(3), 293–301. https://doi.org/10.5194/os-5-293-2009.

Vellinga, M., & Milton, S. F. (2018). Drivers of interannual variability of the East African "Long Rains". *Quarterly Journal of the Royal Meteorological Society, 144*(712), 861–876. https://doi.org/10.1002/qj.3263.

Vicente-Serrano, S. M., Beguería, S., & López-Moreno, J. I. (2010). A multiscalar drought index sensitive to global warming: The standardized precipitation evapotranspiration index. *Journal of Climate, 23*(7), 1696–1718. https://doi.org/10.1175/2009JCLI2909.1.

Wells, N., Goddard, S., & Hayes, M. J. (2004). A self-calibrating Palmer Drought Severity Index. *Journal of Climate, 17*(12), 2335–2351. https://doi.org/10.1175/1520-0442(2004)017<2335:ASPDSI>2.0.CO;2.

Williams, A. P., & Funk, C. (2011). A westward extension of the warm pool leads to a westward extension of the Walker circulation, drying eastern Africa. *Climate Dynamics, 37*(11–12), 2417–2435. https://doi.org/10.1007/s00382-010-0984-y.

Zhang, R., Delworth, T. L., Dixon, K. W., Held, I. M., Ming, Y., Msadek, R., et al. (2013). Have aerosols caused the observed Atlantic multidecadal variability? *Journal of the Atmospheric Sciences, 70*(4), 1135–1144. https://doi.org/10.1175/JAS-D-12-0331.1.

CHAPTER

11

Revisiting the impacts of tropical cyclone Idai in Southern Africa

Collen Mutasa

UNESCO Regional Office for Southern Africa, Harare, Zimbabwe

1 Introduction

Water-related hazards have dominated the list of disasters in terms of both human and economic toll over the past 50 years (WMO, 2021). Between 1970 and 2019, weather, climate and water hazards accounted for 50% of all disasters, 45% of all reported deaths and 74% of all reported economic losses at global level (WMO, 2021). Of the top 10 disasters that occurred during the same period, the hazards that led to the largest human losses have been droughts (650,000 deaths), storms (577,232 deaths), floods (58,700 deaths), and extreme temperatures (55,736 deaths).

Africa accounts for 15% of weather, climate and water-related disasters, 35% of associated deaths and 1% of economic losses reported globally (WMO, 2021). Storms and floods, combined, accounted for the highest economic losses (71%) of the total economic losses recorded in 1970–2019.

Southern Africa is vulnerable to natural hazards, the majority being hydro-meteorological in nature. The most frequent hazards are floods, while recurrent droughts affect the largest number of people and also account for the highest economic cost of damages (UNECA, 2015). Climate extremes, especially precipitation and drought extremes, interacting with exposed and vulnerable human and natural systems can lead to disasters (Liu, Shen, Qi, Wang, & Geng, 2019), resulting in environmental and infrastructural damage, economic losses and even human deaths. Other major hydrometeorological hazards that affect southern Africa include landslides, storms, wildfires and tropical cyclones (TCs) that originate from the South West Indian Ocean.

Tropical cyclones are ranked after drought as the second leading cause of loss of human life due to natural disasters (De, Khole, & Dandekar, 2004). Landfalling TCs are associated with severe weather phenomena such as strong winds, torrential rains and storm surges that cause substantial consequences that include loss of lives and devastation of property and livelihood (Shanko & Camberlin, 1998; Vitart, Anderson, & Stockdale, 2003). Coastal areas are particularly prone to storm surges caused by TCs that make landfall, while widespread flooding can result over some parts of the southern African mainland (Crimp & Mason, 1999; Reason & Keibel, 2004). The TCs also trigger weather extremes such as intense precipitation events which result in landslides and flash floods causing fatalities and huge socioeconomic damages (Chen, Chang, Chiu, Lau, & Lee, 2013; Czajkowski, Villarini, Michel-Kerjan, & Smith, 2013; Peduzzi et al., 2012; Webster, Holland, Curry, & Chang, 2005). However, in some instances TCs provide useful rainfall to communities. For instance, in February 2000, ex-tropical cyclone Eline contributed significantly to the seasonal rainfall total of southern Namibia. It is approximated that 25% of the January–February–March (JFM) 2000 rainfall over southern Namibia was caused by this system. Furthermore, the system resulted in southern Namibia recording its wettest summer since 1976 (Reason & Keibel, 2004).

Worldwide over the 50-year period (1970–2019), TCs were responsible for 9% of all recorded disasters, and accounted for 17% of all reported deaths and 29% of all economic losses (WMO, 2021). During the same period, 86% of disasters, 99% of deaths and 29% of economic losses associated with TCs occurred in developing economies. The number of disasters related to TCs over the period increased, while the number of deaths markedly decreased following a peak in the 1970s. Socioeconomic losses resulting from landfalling TCs have increased sharply worldwide over the last decades (Pielke Jr. et al., 2008; Zhang, Liu, & Wu, 2009).

Climate Impacts on Extreme Weather
https://doi.org/10.1016/B978-0-323-88456-3.00012-5

Results from attribution studies show that human induced climate change is changing the likelihood and intensity of extreme weather and climate events (Carbon Brief, 2021; IPCC AR5, 2013; Murakami et al., 2020), making some extreme events more frequent and/or intense and others less so (Cattiaux, Chauvin, Douville, & Ribes, 2021). Globally, an increasing number of extreme events has been observed over the past few decades, some of them attributed to global warming (Hao, Aghakouchak, & Phillips, 2013; IPCC AR5, 2014; Leonard et al., 2013). Weather, climate and water-related disasters are also on the rise worldwide, causing loss of life and setting back economic and social development by years, if not decades (WMO, 2014). In particular, since 1950, heat waves have increased in frequency, duration and intensity in nearly every part of the world (Perkins-Kirkpatrick & Lewis, 2020), while an intensification of drought patterns, extreme rainfall and an increase in aridity has also been observed in response to global warming (Bonfils, Santer, & Fyfe, 2020; Sun, Zhang, Zwiers, Westra, & Alexander, 2021). An increase in the global average intensity of the strongest TCs since the early 1980s has also been noticed (Knutson et al., 2019; Kossin, Knapp, Olander, & Velden, 2020). The number of TCs has been rising since 1980 in some basins such as the north Atlantic and central Pacific Oceans, while declining in the western Pacific and south Indian Ocean (Murakami et al., 2020).

This chapter gives an overview of TCs in general. It also seeks to establish present and future characteristics of TCs in the South West Indian Ocean (SWIO) basin, determine how TC Idai formed and evolved, understand its characteristics and document the impacts it caused in southern Africa, particularly in Malawi, Mozambique and Zimbabwe. The study also seeks to summarize the lessons derived from TC Idai in order to minimize future impacts from extreme weather events such as TCs.

2 Materials and methods

2.1 Study area

The study area is southern Africa, which for purposes of this study, is taken as that region south of the equator which includes the 16 member countries of the Southern African Development Community (SADC). The region is bordered by the Indian Ocean to the East and the Atlantic Ocean to the West. Fig. 11.1 shows a map of southern Africa. As at 2018, the estimated SADC population stood at 345.2 million, which represents a 2.5% annual population growth rate when compared with the 2017 population of 336.9 million (SADC, 2018).

Agriculture is an important source of livelihood for most of the rural population of southern Africa (Hachigonta, Nelson, Thomas, & Sibanda, 2013; Wamukonya & Rukato, 2001). The sector contributes between 4% and 27% of the Gross Domestic Project (GDP) in the different member states (SADC, 2012). The performance of this sector therefore

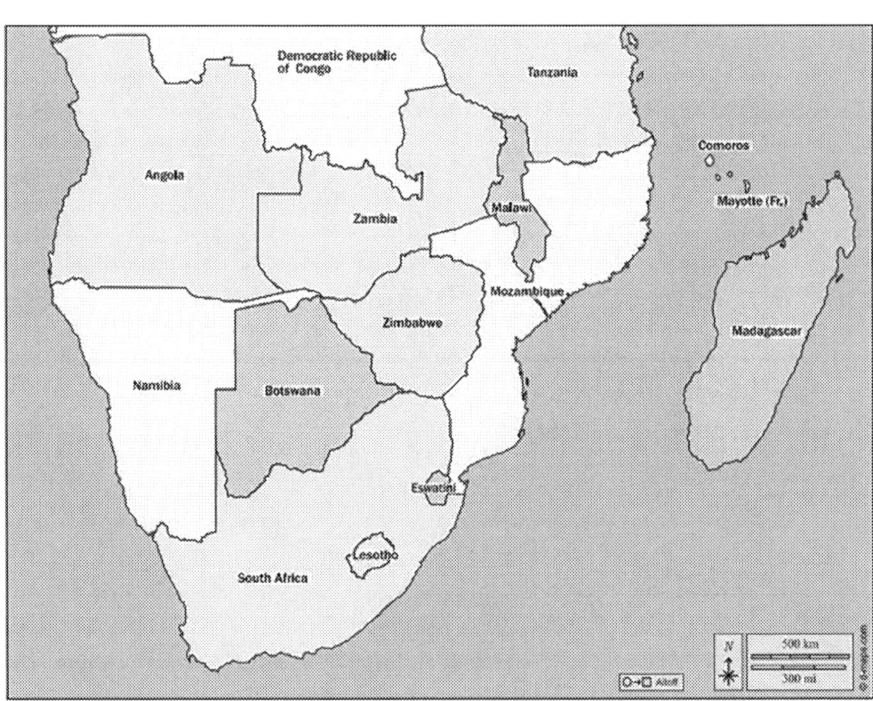

FIG. 11.1 Map of southern Africa. *Source: d-maps.org.*

has a strong influence on food security, economic growth and social stability in the region (SADC, 2012). About 80% of the population in Mozambique and Malawi, and 70% of the population in Zimbabwe rely on rainfed, subsistence agriculture for their livelihoods. Southern Africa is also one of the world's poorest and most vulnerable regions (Bauer & Scholz, 2010) and suffers from poor infrastructure and low socio-economic development. Hence, the consequences of extreme weather or climate anomalies are often devastating to both people and property. The region is prone to frequent droughts and floods, some of the floods being caused by TCs that occasionally make landfall on the Mozambican and South African coastlines, bringing significant rainfall and associated flooding to Mozambique, the northern parts of South Africa, and Zimbabwe (Davis, 2011). One of the notable TCs to have affected the region in the past is TC Eline which struck southern Africa in February 2000 and caused floods in northern South Africa, Mozambique and Zimbabwe, resulting in the death of 600 people and damage to bridges, roads and other infrastructure (Dyson & van Heerden, 2001). Prolonged periods of drought in the summer seasons of 1982–1983 and 1991–1992 also caused severe falls in crop and stock production in many parts of the region (Harsch, 1992; Vogel, 1994).

The climate across southern Africa varies from arid conditions in the west to humid subtropical conditions in the north and east, while much of the central parts are classified as semi-arid (Stringer et al., 2009). Most parts of the region, with the exception of southwest South Africa which experiences Mediterranean conditions, and thus receives winter rains, receives the majority of its rainfall in the summer months, between November and March. Precipitation patterns reveal lower annual rainfall in the south versus higher annual rainfall in the north (Kandji, Verchot, & Mackensen, 2006). The majority of the region receives between 500 and 1500 mm per year, with the more semi-arid regions receiving between 250 and 500 mm per year (Davis, 2011). In terms of temperature, southern Africa has a warm climate with a mean annual minimum temperature ranging from 3 to 25°C while the mean annual maximum temperature ranges from 15 to 36°C.

2.2 Methodology

The study used secondary sources of information to understand the impacts experienced by each of the countries; Malawi, Mozambique and Zimbabwe as a result of TC Idai. These sources included government reports, reports from disaster management institutions, briefings from local and international Non-Governmental Organizations (NGOs), policy briefs, reports from donor agencies, reports from UN agencies such as the United Nations Children's Fund (UNICEF), United Nations Office for the Coordination of Humanitarian Affairs (UNOCHA), International Organization for Migration (IOM), World Food Programme (WFP), and the World Bank, among others. Other information was obtained from researchers' observations and interviews with the affected people in different communities. In addition, peer reviewed journal articles also acted as a source of valuable information particularly on historical and future characteristics of TCs in the SWIO basin. Sustained wind speed (sustained 3 min) in km/hour for TC Idai was calculated from Best Track data from the US National Center for Atmospheric Research (NCAR). Estimated precipitation data was obtained from the US National Aeronautics and Space Administration (NASA) Global Precipitation Mission (GPM) (Probst & Annunziato, 2019). Storm surge data is based on calculations from National Oceanic and Atmospheric Administration (NOAA) Hurricane Weather Research and Forecast (HWRF), a non-hydrostatic coupled ocean–atmosphere model. The model uses observations from satellites, data buoys, and hurricane hunter aircraft (Probst & Annunziato, 2019).

3 Results

3.1 An overview of tropical cyclones

3.1.1 Understanding tropical cyclones

Tropical cyclones are synoptic-scale low pressure systems without fronts, occurring over tropical or subtropical waters with organized thunderstorm activity (Anthes, 1982; Holland & Lander, 1993). They form in seven specific areas around the globe, known as basins, and in particular seasons. The basins, depending on the particular ocean of formation are named; *1. Atlantic Basin, 2. Northeast Pacific Basin, 3. Northwest Pacific Basin, 4. North Indian Basin, 5. Southwest Indian Basin, 6. Southeast Indian/Australian Basin, 7. Australian/Southwest Pacific Basin.* Fig. 11.2 shows the seven basins. Within the basins, the TCs are known by different names; they are called hurricanes in the North Atlantic and Northeastern Pacific oceans, TCs in the South West Indian Ocean, and typhoons in the North West Pacific Ocean. They form in equatorial regions at 5°–30° and rarely within less than 4°–5° from the Equator (Anthes, 1982). Typically, they form where sea surface temperatures (SSTs) are about 26–27°C (Dare & McBride, 2011; Palmén, 1948). Other conditions that favor cyclogenesis are: (i) large values of low level relative vorticity, (ii) weak vertical and

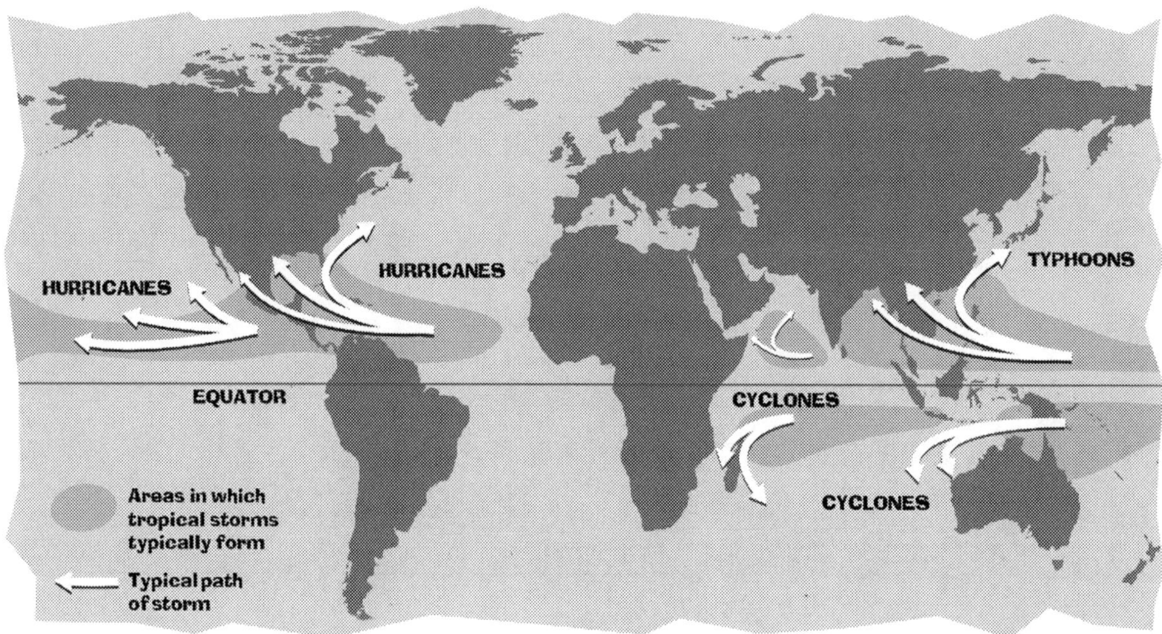

FIG. 11.2 Location of tropical cyclone formation areas and their typical tracks once formed. *Courtesy: NOAA. Retrieved from https://scijinks.gov/hurricane/ on 5 May 2022.*

horizontal wind shear, (iii) conditional instability through a deep atmospheric layer, (iv) large relative humidity in the lower and middle troposphere, and (v) a deep oceanic mixed layer (Henderson-Sellers et al., 1998).

Approximately 69% of TCs occur in the Northern Hemisphere, while only 31% occur in the Southern Hemisphere. About 12% occur in the Atlantic Ocean, 57% occur in the Pacific Ocean and 31% occur in the Indian Ocean (NOAA). An analysis of data from 1968 to 2010 shows that on average 88 tropical storms formed each year across the globe (Shultz, Russell, & Espinel, 2005; Weinkle, Maue, & Pielke Jr., 2012), and of these, 48 acquired TC strength (Doocy, Dick, Daniels, & Kirsch, 2013; Weinkle et al., 2012).

3.1.2 *Tropical cyclones in the SWIO basin*

Countries in southern Africa are affected by TCs that originate from the South West Indian Ocean (SWIO) basin. The SWIO basin lies within 30°E to 90°E and 0°S to 40°S, covering a number of countries; Botswana, Comoros, Eswatini, Kenya, La Réunion, Lesotho, Madagascar, Malawi, Mauritius, Mozambique, Namibia, Seychelles, South Africa, Tanzania and Zimbabwe. The TCs that form in the SWIO tend to move westwards, toward the east coast of Madagascar and sometimes recurve southeast into the central South Indian Ocean. Only 5% of TCs that form in the SWIO basin make landfall over southern Africa (Fitchett & Grab, 2014; Mavume, Rydberg, Rouault, & Lutjeharms, 2010; Reason, 2007) and they can move further westwards into the southern Africa interior, or may be deflected towards the south or north (Mavume et al., 2010). The TC season in the SWIO basin officially starts in November and ends in April, with a peak in January and February, coinciding with southern hemisphere summer. According to Mavume et al. (2010), 85% of the TCs form during this November to April period.

On average, nine TCs occur in the basin annually, and if tropical storms are also included, the number of tropical systems rise to 11 (Jury, 1993). Tropical systems (tropical storms and tropical cyclones) that occur in the SWIO basin account for about 14% of the global total (Ho, Kim, Jeong, Kim, & Chen, 2006; Jury, 1993; Mavume et al., 2010). According to an analysis of 1966 to 2000 data, 13% of the cyclogenesis in the SWIO basin occurs over the Mozambique Channel (Mavume et al., 2010). The majority of TCs in the Mozambique Channel occur in January or February with no formations in November and April while formation in December and March is rare (Mavume et al., 2010). A study of 94 TCs that occurred in the Mozambique Channel between 1948 and 2010 revealed that the conditions ideal for cyclogenesis in the Mozambique Channel include high SSTs (29.6°C on average), a southerly wind shear with height and below-normal geopotential height anomalies at 500 hPa (Matyas, 2015).

In their developmental cycle in the SWIO basin, weather systems go through a series of stages from starting out as tropical disturbances until reaching very intense TC status. Table 11.1 shows the classification of tropical weather

TABLE 11.1 Classification of tropical weather systems in the SWIO basin.

Stage	Wind speed (knots)	Wind speed (km/h)
Tropical disturbance	<28	<51
Tropical depression	28–33	51–63
Moderate tropical storm	34–47	64–88
Severe tropical storm	48–63	89–117
Tropical cyclone	64–89	118–165
Intense tropical cyclone	90–115	166–212
Very intense tropical cyclone	>115	>212

TABLE 11.2 The Saffir-Simpson Hurricane Wind Scale.

Category	Wind speed (knots)	Wind speed (km/h)	Damage potential
1	64–82	119–153	Minimal
2	83–95	154–177	Moderate
3	96–112	178–208	Extensive
4	113–136	209–251	Extreme
5	≥137	≥252	Catastrophic

systems in the SWIO basin. Tropical cyclone intensity is measured by the maximum sustained wind speed at the surface as either 1 min or 10 min averages. In the SWIO the intensity or classification of tropical cyclones is determined by the averaged 10 min wind speeds and systems are given a name when they reach moderate tropical storm stage, corresponding to a wind speed of 64 km/hr (34 knots). The names are chosen from a predetermined list containing both female and male names, which is arranged in alphabetical order, with each TC season starting with the letter "A" (WMO). Names from the list are contributed by all the nations that are members of the World Meteorological Organization's SWIO Tropical Cyclone Committee (WMO). A different scale, the Saffir-Simpson Hurricane Wind Scale (SSHWS), Table 11.2, is used to rate hurricane intensity in the North Eastern Pacific Ocean and North Atlantic Ocean basins (National Geographic, n.d). According to the National Hurricane Center (NHC) and Central Pacific Hurricane Center (n.d), the SSHWS uses a 1 to 5 rating based on a hurricane's sustained wind speed over a 1 min interval, 10 m above the surface and describes estimated potential property damage.

3.1.3 Historical and future characteristics of tropical cyclones in the SWIO basin

Research shows that TC characteristics in individual basins, i.e., frequency and intensity, depend on natural cycles such as the El Niño-Southern Oscillation (ENSO) (Landsea, Pielke Jr., Mesta Nuñez and Knaff, 1999), the Quasi-Biennial zonal wind Oscillation (QBO) (Landsea et al., 1999) and the Madden-Julian Oscillation (MJO) (Madden and Julian, 1994). A number of studies have been carried out to establish the historical trends in TC characteristics in the SWIO basin (Emanuel, 2005; Mavume et al., 2010; Trenberth, 2005; Webster et al., 2005). In the study by Mavume et al. (2010), track data from the Joint Typhoon Warning Center was merged with data from La Réunion Regional Specialized Meteorological Centre to study TC characteristics during the period 1980–2007. This period was split into two-time series, 1980 to 1993 and 1994 to 2007. An increase in the number of intense TCs from 36 during the 1980–1993 period to 56 during the 1994–2007 period was observed. This increase in intense TCs is consistent with results of work by Webster et al. (2005) who used data from 1975 to 2004 and also the work by Kuleshov, Qi, Fawcett, and Jones (2008). Webster et al. (2005) noted in particular, an increase in category 4 and 5 hurricanes in all the basins using 1975–1989 and 1990–2004 data. A minor increase in the total number of TC days was also noticed for the period 1980–2007 (Mavume et al., 2010). When a longer time series (1952–2007) split into two periods, 1952–1979 and 1980–2007 was investigated to analyze trends in frequency, there was a decrease in TC frequency and the number of land falling TCs on both Madagascar and Mozambique (Mavume et al., 2010). The decrease in TC frequency occurred in both the Mozambique Channel and the SWIO in general. Some authors however view records before

1980 when satellite data was unavailable as unreliable due to the scarcity of observational data, particularly wind intensity data (Knapp & Kossin, 2007; Landsea, 2007).

Despite advances in modeling of the global climate, predicting future characteristics of TCs remains a challenge due to a number of factors (Mutasa, 2019). However a few future numerical modeling results of TC characteristics are worth taking note of. By the end of the 21st century, simulation studies under an A2 emission scenario indicate a decrease in the frequency of TCs over the SWIO basin (Caron & Jones, 2008; Malherbe, Engelbrecht, & Landman, 2013; Muthige et al., 2018). A projected decrease ranging between 25% and 50% in the number of TCs making landfall over southern Africa is also predicted (Muthige et al., 2018). Furthermore, there will be an increase in the frequency of the most intense (Category 4–5) TCs (Christensen et al., 2013; Knutson et al., 2015) and an increase in the maximum intensity particularly of the most intense TCs by the late 21st century.

3.1.4 The formation and evolution of tropical cyclone Idai

Tropical cyclone Idai, then a tropical depression, formed on 4 March 2019 in the Mozambique Channel and made landfall for the first time on the same day north of Quilimane, Mozambique (Meteo-France, 2019). It then headed towards Malawi where it weakened to a low pressure system but still causing significant floods in the southern parts of the country and northern Mozambique. The low pressure system recurved between 7 and 8 March and moved back into the Mozambique Channel on 9 March where it re-intensified. The system was subsequently named at 2300 UTC on 9 March, being the 9th 2018–2019 season named system of the SWIO basin, and had wind speeds of ~75 mph (equivalent to category 1) on the SSHWS. TC Idai made its second landfall around 2330 UTC on 14 March near Beira, the fourth largest city in Mozambique with a population of about 530,000 people. The TC had maximum sustained 10-min winds of 160–180 km/h, with higher gusts on landfall (Probst & Annunziato, 2019), making it an intense tropical cyclone or Category 2 storm according to the SSHWS. The TC's strong winds and heavy rain caused extensive damage to the city of Beira in Sofala Province and surrounding areas before heading inland to eastern Zimbabwe.

From 15 March, TC Idai started weakening as it moved inland towards the central parts of Zimbabwe, becoming a tropical depression in the process (WMO, 2019). On the 16th, it further weakened to an area of low pressure, and on the 17th only a wide clockwise circulation remained over Zimbabwe. Despite the weakening, remnants of TC Idai still produced rains that affected the region. On the 19th, it left the mainland and on the 21st of March it dissipated (ECDC, 2019). Fig. 11.3 shows the path of TC Idai.

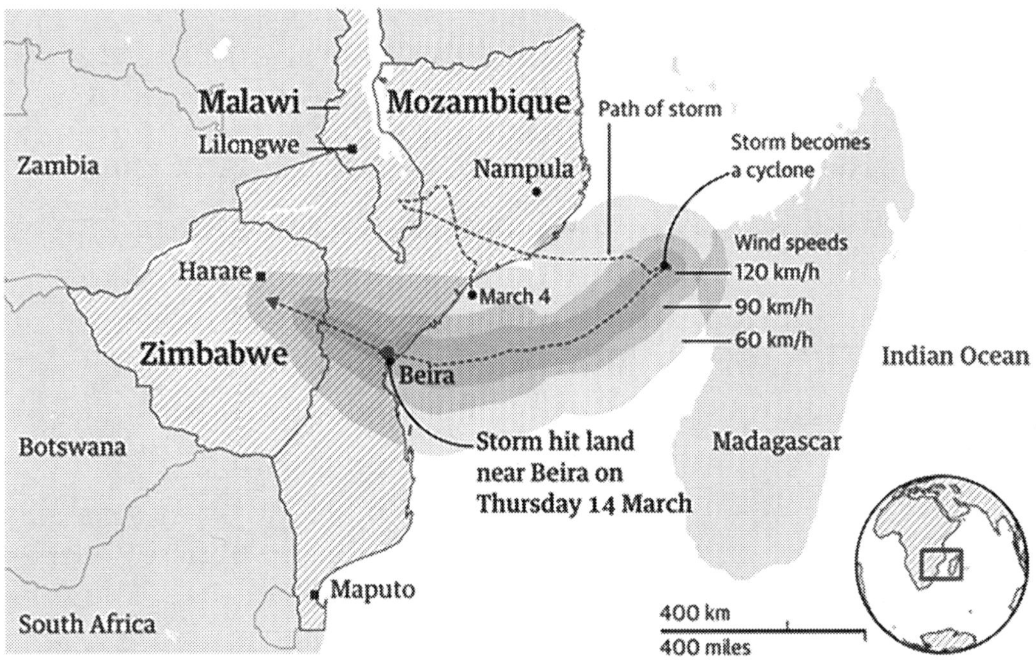

FIG. 11.3 Track of tropical cyclone Idai in Malawi, Mozambique and Zimbabwe. *Source: Global Disaster Alert and Coordination System. Retrieved from https://www.internetgeography.net/weather-and-climate/cyclone-idai-case-study/ on 5 May 2022.*

3.2 An overview of the impacts of tropical cyclone Idai

Tropical Cyclone Idai is one of the strongest tropical cyclones on record to hit Africa and the Southern Hemisphere, and the worst humanitarian disaster caused by a natural hazard event in 2019 (WMO, 2019). The worst affected countries were Mozambique, Zimbabwe and Malawi where strong winds, heavy rainfall and subsequent flooding caused extensive damage to houses, schools, hospitals, roads, bridges, energy infrastructure, communication networks, and disrupted water supplies (Mühr et al., 2019). The TC left over 1000 people dead and affected nearly 3 million people in the three countries. Hundreds of thousands of people were also left homeless and displaced, and thousands of hectares of crops, including maize that was nearing harvesting, were destroyed. The total economic loss caused by TC Idai was estimated to be at least US$2 billion, making it the costliest TC ever in the SWIO basin (World Bank, 2019). The worst affected provinces in Zimbabwe were Manicaland and Masvingo while in Mozambique the worst affected provinces were Manica, Sofala, Tete, and Zambezia. In Malawi the worst affected districts were Nsanje, Chikwawa, and Phalombe which are all in the Southern Region. Estimates from the Integrated Multi-Satellite Retrievals (IMERG), a product of the Global Precipitation Measurement (GPM) mission indicate that from 13 to 20 March, some areas in central Mozambique received cumulative rainfall totals of up to 50 cm , while elsewhere in southern Malawi and Zimbabwe, amounts of up to 25 cm of cumulative rainfall were recorded (NASA, 2019).

Apart from the three countries that were worst affected, there were other countries which were impacted due to some knock-on effects. South Africa, which imports roughly 1000 MW of electricity daily from Mozambique's Cahora Bassa hydroelectric power plant, experienced severe electricity shortages when the TC hit, resulting in the country instituting load shedding (Zimba, Houane, & Chikova, 2020).

The TC also redirected precipitation away from east Africa, resulting in a delay in the onset of the March–May (MAM) "long rains," causing suppressed rainfall between early March and mid-April (FAO, 2019; Ongoma, 2019). The resulting dryness caused a delay and disruption of planting operations, poor crop germination, water stress in early planted crops, deterioration of rangeland conditions and widespread water shortages (FAO, 2019). The water deficit also impacted livestock production (FEWSNET, 2019). The worst affected areas were most of Somalia, southeastern Ethiopia, northern and eastern Kenya, most of Uganda and northeastern areas of Tanzania (FAO, 2019).

3.2.1 Impacts on Malawi

Tropical cyclone Idai caused initial heavy rains and flooding as a tropical depression and low pressure system in Malawi from 4 to 9 March (Probst & Annunziato, 2019). The system made a second landfall as an intense tropical cyclone on March 14 on the Mozambique coast producing strong winds and heavy rainfall, causing further flooding in the southern and, to a lesser extent, central regions of Malawi.

The floods affected 15 out of Malawi's 28 districts, and two cities (IFRC, 2019). The districts affected included Balaka, Blantyre, Chikwawa, Chiradzulu, Machinga, Mangochi, Mulanje, Mwanza, Neno, Nsanje, Phalombe, Thyolo, and Zomba in the Southern Region and Dedza and Ntcheu in the Central Region (IFRC, 2019). The two cities affected were Zomba and Blantyre (GoMw PDNA, 2019). The highest 24-h rainfall totals recorded were 239 mm at Limbuli station on 5 March; 255.5 mm at Mpemba on 6 March and 220.8 mm at Tsangano on 7 March (GoMw PDNA, 2019).

The floods caused injuries and loss of human lives, damaged structures, irrigation and water supply systems, power supply systems, hydrological monitoring stations, and transport system among other devastating impacts (GoMw PDNA, 2019). The number of people affected by TC Idai included an estimated 230,737 women of childbearing age, of whom around 27,693 were expectant, and 18,462 live births were expected within the following 6 months (UNFPA, 2019). Furthermore, the floods washed away livestock and crops in the fields which were almost to be harvested. The districts that suffered most crop losses were Nsanje, Phalombe and Machinga (GoMw PDNA, 2019). Malawi is one of the poorest countries in the world with more than half the population living below the poverty threshold (IMF, 2017). Furthermore, the largest concentration of the poor population is in southern Malawi, which is also the most densely populated region in the country and the most affected by TC Idai. Table 11.3 gives the impacts of TC Idai on different sectors in Malawi.

A floods post disaster needs assessment (GoMw PDNA, 2019) estimated that the total value of the effects of the disaster was US$ 220.2 million, while the total needs for recovery and reconstruction stood at US$ 370.5 million. Following the devastation, the Government of the Republic of Malawi declared a state of disaster on the 8th of March 2019 in all flood-affected areas and appealed for international assistance.

TABLE 11.3 Impacts of tropical cyclone Idai on different sectors in Malawi.

Sector	Impacts
Human and social development	60 people were killed; 975,588 were affected; 86,976 were displaced; 731,879 were in immediate need and 672 were injured
Housing	More than 288,371 houses were partially or totally destroyed
Transport	1841 km of road network (11.9%) of the total, and 129 bridges were destroyed
Agriculture	91, 638 ha of crops and 64 irrigation systems were damaged; 47,899 livestock were washed away and lost
Health	25 health facilities were partially or completely destroyed
Water, Sanitation and Hygiene	396 boreholes were damaged, 140 latrines were damaged
Education	154 school blocks and 81 teachers' houses were partially or totally destroyed; almost 67,347 school textbooks were damaged; more than 5500 desks and 967 learners' toilets were destroyed
Energy	Power generation plants and distribution network infrastructure was destroyed
Fisheries and aquaculture	287 fishing boats; 6589 sets of fishing gears and 189 fish ponds were damaged

Reproduced with permission from Government of Malawi (GoMw) (2019). Malawi 2019 Floods Post Disaster Needs Assessment (PDNA). https://www.unicef.org/malawi/reports/malawi-2019-floods-post-disaster-needs-assessment-report (Accessed on 27 May 2021).

3.2.2 Impacts on Mozambique

The tropical depression that formed in the Mozambique Channel on the 4th of March brought heavy rains to Mozambique, causing flooding in Tete and Zambezia Provinces. After making the second landfall, as an intense TC, Idai also brought heavy rain of up to more than 200 mm in 24 h across the provinces of Sofala, Manica, Zambezia, Tete, and Inhambane, causing rivers to overflow, with flood waters reportedly rising above 10 m (GoM, 2019). The TC also caused a storm surge of 2.5 m (World Meteorological Organization, 2019) in Beira and surrounding areas of Sofala province, although other reports put the storm surge at 4.5 m (e.g. Probst & Annunziato, 2019). The TC induced winds, heavy rain and floods caused massive destruction to houses, roads, bridges, schools, health facilities, water supply systems and power distribution and transmission infrastructure. It was estimated that 90% of the city of Beira was damaged or destroyed. The floods also swept away crops and livestock, in addition to causing human deaths, injuries and displacements.

In response to the disaster, the Government of Mozambique declared a State of Emergency on 19 March 2019. Estimates indicated that TC Idai caused about US$1.4 billion in total damage and US$1.39 billion in economic losses. The total cost of recovery and reconstruction was estimated at US$2.9 billion for the 4 provinces of Sofala, Manica, Tete and Zambezia (GoM, 2019). Table 11.4 summarizes the number of people affected by the TC per province in the main affected provinces, while Table 11.5 lists the main effects of the TC on different sectors.

TABLE 11.4 Total population and number of affected people by province in Mozambique.

	Population (2019)	Affected population (2019)	% of total
Zambezia	5,164,732	6035	0.1
Tete	2,648,941	54,721	2.1
Manica	1,945,994	262,890	13.5
Sofala	2,259,248	1,190,596	52.7
Inhambane	1,488,676	422.0	0.0
Total	13,507,591	1,514,662	11.2

TABLE 11.5 Impacts of tropical cyclone Idai on different sectors in Mozambique.

Sector	Impact
Human and social development	602 people were killed; 1,500,000 people were affected; 400,000 people were displaced; 237,789 families were displaced; 1600 people were injured; 750,000 people were in need of urgent assistance
Housing	110,000 houses were totally destroyed; 128,529 houses were partially destroyed
Education	4222 school buildings were partially or totally destroyed
Water, Sanitation and Hygiene	211,000 people had restricted water access; 71,450 latrines in rural areas were damaged; 118,600 latrines were damaged in urban areas
Health	89 health facilities were partially destroyed; 3 health facilities were completely destroyed
Agriculture	715,378 ha of cultivated land was destroyed
Transport	3490 km of road was damaged; 20 bridges were damaged
Energy	1345 km of transmission lines were damaged; 10,216 km of distribution lines were damaged; 30 substations were damaged; 4000 transformers were damaged

Data from Government of Mozambique. (2019). Mozambique cyclone Idai. Post Disaster Needs Assessment (PDNA). https://www.undp.org/publications/mozambique-cyclone-idai-post-disaster-needs-assessment-pdnadna (Accessed on 27 May 2021).

3.2.3 *Impacts on Zimbabwe*

Tropical Cyclone Idai struck eastern Zimbabwe on 15 March 2019. Although it had weakened from the time it made landfall over Mozambique, it still produced heavy rains, flooding, landslides and strong winds. This resulted in loss of lives, significant damage to property, crops, livestock and infrastructure that included roads, bridges, water installations, power and communication systems. Many homes, schools and community structures were also damaged. The worst affected provinces were Manicaland and Masvingo. Districts that were affected in Manicaland included Buhera, Chimanimani, Chipinge, Makoni, Mutare, Mutasa, and Nyanga (DTM: IOM DISPLACEMENT MATRIX, 2019). More than half of the damages to houses and infrastructure occurred in Chimanimani and Chipinge (GoZ, 2019). About 95% of the road network and at least 10 critical bridges damaged or destroyed were in Chimanimani district (Oxfam, 2019). Chimanimani and Chipinge districts also reported the highest number of deaths, with the former recording 173 while the latter reported 6. Of the total deaths, 158 bodies were swept away to Mozambique (CPU, 2019). The most affected districts in Masvingo Province were Bikita, Gutu, Chiredzi, and Zaka. The Government of Zimbabwe (GoZ) declared a state of emergency on 17 March 2019, activating a government-led response led by the Department of Civil Protection Unit (CPU) in coordination with humanitarian partners, the military, and sub-national flood command centers (GoZ, 2019).

As a result of the TC induced rains, several stations in Manicaland reported heavy rains during the time the TC affected the province. Table 11.6 shows measured 24 h accumulated rainfall totals greater than 100 mm recorded at some selected meteorological stations in Manicaland province during the time when tropical cyclone Idai affected Zimbabwe. In only 4 days, Chisengu, the rainfall measuring station in Chimanimani district, recorded a total of more than 700 mm. Table 11.7 shows the sector impacts of tropical cyclone Idai in Zimbabwe.

TABLE 11.6 Notable 24 h accumulated rainfall amounts from 13 to 17 March 2019 in Manicaland, Zimbabwe.

Place	Rainfall (mm)	Date
Chisengu	153	14 March
Nyanga	145	15 March
Chisengu	407	15 March
Mutare	250	15 March
Mukandi	203	15 March
Chipinge	219	16 March
Chisengu	127	16 March
Buhera	102	16 March

TABLE 11.7 Sector impacts of tropical cyclone Idai in Zimbabwe.

Sector	Impacts
Human and social development	347 people were killed; 50, 906 were internally displaced; 270, 000 were affected; 90, 000 were in need of psychosocial support services; 344 were missing; 183 were injured
Housing	1838 houses were completely destroyed; 193 houses were partially destroyed
Transport	865 km of road was destroyed; 11 major bridges were damaged or destroyed
Health	12 health facilities were damaged
Energy	88.9 km of 33 kV line, 106.2 km of 11 kV line and 33.6 km of mv line, and 44 sub-stations, were adversely affected
Agriculture	18 irrigation schemes were damaged; at least 362 cattle, 514 goats and sheep, 17,000 chicken were lost whilst 86 dipping facilities were damaged. 1.4 million hectares of arable land were destroyed
Education	139 schools were damaged
Water, Sanitation and Hygiene	250 boreholes were damaged; 5000 m of the water distribution network was washed away in Chimanimani town; the main pipeline from the water treatment plant in Chipinge town was washed away. Water supply systems and sanitation (in private households, health facilities, schools, and government buildings) suffered extensive damage or were destroyed

Based on Government of Zimbabwe (GoZ). (2019). Zimbabwe—Rapid Impact and Needs Assessment (RINA). https://www.gfdrr.org/en/publication/zimbabwe-rapid-impact-needs-assessment-2019. (Accessed on 27 May 2021); Civil Protection Unit. CPU. (2019). Cyclone Idai death toll set to double – CPU. https://www.newzimbabwe.com/cyclone-idai-death-toll-set-to-double-cpu/ (Accessed on 12 May 2021).

Many people in the affected communities lost civil registration documents that ensure identification, protection and access to entitlements, such as birth, death and marriage certificates (OCHA, 2019). The TC also caused soil erosion, landslides, and the formation of gullies, while damaging 1.17 million hectares of forest. Wildlife habitats were also lost due to loss of forest cover (Oxfam, 2019).

The full cost of the impacts of tropical cyclone Idai are difficult to quantify. However, the cost of direct damages was estimated to be around $622 million while the cost of recovery was put at US$584-803 million (GoZ, 2019).

4 Discussion and conclusion

This work examined the formation and evolution of TC Idai and the impacts it caused in southern Africa. It also looked at the current and future characteristics of TCs in the SWIO basin. Government reports and reports from donor agencies, disaster management institutions, individual consultants, non-governmental organizations, international organizations and United Nations Agencies, as well as technical reports, policy briefs and peer reviewed journal articles were the sources of information. Strong winds, heavy rain and floods caused by TCs usually damage meteorological measuring instruments, hence where observational data could not be obtained numerical model output was used as an estimate of observations. In particular, estimated precipitation data was obtained from NASA Global Precipitation Mission (GPM), while storm surge data is based on calculations from NOAA Hurricane Weather Research and Forecast (HWRF) model (Probst & Annunziato, 2019). Sustained 3-min wind speed in km/hour was calculated from Best Track data from NCAR.

Results indicate that TC Idai started off as a tropical depression that formed in the Mozambique Channel on the 4th of March in 2019. It then made its first landfall later on the same day and caused flooding in southern Malawi and in Tete and Zambezia Provinces of Mozambique. The depression re-entered the Mozambique Channel on the 9th of March where it intensified and later made a second landfall on the 14th near Beira as an intense TC named Idai. At its most intense stage, just before making the second landfall, TC Idai reached 10-min maximum sustained wind speeds of 195 km/h and a minimum central pressure of 940 hPa (Meteo-France, 2019). When it made landfall, its 10-min sustained wind speed had decreased to 167 km/hour (90 knots), a Category 2 storm, with a central pressure of 960 hPa (Meteo-France, 2019). After making the second landfall, TC Idai weakened to a depression and then to a remnant low as it moved inland to the west on the 15th and 16th. On the 17th the remnant low turned eastward, and on the 19th re-entered the Mozambique Channel (Météo-France La Réunion, 2019) before eventually dissipating on the 21st of March. Tropical cyclone Idai was one of 9 intense TCs that formed in the SWIO basin during the 2018–2019 season, consistent with results from other studies which show that globally the average proportion of intense TC occurrence (Category 3 or higher) has increased since 1979 (Holland & Bruyère, 2014; Knutson et al., 2021; Kossin et al., 2020), and so has the intensity (Kossin et al., 2020). Since the start of reliable satellite coverage, the largest number of intense TCs to have occurred in the SWIO basin during one season was 6 which occurred during the 2006–2007 season (Wiki, 2019).

Tropical cyclone Idai caused torrential rainfall, flooding and landslides in the affected countries, resulting in severe damage to road networks and bridges, telecommunication infrastructures, health and education facilities, agricultural crops and livestock, as well as loss of human lives. The worst affected countries were Mozambique, Zimbabwe and Malawi which collectively suffered economic losses of over US$2 billion, over 1000 deaths, with thousands missing, while hundreds of thousands were displaced (World Bank, 2019). Idai was the deadliest and costliest TC to affect the SWIO basin (World Bank, 2019) and one of the strongest to hit the southern hemisphere (WMO, 2019). The huge cost of the damages is in agreement with observations which indicate that the economic cost of water and climate related extreme weather events is increasing (WMO, 2014). The displacement, loss of property, and deaths of family members left a lot of people in need of psychosocial support to help them cope with the impact of the disaster on their families (Oxfam, 2019). The floods and resultant damages caused by TC Idai increased the proportion of highly vulnerable people such as the chronically ill, the elderly and the disabled (GoMw PDNA, 2019), while poverty levels and food insecurity were also expected to increase in Mozambique and Malawi (GoM, 2019; GoMw PDNA, 2019). In Zimbabwe, losses in agricultural production and losses in trade related activities due to damages to electricity distribution networks, communication infrastructure and people losing employment opportunities resulted in projections of the economy shrinking by 3.1% (GoZ, 2019). In Mozambique and Malawi, destruction of productive capacity in sectors such as agriculture, transport, trade, manufacturing and services negatively affected the economic growth and the GDP per capita was expected to contract. Increases were also projected in annual inflation and imports in Mozambique. The fishing and aquaculture sector contributes about 4% of Malawi's national GDP hence the destruction of fishing infrastructure negatively affected income and household livelihoods since the sector also employs 60, 000 fishers (GoMw PDNA, 2019). Displaced people were housed in makeshift shelters provided by humanitarian agencies, and in schools (GoMw PDNA, 2019), and some of the makeshift shelters lacked proper sanitation facilities and safe water, thereby posing some health risks (DTM, 2019). As a result of displacements, families were sometimes separated, exposing women and girls to sexual exploitation and other gender based violence (GoZ, 2019). The destruction of crops, some ready for harvesting, worsened the food security situation which was already critical due to an ongoing drought that was affecting southern Africa when TC Idai occurred. In Mozambique 1,359,159 people were left in need of food aid due to the destruction of crops (GoZ, 2019) and by July 2019, 2420 children were reported to be suffering from severe malnutrition (UNICEF, 2019). In Zimbabwe, destruction of 1.4 million hectares of arable land affected 50,000 mainly smallholder farmers worsening the already high levels of malnutrition (Oxfam, 2020). Damages to roads and bridges made some roads impassable thereby hampering rescue and recovery efforts, and in Malawi there was also restricted access to health facilities and work places by people, resulting in a decline in incomes and deepening poverty and reduced quality of life (GoMw PDNA, 2019). Damages to electricity and communication infrastructure resulted in power disruptions and negative effects on the economies of the three countries, and in Malawi this caused commodity prices to increase by around 20% in the post disaster period (GoMw PDNA, 2019). The port of Beira is a strategic port connecting Malawi, Zambia and Zimbabwe to the Indian Ocean for trade purposes, and damage to railroads significantly affected trade. Damaged schools' infrastructure disrupted learning and teaching services resulting in a decline in attendance in the three countries. Health facilities also suffered massive damages and this disrupted healthcare services, negatively affecting vulnerable populations such as people living with HIV, children and pregnant women. Destruction and damage to pit latrines and piped water systems caused a spillage of human waste into water, causing an increase in the reported number of infectious diseases such as coughs, malaria, diarrhea and cholera (GoMw PDNA, 2019). In Mozambique, the districts most affected by cholera were Beira, Dondo, Nhamatanda, and Búzi in Sofala Province, where 6627 cases and 8 deaths were reported, while 14,863 cases of malaria were also recorded in Sofala province (GoM, 2019).

Numerical model guidance indicate that the frequency of the most intense TCs and the maximum intensity, particularly of the most intense TCs is expected to increase (Christensen et al., 2013; Knutson et al., 2015). Larger and more intense TCs tend to cause more damage than smaller, weaker storms (Knutson et al., 2021). Models also project an increase in TC precipitation rates by the end of the 21st century, which could further elevate the risk of flooding (Knutson et al., 2021).

A number of factors contributed to the extreme destruction caused by TC Idai, as well as the human death toll and economic loss. A lot of people in Mozambique, Zimbabwe and Malawi survive on low incomes and live in geographically vulnerable areas (Reall, 2019), such as along or close to river banks as well as on mountain slopes, thereby increasing their vulnerability. In addition, houses, particularly in the informal settlements, were constructed using poor construction materials (Reall, 2019). The three countries are also heavily reliant on rain-fed subsistence agriculture for their livelihoods. These socioeconomic, development and environmental challenges worsened the impacts of TC Idai. There was also too much reliance on external aid agencies in the rescue phase and local disaster management institutions, mainly headquartered in the big cities, were under resourced to deal with a disaster of the magnitude of TC Idai. The slow speed, less than 10 mph of TC Idai at landfall, also contributed to its highly destructive rains (Henson, 2019).

Tropical cyclone Idai exposed the vulnerability of communities and certain sectors in the affected countries to extreme weather conditions. A number of lessons can be drawn from the damage and destruction caused by the TC in Malawi, Mozambique and Zimbabwe so as to reduce the impacts in future, be better prepared and build resilience to TCs or other weather related extreme events. The lessons are derived from interviews conducted by researchers with affected people, government officials, representatives of donor agencies, UN agencies, NGOs, academics, humanitarian response agents and a review of secondary sources of literature. The lessons are generally applicable to all the three most affected countries and can be enumerated as follows: (a) Governments need to improve disaster preparedness and strengthen early warning systems (ISET, 2019), which must be detailed enough and reach the affected communities and individual households. (b) The messages conveyed by the early warnings must be understandable and actionable (Norton, MacClune, & Szönyi, 2020). (c) It is important that countries prone to natural disasters do hazard risk mapping and proper land use planning to demarcate areas where settlements and cultivation can be allowed, and desist from establishing settlements in areas prone to flooding, landslides and on wetlands to avoid or minimize properties being destroyed by floods in future (ISET, 2019). (d) Disaster management institutions need to be strengthened and decentralized to local authority and district levels, in addition to being financially capacitated (ISET, 2019; Norton et al., 2020). (e) When coordinating disaster management after a disaster has occurred, the roles of stakeholders such as governments, NGOs, international organizations, private citizens and other players should be clearly spelt out (Norton et al., 2020). (f) Governments must strengthen the capacity to manage disasters at national and local levels and avoid relying on external organizations. (g) National Meteorological and Hydrological institutions must be adequately financed so that they have enough functional stations to carry out flood monitoring purposes. (h) Meteorological institutes must invest in modern equipment to improve flood forecasting capabilities (Norton et al., 2020). (i) Governments need to incorporate disaster risk reduction into recovery and reconstruction plans, as well as into existing or new policies and practices, through for example implementing hazard resistant building codes (Norton et al., 2020), so that buildings can withstand strong winds, heavy rains and floods. (j) It is important for communities to have multi-purpose fortified infrastructure that serve as shelters and accommodation in times of TCs and during floods (ISET, 2019). (k) Communities need to be trained in good agricultural practices, diversifying income streams and production of higher-value end products and strengthening markets and supply chains (Norton et al., 2020).

Although a total of 15 moderate tropical storms, tropical storms, tropical cyclones and intense tropical cyclones (Wiki, 2019) occurred during the 2018–2019 TC season in the SWIO basin, this study only considered the impacts caused by TC Idai. Further studies could consider the total impacts caused by all the 2018–2019 season SWIO basin TCs and compare that with other seasons in terms of intensities of the TCs and the damages they caused. The total number of human deaths and displacements should be taken as estimates, as ascertaining the number of deaths or displacements in a disaster is difficult since the figures keep changing. The total cost of damages due to TC Idai could also be much higher than the quoted values since this study only considered the cost of damages in Mozambique, Zimbabwe and Malawi.

References

Anthes, R. A. (1982). Tropical cyclones: Their evolution, structure and effects. *American Meteorological Society Meteorological Monographs, 19*. 208 pp.

Bauer, S., & Scholz, I. (2010). Adaptation to climate change in southern Africa: New boundaries for sustainable development? *Climate and Development, 2*(2), 83–93. https://doi.org/10.3763/cdev.2010.0040.

Bonfils, C. J. W., Santer, B. D., & Fyfe, J. C. (2020). Human influence on joint changes in temperature, rainfall and continental aridity. *Nature Climate Change, 10*, 726–731. https://doi.org/10.1038/s41558-020-0821-1.

Carbon Brief. (2021). *Mapped: How climate change affects extreme weather around the world*. https://www.carbonbrief.org/mapped-how-climate-change-affects-extreme-weather-around-the-world. (Accessed 13 July 2021).

Caron, L. P., & Jones, G. J. (2008). Analysing present, past and future tropical cyclone activity as inferred from an ensemble of coupled global climate models. *Tellus, 60*(1), 80–96.

Cattiaux, J., Chauvin, F., Douville, H., & Ribes, A. (2021). *Weather extremes and climate change, encyclopedia of the environment*. https://www.encyclopedie-environnement.org/en/climate/extreme-weather-events-and-climate-change/. (Accessed 16 May 2021).

Chen, Y. C., Chang, K. T., Chiu, Y. J., Lau, S. M., & Lee, H. Y. (2013). Quantifying rainfall controls on catchment-scale landslide erosion in Taiwan. *Earth Surface Processes and Landforms, 38*, 372–382.

Christensen, J. H., Krishna Kumar, K., Aldrian, E., An, S.-I., Cavalcanti, I. F. A., de Castro, M., … Zhou, T. (2013). Climate phenomena and their relevance for future regional climate change. In T. F. Stocker, D. Qin, G.-K. Plattner, M. Tignor, S. K. Allen, J. Boschung, … P. M. Midgley (Eds.), *Climate change 2013: The physical science basis. Contribution of working group I to the fifth assessment report of the intergovernmental panel on climate change*. Cambridge, United Kingdom and New York, NY: Cambridge University Press.

Civil Protection Unit (CPU). (2019). *Cyclone Idai death toll set to double – CPU*. https://www.newzimbabwe.com/cyclone-idai-death-toll-set-to-double-cpu/. (Accessed 12 May 2021).

Crimp, S. J., & Mason, S. J. (1999). The extreme precipitation event of 11 to 16 February 1996 over South Africa. *Meteorology and Atmospheric Physics*, *70*, 29–42. https://doi.org/10.1007/s007030050023.

Czajkowski, J., Villarini, G., Michel-Kerjan, E., & Smith, J. A. (2013). Determining tropical cyclone inland flooding loss on a large scale through a new flood peak ratio-based methodology. *Environmental Research Letters*, *8*, 44056. https://doi.org/10.1088/1748-9326/8/4/044056.

Dare, R. A., & McBride, J. L. (2011). The threshold sea surface temperature condition for tropical cyclogenesis. *Journal of Climate*, *24*, 4570–4576. https://doi.org/10.1175/JCLI-D-14-00637.1.

Davis, C. L. (2011). *Climate risk and vulnerability: A handbook for southern Africa* (p. 92). Pretoria, South Africa: Council for Scientific and Industrial Research.

De, U. S., Khole, M., & Dandekar, M. M. (2004). Natural hazards associated with meteorological extreme events. *Natural Hazards*, *31*, 487–497.

Doocy, S., Dick, A., Daniels, A., & Kirsch, T. D. (2013). The human impact of tropical cyclones: A historical review of events 1980-2009 and systematic literature review. *PLoS Currents*, *5*. https://doi.org/10.1371/currents.dis.2664354a5571512063ed29d25ffbce74, ecurrents.dis.2664354a5571512063ed29d25ffbce74.

DTM. (2019). IOM displacement tracking matrix. In *Zimbabwe - Tropical cyclone IDAI mobility tracking rounds 1–3*. https://displacement.iom.int/system/tdf/reports/MT%20R1-3%20Zimbabwe%2015-5_0.pdf?file=1&type=node&id=5717. (Accessed 12 March 2021).

Dyson, L. L., & van Heerden, J. (2001). The heavy rainfall and floods over the north-eastern interior of South Africa during February 2000. *South African Journal of Science*, *97*(3–4), 80–86.

Emanuel, K. A. (2005). Increasing destructiveness of tropical cyclones over the past 30 years. *Nature*, *436*, 686–688.

European Centre for Disease Prevention and Control. (2019). *Cyclone Idai: Risk of communicable diseases in southern Africa – 10 April 2019*. Stockholm: ECDC. https://www.ecdc.europa.eu/sites/default/files/documents/RRA-cyclone-idai-10-Apr-2019.pdf. (Accessed 15 June 2021).

Famine Early Warning Systems Network (FEWSNET). (2019). *EAST AFRICA seasonal monitor*. https://reliefweb.int/report/somalia/east-africa-seasonal-monitor-april-26-2019. (Accessed 21 July 2021).

Fitchett, J. M., & Grab, S. W. (2014). A 66-year-old tropical cyclone record for Southeast Africa: Temporal trends in a global context. *International Journal of Climatology*, *34*, 3604–3615.

Food and Agriculture Organization of the United Nations (FAO). (2019). *Global Information and Early Warning System on Food and Agriculture (GIEWS) special alert no. 345, East Africa*. http://www.fao.org/3/ca4301EN/ca4301en.pdf. (Accessed 21 July 2021).

Government of Malawi (GoMw). (2019). *Malawi 2019 Floods Post Disaster Needs Assessment (PDNA)*. https://www.unicef.org/malawi/reports/malawi-2019-floods-post-disaster-needs-assessment-report. (Accessed 27 May 2021).

Government of Mozambique. (2019). *Mozambique cyclone Idai. Post Disaster Needs Assessment (PDNA)*. https://www.undp.org/publications/mozambique-cyclone-idai-post-disaster-needs-assessment-pdnadna. (Accessed 27 May 2021).

Government of Zimbabwe (GoZ). (2019). *Zimbabwe – Rapid Impact and Needs Assessment (RINA)*. https://www.gfdrr.org/en/publication/zimbabwe-rapid-impact-needs-assessment-2019. (Accessed 27 May 2021).

Hachigonta, S., Nelson, G. C., Thomas, T. S., & Sibanda, L. M. (2013). *Southern african agriculture and climate change: A comprehensive analysis*. Washington, DC: International Food Policy Research Institute (IFPRI). https://doi.org/10.2499/9780896292086.

Hao, Z., Aghakouchak, A., & Phillips, T. J. (2013). Changes in concurrent monthly precipitation and temperature extremes. *Environmental Research Letters*, *8*, 1402–1416.

Harsch. (1992). Drought devastates Southern Africa. *Drought Network News*, *4*, 17–19 (Reprint from Africa Recovery, April 1992).

Henderson-Sellers, Zhang, H., Berz, G., Emanuel, K., Gray, W., Landsea, C., … Webster, P. (1998). Tropical cyclones and global climate change: A post IPCC assessment. *Bulletin of the American Meteorological Society*, *79*, 19–38.

Henson, B. (2019). *Category 4 Kenneth crashes ashore in Mozambique; Devastating Rains Still to Come*. https://www.wunderground.com/cat6/Category-4-Kenneth-Crashes-Ashore-Mozambique-Devastating-Rains-Still-Come. (Accessed 15 May 2021).

Ho, C., Kim, H., Jeong, J. H., Kim, H. S., & Chen, D. L. (2006). Variation of tropical cyclone activity in the South Indian Ocean: El Nino-southern oscillation and Madden-Julian oscillation effects. *Journal of Geophysical Research – Atmospheres*, *111*, D22101. https://doi.org/10.1029/2006JD007289.

Holland, G., & Bruyère, C. L. (2014). Recent intense hurricane response to global climate change. *Climate Dynamics*, *42*, 617–627. https://doi.org/10.1007/s00382-013-1713-0.

Holland, G. J., & Lander, M. (1993). The meandering nature of tropical cyclone tracks. *Journal of the Atmospheric Sciences*, *50*, 1254–1266. https://doi.org/10.1175/1520-0469(1993)050<1254:TMNOTC>2.0.CO;2.

IMF. (2017). *Malawi economic development document. Country report no:17/184*. https://doi.org/10.5089/9781484307311.002.

Institute for Social and Environmental Transition - International (ISET). (2019). *Executive summary: Learning from 2019 cyclone Idai and cyclone Kenneth Malawi, Mozambique, and Zimbabwe*. https://www.preventionweb.net/publication/learning-2019-cyclone-idai-and-cyclone-kenneth-executive-summary. (Accessed 26 April 2021).

International Federation of Red Cross and Red Crescent Societies (IFRC). (2019). *Emergency plan of action operation update Malawi: Floods*. https://reliefweb.int/sites/reliefweb.int/files/resources/MDRMW014ou1_0.pdf. (Accessed 4 April 2021).

IPCC. (2013). Climate change 2013: The physical science basis. In T. F. Stocker, D. Qin, G.-K. Plattner, M. Tignor, S. K. Allen, J. Boschung, … P. M. Midgley (Eds.), *Contribution of working group I to the fifth assessment report of the intergovernmental panel on climate change*. Cambridge, United Kingdom and New York, NY: Cambridge University Press. 1535 pp.

IPCC. (2014). Climate change 2014: Synthesis report. In Core Writing Team, R. K. Pachauri, & L. A. Meyer (Eds.), *Contribution of working groups I, II and III to the fifth assessment report of the intergovernmental panel on climate change*. Geneva, Switzerland: IPCC. 151 pp.

Jury, M. R. (1993). A preliminary study of climatological associations and characteristics of tropical cyclones in the SW Indian Ocean. *Meteorology and Atmospheric Physics*, *51*, 101–115. https://doi.org/10.1007/BF01080882.

Kandji, S. T., Verchot, L., & Mackensen, J. (2006). *Climate change and variability in southern Africa: Impacts and adaptation in the agricultural sector.* World Agroforestry Centre. http://apps.worldagroforestry.org/downloads/Publications/PDFS/B14549.pdf. (Accessed 19 June 2021).

Knapp, K. R., & Kossin, J. P. (2007). New global tropical cyclone data set from ISCCP B1 geostationary satellite data. *Journal of Applied Remote Sensing, 1*(1). https://doi.org/10.1117/1.2712816, 013505.

Knutson, T. R., Sirutis, J. J., Zhao, M., Tuleya, R. E., Bender, M., Vecchi, G. A., et al. (2015). Global projections of intense tropical cyclone activity for the late twenty-first century from dynamical downscaling of CMIP5/RCP4.5 scenarios. *Journal of Climate, 28*(18), 7203–7224. https://doi.org/10.1175/JCLI-D-15-0129.1.

Knutson, T., Camargo, S. J., Chan, J. C. L., Emanuel, K., Ho, C., Kossin, J., … Wu, L. (2019). Tropical cyclones and climate change assessment: Part I: Detection and attribution. *Bulletin of the American Meteorological Society, 100*(10), 1987–2007. https://doi.org/10.1175/BAMS-D-18-0189.1.

Knutson, T. R., Chung, M. V., Vecchi, G., Sun, J., Hsieh, T.-L., & Smith, A. J. P. (2021). ScienceBrief review: Climate change is probably increasing the intensity of tropical cyclones. In C. Le Quéré, P. Liss, & P. Forster (Eds.), *Critical issues in climate change science.* https://doi.org/10.5281/zenodo.4570334.

Kossin, J., Knapp, K., Olander, T., & Velden, C. (2020). Global increase in major tropical cyclone exceedance probability over the past four decades. *Proceedings of the National Academy of Sciences of the United States of America, 117,* 201920849. https://doi.org/10.1073/pnas.1920849117.

Kuleshov, Y., Qi, L., Fawcett, R., & Jones, D. (2008). On tropical cyclone activity in the southern hemisphere: Trends and the ENSO connection. *Geophysical Research Letters, 35,* L14S08. https://doi.org/10.1029/2007GL032983.

Landsea, C. W. (2007). Counting Atlantic tropical cyclones back to 1900. *Eos, Transactions of the American Geophysical Union, 88,* 197–208.

Landsea, C. W., Pielke, R. A., Jr., Mesta Nuñez, A. M., & Knaff, J. A. (1999). Atlantic basin hurricanes: Indices of climatic changes. *Climatic Change, 42,* 89–129.

Leonard, M., Westra, S., Phatak, A., Lambert, M., Hurk, B. V. D., Mcinnes, K., … Stafford-Smith, M. (2013). A compound event framework for understanding extreme impacts. *Wiley Interdisciplinary Reviews: Climate Change, 5,* 113–128.

Liu, M., Shen, Y., Qi, Y., Wang, Y., & Geng, X. (2019). Changes in precipitation and drought extremes over the past half century in China. *Atmosphere, 10*(4), 203. https://doi.org/10.3390/atmos10040203.

Madden, R. A., & Julian, P. R. (1994). Observations of the 40-50 day tropical oscillation—A review. *Monthly Weather Review, 122,* 814–837.

Malherbe, J., Engelbrecht, F. A., & Landman, W. A. (2013). Projected changes in tropical cyclone climatology and landfall in the Southwest Indian Ocean region under enhanced anthropogenic forcing. *Climate Dynamics, 40,* 2867–2886.

Matyas, C. J. (2015). Tropical cyclone formation and motion in the Mozambique Channel. *International Journal of Climatology, 35*(3), 375–390. https://doi.org/10.1002/joc.3985.

Mavume, A., Rydberg, L., Rouault, M., & Lutjeharms, J. (2010). Climatology and landfall of tropical cyclones in the south- west indian ocean. *Western Indian Ocean Journal of Marine Science, 8*(1), 15–36. https://doi.org/10.4314/wiojms.v8i1.56672.

Météo-France La Réunion. (2019). Bulletin for cyclonic activity and significant tropical weather in the Southwest Indian Ocean. In *AWIO20 FMEE 191257.* http://www.meteo.fr/temps/domtom/La_Reunion/webcmrs9.0/anglais/activiteope/bulletins/zcit/ZCITA_201903191255.pdf. Accessed 24 April 2021.

Meteo-France. (2019). *Intense tropical cyclone Idai warning number 26/11/20182019.* http://www.meteo.fr/temps/domtom/La_Reunion/webcmrs9.0/anglais/activiteope/bulletins/cmrs/CMRSA_201903150000_IDAI.pdf. (Accessed 23 April 2021).

Mühr, B., Daniell, J., Schaefer, A., Brand, J., Barta, T., Neuweiler, A., … Kunz, M. (2019). *Tropical cyclone IDAI (southern Africa). Report no. 1. Center for Disaster Management and Risk Reduction Technology (CEDIM).* Forensic Disaster Analysis Group (FDA). https://www.cedim.kit.edu/download/CEDIM_FDA_IDAI_2019.pdf. (Accessed 2 September 2021).

Murakami, H., Delworth, T. L., Cooke, W. F., Zhao, M., Xiang, B., & Hsu, P. C. (2020). Detected climatic change in global distribution of tropical cyclones. *Proceedings of the National Academy of Sciences of the United States of America, 117*(20), 10706–10714. https://doi.org/10.1073/pnas.1922500117.

Mutasa, C. (2019). Zimbabwe's climate: Past, present and future trends. In T. Murombo, M. Dhliwayo, & T. Dhlakama (Eds.), *Climate change law in Zimbabwe: Concepts and insights.* https://www.kas.de/documents/277198/0/Climate+Change+Law+in+Zimbabwe.pdf/81decd91-6699-708e-b80a-8a04f199b50d?t=1573136985486. (Accessed 12 September 2021).

Muthige, M. S., Malherbe, J., Englebrecht, F. A., Grab, S., Beraki, A., Maisha, T. R., & Van der Merwe, J. (2018). Projected changes in tropical cyclones over the south West Indian Ocean under different extents of global warming. *Environmental Research Letters, 13*(6). https://doi.org/10.1088/1748-9326/aabc60, 065019.

National Aeronautics and Space Administration (NASA). (2019). *Devastation in Mozambique.* https://earthobservatory.nasa.gov/images/144712/devastation-in-mozambique. (Accessed 13 April 2021).

National Geographic. *The Saffir-Simpson Hurricane wind scale.* https://www.nationalgeographic.org/encyclopedia/saffir-simpson-hurricane-wind-scale/. (Accessed 20 June 2021).

National Hurricane Center (NHC) and Central Pacific Hurricane Center. *Saffir-Simpson Hurricane wind scale.* https://www.nhc.noaa.gov/aboutsshws.php. (Accessed 16 July 2021).

Norton, R., MacClune, K., & Szönyi, M. (2020). *When the unprecedented becomes precedented: Learning from cyclones Idai and Kenneth.* Boulder, CO: ISET International and the Zurich Flood Resilience Alliance. https://2eac3a3b-5e23-43c7-b33c-f17ad8fd3011.filesusr.com/ugd/558f8a_753e1b7efa6148f3bd09d243c3c66b58.pdf. (Accessed 6 June 2021).

OCHA. (2019). *Zimbabwe flash appeal January–June 2019* (revised following cyclone Idai, March 2019) https://reliefweb.int/sites/reliefweb.int/files/resources/ROSEA_Zimbabwe_FlashAppeal_05042019.pdf. (Accessed 6 June 2021).

Ongoma, V. (2019). *Why Kenya's seasonal rains keep failing and what needs to be done.* The Conversation. https://theconversation.com/why-kenyas-seasonal-rains-keep-failing-and-what-needs-to-be-done-115635. (Accessed 21 July 2021).

Oxfam briefing paper. (2019). *Cyclone Idai in Zimbabwe: An analysis of policy implications for post-disaster institutional development to strengthen disaster risk management.* https://policy-practice.oxfam.org/resources/cyclone-idai-in-zimbabwe-an-analysis-of-policy-620892/. (Accessed 10 July 2021).

Oxfam media briefing. (2020). *After the storm Barriers to recovery one year on from Cyclone Idai.* https://reliefweb.int/report/mozambique/after-storm-barriers-recovery-one-year-cyclone-idai-10-march-2020. (Accessed 26 July 2021).

Palmén, E. H. (1948). On the formation and structure of tropical cyclones. *Geophysica, 3,* 26–38.

Peduzzi, P., Chatenoux, B., Dao, H., De Bono, A., Herold, C., Kossin, J., … Nordbeck, O. (2012). Global trends in tropical cyclone risk. *Nature Climate Change*, *2*, 289–294. https://doi.org/10.1038/nclimate1410.

Perkins-Kirkpatrick, S. E., & Lewis, S. C. (2020). Increasing trends in regional heatwaves. *Nature Communications*, *2020*, 11 (1) https://doi.org/10.1038/s41467-020-16970-7.

Pielke, R. A., Jr., Gratz, J., Landsea, C. W., Collins, D., Saunders, M. A., & Musulin, R. (2008). Normalized hurricane damages in the United States: 1900-2005. *Natural Hazards Review*, *9*, 29–42.

Probst, P., & Annunziato, A. (2019). *Tropical cyclone Idai: Analysis of the wind, rainfall and storm surge impact*. European Commission. Joint Research Centre. https://reliefweb.int/sites/reliefweb.int/files/resources/joint_research_centre_analysis_of_wind_rainfall_and_storm_surge_impact_09_april_2019.pdf. (Accessed 12 June 2021).

Reall. (2019). *Affordable and climate-resilient: how casa real's homes withstood cyclone idai*. https://www.reall.net/wp-content/uploads/2019/09/Affordable-Homes-Cycolne-Idai.pdf. (Accessed 24 June 2021).

Reason, C. J. C. (2007). Tropical cyclone Dera, the unusual 2000/01 tropical cyclone season in the Southwest Indian Ocean and associated rainfall anomalies over southern Africa. *Meteorology and Atmospheric Physics*, *97*, 181–188.

Reason, C. J., & Keibel, A. (2004). Tropical Cyclone Eline and its unusual penetration and impacts over the southern African mainland. *Weather and Forecasting*, *19*, 789–805.

SADC. (2012). *SADC facts & figures*. http://www.sadc.int/aboutsadc/overview/sadc-facts-figures/. (Accessed 30 March 2021).

SADC. (2018). *Southern African development community, towards a common future. Selected economic and social indicators 2018*. https://www.sadc.int/files/6215/6630/2592/SADC_Selected_Indicators_2018.pdf. (Accessed 24 June 2021).

Shanko, D., & Camberlin, P. (1998). The effects of the Southwest Indian Ocean tropical cyclones on Ethiopian drought. *International Journal of Climatology*, *18*, 1373–1388.

Shultz, J. M., Russell, J., & Espinel, Z. (2005). Epidemiology of tropical cyclones: The dynamics of disaster, disease, and development. *Epidemiologic Reviews*, *27*(1), 21–35. https://doi.org/10.1093/epirev/mxi011.

Stringer, L. C., Dyer, J. C., Reed, M. S., Dougill, A. J., Twyman, C., & Mkwambisi, D. (2009). Adaptations to climate change, drought and desertification: Local insights to enhance policy in southern Africa. *Environmental Science & Policy*, *12*(7), 748–765.

Sun, Q., Zhang, X., Zwiers, F., Westra, S., & Alexander, L. V. (2021). A global, continental, and regional analysis of changes in extreme precipitation. *Journal of Climate*, *34*(1), 243–258.

Trenberth, K. E. (2005). Uncertainty in hurricanes and global warming. *Science*, *308*, 1753–1754.

UNICEF. (2019). *Mozambique humanitarian situation report*. https://www.unicef.org/media/75236/file/Mozambique-SitRep-Cyclone-Idai-1-May-2019.pdf. (Accessed 18 April 2021).

United Nations Economic Commission for Africa (UNECA). (2015). *Assessment report on mainstreaming and implementing disaster risk reduction in southern Africa*. Addis Ababa https://repository.uneca.org/bitstream/handle/10855/23280/b11569207.pdf?sequence=1&isAllowed=y. (Accessed 18 March 2021).

United Nations Population Fund (UNFPA) East and Southern Africa. (2019). *Cyclone Idai flood appeal*. https://reliefweb.int/sites/reliefweb.int/files/resources/UNFPA%20Flood%20Appeal.pdf. (Accessed 17 April 2021).

Vitart, F., Anderson, D., & Stockdale, T. (2003). Seasonal forecasting of tropical cyclone landfall over Mozambique. *Journal of Climate*, *16*, 3932–3945.

Vogel, C. H. (1994). South Africa. In M. H. Glantz (Ed.), *Drought follows the plow*. Cambridge: Cambridge University Press.

Wamukonya, N., & Rukato, H. (2001). *Climate change implications for southern Africa: A gendered perspective. Background paper prepared for the southern African gender and energy network*. South Africa: MEPC. https://www.energia.org/assets/2015/06/16-Climate-change-implications-for-Southern-Africa.pdf (Accessed 22 February 2021).

Webster, P. J., Holland, G. J., Curry, J. A., & Chang, H.-R. (2005). Changes in tropical cyclone number, duration, and intensity in a warming environment. *Science*, *309*, 1844–1846. https://doi.org/10.1126/science.1116448.

Weinkle, J., Maue, R., & Pielke, R., Jr. (2012). Historical global tropical cyclone landfalls. *Journal of Climate*, *25*(13), 4729–4735. https://doi.org/10.1175/JCLI-D-11-00719.1.

Wiki. (2019). *2018–19 South-West Indian Ocean cyclone season*. https://wikivisually.com/wiki/2018-19_South-West_Indian_Ocean_cyclone_season.9. (Accessed 28 June 2021).

WMO mission report following tropical cyclone IDAI (29 April to 7 May). (2019). *Reducing vulnerability to extreme hydro-meteorological hazards in Mozambique after Cyclone IDAI*. https://library.wmo.int/doc_num.php?explnum_id=6259. (Accessed 19 July 2021).

World Bank. (2019). *Statement on high-level meeting on humanitarian and recovery efforts following cyclone Idai*. https://www.worldbank.org/en/news/statement/2019/04/11/statement-on-high-level-meeting-on-humanitarian-and-recovery-efforts-following-cyclone-idai. (Accessed 25 June 2021).

World Meteorological Organization (WMO). (2014). *Atlas of mortality and economic losses from weather, climate and water extremes (1970–2012)*. https://library.wmo.int/doc_num.php?explnum_id=7839. (Accessed 12 May 2021).

World Meteorological Organization (WMO). (2021). *Water-related hazards dominate disasters in the past 50 years*. https://public.wmo.int/en/media/press-release/water-related-hazards-dominate-disasters-past-50-years. (Accessed 28 July 2021).

Zhang, Q., Liu, Q., & Wu, L. (2009). Tropical cyclone damages in China 1983-2006. *Bulletin of the American Meteorological Society*, *90*, 489–495.

Zimba, S. K., Houane, M. J., & Chikova, A. M. (2020). Impact of tropical cyclone Idai on the southern African electric power grid. In *2020 IEEE PES/IAS PowerAfrica* (pp. 1–5). https://doi.org/10.1109/PowerAfrica49420.2020.9219944.

CHAPTER

12

Impacts of climate extremes over Arctic and Antarctic

Masoud Irannezhad[a,b], *Behzad Ahmadi*[c], *and Hannu Marttila*[a]

[a]Water, Energy and Environmental Engineering Research Unit, Faculty of Technology, University of Oulu, Oulu, Finland
[b]School of Environmental Science and Engineering, Southern University of Science and Technology (SUSTech), Shenzhen, People's Republic of China
[c]WSP, Portland, OR, United States

1 Introduction

Mankind is promptly fascinated by the polar regions on Earth during the 21st century more than ever. The terrestrial and aquatic ecosystems of both Arctic and Antarctic areas are generally characterized by biting cold climate, darkness through half of the year, ice, and snow as well as survival organisms that could adapt to very extreme weather conditions. Nowadays, there are undoubtedly no other geographical landscapes than such coldest, harshest, most sparsely populated distant land, and marine polar regions that folks eagerly want to see for themselves. For researchers, on the other hand, most of the practically inaccessible wide expanses of ice and snow throughout the Arctic and Antarctic are today viewed as frontier parts of the world, with numerous fundamental scientific questions; e.g., what is exactly hidden under the Antarctic's kilometer-thick ice sheet?

The Arctic and Antarctic play a critical role in the climate system of Earth by acting as cooling chambers, besides fascination. Small alterations in their complex structures, particularly in response to global warming, can significantly influence the patterns of airflow and ocean currents around the world, with extreme impacts. It is particularly true for the ice sheets of Antarctica and Greenland, which both together contain 99% of the solid water on our planet. If these two ice covers melt, the global sea level will rise about 70m. As a particular consequence, different long stretches of worldwide coastlines would be flooded. Hence, scientists are interested in following such changes in real time, by utilizing observations recorded by measuring networks or satellites.

1.1 Polar regions: Similar but fundamentally different

The polar regions (Fig. 12.1) are the areas located between the North or South Pole and the Arctic or Antarctic Circles marked by dashed lines at 66°33′ north or south latitude on the world maps, respectively. The distance between the Arctic (Antarctic) Circle and the North (South) Pole is about 2602km. The Arctic or Northern polar region geographically contains the Arctic Ocean and some parts of surrounding lands. The Antarctic or Southern polar region encompasses Antarctica and some areas of the surrounding Southern Ocean. At first glance, the Arctic and Antarctic appear to have many striking parallels, but there are fundamental differences (e.g., geographical characteristics, ice formation history, and conquest by humanity) between these two polar regions.

In the Northern polar region, the Arctic Ocean (or Sea) is naturally centered on the pole and surrounded by different landmasses. As the smallest ocean on Earth with an area of 14 million km^2, the Arctic Sea is linked to the oceans around the world by only a limited number of waterways. In the Arctic Ocean, there is a permanent sea-ice cover with seasonally different areas. Its largest (smallest) extent is generally seen at the end of winter (summer). Since 1979, the area of summertime ice cover in the Arctic Ocean has shrunk by almost 3 million km^2 (Petty et al., 2018; Wu & Wang, 2019;

Climate Impacts on Extreme Weather
https://doi.org/10.1016/B978-0-323-88456-3.00004-6

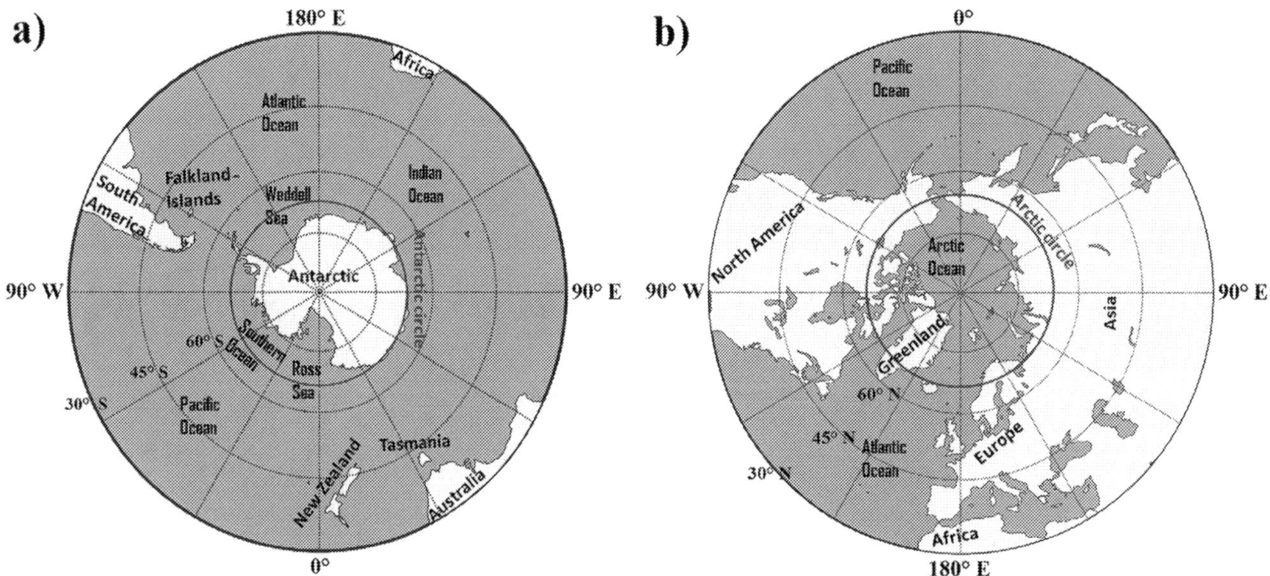

FIG. 12.1 The geographical locations of the Arctic (A) and the Antarctic (B).

Zhao et al., 2018). As Europe, North America, and Asia outspread far to the Northern polar region, the Arctic has earlier and more successfully than the Antarctic been settled by people, animals, and plants. Historically, the first aboriginal persons were moving to the coastal areas of the Arctic Sea around 45,000 years ago (Pitulko et al., 2016). However, more than 4 million people are living within the Arctic Circle today (NSIDC, 2019a; Nymand Larsen, 2014).

Mostly (98%) covered by ice (up to 4.7 km), the Antarctic is a vast land (14.2 million km^2) in the far south of Earth (Gonzalez & Vasallo, 2020). Physically, the Antarctic (or Southern) Ocean separates this continent from other parts of the world. The clockwise water current in this ocean also insulates the Antarctic continent climatically (Orsi, Whitworth, & Nowlin, 1995). In general, the annual average surface air temperature over the Antarctic (−49.3°C) is much colder than over the Arctic (−18°C) (Fig. 12.2, based on (Hersbach et al., 2020)). Furthermore, the Antarctic is known as the driest and windiest region on our planet (Turner, 2003). Such very extreme climatic conditions principally explain why only a very few plants and animal species have been able to establish and survive themselves on this frozen and remote landscape. Today, different research stations are the only permanent human settlements on the Antarctic, while the number of tourists coming to visit this continent for a short time is increasing.

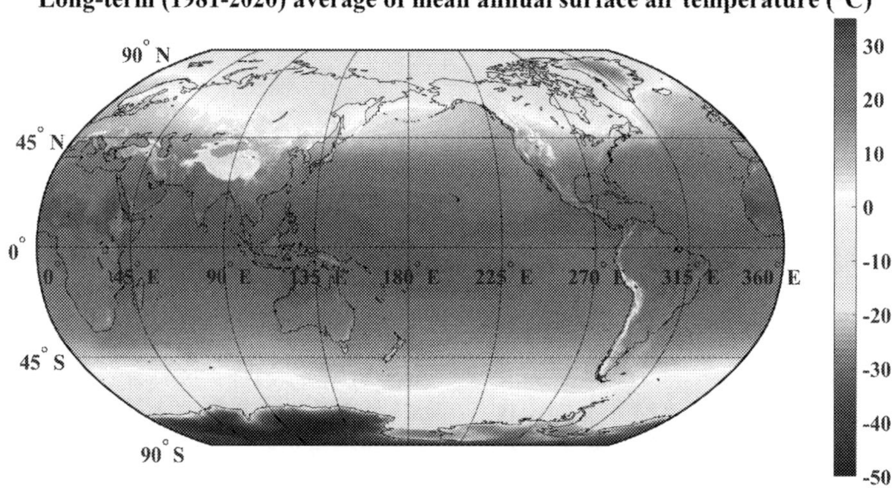

FIG. 12.2 The spatial distribution map of the long-term average of mean annual surface air temperature during 1981–2020. *Based on Hersbach, H., Bell, B., Berrisford, P., Hirahara, S., Horányi, A., Muñoz-Sabater, J., et al. (2020). The ERA5 global reanalysis.* Quarterly Journal of the Royal Meteorological Society, 146*(730), 1999–2049. https://doi.org/10.1002/qj.3803.*

1.2 Arctic and Antarctic: The cooling champers of earth

The extremely cold climate in both Arctic and Antarctic regions is basically related to a self-reinforcing process (Goosse et al., 2018). The process involves the following factors:

Solar energy (Crook, Forster, & Stuber, 2011; Pithan & Mauritsen, 2014; Taylor et al., 2013)—The surface of Earth naturally receives much less solar energy in both polar regions than at the equator (Fig. 12.3). The key reasons are: (1) the low angle of arriving sunlight, (2) the slope of the Earth's axis, and (3) the orbit of this planet around the sun. In combination, all these fundamentally result in the undersupply of solar energy throughout both the Arctic and the Antarctic compared to other parts of the world. In contrast, a great fraction of solar energy reaches the tropical areas on Earth. This huge difference in temperatures between the polar regions and the tropics principally generates the large airflows and ocean currents, which distribute the heat and thereby determine weather conditions around the world. Without such cold regions in the North and South Pole of Earth, in fact, there would be very limited or no global circulation patterns in both air and water masses.

Wind (Rainville, Lee, & Woodgate, 2011)—The differential heating between the poles and the equator as well as the rotation of Earth (Coriolis effect) are the main two causes of large-scale wind circulation around the world (Fig. 12.3). In both hemispheres, such strong winds act as shields for preventing tropical heat to deeply reach the Arctic and Antarctic. Hence, the climate in both northern and southern polar regions is so much colder than in the rest of the world.

The high albedo of white snow and ice (Hall, 2004; Winton, 2006)—The freezing climatic conditions in the polar regions not only forces precipitation to be in the form of snow but also surface water to be frozen, particularly during the winter season. Due to the high albedo of white snow and ice covers, the Earth's surface in both Arctic and Antarctic reflects a large proportion of incoming solar radiation, and thus, amplifies cooling conditions (Fig. 12.4). Such effect is known by scientists as positive feedback.

Humidity (Bodas-Salcedo, Andrews, Karmalkar, & Ringer, 2016; Dessler, Zhang, & Yang, 2003; Gordon, Jonko, Forster, & Shell, 2013)—In both Arctic and Antarctic regions, the capacity of air to hold water vapor is basically very low due to the extremely cold climate conditions. This also facilitates the low temperatures over the polar regions. This is particularly true for the central Antarctic, where the air masses without such important heat reservoirs as well as

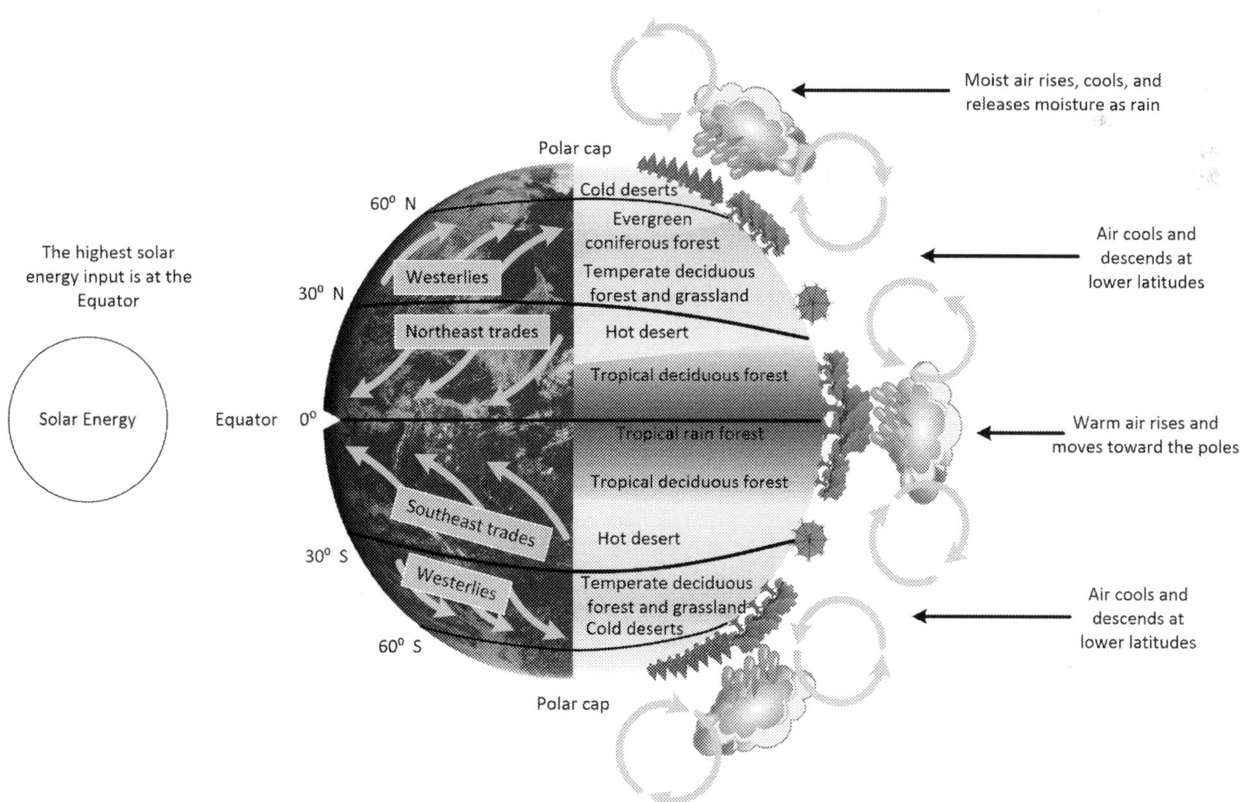

FIG. 12.3 The uneven distribution of solar radiation creates different temperatures and trade wind zones around the world.

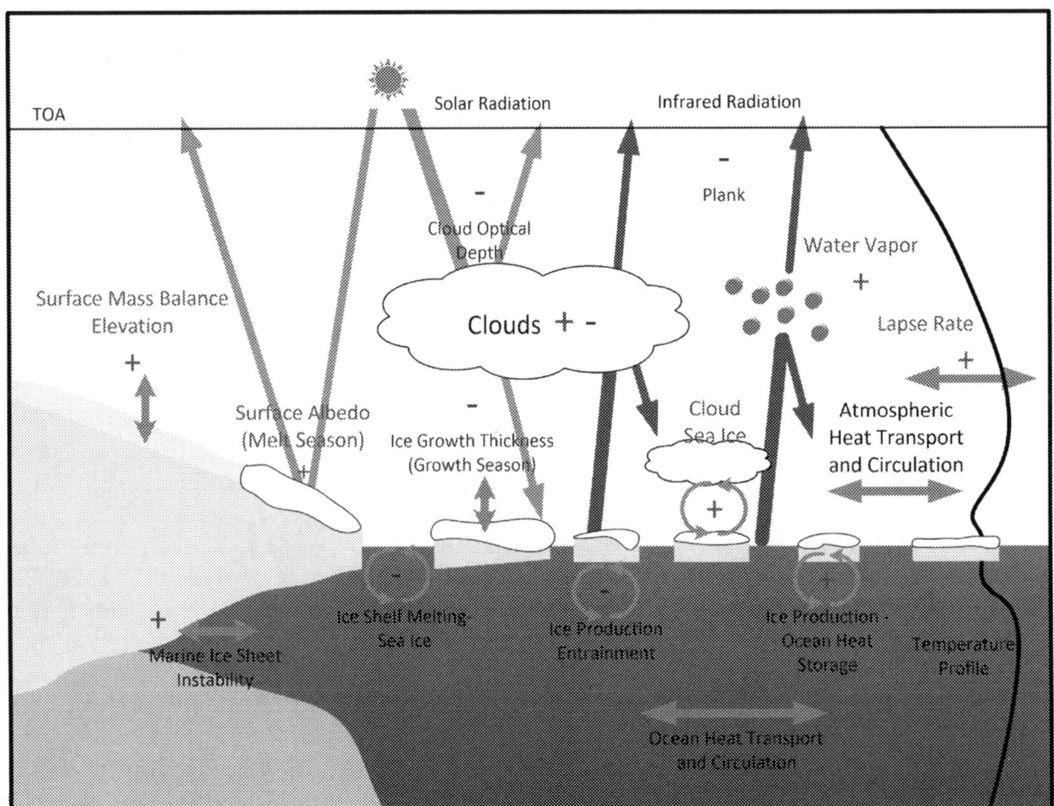

FIG. 12.4 A schematic of key radiative and nonradiative feedbacks throughout the polar regions involving the atmosphere, the ocean, sea ice, and ice sheets. TOA is the top of the atmosphere. Solar radiation in *yellow (gray in the printed version)* and Infrared Radiation in *red (dark gray in the printed version)* represent the shortwave (solar) and longwave (infrared) radiation exchanges. A *red (dark gray in the printed version)* (*blue (light gray in the printed version)*) plus (negative) sign shows the positive (negative) feedback. The *black line* on the right corresponds to the simplified temperature profile in polar regions for both of atmosphere and ocean. *Adopted from Goosse, H., Kay, J. E., Armour, K. C., Bodas-Salcedo, A., Chepfer, H., Docquier, D., et al. (2018). Quantifying climate feedbacks in polar regions.* Nature Communications, *9(1). https://doi.org/10.1038/s41467-018-04173-0.*

thick cloud cover amplify the cooling processes. Besides the other factors, this helps in generating favorable conditions for the development of enormous sea-ice covers, ice sheets, and glaciers in the central Antarctic.

The extremely low temperatures in the Arctic and the Antarctic cool down the warm air and ocean currents from the tropical regions and send them back as floating ice to the equator because of absorbing a little fraction of incoming solar energy. Hence, with such a crucial role in the heat distribution around our planet, the polar regions act as the cooling chambers of Earth. This smooth functionality will only continue if the atmosphere–ocean-glaciers-sea ice interactions do not change. Due to the very high sensitivity of ice masses to heat, however, increases in temperature pose serious challenges to such interactions, particularly throughout both polar regions.

2 The new face of Arctic and Antarctic

In the current era of global warming (Scheffer, Brovkin, & Cox, 2006; Zahn, 2009), the polar regions are more rapidly and evidently undergoing a rapid transformation than in most parts of the world. The consequences of such a warmer climate have so far been most revealed in the Arctic region, where the seawater temperature is significantly increasing, and the sea ice, snow cover, and glaciers are tremendously melting. On the other hand, such changes are spatially different throughout the Antarctic. Hence, this chapter reviews the extreme impacts of climate change on both the Arctic and Antarctic regions.

2.1 Polar climate warming

On 29th December 2015, the surface air temperature warmed from −26.8°C to −0.8°C at the North Pole (Moore, 2016). According to the meteorological records in Ny-Ålesund, Spitsbergen, a storm brought warm airflow from the North Atlantic to the North Pole during the middle of Arctic winter (Moore, 2016). This was confirmed by sea-ice buoys drifting that showed a positive average temperature of 0.7°C at the 85°N throughout the Arctic Ocean (Grosfeld et al., 2016). Besides, the North Pole was warmer than some other areas of Central Europe on December 30, 2015 (Boisvert, Petty, & Stroeve, 2016). Such a considerable heat invasion into the northern polar region would have been an exceptional extreme event two decades ago, while becoming more frequent, particularly during winter (Moore, 2016). In February 2017, for example, there was a rainfall event over Ny-Ålesund, where the surface air temperature was positive (Graham et al., 2017). A year later (February 2018), the combination of strong offshore winds and above-average surface air temperature over the north coast of Greenland led to breaking off the old sea ice frozen to the coast to form a large polynya (Moore, Schweiger, Zhang, & Steele, 2018). The daily high temperature was about 6.1°C across the northernmost part of Greenland on February 24, 2018, when the polynya experienced its maximum width (Moore et al., 2018). In fact, at the end of February 2018, there was an exceptional atmospheric circulation over the Arctic. Due to the splitting of the polar vortex, the power of the jet stream weakened, and consequently, warm airflow penetrated far into the Arctic region. The air temperatures warmed to as high as 15°C above normal in the Siberian area and in the Labrador Sea and of the Arctic Ocean, while Central Europe was simultaneously experiencing spells of extreme cold (Moore et al., 2018).

In general, the annual mean surface air temperature has significantly warmed by 2.7°C across the Arctic during 1971–2017 (AMAP, 2019). In particular, the Atlantic sector of the northern polar region has prominently experienced warmer surface air temperature during winter (AMAP, 2019). In 1971–2017, the Arctic wintertime surface air temperature increased by 3.1°C (AMAP, 2019). On the other hand, the warmer surface air temperature in the Arctic during summer was less markedly. For example, in Ny-Ålesund, the increase in temperature during the warm months was only about 1.4°C (AMAP, 2019).

Previous studies including almost all different areas of the northern polar region express one central message: the Arctic has been warming more than twice as other parts of the world during recent decades (Cohen et al., 2014; Gjelten et al., 2016; Serreze & Barry, 2011) (Fig. 12.5). This upward trend in the Arctic air temperature is still on (Overland et al., 2019). Scientists have reported the highest warming rate of surface air temperature across the Arctic during winter (Bekryaev, Polyakov, & Alexeev, 2010; Boisvert & Stroeve, 2015; Overland et al., 2019). In winter 2015–2016, for example, the temperature at the latitudes higher than 66°N was 5°C warmer than the average monthly value for the period 1981–2010 (Overland & Wang, 2016). All over the Arctic, the surface air temperature was also about 1.7°C higher in October 2017–September 2018, compared to the reference period 1981–2010 (AMAP, 2019).

While Arctic warming appeared as early as the 1830s, the surface air temperatures remained steady in Australia and South America through the turn of the century (Abram, McGregor, & Tierney, 2016). During the 1950s, significantly warmer temperatures were initially experienced on the Antarctic Peninsula and in the West Antarctic (Abram et al., 2013; Steig et al., 2009). However, slightly warmer local temperature is not necessarily translated into a sign of climate

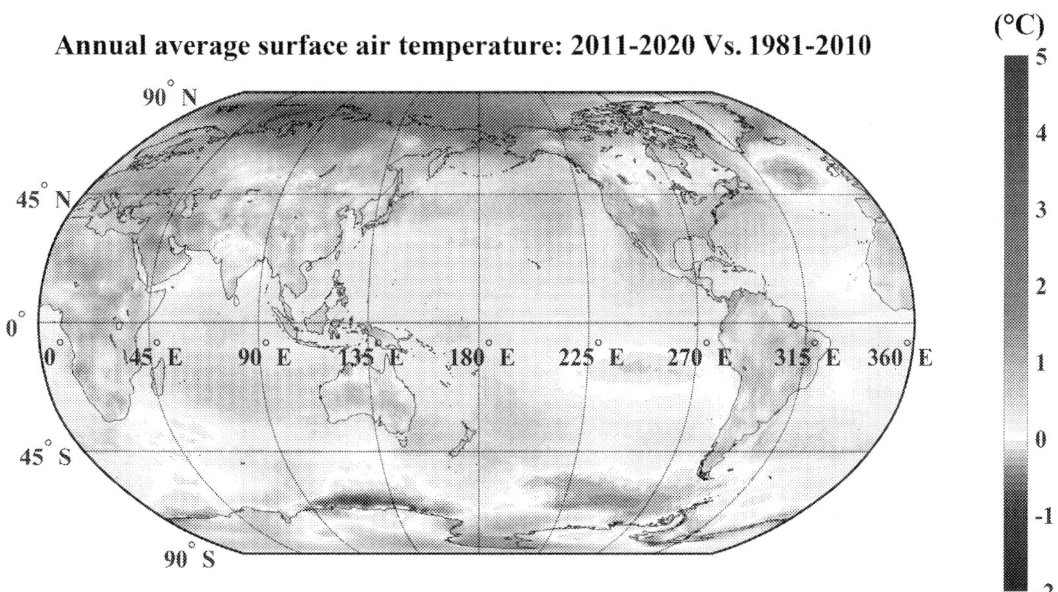

FIG. 12.5 The spatial distribution map of annual average surface air temperature during 2011–2020 versus 1981–2010. *Based on Hersbach, H., Bell, B., Berrisford, P., Hirahara, S., Horányi, A., Muñoz-Sabater, J., et al. (2020). The ERA5 global reanalysis.* Quarterly Journal of the Royal Meteorological Society, *146(730), 1999–2049. https://doi.org/10.1002/qj.3803.*

change. According to Intergovernmental Panel on Climate Change (IPCC, 2013), climate change is commonly used when a clear and sustained surface air temperature curve during a period of at least 30 years surpasses the margins that were previously defined by naturally occurring variations in climatic conditions. This became clear in the northern polar region during the 1930s, earlier than in any other part of the planet (Abram, McGregor, & Tierney, 2016). This was followed by the tropical regions and the mid-latitude areas throughout the northern hemisphere, where the different warming signal was first seen during the 1950s (Levitus et al., 2012). Later, the mounting evidence for climate change was detected in Australia and Southeast Asia about 60 years ago. Since the beginning of this century, surface air temperature warming has been evident in all regions around the world, with the exception of central Antarctica (PAGES 2k Consortium, 2013).

The Antarctic is not experiencing a uniform warming pattern that has been recorded over the Arctic. This is generally due to (1) the cooling role of high-albedo continental ice masses, and (2) the protecting effects of Circumpolar Currents within the Antarctic region (King, Turner, Marshall, Connolley, & Lachlan-Cope, 2003; Morris & Vaughan, 2003). The marine-dominated coasts also show significant regional differences with the continental climatic conditions in central Antarctica con (Convey, 2003). Throughout the Antarctic Peninsula and Pacific-West Antarctica, an acceleration in movements of glaciers as well as in sea ice diminishing have been detecting during recent decades (Cook & Vaughan, 2010). Such developments are basically explained by alterations in (1) atmospheric circulation that transport more heat and humidity toward the South Pole, and (2) oceanic currents that bring warmer water into coastal parts (King et al., 2003). The increases in atmospheric heat are related to the westerly winds throughout the Southern Ocean, strengthened toward the South Pole in response to significant rises in atmospheric greenhouse gas emissions (Fyfe, 2015; Swart, Gille, Fyfe, & Gillett, 2018) as well as to sustainable increases in ozone depletion across Antarctica (Sigmond, Reader, Fyfe, & Gillett, 2011; Solomon, Polvani, Smith, & Abernathey, 2015; Swart et al., 2018; Thompson et al., 2011). In fact, the two processes have resulted in greater temperature differences between the tropical areas and the Antarctic Region, forcing considerably stronger winds (Armour, Marshall, Scott, Donohoe, & Newsom, 2016; Turner, Hosking, Bracegirdle, Marshall, & Phillips, 2015).

2.2 The role of ozone hole in Antarctic climate

An atmospheric layer composed of ozone, the stratosphere ozone the absorbs the shortest (i.e., highest energy) rays of the sun (Rowland, 2006). This prevents ultraviolet radiation (UV rays), which is invisible to humans, from reaching the surface of Earth. Without this natural protective screen, life on our planet would not be possible (van der Leun, Tang, & Tevini, 1995).

Ozone is a highly reactive gas and has a higher concentration in the Earth's atmosphere above an altitude of 12 km (with the highest at an altitude of 30–35 km) (Dameris, 2010). Nonetheless, the total proportion of ozone in the atmosphere is significantly low compared to other gases. This makes the effect of the ozone layer on the Earth's climate more notable. In fact, the stratospheric ozone not only absorbs the passing UV rays through the mesosphere layer. Tropospheric ozone is a greenhouse gas (GHG) it also absorbs heat energy that is radiated from the Earth (World Meteorological Organization, 2007). Thus, more tropospheric ozone in the atmosphere leads to more absorption of GHGs, resulting in warming. On the contrary, the atmosphere would cool down if the ozone concentration in the stratosphere declines (Randel & Wu, 1999).

Scientists have been monitoring the ozone layer thickness over the Antarctic, which began to thin out regularly at the end of (southern) winter and the ozone hole began to appear in September and October (World Meteorological Organization, 2018). Man-made gases (chlorofluorocarbons and brominated hydrocarbons) that have been used—or are still being used—as propellants, refrigerants, or solvents, and contain chlorine or bromine compounds can destroy ozone (World Meteorological Organization, 2011). These gases require special conditions to activate that are only present during the long, dark winters in the polar regions. Therefore, ozone holes can only occur in the Antarctic or, in some exceptional cases, also in the Arctic (World Meteorological Organization, 2011).

An ozone hole is created when the concentration of ozone in the stratosphere falls below 220 Dobson Units which denotes 220 ozone molecules, corresponding to a pure ozone layer with a thickness of 2.2 mm (Stolarski, Schoeberl, Newman, McPeters, & Krueger, 1990). The average ozone concentration was 250–350 Dobson Units in the Antarctic before its first ozone hole experience (World Meteorological Organization, 2014). Today, it regularly sinks to about 100 Dobson Units during the Antarctic springtime (Bednarz et al., 2016; Langematz et al., 2014). In response to such thinning of the ozone, the air layers in the lower stratosphere over the Antarctic are now 10°C cooler than in the 1990s (World Meteorological Organization, 2018). Due to the ozone depletion throughout the stratosphere, the underlying troposphere also cools down. This is accordingly the key reason for the slight drop of surface temperatures in Central Antarctica during recent decades (World Meteorological Organization, 2018).

The sustainable cold air in the lower stratosphere not only prevents the polar vortex from a timely collapse but also increases its lifespan, and consequently lengthens the ozone depletion period (Sheshadri & Plumb, 2016; Zhang, Li and

Zhou, 2017). Simultaneously, it amplifies the contracts between the tropics and the southern polar region and thereby changes the atmospheric circulation patterns around the world. Along with such stronger winds in the stratosphere, the descending of tropopause above Antarctica directly changes the southern polar weather patterns by influencing both line-up and expansion of high- and low-pressure areas (Sheshadri, Alan Plumb, & Domeisen, 2014) (Fig. 12.6). Thus, the band of westerly has shifted further toward the South Pole across the Southern Ocean (Hogg, Meredith, Blundel, & Wilson, 2008), while the temperature has changed across some coastal parts within Antarctica, particularly

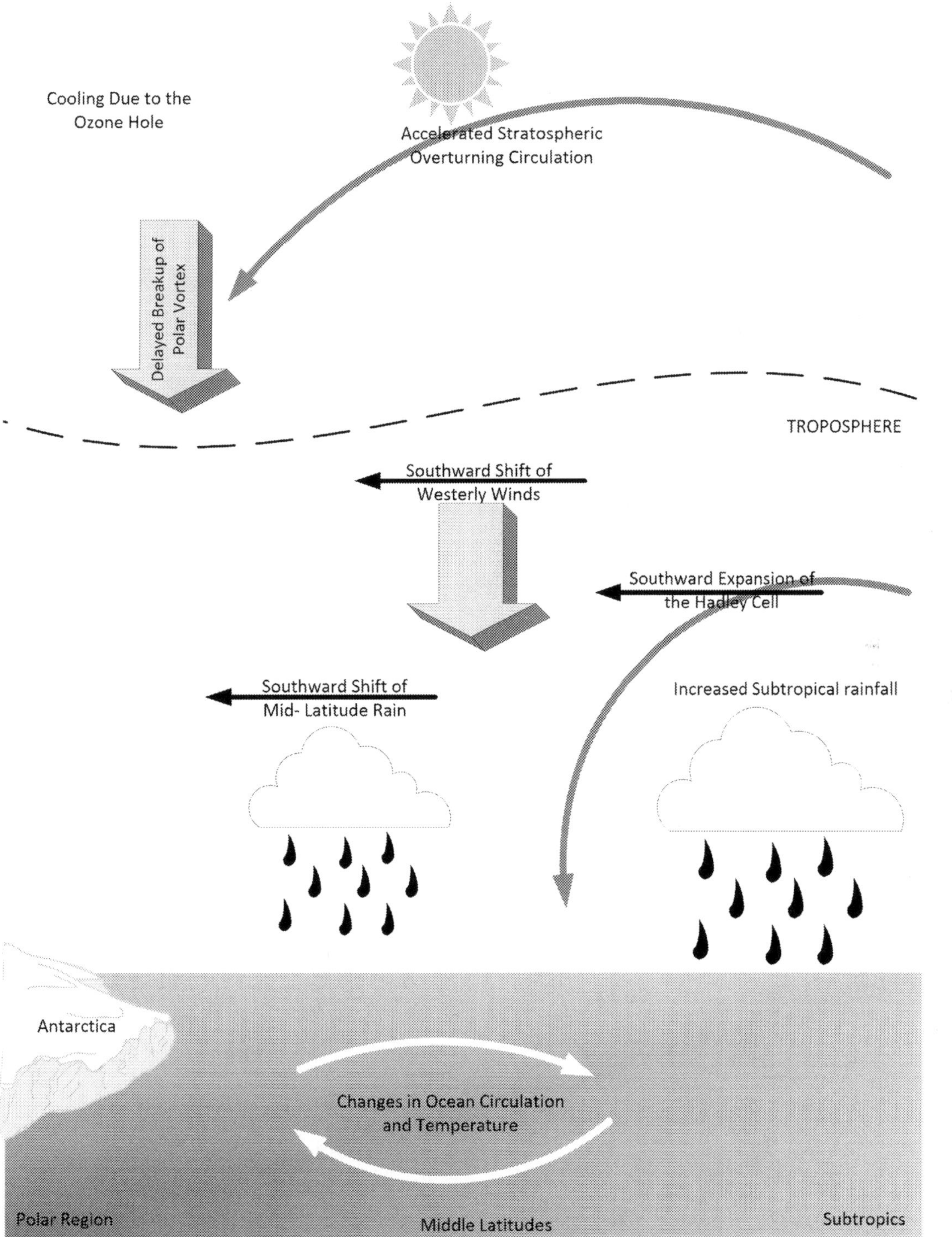

FIG. 12.6 In recent decades, more frequent ozone depletion across Antarctica has significantly affected the climate of the southern polar region. The lower stratosphere has cooled down, leading to, for example, southward shifts of wind and rain systems.

during summer (Seviour, Gnanadesikan, Waugh, & Pradal, 2017). The Antarctic Peninsula has accordingly experienced warmer summer seasons since 1985 when the ozone hole was discovered (Turner et al., 2016). In the Bellingshausen Sea and the waters to the west and northeast of the Antarctic Peninsula, significantly shorter periods of sea-ice cover have also been recorded during the last 30 years (Hobbs et al., 2016; Turner et al., 2015).

2.3 Heat tracking in the polar seas

The comparatively moderate increase (1°C) in global temperature can basically be attributed to the existence of oceans around the world (IPCC, 2013). Absorbing about 30% of atmospheric CO_2 in the past, the oceans considerably buffered the impacts of GHG emissions (IPCC, 2014). The oceans also have a massive capacity for heat saving through the physical features of saltwater as well as the sheer magnitude of oceanic water. Besides, the very sluggishly reaction of oceans to the environmental changes is fundamentally related to the water temperature cooling down processes through the circulation and passing of oceans throughout the polar regions (Marshall, 2014). Hence, it generally takes about 10 years (centuries to millennia) for the ocean's surface (deep) water to adjust to global warming (IPCC, 2014).

Since 1871, the oceans around the world have absorbed about 436 zettajoules of heat energy, a thousand times the energy amount that humans currently consume each year (Zanna, Khatiwala, Gregory, Ison, & Heimbach, 2019). For decades, hence, all oceans have been continuously becoming warmer. The heat energy is mostly retained in the water masses upper than 700 m depth. The water layers at depths from 700 to 2000 m are also warming, with potentially serious impacts on the global ocean-current conveyor belt (Cazenave et al., 2018) (Fig. 12.7). Accordingly, ocean warming can weaken the thermohaline circulation in two different ways (Beszczynska-Möller, Woodgate, Lee, Melling, & Karcher, 2011). First, additional heat reduces the water density because of thermal extension, and consequently, the water turns out to be lighter. Second, the seawater is diluted with freshwater from more melting of glaciers (particularly in Greenland and Antarctica) or rainfall, and thus, the density of water in oceans is similarly lowered. Substantial increases in both water temperature and freshwater influx prevent the water masses in the Southern Ocean and the North Atlantic from sinking and thereby extinguish the influential drivers for thermohaline circulation (Swart et al., 2018; Williams, Roussenov, Smith, & Lozier, 2014).

Sea surface temperature (SST) (Fig. 12.8) is crucially important for the polar regions. Today, the ocean currents flowing toward the poles transport more heat to the Arctic and Antarctic regions than in the past. Since the early 1990s, for example, the Atlantic water flowing into the Arctic Ocean has become verifiably warmer (Holliday et al., 2008; Schauer et al., 2008). German and Norwegian Scientists tracked such heat pathways into the Arctic Ocean by setting up a transect of oceanographic survey sites across the Fram Strait (79°N) in 1997, from the west coast of Spitsbergen to Greenland's northeastern coasts (Schauer et al., 2008). At 16 sites, all temperature, current speed, and salinity of both inflowing and outflowing water masses are measured throughout the water column. These records indicate that the water of the West Spitsbergen Current arriving from the North Atlantic is today 1°C warmer when moving

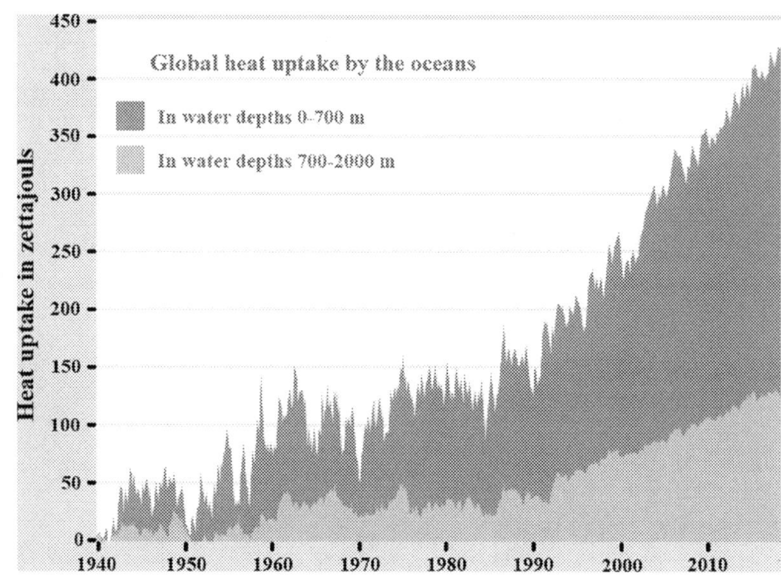

FIG. 12.7 The massive amounts of heat energy are continuously absorbed by the world's oceans. Although such heat was initially stored in the upper layers of water, it has now reached deeper levels. *Based on Cazenave, A., Meyssignac, B., Ablain, M., Balmaseda, M., Bamber, J., Barletta, V., et al. (2018). Global sea-level budget 1993-present.* Earth System Science Data, 10*(3), 1551–1590. https://doi.org/10.5194/essd-10-1551-2018.*

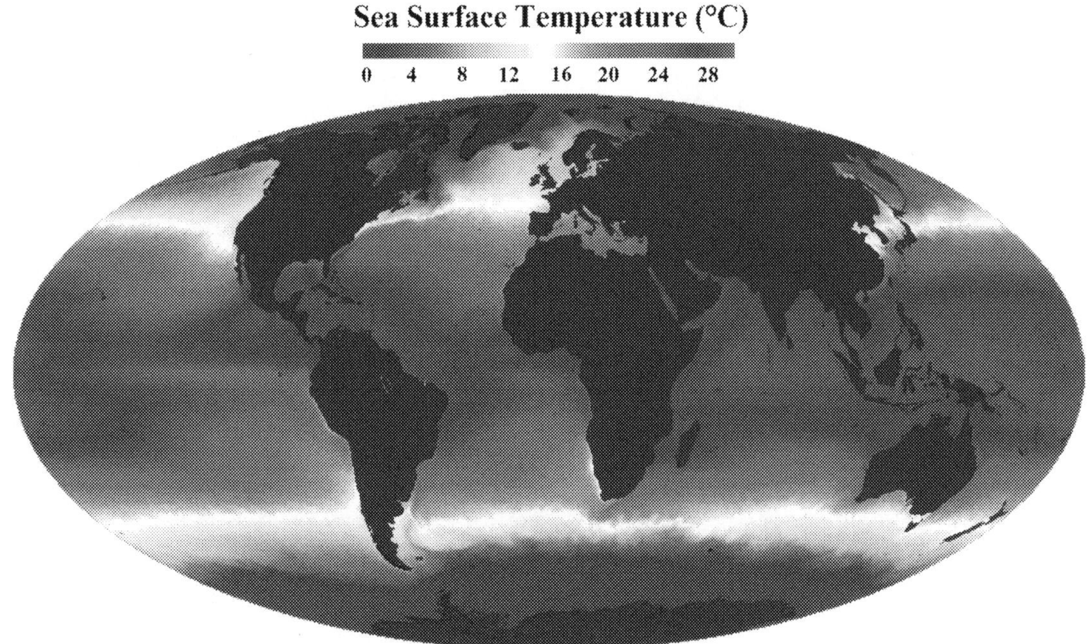

FIG. 12.8 The spatial distribution map of average sea surface temperature (SST) in °C during 2003–2011 base on the MODIS Aqua Data. There is a strong contrast in ocean surface temperatures between the warm equatorial areas and cold polar regions.

through the Fram Strait toward the Arctic Ocean, compared to the long-term measurements during the last 20 years. Such warmer water is already evident within the entire Eurasian Basin (Schauer et al., 2008).

In the ice-free parts of the Arctic Ocean, the SSTs have also increased (Dethloff, Handorf, Jaiser, Rinke, & Klinghammer, 2019; Jahn, Kay, Holland, & Hall, 2016). This explains why, not only the sea in the Arctic Ocean freezes over later through the year, but also the sea ice melts earlier. Hence, there would be different large ice-free areas throughout the Arctic Ocean for longer periods during summers. This facilitates the absorption of solar energy and thereby exacerbates further temperature warming (Jaiser, Dethloff, Handorf, Rinke, & Cohen, 2012; Semmler et al., 2016).

The Southern Ocean plays an important role in the global climate system (Brook & Buizert, 2018). Both cooling and overturning of water masses throughout Antarctica enable the oceans to store considerable amounts of heat (Fig. 12.9) and GHG (Armour et al., 2016; Swart et al., 2018). The sinking of heavy water leads to the possibility of heat and CO_2 transportation from the upper to deeper water layers for long periods (Durack & Wijffels, 2010; Frölicher et al., 2015).

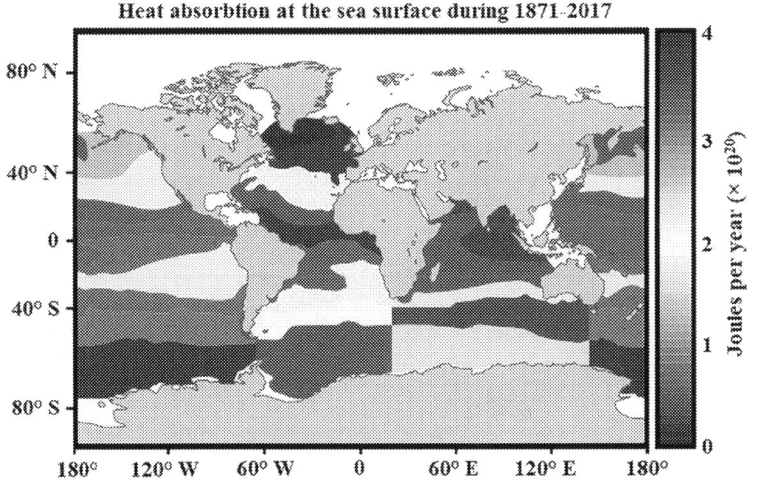

FIG. 12.9 Long-term (1871–2017) average value of annual heat absorption at the sea surface on Earth. *Adopted from Zanna, L., Khatiwala, S., Gregory, J. M., Ison, J., & Heimbach, P. (2019). Global reconstruction of historical ocean heat storage and transport.* Proceedings of the National Academy of Sciences of the United States of America, 116(4), 1126–1131. *https://doi.org/10.1073/pnas.1808838115.*

The scale of this overturning process is much larger in the Antarctic than in the North Atlantic. Accordingly, a warming trend has been determined in water temperature all over the Southern Ocean since the 1950s (Gille, 2002, 2008). Its slope indicates that the sea south of the 40th parallel has absorbed considerably more heat from the atmosphere than all other oceanic areas combined (Swart et al., 2018).

The heat storage in large magnitudes over several decades has other consequences. According to long-term records along the prime meridian, the entire water column throughout the Weddell Sea, and particularly the deepest layer of water known as the Antarctic Bottom Water, has been experiencing a warming trend since the 1990s (Jullion et al., 2013; Purkey & Johnson, 2010, 2012). Similarly, in other marine regions of Antarctica, the Southern Ocean water temperature warming at depths between 700 and 1100m after 1950 was double the worldwide oceanic average (Gille, 2002; IPCC, 2014; Levitus et al., 2012; Purkey & Johnson, 2010).

Scientists are not sure if this water warming is primarily caused by the increasing air temperature across the Southern Ocean? (IPCC, 2014). When the air warms, the seawater is not able to release as much of its own heat to the surrounding atmosphere. Feasible changes in wind conditions across the sea can also alter the speed of certain ocean currents and thereby influence deep-water development. The inflowing of warmer waters into the Southern Ocean can potentially be another reason for observing such increases in water temperature at deep layers. However, researchers can today track the Antarctic-wide warming of deepwater northward to beyond the equator, thanks to the new remarkable advances (Zanna et al., 2019). Accordingly, the heavy water masses flow there, after they have filled up the deepest water layers of the Southern Ocean.

2.4 Paradox in Antarctic sea ice

Since the beginning of satellite measurements, the seasonal sea ice has continued to develop/shrink in Antarctica/the Arctic (National Academies of Sciences, Engineering, and Medicine, 2017) (Fig. 12.10, adopted from Perovich et al., 2018). In the Arctic, the sea ice is thinner today than in 1980, and its larger portions are covered by meltwater pools (Fig. 12.10A). These two have increased solar energy absorption by the ice cover and its ocean below, and thereby reinforced the sea ice declines. Hence, the areas of seasonal sea-ice cover are shrinking in the Arctic. Comparing to the long-term (1981–2010) average value, the total loss of Arctic sea-ice cover is greater in summer or September than in winter or March (Fig. 12.10B). In 2014, the record-breaking of ice cover (20.1 million km^2) throughout the southern polar region was reported (NSIDC, 2014). This was mainly related to the development of sea ice in Antarctica, while the decline in ice cover in the far north was simultaneously pronounced with increasing global temperatures (NSIDC, 2014). Such expansions in the Antarctic Sea ice posed a mystery for humanity: the Antarctic region might be spared the impacts of global warming.

In the Antarctic, the sea ice was only growing in certain parts (Hobbs et al., 2016; Parkinson & Cavalieri, 2012). For example, the highest rate of significant increasing (decreasing) trend in ice surface was recorded in the Ross Sea (Bellingshausenhas and Amundsen Seas) in recent decades (Holland, 2014; Lecomte et al., 2017; Parkinson, 2019; Turner et al., 2015). On the other hand, there were also conflicts in the duration of sea ice developments. The sea ice melted much later during the summer season in the Ross Sea, while returned earlier and earlier throughout the marine areas of West Antarctica (Jenkins et al., 2016; Mouginot, Rignot, & Scheuchl, 2014; Stammerjohn, Massom, Rind, & Martinson, 2012; Turner et al., 2017). In the Weddell Sea, the sea ice expanded in some parts, while shrank in others (Hobbs et al., 2016). In 2014, the growth of sea ice was relatively enough to turn the balance of Antarctic ice cover into a positive variation (NSIDC, 2014). Two years later (winter 2016), however, the Antarctic Sea ice reached only 18.5 million km^2, and then, showed a shrinking trend (NSIDC, 2016; Stuecker, Bitz, & Armour, 2017). On 1st January 2019, the new record of 5.47 million km^2 was reported as the smallest area of January ice cover during the last 40 years (NSIDC, 2019b; Nymand Larsen, 2014). With a typically heavy ice cover, the Ross Sea even showed many parts of ice-free water in January 2019 (NSIDC, 2019b).

Today, intensive researches (Matear, O'Kane, Risbey, & Chamberlain, 2015; Purich et al., 2016; Simpkins, Ciasto, & England, 2013) are focusing on both decreasing of sea ice in the southern polar region during winter and speedy melting in the summertime. US researchers propose a new theory: the existence of natural current variations throughout the Southern Ocean in a 30-year cycle (Sheshadri & Plumb, 2016; Zhang et al., 2017). Their study concluded that both convection and deep-water formation weakened in different parts of the Antarctic during 1980–2000. Accordingly, the intermediate water heat was trapped throughout the deep parts of the ocean and cannot reach its surface. The surface water simultaneously cooled down, providing supreme conditions for the sea ice formation despite global warming.

There is now evidence that the water masses are increasingly overturning in the Southern Ocean again (Caesar, Rahmstorf, Robinson, Feulner, & Saba, 2018; Delworth & Zeng, 2018; Zhang, Delworth, & Zeng, 2017). This may cause

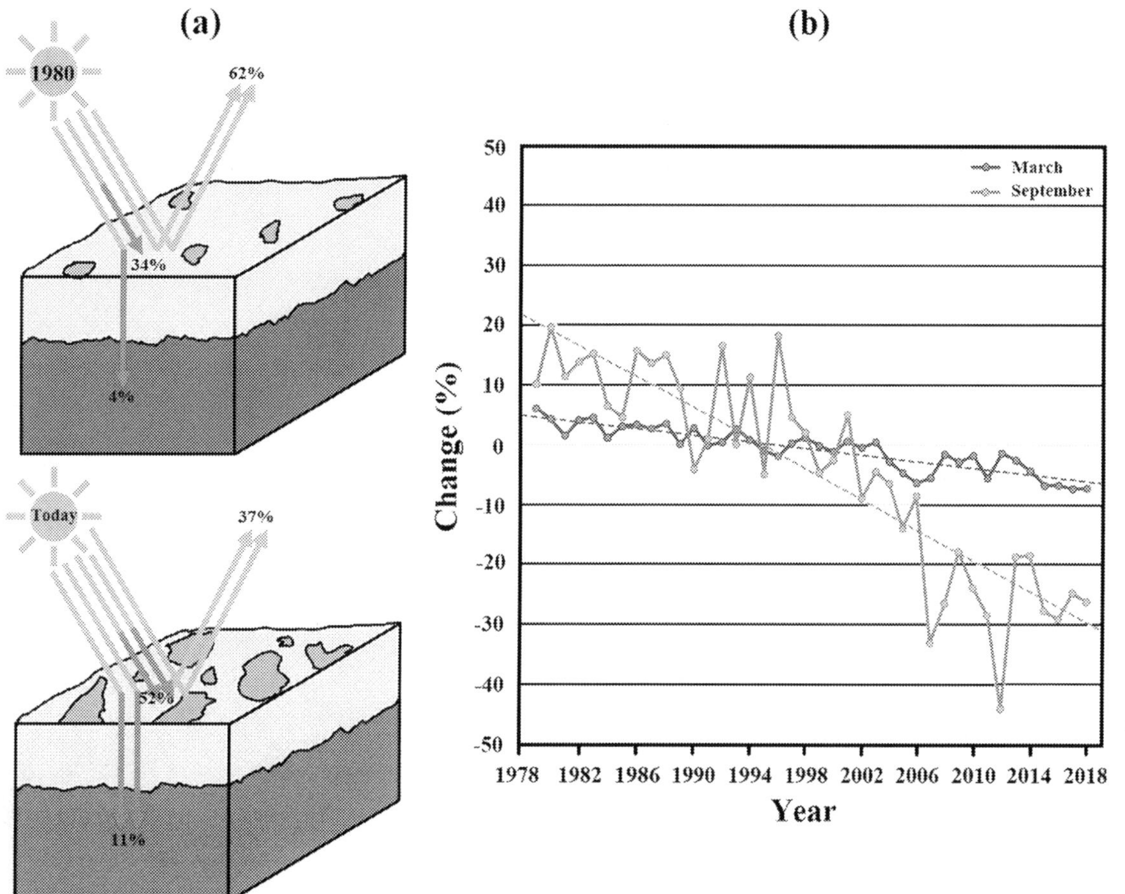

FIG. 12.10 (A) Thinner sea ice with more meltwater pools has already increased solar energy absorption by the Arctic ice cover and its ocean below, and (B) the total loss of Arctic sea-ice cover in summer/September and winter/March during 1981–2010. *Adopted from Perovich, D., Meier, W., Tschudi, M., Farrell, S., Hendricks, S., Gerland, S., et al. (2018). Sea ice. In* Arctic report card: Update for 2018. *https://arctic.noaa.gov/Report-Card/Report-Card-2018/ArtMID/7878/ArticleID/780/SeanbspIce.*

the intermediate water heat to move toward the surface and thereby retreat the ice again. If this theory comes into reality, the sea-ice cover would decrease more in response to the warmer intermediate water (Zhang, Delworth, Cooke, & Yang, 2019).

The role of the atmosphere in the disappearance of sea ice throughout the Antarctic cannot be neglected. Cold winds, which caused wide-scale surface water freezing in the past, are weakening over the Southern Ocean and thereby reduce the sea ice throughout the Bellingshausen and Amundsen Seas (Lee et al., 2017; Matear et al., 2015). However, up to 2014 changes in wind patterns could also result in the inconsistent increase in sea ice development in the western part of the Ross Sea. On the other hand, an ongoing discussion is related to the influences of melting ice sheets and glaciers. Has it possibly contributed to the cooling processes of surface waters during recent decades? Has such meltwater continuously diluted the surface waters in the Southern Ocean? However, there are yet no clear responses for these questions, partially due to comparatively limited available measurement records at the Southern Ocean. Moreover, for the Antarctic region, climate models have not been able to appropriately simulate the sea-ice development patterns yet. Hence, the prediction of sea-ice fate in the southern polar region during the future is still lacking. Even, the Intergovernmental Panel on Climate Change (IPCC) special report on global warming of 1.5°C (IPCC, 2018) deliberately avoided providing any projection about the future of sea ice in the Antarctic in response to the 1.5–2°C warmer global average temperature.

2.5 Changes in different glaciers and ice sheets

A simple mass balance equation can reveal the response of glaciers or ice sheets to both global warming and climate change (Cogley et al., 2011). This balance turns positive if the amount of snow falling on ice surpasses the mass that it

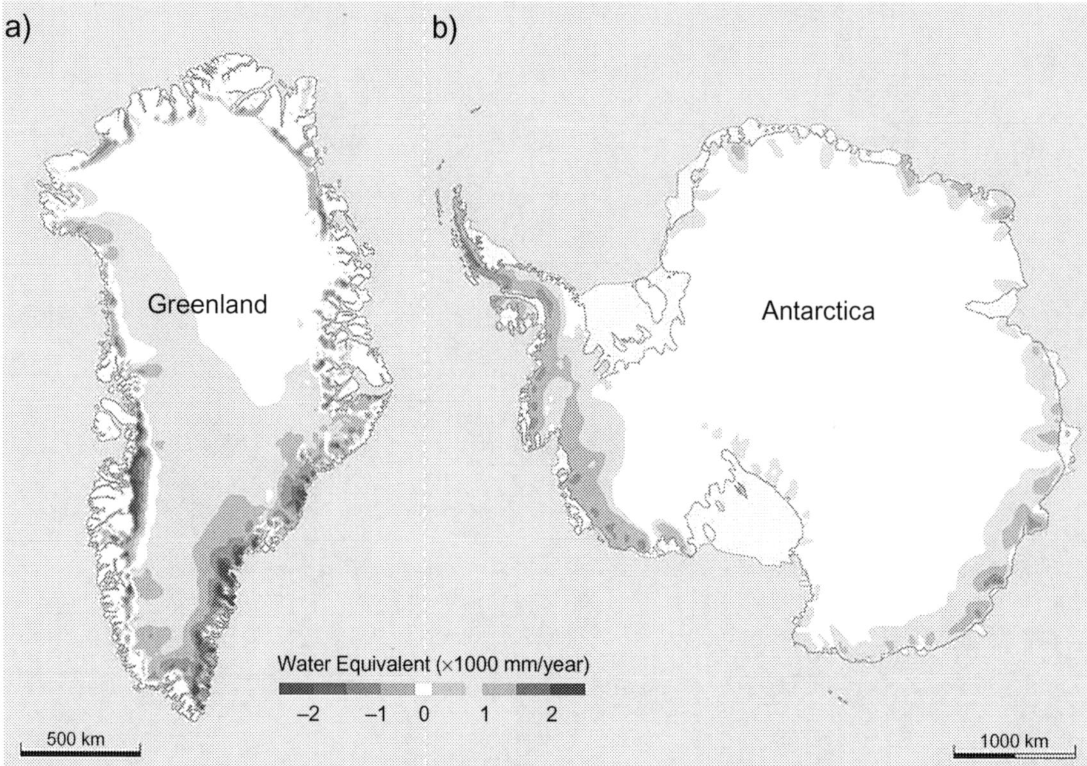

FIG. 12.11 Average annual surface mass balance in water equivalent (×103 mm/year) for (A) Greenland during 1958–2016, and (B) Antarctica during 1979–2017. A positive or negative surface mass balance in an ice sheet depends on snowfall, melting processes, wind transport, and sublimation. *Adopted from World Ocean Review 6. (2019).* The Arctic and Antarctic-extreme, climatically crucial and in crisis. *Hamburg, Germany: maribus gGmbH.*

loses in several ways. But it tends to be negative when the ice mass loses more than it receives through precipitation. Hence, such positive and negative values indicate the growing and shrinking status of the ice shield, respectively.

In 2002, the satellite mission GRACE (Gravity Recovery and Climate Experiment) to minimize the uncertainty in the mass balance estimations (Fig. 12.11, adopted from World Ocean Review 6, 2019) for glaciers and ice sheets around the world was begun (Tapley, Bettadpur, Watkins, & Reigber, 2004). At the elevation of 490 km, two identical satellites of GRACE circle the Earth, one behind the other in a near-polar orbit. Hence, GRACE can provide us with what is not possible via ground-based observations: The system for measuring the total gravitational field of the Earth during a single month. In fact, the satellites generally record modifications in mass on the Earth and measure the water redistribution among all oceans, continents, and ice sheets (Chambers, 2006; Luthcke et al., 2013; Velicogna, 2009). Such remote sensing data sets, thus, help in answering two of the most crucial questions: How much ice are the ice sheets and glaciers of Greenland and Antarctica losing in response to global warming and climate change? And in which parts around the world is sea level rising consequently?

The first GRACE mission continued between 2002 and 2017 (Fig. 12.12). As it was impressively successful, the follow-up satellite GRACE-FO was launched into space in May 2018 (Kornfeld et al., 2019). It is providing climate scientists with reliable data on the development and reduction of the ice sheets for another 10 years. Under favorable solar conditions, however, it should receive an optimal energy supply for even next 30 years.

Most previous studies have been limited to the GRACE mission data. Accordingly, an average of 286 billion tons of ice has annually been lost in Greenland during 2010–2018 (Mouginot et al., 2019). This is basically due to the warmer air over Greenland during recent decades (Box, Yang, Bromwich, & Bai, 2009), causing the melting season to be longer and more intensive (Vaughan, Carrasco, Kaser, et al., 2013). Because of such surface melting processes, Greenland today loses ice about twice each year, compared to the period 1960–1990 (Sasgen et al., 2020). According to the subsequent model simulations, the total ice loss and development were evenly balanced at that time. To the present, however, the ice mass losses also increased by 25% mainly due to the iceberg break-off (Enderlin et al., 2014; Truffer & Motyka, 2016). Nowadays, Greenland has the greatest share (0.7 mm/year) of meltwater contributing to sea-level rise (3.34 mm/year) around the world (Cazenave et al., 2018).

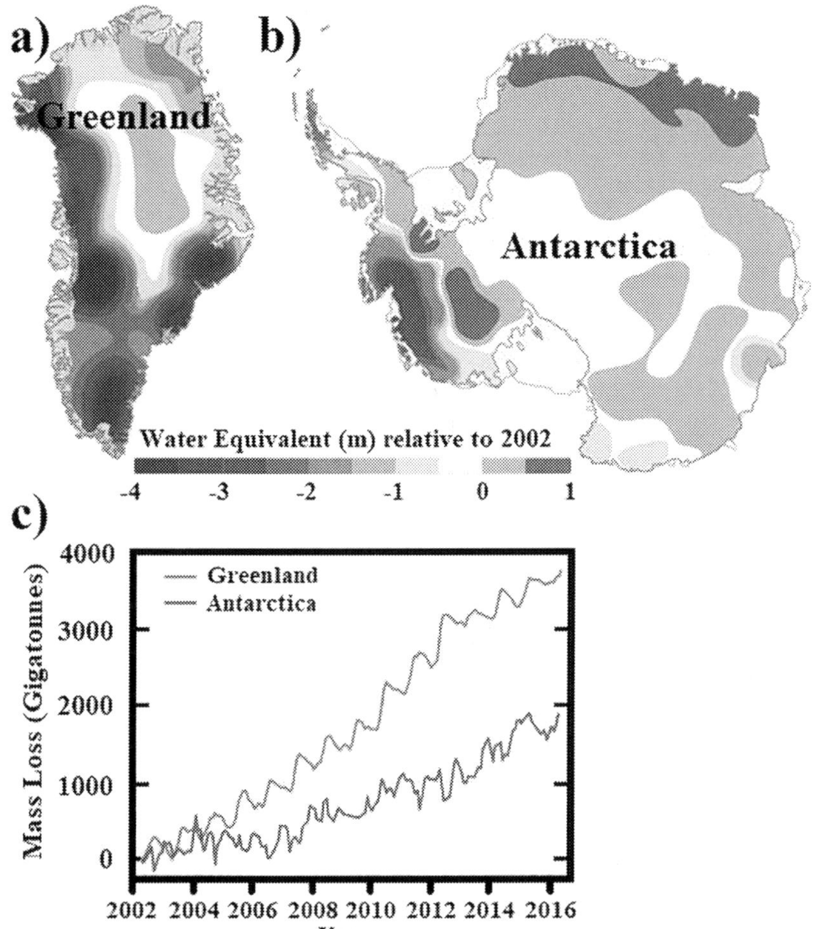

FIG. 12.12 The GRACE satellite system could reliably measure the mass balance of the ice sheets in (A) Greenland and (B) Antarctica, for the first time. (C) Average mass loss is about 218 and 125 gigatonnes per year for Greenland and Antarctica, respectively. During the measurement period, both of these ice shields lost more ice than they gained. *Adopted from World Ocean Review 6. (2019).* The Arctic and Antarctic-extreme, climatically crucial and in crisis. *Hamburg, Germany: maribus gGmbH.*

In Greenland, as a warming hotspot (Rogozhina et al., 2016), the ice cap has been principally lessening from above since the 1990s due to the upper surface melting under simultaneous summer surface air temperature warming by 2°C (Vaughan et al., 2013). The lessening rate of ice was extremely high in summer 2012 (Hanna et al., 2014) when exceptionally warm air (Hanna et al., 2014) and persistent cloudless skies (Bennartz et al., 2013) resulted in the surface melting over 97% of Greenland's ice-sheet area (Barry & Hall-McKim, 2018). Only in July 2012, the ice loss was about 400 to 500 billion tons (Nghiem et al., 2012; Tedesco et al., 2013). Based on the ice core data, such scale of ice melting event has only happened two times during the recent climate history of Greenland. One of these events was in 1889 (Clausen, Gundestrup, Johnsen, Bindschadler, & Zwally, 1988), and the other was 7 centuries earlier during the Medieval Climatic Anomaly (Meese et al., 1994). Surprisingly, the total ice lost by both melting and disintegrating of icebergs in 2012 was almost offset by the new snow accumulation during 2013, as a comparatively cold year (Bevis et al., 2019; Rajewicz & Marshall, 2014). However, Greenland is generally experiencing significant warming and losing ice trends from the 1990s, while the GRACE data also show large variability in the ice sheet's mass balance (Vaughan et al., 2013).

Significant increases in atmospheric GHG concentrations play a crucial role in the current continued warming trends in summertime surface air temperature across Greenland (Box, 2013; Hanna, Mernild, Cappelen, & Steffen, 2012). Since 2003, the frequency of warmer surface air arriving from the south to Greenland, particularly its western parts, has been increased (Fettweis et al., 2013). As the surface air temperature gradient between the middle latitudes and the Arctic is decreasing, the strength of the jet stream weakens. Hence, the blocking high-pressure zone can develop across Greenland (Hanna, Cropper, Hall, & Cappelen, 2016). This consequently results in cloudless skies, high incoming solar energy, minimum snowfall, and a warm influx of airflow from the south of Greenland (van den Broeke et al., 2017).

In a complex way, clouds can also play an important role in the intensity of surface ice melting in Greenland (Bennartz et al., 2013; Van Tricht et al., 2016). For example, the snow cover on the highest plateau of the Greenland Ice Sheet sometimes melts during summertime (Ryan et al., 2019), when clouds consisting of a particular combination of ice crystal drift and water drops (Shupe et al., 2013). This intensifies the absorption features and optical density of the atmosphere, and thereby amplifies the long-wave heat radiation that accelerates ice melting processes (Bennartz et al., 2013). On the other hand, recent studies indicated that the absence of clouds lets the rays of the Sun fall on the Greenland Ice Sheet almost unobstructed. This accordingly empowers the ice melting in Greenland, particularly throughout the marginal parts with lower albedo (Box et al., 2012; Leeson, Shepherd, Palmer, Sundal, & Fettweis, 2012).

Almost 90% of Greenland's surface is naturally covered by snow. A fresh, fine-grained, and dry layer of snow can reflect up to 85% of short-wave solar energy (Konzelmann & Ohmura, 1995). Hence, the Sun's radiation heat poses very little threat to the Greenland Ice Sheet in the presence of snow cover (van den Broeke et al., 2017). When the snow merely starts to melt, its crystals become wet, clump together and grow. This changes the optical properties of snow crystals and consequently reduces the power of snow cover reflectivity. Accordingly, the snow grains absorb higher light energy in the long-wave infrared range (Box et al., 2012). Hence, such added solar heat energy develops the clumping of wet snow crystals, darkens the snow cover, increases the absorption of warmth, and consequently intensifies the rate of melting.

Even with an annual snowfall of about 2000 gigatons over the Antarctic glaciers and ice sheets, the two ice sheets in Antarctica are losing their masses (Gardner et al., 2018; McMillan et al., 2014; Shepherd et al., 2012). About 10% of such snow is generally lost because of surface melting, evaporation, wind transport, and sublimation. The other 90% is compressed to firn and consequently to ice (Gardner et al., 2018). In 2018, scientists reported that the Antarctic annually lost about 76 gigatons of ice during 1992–2011 (Shepherd, Fricker, & Farrell, 2018; Shepherd, Ivins, et al., 2018). It added about 0.2 (mm/year) to the global sea-level rising trend (Shepherd, Fricker, et al., 2018; Shepherd, Ivins, et al., 2018). However, since 2012, the annual ice loss throughout the Antarctic has almost tripled (219 gigatons) (Shepherd, Fricker, et al., 2018; Shepherd, Ivins, et al., 2018). During 1992–1997, the glaciers and ice streams in West Antarctica annually transported about 53 gigatons more ice into the Southern Ocean than was being developed by snowfall on the ice sheet (Shepherd, Fricker, et al., 2018; Shepherd, Ivins, et al., 2018). However, this amount was tripled (159 gigatons/year) during 2012–2017 (Shepherd, Fricker, et al., 2018; Shepherd, Ivins, et al., 2018).

The ice mass in West Antarctica has largely been lost since the late 2000s when both Pine Island Glacier and Thwaites Glacier started to flow much faster into the Amundsen Sea (Mouginot et al., 2014). Hence, the warm ocean currents rising from below in this sea melt the ice shelves in front of the glaciers (Shepherd, Ivins, et al., 2018). In the Antarctic Peninsula, the northernmost part of Antarctica, about one-third of ice shelves have collapsed during recent decades (Shepherd, Ivins, et al., 2018). Three ice shelves have lost as much as 70% of their area (Shepherd, Ivins, et al., 2018). Accordingly, the rate of ice-mass loss on the Antarctic Peninsula increased to about 25 gigatons per year. Contrarily, the balance between ice growth and loss (5 ± 46 gigatons/year) was almost neutral in East Antarctica during 1992–2017 (Shepherd, Ivins, et al., 2018).

The analysis of GRACE data determined that the loss rate of ice in the Antarctic continent is now about 127 gigatons per year (Wang, Davis, & Howat, 2021). The western area of the Antarctic Peninsula, the coastal regions in West Antarctica, and both Wilkes Land and Adélie Land in East Antarctica (Fig. 12.13) are experiencing the greatest ice losses (Groh et al., 2019; Wang et al., 2021). The growth of the ice sheet is however recording in the southern reaches of West Antarctica and the northern areas of Queen Maud Land (Sasgen et al., 2013).

In Antarctica, significant increases in ice losses (Hogg & Gudmundsson, 2017) are directly related to the thinning or even complete vanishing of the ice-shelf areas (Pritchard et al., 2012; Pritchard, Arthern, Vaughan, & Edwards, 2009; Scambos et al., 2009). The once-massive ice tongues are becoming lighter, shorter, and narrower, and consequently less able to resist the push of the inland ice from behind. The ice sheets are also more unstable today, particularly due to two different processes (Adusumilli et al., 2018; Bevan et al., 2017; Lai et al., 2020): (1) basal melting in response to significant increases in the temperature of the ocean currents that influence the underside parts of the ice shelf, and (2) melting on the upper part of ice surface due to warmer air temperatures. Such surface meltwaters can deepen different cracks and crevices in the ice mass, and thereby increase the risk of icebergs breaking off (Liu et al., 2015).

Until the 1990s, the Larsen B Ice Shelf was one of five almost contiguous ice shelves extended about 200 km (Larsen C) into the Weddell Sea throughout the eastern coasts of the Antarctic Peninsula. In 1995, the Prince Gustav Ice Shelf and the Larsen A Ice Shelf were broken up (Pritchard et al., 2009, 2012; Scambos et al., 2009). These two northernmost segments were followed 6 years later (2002) by the Larsen B Ice Shelf in the northwest of Weddell Sea (Banwell et al., 2014; Glasser & Scambos, 2008; Scambos, Hulbe, & Fahnestock, 2003). A 3250 km^2 floating ice broke into millions of smaller pieces and consequently disappeared within a single month during the Antarctic summertime of 2002 (Pritchard et al., 2009, 2012; Scambos, Bohlander, Shuman, & Skvarca, 2004). The two southernmost segments (Larsen C and D) still exist today (Baumhoer, Dietz, Kneisel, Paeth, & Kuenzer, 2021), while a huge iceberg (5800 km^2) broke off from Larsen C in July 2017 (Benn & Åström, 2018).

Along the Antarctic Peninsula, there were originally 12 ice shelves fed off the glaciers, forming in the mountainous regions (Cook & Vaughan, 2010). For these ice masses, the precipitation over the peninsula was the only source. Such

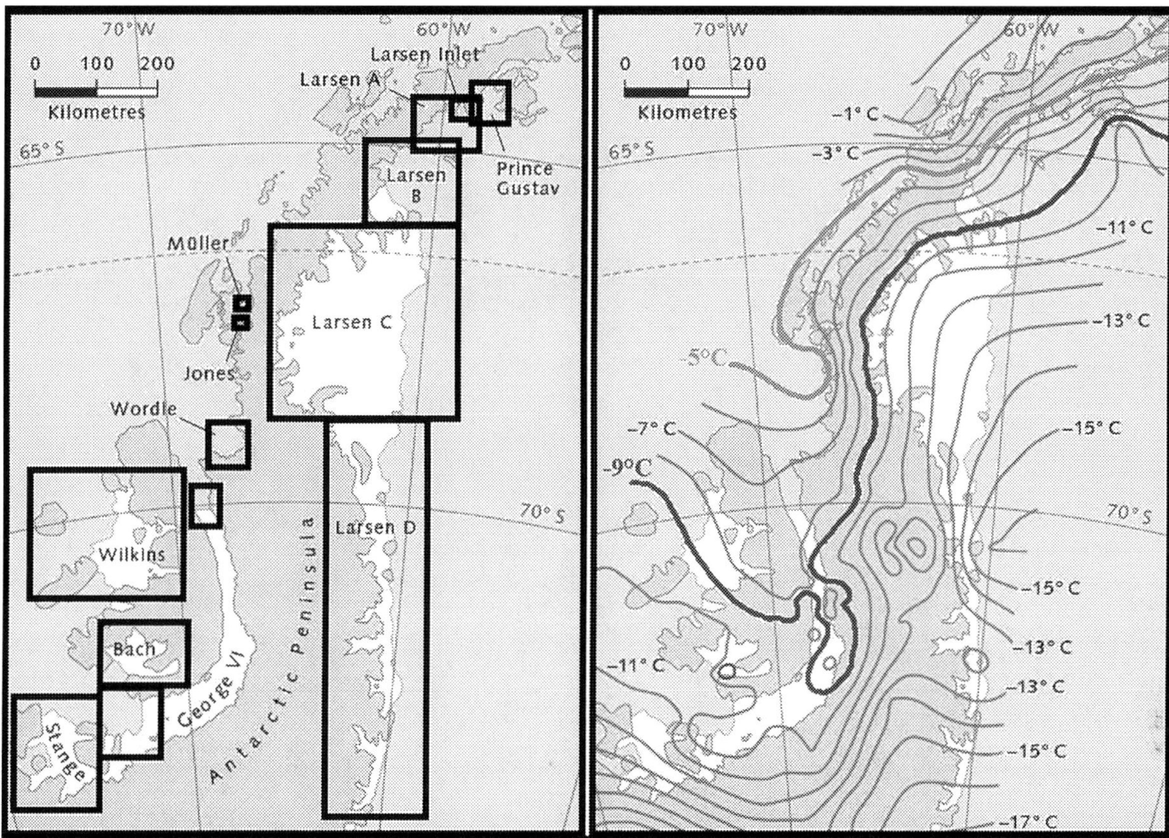

FIG. 12.13 Since the 1960s, 7 of 12 ice shelves along the Antarctic Peninsula have experienced substantially large ice depletions. The Jones, Wordie, Prince Gustav, Larsen A ice shelves have completely been disintegrated. Their disappearance was mainly triggered by significantly warmer surface air temperatures (up to 3°C). Since that time, ice shelves only occur in parts of the peninsula where the average annual temperature is less than −9°C. *Adopted from Morris, E., & Vaughan, D. (2003). Spatial and temporal variation of surface temperature on the antarctic peninsula and the limit of viability of ice shelves. In* Antarctic research series of the American Geophysical Union *(Vol. 79, pp. 61–68). https://doi.org/10.1029/AR079p0061.*

glaciers, and those that remain today, were not connected to the West or East Antarctic Ice Sheets in any way. On average, the thickness of ice shelves throughout the Antarctic Peninsula is about 200 to 250 m (Scambos, Hulbe, Fahnestock, & Bohlander, 2000). It is considerably thinner than the Filchner-Ronne or Ross Ice Shelves that both transport ice masses out of the inner parts of Antarctica. Besides, the Antarctic Peninsula is the northernmost, and thus, warmest region of Antarctica. Its surface air temperature has averagely increased by 3.5 ± 0.8°C per century since 1958 (Steig et al., 2009; Turner et al., 2005; Vaughan et al., 2003). In response to such regional warming, the stability of ice shelves in the Antarctic Peninsula has significantly been impacted since the 1950s (Mulvaney et al., 2012; Rignot, Jacobs, Mouginot, & Scheuchl, 2013).

During the 1970s, the British glaciologist John H. Mercer recognized that the ice shelves only exist throughout those parts of Antarctica where the average annual surface air temperature is not warmer than −5°C (Mercer, 1978). For a long time, in fact, there was such a −5°C isotherm for different ice shelves throughout Antarctica (Cook & Vaughan, 2010). Under regional warming, the northern boundary of these ice shelves has steadily been shifting further to the southern parts of the Antarctic Peninsula during recent decades, seriously impacting the ice-shelf areas that today (previously) lied on the north (south) of −5°C isothermal (Cook & Vaughan, 2010). Hence, such areas started to melt persistently due to the warmer summers and foehn winds (Cape et al., 2015; Gagliardini, 2018; Lenaerts et al., 2017). The consequential meltwater flows into the cracks and crevices in the ice shelves, increases the weight of water, and thereby intensifies the hydrostatic pressure at the bottom of the individual fractures and consequently deepens them (Veen., 2007). In winter, the meltwater in the fractures froze again, eventually permeating the whole ice shelf, and consequently increase the risk of the iceberg breaking off (Bell, Banwell, Trusel, & Kingslake, 2018; Macayeal & Sergienko, 2013).

The stability of different ice shelves was also affected by a load of returning meltwater lakes (Bell et al., 2017) (Fig. 12.14). Based on measurements, the collection of meltwaters in a depression on the ice shelf can bend the underlying ice downward by as much as 1 m by the massive water weight. In summer, however, the lake suddenly empties, and thus

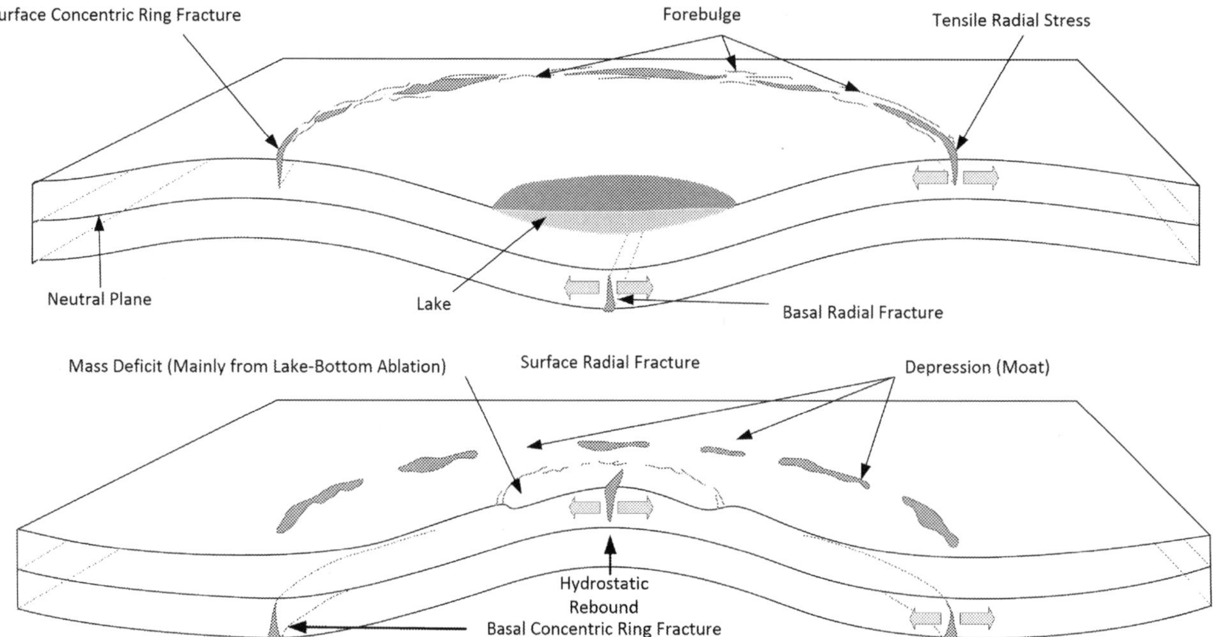

FIG. 12.14 There would be a pond or a lake when meltwater is collected on an ice shelf. The ice surface slowly yields under its weight and forms a depression. By the drainage of water in the lake, possibly through the ice fractures, the depression rebounds. This develops a ring of concentric fractures around the lake margin and facilitates the breakup of the ice shelf over time. *Adopted from Banwell, A. F., MacAyeal, D. R., & Sergienko, O. V. (2013). Breakup of the Larsen B Ice Shelf triggered by chain reaction drainage of supraglacial lakes.* Geophysical Research Letters, *40(22), 5872–5876. https://doi.org/10.1002/2013GL057694.*

the bulged part of the ice shelf rapidly returns to its original position (Banwell & MacAyeal, 2015; Bell et al., 2018). This motion forms a small ring-shaped fracture throughout the ice around the lake and its center. If water from lakes subsequently leaks into this fracture, it would continue to develop, and thereby increase the breaking risk (Banwell & MacAyeal, 2015; Bell et al., 2018; Macayeal & Sergienko, 2013). During the Antarctic summer 2001/2002, such domino reactions of melting, bending, cracking, and deepening probably resulted in the drainage of more than 2000 closely lying meltwater lakes (Glasser & Scambos, 2008) throughout the Larsen B Ice Shelf (Banwell, MacAyeal, & Sergienko, 2013).

But the warming surface air temperature has fundamentally played the most important role in the destruction of ice shelves throughout the Antarctic Peninsula (Mulvaney et al., 2012; Scambos et al., 2000; Steig et al., 2009). The ice shelves in the northern areas of the peninsula initially retreated for several decades, but then, collapsed one by one coo (Cook & Vaughan, 2010; Rignot et al., 2013; Shepherd, Ivins, et al., 2018). First, the Wordie Ice Shelf on the western coastal areas of the peninsula was collapsed during the 1980s (Cook & Vaughan, 2010). Later, the Prince Gustav Ice Shelf on the east coast was demolished in 1995, followed by Larsen A in 1995 (Scambos et al., 2000) and Larsen B in 2002 (Banwell et al., 2014; Glasser & Scambos, 2008; Scambos et al., 2003). Recently, the Jones Ice Shelf near the Arrowsmith Peninsula (2003) (Fox & Vaughan, 2005) and large parts of the Wilkins Ice Shelf (2008) (Braun & Humbert, 2009; Braun, Humbert, & Moll, 2009) were devastated. Based on the satellite images, there were numerous blue meltwater ponds throughout the Larsen B ice shelf shortly before its collapse (Banwell et al., 2013, 2014). It can confirm the connections between the warmer surface air temperature and the disintegration of ice shelves.

One year after the destruction of the Larsen B ice shelf, researchers analyzed the surface air temperature over the Antarctic Peninsula at the time. Accordingly, no large-scale surface melting was only observed throughout the ice shelves laying south of −9°C isotherm. To the north of this isothermal line, all ice shelves had either shrunk expansively or even disintegrated completely. Since that time, hence, the isotherm defining an annual mean of −9°C has been considered as the new northern limit for the presence of ice shelves along the Antarctic Peninsula (Morris & Vaughan, 2003).

2.6 The meltwater pathways

Since 1908, in the expedition of Ernest Shackelton's Nimrod, the meltwater streams flowing through Nansen Ice Shelf were witnessed by the explorers (David & Priestley, 1909). It helped humans to understand that ice shelves and glaciers in the Antarctic sometimes melt at the surface. In 1912, while the members of a British expedition mapping the Nansen Ice Shelf, they were frequently forced to paddle through meltwater streams (Campbell, 1988; Priestley, 1915). On a few occasions, their tents were also flooded (Levick, 1912).

According to the satellite observations, aerial photographs, and mass-balance models, the significance of ice melting and loss in Antarctica is higher than what humanity previously supposed. Nearly 700 different networks of meltwater lakes and streams were transporting liquid water throughout all ice shelves of Antarctica in 2017 (Kingslake, Ely, Das, & Bell, 2017). The highest surface-melting rates on the present ice shelves are today being recorded along the Antarctic Peninsula (Banwell et al., 2013; Glasser & Scambos, 2008). The meltwater is also changing the surface of ice shelves in the southern part of East Antarctica (Lenaerts, Vizcaino, Fyke, van Kampenhout, & van den Broeke, 2016; Trusel, Frey, & Das, 2012; Trusel, Frey, Das, Munneke, & Van Den Broeke, 2013). On the Amery and King Baudouin Ice Shelves, the intensity of ice melting during the summers is at the level that networks of meltwater lakes and streams are observable over great distances (Kingslake et al., 2017; Phillips, 1998; Stokes, Sanderson, Miles, Jamieson, & Leeson, 2019). On the other hand, only minor evidence of melting has so far been seen on the Ross and Filchner-Ronne Ice Shelves (Kingslake et al., 2017). In Antarctica, melting events are experiencing in all the ice areas at elevations below 1400 m (Nicolas, 2016; Trusel et al., 2013). This is basically related to the fact that the temperature is still below the melting point in the higher elevations than 1400 m throughout Antarctica.

On Greenland, meltwater lakes are draining almost vertically into the interior parts of the ice sheet through moulins. Then, they flow toward the sea along the bottom of the ice body. Such a phenomenon has not yet been observed in Antarctica (Bell et al., 2018). The emptying of meltwater lakes is only known to occur on the floating ice masses (MacDonald, Banwell, & Macayeal, 2018). The empty lake basins look rather like large craters. The meltwater lakes that form directly on glaciers above a land surface normally freeze again during wintertime and are covered by snow. Such hidden lakes can completely freeze and form huge ice lenses if the surface air temperature becomes cold enough (Bell et al., 2018).

In response to global warming, 2–3 times more meltwater will be generated on Antarctic glaciers and ice shelves by 2050 than is the case today (Trusel et al., 2015). Such amount of liquid water will possibly impact the ice mass balance throughout the Antarctic in three ways (Bell et al., 2018) (Fig. 12.15):

If the surface ice melts, the liquid water will typically runoff, and thereby resulting in a thinner ice body.

The surface meltwater may also leak into the snow-firn layer of an ice sheet, forming water lenses below the ice surface that fundamentally change the dynamics of the Antarctic glaciers and ice sheets.

Due to the warmer temperatures and winds, even the ice shelves that have previously not been significantly influenced (Fig. 12.16) will commence producing more surface meltwater. This can substantially contribute to the disintegration of ice shelves, particularly in southern parts of the Antarctic.

For a long time, thus, the amount and importance of meltwater in the Antarctic and its ice sheets, respectively, will increase. In the future, hence, the already negative balance of ice mass in the southern polar region will therefore become even stronger (Trusel et al., 2015).

3 Conclusion

Global mean surface air temperature is significantly warming basically due to the considerable increases in atmospheric GHGs emissions during recent decades. The Earth's oceans are also warming following the absorption of 93% of additional heat. In both polar regions, such warmer air and water are particularly causing principal changes that are happening earlier and more evident throughout the Arctic than in the Antarctic. The warming trend is twice in the Arctic as compared to other parts of the world, primarily due to the closely interrelated interactions among atmosphere, land, sea, and ice throughout this region. This effect is commonly called "Arctic amplification," and is particularly predominant during the winter season.

Many parts of the Arctic region are experiencing substantially less snowfall, sea ice formation, and ice cover extent. The warmer climate is also thawing permanently frozen soils in the Arctic, particularly during summertime. Throughout the Arctic, the ice masses (Greenland Ice Sheet as well as the glaciers in Canada and Alaska) on land are losing more ice than is being substituted by new snowfall. Its primary reason is the melting of ice throughout the upper surfaces of ice masses as well as the underside of the ice tongues contacted with seawater.

In the southern polar region, the warmer climate was only pronounced by the breaking up of the ice shelves and decreasing of sea ice on the northern and western sides of the Antarctic Peninsula, respectively, during the end of the 21st century. No significant increases in temperature across all other parts of the southern polar region are principally explained by the cooling effects of the Antarctic ozone hole. However, the warming trends in the Southern Ocean have

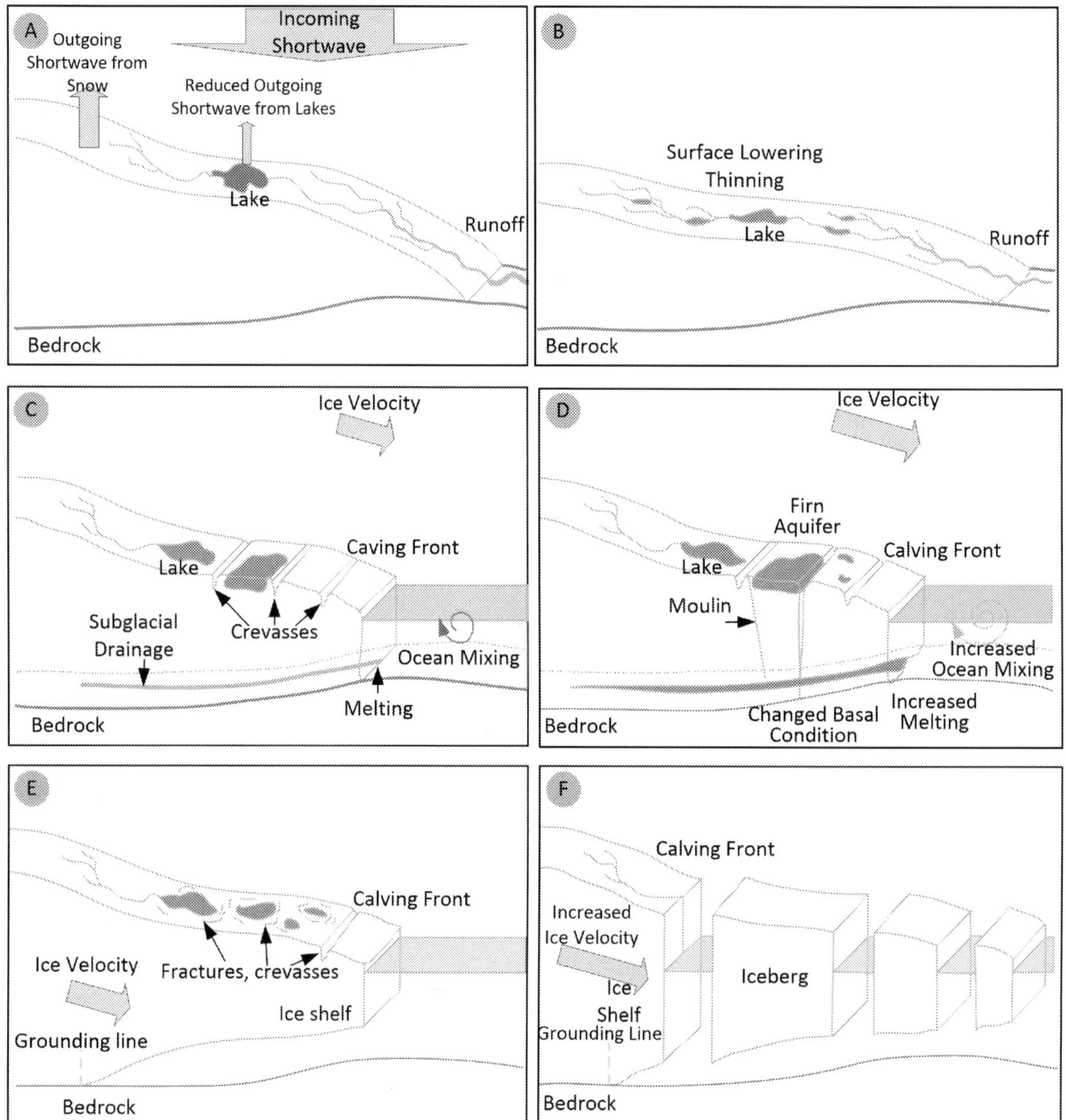

FIG. 12.15 If upper surfaces of glaciers and ice shelves start to melt, the presence of meltwater strengthens the further ice mass loss in three ways. (A and B) The darker surfaces of streams and lakes absorb more solar radiation heat than the brighter ice. Consequently, the water warms up and increases the rate of ice melting on the surface of the glacier. Hence, the ice body thins from its surface. (C and D) Meltwater might find its way toward the glacier's underside and thereby increases the ice mass flow velocity. (E and F) Meltwater on the surface of the ice shelf deepens existing fractures and develops new ones, increasing the risk of icebergs break off. *Adopted from Bell, R. E., Banwell, A. F., Trusel, L. D., & Kingslake, J. (2018). Antarctic surface hydrology and impacts on ice-sheet mass balance.* Nature Climate Change, *8(12), 1044–1052. https://doi.org/10.1038/s41558-018-0326-3.*

caused the loss of ice throughout West Antarctica. The ice shelf in the Amundsen Sea is melting from below in response to the penetration of warmer water from the Circumpolar Current. Today, a similar development is evident in East Antarctica, where the contact of the Totten Glacier with the bottom is losing and the warmer Weddell Sea is melting the 2nd largest ice shelf in the southern polar region. Since 2012, the rate of loss of ice mass has tripled in the Antarctic region. Although this has increased the contribution of the southern polar region to the global sea-level rise, its most important driver is the melting of the Greenland ice sheet and the glaciers outside of Antarctica. In general, such rising water levels are threatening the coastal parts of Earth.

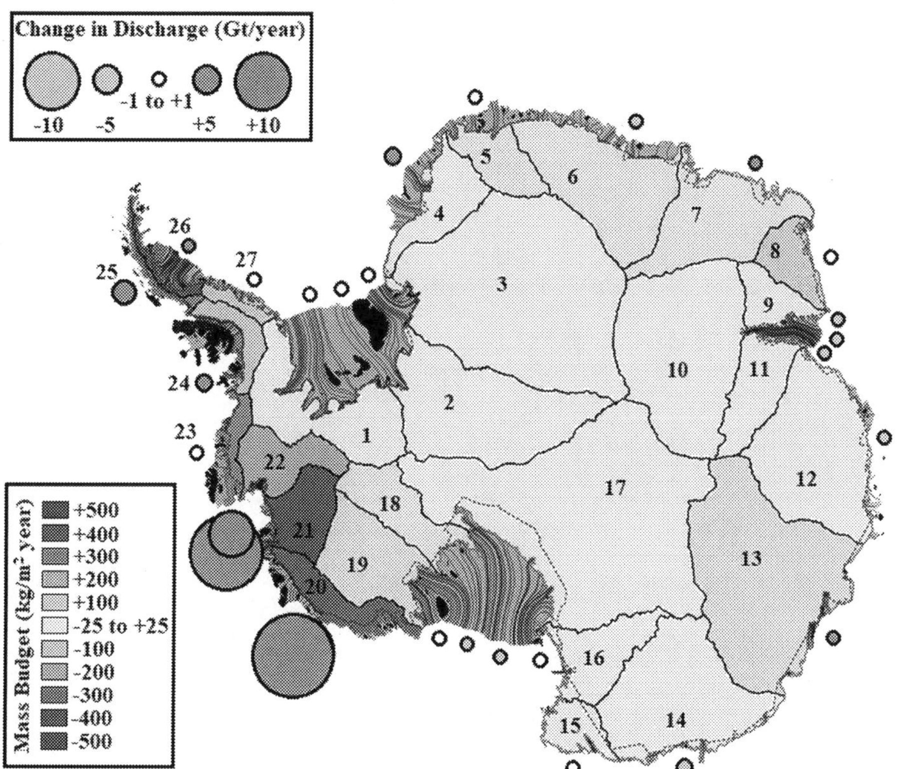

FIG. 12.16 During 2008–2015, the glaciers and ice shelves of West Antarctica lost substantially large amounts of ice because of both basal melting and calving of icebergs, while the loss of ice is not so discouraging in East Antarctica. In some areas, more snowfall could actually offset the ice lost by calving or melting. The Totten Glacier was the only exception. *Adopted from Gardner, A. S., Moholdt, G., Scambos, T., Fahnstock, M., Ligtenberg, S., Van Den Broeke, M., & Nilsson, J. (2018). Increased West Antarctic and unchanged East Antarctic ice discharge over the last 7 years.* Cryosphere, *12(2), 521–547. https://doi.org/10.5194/tc-12-521-2018.*

Acknowledgments

This work was partially financed by the Arctic Interactions research profile action supported by the University of Oulu (Finland) and the Academy of Finland PROFI4 (Grant No. 318930).

References

Abram, N., McGregor, H., Tierney, J., et al. (2016). Early onset of industrial-era warming across the oceans and continents. *Nature, 536*, 411–418. https://doi.org/10.1038/nature19082.

Abram, N., Mulvaney, R., Wolff, E., et al. (2013). Acceleration of snow melt in an Antarctic Peninsula ice core during the twentieth century. *Nature Geoscience, 6*, 404–411. https://doi.org/10.1038/ngeo1787.

Adusumilli, S., Fricker, H. A., Siegfried, M. R., Padman, L., Paolo, F. S., & Ligtenberg, S. R. M. (2018). Variable basal melt rates of Antarctic peninsula ice shelves, 1994–2016. *Geophysical Research Letters, 45*(9), 4086–4095. https://doi.org/10.1002/2017GL076652.

PAGES 2k Consortium. (2013). Continental-scale temperature variability during the past two millennia. *Nature Geosciences, 6*, 339–346. https://doi.org/10.1038/ngeo1797.

AMAP. (2019). *Arctic climate change update 2019*. Arctic Monitoring and Assessment Programme.

Armour, K. C., Marshall, J., Scott, J. R., Donohoe, A., & Newsom, E. R. (2016). Southern Ocean warming delayed by circumpolar upwelling and equatorward transport. *Nature Geoscience, 9*(7), 549–554. https://doi.org/10.1038/ngeo2731.

Banwell, A. F., Caballero, M., Arnold, N. S., Glasser, N. F., Cathles, L. M., & MacAyeal, D. R. (2014). Supraglacial lakes on the Larsen B ice shelf, Antarctica, and at Paakitsoq, West Greenland: A comparative study. *Annals of Glaciology, 55*(66), 1–8. https://doi.org/10.3189/2014AoG66A049.

Banwell, A. F., & MacAyeal, D. R. (2015). Ice-shelf fracture due to viscoelastic flexure stress induced by fill/drain cycles of supraglacial lakes. *Antarctic Science, 27*(6), 587–597. https://doi.org/10.1017/S0954102015000292.

Banwell, A. F., MacAyeal, D. R., & Sergienko, O. V. (2013). Breakup of the Larsen B ice shelf triggered by chain reaction drainage of supraglacial lakes. *Geophysical Research Letters, 40*(22), 5872–5876. https://doi.org/10.1002/2013GL057694.

Barry, R. G., & Hall-McKim, E. A. (2018). *Polar environments and global change*. Cambridge: Cambridge University Press.

Baumhoer, C. A., Dietz, A. J., Kneisel, C., Paeth, H., & Kuenzer, C. (2021). Environmental drivers of circum-Antarctic glacier and ice shelf front retreat over the last two decades. *The Cryosphere, 15*(5), 2357–2381. https://doi.org/10.5194/tc-15-2357-2021.

Bednarz, E. M., Maycock, A. C., Abraham, N. L., Braesicke, P., Dessens, O., & Pyle, J. A. (2016). Future Arctic ozone recovery: The importance of chemistry and dynamics. *Atmospheric Chemistry and Physics, 16*(18), 12159–12176. https://doi.org/10.5194/acp-16-12159-2016.

Bekryaev, R. V., Polyakov, I. V., & Alexeev, V. A. (2010). Role of polar amplification in long-term surface air temperature variations and modern arctic warming. *Journal of Climate*, *23*(14), 3888–3906. https://doi.org/10.1175/2010JCLI3297.1.

Bell, R. E., Banwell, A. F., Trusel, L. D., & Kingslake, J. (2018). Antarctic surface hydrology and impacts on ice-sheet mass balance. *Nature Climate Change*, *8*(12), 1044–1052. https://doi.org/10.1038/s41558-018-0326-3.

Bell, R., Chu, W., Kingslake, J., et al. (2017). Antarctic ice shelf potentially stabilized by export of meltwater in surface river. *Nature*, *544*, 344–348. https://doi.org/10.1038/nature22048.

Benn, D. I., & Åström, J. A. (2018). Calving glaciers and ice shelves. *Advances in Physics: X*, *3*(1), 1048–1076. https://doi.org/10.1080/23746149.2018.1513819.

Bennartz, R., Shupe, M., Turner, D., et al. (2013). July 2012 Greenland melt extent enhanced by low-level liquid clouds. *Nature*, *496*, 83–86. https://doi.org/10.1038/nature12002.

Beszczynska-Möller, A., Woodgate, R., Lee, C., Melling, H., & Karcher, M. (2011). A synthesis of exchanges through the main oceanic gateways to the Arctic Ocean. *Oceanography*, 82–99. https://doi.org/10.5670/oceanog.2011.59.

Bevan, S. L., Luckman, A., Hubbard, B., Kulessa, B., Ashmore, D., Kuipers Munneke, P., et al. (2017). Centuries of intense surface melt on Larsen C Ice Shelf. *The Cryosphere*, *11*, 2743–2753. https://doi.org/10.5194/tc-11-2743-2017.

Bevis, M., Harig, C., Khan, S. A., Brown, A., Simons, F. J., Willis, M., et al. (2019). Accelerating changes in ice mass within Greenland, and the ice sheet's sensitivity to atmospheric forcing. *Proceedings of the National Academy of Sciences*, *116*(6), 1934–1939. https://doi.org/10.1073/pnas.1806562116.

Bodas-Salcedo, A., Andrews, T., Karmalkar, A. V., & Ringer, M. A. (2016). Cloud liquid water path and radiative feedbacks over the Southern Ocean. *Geophysical Research Letters*, *43*(20), 10938–10946. https://doi.org/10.1002/2016gl070770.

Boisvert, L. N., Petty, A. A., & Stroeve, J. C. (2016). The impact of the extreme winter 2015/16 arctic cyclone on the Barents-Kara seas. *Monthly Weather Review*, *144*(11), 4279–4287. https://doi.org/10.1175/MWR-D-16-0234.1.

Boisvert, L. N., & Stroeve, J. C. (2015). The Arctic is becoming warmer and wetter as revealed by the atmospheric infrared sounder. *Geophysical Research Letters*, *42*(11), 4439–4446. https://doi.org/10.1002/2015GL063775.

Box, J. E. (2013). Greenland ice sheet mass balance reconstruction. Part II: Surface mass balance (1840-2010). *Journal of Climate*, *26*(18), 6974–6989. https://doi.org/10.1175/JCLI-D-12-00518.1.

Box, J. E., Fettweis, X., Stroeve, J. C., Tedesco, M., Hall, D. K., & Steffen, K. (2012). Greenland ice sheet albedo feedback: Thermodynamics and atmospheric drivers. *The Cryosphere*, *6*(4), 821–839. https://doi.org/10.5194/tc-6-821-2012.

Box, J. E., Yang, L., Bromwich, D. H., & Bai, L. S. (2009). Greenland ice sheet surface air temperature variability: 1840-2007. *Journal of Climate*, *22*(14), 4029–4049. https://doi.org/10.1175/2009JCLI2816.1.

Braun, M., & Humbert, A. (2009). Recent retreat of wilkins ice shelf reveals new insights in ice shelf breakup mechanisms. *IEEE Geoscience and Remote Sensing Letters*, *6*(2), 263–267. https://doi.org/10.1109/LGRS.2008.2011925.

Braun, M., Humbert, A., & Moll, A. (2009). Changes of Wilkins ice shelf over the past 15 years and inferences on its stability. *The Cryosphere*, *3*(1), 41–56. https://doi.org/10.5194/tc-3-41-2009.

Brook, E. J., & Buizert, C. (2018). Antarctic and global climate history viewed from ice cores. *Nature*, *558*(7709), 200–208. https://doi.org/10.1038/s41586-018-0172-5.

Caesar, L., Rahmstorf, S., Robinson, A., Feulner, G., & Saba, V. (2018). Observed fingerprint of a weakening Atlantic Ocean overturning circulation. *Nature*, *556*(7700), 191–196. https://doi.org/10.1038/s41586-018-0006-5.

Campbell, V. (1988). *The wicked mate: The Antarctic diary of Victor Campbell*. Bluntisham Books.

Cape, M. R., Vernet, M., Skvarca, P., Marinsek, S., Scambos, T., & Domack, E. (2015). Foehn winds link climate-driven warming to ice shelf evolution in Antarctica. *Journal of Geophysical Research-Atmospheres*, *120*(21), 11037–11057. https://doi.org/10.1002/2015jd023465.

Cazenave, A., Meyssignac, B., Ablain, M., Balmaseda, M., Bamber, J., Barletta, V., et al. (2018). Global sea-level budget 1993-present. *Earth System Science Data*, *10*(3), 1551–1590.

Chambers, D. P. (2006). Observing seasonal steric sea level variations with GRACE and satellite altimetry. *Journal of Geophysical Research, Oceans*, *111*(3). https://doi.org/10.1029/2005JC002914.

Clausen, H. B., Gundestrup, N. S., Johnsen, S. J., Bindschadler, R., & Zwally, J. (1988). Glaciological investigations in the Crête area, Central Greenland: A search for a new deep-drilling site. *Annals of Glaciology*, 10–15. https://doi.org/10.1017/S0260305500004080.

Cogley, J. G., Hock, R., Rasmussen, L. A., Arendt, A. A., Bauder, A., Braithwaite, R. J., et al. (2011). Recent Arctic amplification and extreme mid-latitude weather. *Arctic, Antarctic, and Alpine Research*, *44*(2), 627–637.

Cohen, J., Screen, J., Furtado, J., et al. (2014). Recent Arctic amplification and extreme mid-latitude weather. *Nature Geosciences*, *7*, 627–637. https://doi.org/10.1038/ngeo2234.

Convey, P. (2003). Maritime Antarctic climate change: Signals from terrestrial Biology. In: E.W. Domack, A. Leventer, A. Burnett, R. Bindschadler, P. Convey & W. Kirby (Eds.), *Antarctic Peninsula climate variability: Historical and palaeoenvironmental perspectives. Antarctic Research Series: Vol. 79* (pp. 145–158). Washington, D.C.: American Geophysical Union.

Cook, A. J., & Vaughan, D. G. (2010). Overview of areal changes of the ice shelves on the Antarctic Peninsula over the past 50 years. *The Cryosphere*, *4*(1), 77–98. https://doi.org/10.5194/tc-4-77-2010.

Crook, J. A., Forster, P. M., & Stuber, N. (2011). Spatial patterns of modeled climate feedback and contributions to temperature response and polar amplification. *Journal of Climate*, *24*(14), 3575–3592. https://doi.org/10.1175/2011JCLI3863.1.

Dameris, M. (2010). Depletion of the ozone layer in the 21st century. *Angewandte Chemie, International Edition*, *49*(3), 489–491. https://doi.org/10.1002/anie.200906334.

David, T. W. E., & Priestley, R. E. (1909). *Geological observations in Antarctica by the British Antarctic expedition*. JP Lippincott.

Delworth, T. L., & Zeng, F. (2018). Simulated impact of altered southern hemisphere winds on the Atlantic meridional overturning circulation. *Geophysical Research Letters*, *35*.

Dessler, A. E., Zhang, Z., & Yang, P. (2003). Water–vapor climate feedback inferred from climate fluctuations. *Geophysical Research Letters*, *35*.

Dethloff, K., Handorf, D., Jaiser, R., Rinke, A., & Klinghammer, P. (2019). Dynamical mechanisms of Arctic amplification. *Annals of the New York Academy of Sciences*, *1436*(1), 184–194. https://doi.org/10.1111/nyas.13698.

Durack, P. J., & Wijffels, S. E. (2010). Fifty-year trends in global ocean salinities and their relationship to broad-scale warming. *Journal of Climate, 23*(16), 4342–4362. https://doi.org/10.1175/2010JCLI3377.1.

Enderlin, E. M., Howat, I. M., Jeong, S., Noh, M. J., Van Angelen, J. H., & Van Den Broeke, M. R. (2014). An improved mass budget for the Greenland ice sheet. *Geophysical Research Letters, 41*(3), 866–872. https://doi.org/10.1002/2013GL059010.

Fettweis, X., Hanna, E., Lang, C., Belleflamme, A., Erpicum, M., & Gallée, H. (2013). Brief communication important role of the mid-tropospheric atmospheric circulation in the recent surface melt increase over the Greenland ice sheet. *The Cryosphere, 7*(1), 241–248. https://doi.org/10.5194/tc-7-241-2013.

Fox, A. J., & Vaughan, D. G. (2005). The retreat of Jones ice shelf, Antarctic peninsula. *Journal of Glaciology, 51*(175), 555–560. https://doi.org/10.3189/172756505781829043.

Frölicher, T. L., Sarmiento, J. L., Paynter, D. J., Dunne, J. P., Krasting, J. P., & Winton, M. (2015). Dominance of the Southern Ocean in anthropogenic carbon and heat uptake in CMIP5 models. *Journal of Climate, 28*(2), 862–886. https://doi.org/10.1175/JCLI-D-14-00117.1.

Fyfe, J. (2015). Southern Ocean warming due to human influence. *Geophysical Research Letters, 33.*

Gagliardini, O. (2018). The health of Antarctic ice shelves. *Nature Climate Change, 8.*

Gardner, A. S., Moholdt, G., Scambos, T., Fahnstock, M., Ligtenberg, S., Van Den Broeke, M., et al. (2018). Increased West Antarctic and unchanged East Antarctic ice discharge over the last 7 years. *The Cryosphere, 12*(2), 521–547. https://doi.org/10.5194/tc-12-521-2018.

Gille, S. T. (2002). Warming of the Southern Ocean since the 1950s. *Science, 295*(5558), 1275–1277. https://doi.org/10.1126/science.1065863.

Gille, S. T. (2008). Decadal-scale temperature trends in the Southern Hemisphere ocean. *Journal of Climate, 21*(18), 4749–4765. https://doi.org/10.1175/2008JCLI2131.1.

Gjelten, H. M., Nordli, Ø, Isaksen, K., Førland, E. J., Sviashchennikov, P. N., Wyszyński, P., et al. (2016). Air temperature variations and gradients along the coast and fjords of western Spitsbergen. *Polar Research, 35*, 29878. https://doi.org/10.3402/polar.v35.29878.

Glasser, N. F., & Scambos, T. A. (2008). A structural glaciological analysis of the 2002 Larsen B ice-shelf collapse. *Journal of Glaciology, 54*(184), 3–16. https://doi.org/10.3189/002214308784409017.

Gonzalez, S., & Vasallo, F. (2020). *Antarctic Climates* (pp. 595–605). Elsevier BV. https://doi.org/10.1016/b978-0-12-409548-9.11876-7.

Goosse, H., Kay, J. E., Armour, K. C., et al. (2018). Quantifying climate feedbacks in polar regions. *Nature Communications, 9*, 1919. https://doi.org/10.1038/s41467-018-04173-0.

Gordon, N. D., Jonko, A. K., Forster, P. M., & Shell, K. M. (2013). An observationally based constraint on the water-vapor feedback. *Journal of Geophysical Research-Atmospheres, 118*(22), 12,435–12,443. https://doi.org/10.1002/2013jd020184.

Graham, R. M., Cohen, L., Petty, A. A., Boisvert, L. N., Rinke, A., Hudson, S. R., et al. (2017). Increasing frequency and duration of Arctic winter warming events. *Geophysical Research Letters, 44*(13), 6974–6983.

Groh, A., Horwath, M., Horvath, A., Meister, R., Sørensen, L. S., Barletta, V. R., et al. (2019). Evaluating GRACE mass change time series for the Antarctic and Greenland Ice Sheet—Methods and results. *Geosciences, 9*, 415. https://doi.org/10.3390/geosciences9100415.

Grosfeld, K., Treffeisen, R., Asseng, J., Bartsch, A., Bräuer, B., Fritzsch, B., et al. (2016). Online sea-ice knowledge and data platform <www.meereisportal.de>, Polarforschung, Bremerhaven. *Alfred Wegener Institute for Polar and Marine Research & German Society of Polar Research, 85*(2), 143–155. https://doi.org/10.2312/polfor.2016.011.

Hall, A. (2004). The role of surface albedo feedback in climate. *Journal of Climate, 17*(7), 1550–1568. https://doi.org/10.1175/1520-0442(2004)017<1550:TROSAF>2.0.CO;2.

Hanna, E., Cropper, T. E., Hall, R. J., & Cappelen, J. (2016). Greenland blocking index 1851–2015: A regional climate change signal. *International Journal of Climatology, 36*(15), 4847–4861. https://doi.org/10.1002/joc.4673.

Hanna, E., Fettweis, X., Mernild, S. H., Cappelen, J., Ribergaard, M. H., Shuman, C. A., et al. (2014). Atmospheric and oceanic climate forcing of the exceptional Greenland ice sheet surface melt in summer 2012. *International Journal of Climatology, 34*, 1022–1037. https://doi.org/10.1002/joc.3743.

Hanna, E., Mernild, S. H., Cappelen, J., & Steffen, K. (2012). Recent warming in Greenland in a long-term instrumental (1881–2012) climatic context: I. Evaluation of surface air temperature records. *Environmental Research Letters, 7*(4). https://doi.org/10.1088/1748-9326/7/4/045404, 045404.

Hersbach, H., Bell, B., Berrisford, P., Hirahara, S., Horányi, A., Muñoz-Sabater, J., et al. (2020). The ERA5 global reanalysis. *Quarterly Journal of the Royal Meteorological Society, 146*(730), 1999–2049.

Hobbs, W. R., Massom, R., Stammerjohn, S., Reid, P., Williams, G., & Meier, W. (2016). A review of recent changes in Southern Ocean sea ice, their drivers and forcings. *Global and Planetary Change, 143*, 228–250. https://doi.org/10.1016/j.gloplacha.2016.06.008.

Hogg, A., Meredith, M. P., Blundel, J. R., & Wilson, C. (2008). Eddy heat flux in the Southern Ocean: Response to variable wind forcing. *Journal of Climate, 21*(4), 608–620. https://doi.org/10.1175/2007JCLI1925.1.

Hogg, A. E., & Gudmundsson, G. H. (2017). Impacts of the Larsen-C ice shelf calving event. *Nature Climate Change*, 540–542. https://doi.org/10.1038/nclimate3359.

Holland, P. R. (2014). The seasonality of Antarctic sea ice trends. *Geophysical Research Letters, 41*(12), 4230–4237. https://doi.org/10.1002/2014GL060172.

Holliday, N. P., Hughes, S., Bacon, S., Beszczynska-Möller, A., Hansen, B., Lavin, A., et al. (2008). Reversal of the 1960s to 1990s freshening trend in the northeast north atlantic and nordic seas. *Geophysical Research Letters, 35*. https://doi.org/10.1029/2007GL032675.

IPCC. (2013). Climate change 2013: The physical science basis. In T. F. Stocker, D. Qin, G.-K. Plattner, M. Tignor, S. K. Allen, J. Boschung, et al. (Eds.), *Contribution of working group I to the fifth assessment report of the intergovernmental panel on climate change* (1535 pp). Cambridge, UK/New York, NY: Cambridge University Press.

IPCC. (2014). Climate change 2014: Impacts, adaptation, and vulnerability. Part B: Regional aspects. In V. R. Barros, C. B. Field, D. J. Dokken, M. D. Mastrandrea, K. J. Mach, T. E. Bilir, et al. (Eds.), *Contribution of working group II to the fifth assessment report of the intergovernmental panel on climate change* (p. 688). Cambridge, UK/New York, NY: Cambridge University Press.

IPCC. (2018). Global warming of 1.5°C. In V. Masson-Delmotte, P. Zhai, H.-O. Pörtner, D. Roberts, J. Skea, P.R. Shukla, et al. *An IPCC Special Report on the impacts of global warming of 1.5°C above pre-industrial levels and related global greenhouse gas emission pathways, in the context of strengthening the global response to the threat of climate change, sustainable development, and efforts to eradicate poverty* (in press).

Jahn, A., Kay, J. E., Holland, M. M., & Hall, D. M. (2016). How predictable is the timing of a summer ice-free Arctic? *Geophysical Research Letters, 43*(17), 9113–9120. https://doi.org/10.1002/2016GL070067.

Jaiser, R., Dethloff, K., Handorf, D., Rinke, A., & Cohen, J. (2012). Impact of sea ice cover changes on the northern hemisphere atmospheric winter circulation. *Tellus Series A: Dynamic Meteorology and Oceanography*, *64*(1). https://doi.org/10.3402/tellusa.v64i0.11595.

Jenkins, A., Dutrieux, P., Jacobs, S., Steig, E. J., Gudmundsson, G. H., Smith, J., & Heywood, K. J. (2016). Decadal ocean forcing and Antarctic ice sheet response: Lessons from the Amundsen Sea. *Oceanography*, *29*(4), 106–117. https://doi.org/10.5670/oceanog.2016.103.

Jullion, L., Garabato, A. C. N., Meredith, M. P., Holland, P. R., Courtois, P., & King, B. A. (2013). Decadal freshening of the antarctic bottom water exported from the weddell sea. *Journal of Climate*, *26*(20), 8111–8125. https://doi.org/10.1175/JCLI-D-12-00765.1.

King, J., Turner, J., Marshall, G., Connolley, W., & Lachlan-Cope, T. (2003). Antarctic peninsula climate variability and its causes as revealed by instrumental records. In *Vol. 79. AGUAdv. Res. Series* (pp. 17–30). https://doi.org/10.1029/AR079p0017.

Kingslake, J., Ely, J. C., Das, I., & Bell, R. E. (2017). Widespread movement of meltwater onto and across Antarctic ice shelves. *Nature*, *544*(7650), 349–352. https://doi.org/10.1038/nature22049.

Konzelmann, T., & Ohmura, A. (1995). Radiative fluxes and their impact on the energy balance of the Greenland ice sheet. *Journal of Glaciology*, *41*(139), 490–502. https://doi.org/10.1017/S0022143000034833.

Kornfeld, R. P., Arnold, B. W., Gross, M. A., Dahya, N. T., Klipstein, W. M., Gath, P. F., & Bettadpur, S. (2019). GraCE-FO: The gravity recovery and climate experiment follow-on mission. *Journal of Spacecraft and Rockets*, *56*(3), 931–951. American Institute of Aeronautics and Astronautics Inc https://doi.org/10.2514/1.A34326.

Lai, C. Y., Kingslake, J., Wearing, M. G., et al. (2020). Vulnerability of Antarctica's ice shelves to meltwater-driven fracture. *Nature*, *584*, 574–578. https://doi.org/10.1038/s41586-020-2627-8.

Langematz, U., Meul, S., Grunow, K., Romanowsky, E., Oberländer, S., Abalichin, J., & Kubin, A. (2014). Future arctic temperature and ozone: The role of stratospheric composition changes. *Journal of Geophysical Research*, *119*(5), 2092–2112. https://doi.org/10.1002/2013JD021100.

Lecomte, O., Goosse, H., Fichefet, T., De Lavergne, C., Barthélemy, A., & Zunz, V. (2017). Vertical ocean heat redistribution sustaining sea-ice concentration trends in the Ross Sea. *Nature Communications*, *8*(1). https://doi.org/10.1038/s41467-017-00347-4.

Lee, S. K., Volkov, D. L., Lopez, H., Cheon, W. G., Gordon, A. L., Liu, Y., & Wanninkhof, R. (2017). Wind-driven ocean dynamics impact on the contrasting sea-ice trends around West Antarctica. *Journal of Geophysical Research, Oceans*, *122*(5), 4413–4430. https://doi.org/10.1002/2016JC012416.

Leeson, A. A., Shepherd, A., Palmer, S., Sundal, A., & Fettweis, X. (2012). Simulating the growth of supraglacial lakes at the western margin of the Greenland ice sheet. *The Cryosphere*, *6*(5), 1077–1086. https://doi.org/10.5194/tc-6-1077-2012.

Lenaerts, J., Lhermitte, S., Drews, R., et al. (2017). Meltwater produced by wind–albedo interaction stored in an East Antarctic ice shelf. *Nature Climate Change*, *7*, 58–62. https://doi.org/10.1038/nclimate318.

Lenaerts, J. T. M., Vizcaino, M., Fyke, J., van Kampenhout, L., & van den Broeke, M. R. (2016). Present-day and future Antarctic ice sheet climate and surface mass balance in the community earth system model. *Climate Dynamics*, *47*(5–6), 1367–1381. https://doi.org/10.1007/s00382-015-2907-4.

Levick, G. M. (1912). *British Antarctic expedition journal (Scott polar research institute archive catalogue No. MS1423/1-4)*.

Levitus, S., Antonov, J. I., Boyer, T. P., Baranova, O. K., Garcia, H. E., Locarnini, R. A., et al. (2012). World ocean heat content and thermosteric sea level change (0–2000 m), 1955–2010. *Geophysical Research Letters*, *39*(10).

Liu, Y., Moore, J. C., Cheng, X., Gladstone, R. M., Bassis, J. N., Liu, H., et al. (2015). Ocean-driven thinning enhances iceberg calving and retreat of Antarctic ice shelves. *Proceedings of the National Academy of Sciences*, *112*(11), 3263–3268.

Luthcke, S. B., Sabaka, T. J., Loomis, B. D., Arendt, A. A., McCarthy, J. J., & Camp, J. (2013). Antarctica, Greenland and Gulf of Alaska land-ice evolution from an iterated GRACE global mascon solution. *Journal of Glaciology*, *59*(216), 613–631. https://doi.org/10.3189/2013JoG12J147.

Macayeal, D. R., & Sergienko, O. V. (2013). The flexural dynamics of melting ice shelves. *Annals of Glaciology*, *54*(63), 1–10. https://doi.org/10.3189/2013AoG63A256.

MacDonald, G. J., Banwell, A. F., & Macayeal, D. R. (2018). Seasonal evolution of supraglacial lakes on a floating ice tongue, Petermann Glacier, Greenland. *Annals of Glaciology*, *59*(76), 56–65. https://doi.org/10.1017/aog.2018.9.

Marshall, J. (2014). The ocean's role in polar climate change: Asymmetric arctic and Antarctic responses to greenhouse gas and ozone forcing. *Philosophical Transactions of The Royal Society A Mathematical Physical and Engineering Sciences*, *372*, 20130040.

Matear, R. J., O'Kane, T. J., Risbey, J. S., & Chamberlain, M. (2015). Sources of heterogeneous variability and trends in Antarctic sea-ice. *Nature Communications*, *6*. https://doi.org/10.1038/ncomms9656.

McMillan, M., Shepherd, A., Sundal, A., Briggs, K., Muir, A., Ridout, A., et al. (2014). Increased ice losses from Antarctica detected by CryoSat-2. *Geophysical Research Letters*, *41*, 3899–3905. https://doi.org/10.1002/2014gl060111.

Meese, D. A., Gow, A. J., Grootes, P., Stuiver, M., Mayewski, P. A., Zielinski, G. A., et al. (1994). The accumulation record from the GISP2 core as an indicator of climate change throughout the Holocene. *Science*, *266*, 1680–1682.

Mercer, J. H. (1978). West Antarctic ice sheet and CO2 greenhouse effect: A threat of disaster. *Nature*, *271*(5643), 321–325. https://doi.org/10.1038/271321a0.

Moore, G. W. K. (2016). The December 2015 North Pole warming event and the increasing occurrence of such events. *Scientific Reports*, *6*. https://doi.org/10.1038/srep39084.

Moore, G. W. K., Schweiger, A., Zhang, J., & Steele, M. (2018). What caused the remarkable February 2018 North Greenland polynya? *Geophysical Research Letters*, *45*(24), 13–350. https://doi.org/10.1029/2018GL080902.

Morris, E., & Vaughan, D. (2003). Spatial and temporal variation of surface temperature on the Antarctic peninsula and the limit of viability of ice shelves. *Antarctic Research Series of the American Geophysical Union: Vol. 79* (pp. 61–68). https://doi.org/10.1029/AR079p0061.

Mouginot, J., Rignot, E., Bjørk, A. A., Van den Broeke, M., Millan, R., Morlighem, M., et al. (2019). Forty-six years of Greenland Ice Sheet mass balance from 1972 to 2018. *Proceedings of the National Academy of Sciences*, *116*(19), 9239–9244.

Mouginot, J., Rignot, E., & Scheuchl, B. (2014). Sustained increase in ice discharge from the Amundsen Sea embayment, West Antarctica, from 1973 to 2013. *Geophysical Research Letters*, *41*(5), 1576–1584. https://doi.org/10.1002/2013GL059069.

Mulvaney, R., Abram, N., Hindmarsh, R., et al. (2012). Recent Antarctic Peninsula warming relative to Holocene climate and ice-shelf history. *Nature*, *489*, 141–144. https://doi.org/10.1038/nature11391.

National Academies of Sciences, Engineering, and Medicine. (2017). *Antarctic sea ice variability in the southern ocean-climate system: Proceedings of a workshop*. Washington, DC: The National Academies Press. https://doi.org/10.17226/24696.

Nghiem, S. V., Hall, D. K., Mote, T. L., Tedesco, M., Albert, M. R., Keegan, K., et al. (2012). The extreme melt across the Greenland Ice Sheet in 2012. *Geophysical Research Letters*, *39*, L20502. https://doi.org/10.1029/2012GL053611.

Nicolas, J. P. (2016). Extensive summer melt in West Antarctica favoured by strong El Niño. *Nature Communications, 8*.

NSIDC. (2014). Newsroom. In *Arctic sea ice continues low; Antarctic ice it's a new high*. https://nsidc.org/news/newsroom/arctic-sea-ice-continues-low-while-antarctic-reaches-new-record-high.

NSIDC. (2016). *Antarctic sea ice reaches winter maximum on a record early date*. http://nsidc.org/arcticseaicenews/2016/.

NSIDC. (2019a). *Arctic people*. The National Snow and Ice Data Center.

NSIDC. (2019b). *A record-low start to the new year in Antarctica*. http://nsidc.org/arcticseaicenews/2019/01/a-record-low-start-to-the-new-year-in-antarctica/.

Nymand Larsen, J. (2014). *Arctic human development report: Regional processes and global linkages*. Nordic Council of Ministers.

Orsi, A. H., Whitworth, T., & Nowlin, W. D. (1995). On the meridional extent and fronts of the Antarctic circumpolar current. *Deep-Sea Research Part I, 42*(5), 641–673. https://doi.org/10.1016/0967-0637(95)00021-W.

Overland, J., Dunlea, E., Box, J. E., Corell, R., Forsius, M., Kattsov, V., et al. (2019). The urgency of Arctic change. *Polar Science, 21*, 6–13.

Overland, J. E., & Wang, M. (2016). Recent extreme arctic temperatures are due to a split polar vortex. *Journal of Climate, 29*(15), 5609–5616. https://doi.org/10.1175/JCLI-D-16-0320.1.

Parkinson, C. L. (2019). A 40-y record reveals gradual Antarctic sea ice increases followed by decreases at rates far exceeding the rates seen in the Arctic. *Proceedings of the National Academy of Sciences of the United States of America, 116*(29), 14414–14423. https://doi.org/10.1073/pnas.1906556116.

Parkinson, C. L., & Cavalieri, D. J. (2012). Antarctic sea ice variability and trends, 1979–2010. *The Cryosphere, 6*(4), 871–880. https://doi.org/10.5194/tc-6-871-2012.

Perovich, D., Meier, W., Tschudi, M., et al. (2018). Sea ice. In *Arctic report card: Update for 2018*. https://arctic.noaa.gov/Report-Card/Report-Card-2018/ArtMID/7878/ArticleID/780/SeanbspIce.

Petty, A. A., Stroeve, J. C., Holland, P. R., Boisvert, L. N., Bliss, A. C., Kimura, N., & Meier, W. N. (2018). The Arctic sea ice cover of 2016: A year of record-low highs and higher-than-expected lows. *The Cryosphere, 12*(2), 433–452. https://doi.org/10.5194/tc-12-433-2018.

Phillips, H. A. (1998). Surface meltstreams on the Amery ice shelf, East Antarctica. *Annals of Glaciology, 27*, 177–181. https://doi.org/10.3189/1998AoG27-1-177-181.

Pithan, F., & Mauritsen, T. (2014). Arctic amplification dominated by temperature feedbacks in contemporary climate models. *Nature Geoscience, 7*(3), 181–184. https://doi.org/10.1038/ngeo2071.

Pitulko, V. V., Tikhonov, A. N., Pavlova, E. Y., Nikolskiy, P. A., Kuper, K. E., & Polozov, R. N. (2016). Early human presence in the Arctic: Evidence from 45,000-year-old mammoth remains. *Science, 351*(6270), 260–263. https://doi.org/10.1126/science.aad0554.

Priestley, R. E. (1915). *Antarctic adventure: Scott's northern party*. E.P. Dutton. 382 pp.

Pritchard, H. D., Arthern, R. J., Vaughan, D. G., & Edwards, L. A. (2009). Extensive dynamic thinning on the margins of the Greenland and Antarctic ice sheets. *Nature, 461*(7266), 971–975. https://doi.org/10.1038/nature08471.

Pritchard, H. D., Ligtenberg, S. R. M., Fricker, H. A., Vaughan, D. G., Van Den Broeke, M. R., & Padman, L. (2012). Antarctic ice-sheet loss driven by basal melting of ice shelves. *Nature, 484*(7395), 502–505. https://doi.org/10.1038/nature10968.

Purich, A., England, M. H., Cai, W., Chikamoto, Y., Timmermann, A., Fyfe, J. C., et al. (2016). Tropical Pacific SST drivers of recent Antarctic sea ice trends. *Journal of Climate, 29*(24), 8931–8948.

Purkey, S. G., & Johnson, G. C. (2010). Warming of global abyssal and deep Southern Ocean waters between the 1990s and 2000s: Contributions to global heat and sea level rise budgets. *Journal of Climate, 23*(23), 6336–6351. https://doi.org/10.1175/2010JCLI3682.1.

Purkey, S. G., & Johnson, G. C. (2012). Global contraction of Antarctic bottom water between the 1980s and 2000s. *Journal of Climate, 25*(17), 5830–5844. https://doi.org/10.1175/JCLI-D-11-00612.1.

Rainville, L., Lee, C. M., & Woodgate, R. A. (2011). Impact of wind-driven mixing in the Arctic Ocean. *Oceanography, 24*(3), 136–145. https://doi.org/10.5670/oceanog.2011.65.

Rajewicz, J., & Marshall, S. J. (2014). Variability and trends in anticyclonic circulation over the Greenland ice sheet, 1948-2013. *Geophysical Research Letters, 41*(8), 2842–2850. https://doi.org/10.1002/2014GL059255.

Randel, W. J., & Wu, F. (1999). Cooling of the Arctic and Antarctic polar stratospheres due to ozone depletion. *Journal of Climate, 12*(5), 1467–1479. https://doi.org/10.1175/1520-0442(1999)012<1467:cotaaa>2.0.co;2.

Rignot, E., Jacobs, S., Mouginot, J., & Scheuchl, B. (2013). Ice-shelf melting around Antarctica. *Science, 341*(6143), 266–270. https://doi.org/10.1126/science.1235798.

Rogozhina, I., Petrunin, A., Vaughan, A., et al. (2016). Melting at the base of the Greenland ice sheet explained by Iceland hotspot history. *Nature Geoscience, 9*, 366–369. https://doi.org/10.1038/ngeo2689.

Rowland, F. S. (2006). Stratospheric ozone depletion. *Philosophical Transactions of the Royal Society, B: Biological Sciences, 361*(1469), 769–790. https://doi.org/10.1098/rstb.2005.1783.

Ryan, J., Smith, L., van As, D., Cooley, S., Cooper, M., Pitcher, L., & Hubbard, A. (2019). Greenland ice sheet surface melt amplified by snowline migration and bare ice exposure. *Science Advances, 5*, eaav3738. https://doi.org/10.1126/sciadv.aav3738.

Sasgen, I., Konrad, H., Ivins, E. R., Van den Broeke, M. R., Bamber, J. L., Martinec, Z., & Klemann, V. (2013). Antarctic ice-mass balance 2003 to 2012: Regional reanalysis of GRACE satellite gravimetry measurements with improved estimate of glacial-isostatic adjustment based on GPS uplift rates. *The Cryosphere, 7*(5), 1499–1512. https://doi.org/10.5194/tc-7-1499-2013.

Sasgen, I., Wouters, B., Gardner, A. S., et al. (2020). Return to rapid ice loss in Greenland and record loss in 2019 detected by the GRACE-FO satellites. *Communications Earth & Environment, 1*, 8. https://doi.org/10.1038/s43247-020-0010-1.

Scambos, T., Fricker, H. A., Liu, C. C., Bohlander, J., Fastook, J., Sargent, A., et al. (2009). Ice shelf disintegration by plate bending and hydro-fracture: Satellite observations and model results of the 2008 Wilkins ice shelf break-ups. *Earth and Planetary Science Letters, 280*(1-4), 51–60.

Scambos, T. A., Bohlander, J. A., Shuman, C. A., & Skvarca, P. (2004). Glacier acceleration and thinning after ice shelf collapse in the Larsen B embayment, Antarctica. *Geophysical Research Letters, 31*(18), L18402–L18404. https://doi.org/10.1029/2004GL020670.

Scambos, T. A., Hulbe, C., Fahnestock, M., & Bohlander, J. (2000). The link between climate warming and break-up of ice shelves in the Antarctic Peninsula. *Journal of Glaciology, 46*(154), 516–530. https://doi.org/10.3189/172756500781833043.

Scambos, T., Hulbe, C., & Fahnestock, M. (2003). Climate-induced ice shelf disintegration in the Antarctic Peninsula. In E. Domack, A. Levente, A. Burnet, R. Bindschadler, P. Convey, & M. Kirby (Eds.), *Antarctic peninsula climate variability: Historical and paleoenvironmental perspectives*. Washington, D.C.: American Geophysical Union. https://doi.org/10.1029/AR079p0079.

Schauer, U., Beszczynska-Möller, A., Walczowski, W., Fahrbach, E., Piechura, J., & Hansen, E. (2008). *Variation of measured heat flow through the Fram strait between 1997 and 2006* (pp. 65–85). Springer Science and Business Media LLC. https://doi.org/10.1007/978-1-4020-6774-7_4.

Scheffer, M., Brovkin, V., & Cox, P. M. (2006). Positive feedback between global warming and atmospheric CO2 concentration inferred from past climate change. *Geophysical Research Letters*, *33*(10). https://doi.org/10.1029/2005GL025044.

Semmler, T., Stulic, L., Jung, T., Tilinina, N., Campos, C., Gulev, S., & Koracin, D. (2016). Seasonal atmospheric responses to reduced arctic sea ice in an ensemble of coupled model simulations. *Journal of Climate*, *29*(16), 5893–5913. https://doi.org/10.1175/JCLI-D-15-0586.1.

Serreze, M. C., & Barry, R. G. (2011). Processes and impacts of Arctic amplification: A research synthesis. *Global and Planetary Change*, *77*(1–2), 85–96. https://doi.org/10.1016/j.gloplacha.2011.03.004.

Seviour, W. J. M., Gnanadesikan, A., Waugh, D., & Pradal, M. A. (2017). Transient response of the Southern Ocean to changing ozone: Regional responses and physical mechanisms. *Journal of Climate*, *30*(7), 2463–2480. https://doi.org/10.1175/JCLI-D-16-0474.1.

Shepherd, A., Fricker, H. A., & Farrell, S. L. (2018). Trends and connections across the Antarctic cryosphere. *Nature*, *558*(7709), 223–232. https://doi.org/10.1038/s41586-018-0171-6.

Shepherd, A., Ivins, E. R., Geruo, A., Barletta, V. R., Bentley, M. J., Bettadpur, S., et al. (2012). A reconciled estimate of ice-sheet mass balance. *Science*, *338*(6111), 1183–1189. https://doi.org/10.1126/science.1228102.

Shepherd, A., Ivins, E., Rignot, E., Smith, B., Van Den Broeke, M., Velicogna, I., et al. (2018). Mass balance of the Antarctic Ice Sheet from 1992 to 2017. *Nature*, *558*, 219–222.

Sheshadri, A., Alan Plumb, R., & Domeisen, D. I. V. (2014). Can the delay in antarctic polar vortex breakup explain recent trends in surface westerlies? *Journal of the Atmospheric Sciences*, *71*(2), 566–573. https://doi.org/10.1175/JAS-D-12-0343.1.

Sheshadri, A., & Plumb, R. A. (2016). Sensitivity of the surface responses of an idealized AGCM to the timing of imposed ozone depletion-like polar stratospheric cooling. *Geophysical Research Letters*, *43*(5), 2330–2336. https://doi.org/10.1002/2016GL067964.

Shupe, M. D., Persson, P. O. G., Brooks, I. M., Tjernström, M., Sedlar, J., Mauritsen, T., Sjogren, S., & Leck, C. (2013). Cloud and boundary layer interactions over the Arctic sea ice in late summer. *Atmos. Chem. Phys.*, *13*, 9379–9399. https://doi.org/10.5194/acp-13-9379-2013.

Sigmond, M., Reader, M. C., Fyfe, J. C., & Gillett, N. P. (2011). Drivers of past and future Southern Ocean change: Stratospheric ozone versus greenhouse gas impacts. *Geophysical Research Letters*, *38*(12). https://doi.org/10.1029/2011GL047120.

Simpkins, G. R., Ciasto, L. M., & England, M. H. (2013). Observed variations in multidecadal Antarctic sea ice trends during 1979-2012. *Geophysical Research Letters*, *40*(14), 3643–3648. https://doi.org/10.1002/grl.50715.

Solomon, A., Polvani, L. M., Smith, K. L., & Abernathey, R. P. (2015). The impact of ozone depleting substances on the circulation, temperature, and salinity of the Southern Ocean: An attribution study with CESM1(WACCM). *Geophysical Research Letters*, *42*(13), 5547–5555. https://doi.org/10.1002/2015GL064744.

Stammerjohn, S., Massom, R., Rind, D., & Martinson, D. (2012). Regions of rapid sea ice change: An inter-hemispheric seasonal comparison. *Geophysical Research Letters*, *39*(6). https://doi.org/10.1029/2012GL050874.

Steig, E. J., Schneider, D. P., Rutherford, S. D., Mann, M. E., Comiso, J. C., & Shindell, D. T. (2009). Warming of the Antarctic ice-sheet surface since the 1957 international geophysical year. *Nature*, *457*(7228), 459–462. https://doi.org/10.1038/nature07669.

Stokes, C. R., Sanderson, J. E., Miles, B. W. J., Jamieson, S. S. R., & Leeson, A. A. (2019). Widespread distribution of supraglacial lakes around the margin of the East Antarctic ice sheet. *Scientific Reports*, *9*(1). https://doi.org/10.1038/s41598-019-50343-5.

Stolarski, R. S., Schoeberl, M. R., Newman, P. A., McPeters, R. D., & Krueger, A. J. (1990). The 1989 Antarctic ozone hole as observed by TOMS. *Geophysical Research Letters*, *17*(9), 1267–1270. https://doi.org/10.1029/GL017i009p01267.

Stuecker, M. F., Bitz, C. M., & Armour, K. C. (2017). Conditions leading to the unprecedented low Antarctic sea ice extent during the 2016 austral spring season. *Geophysical Research Letters*, *44*(17), 9008–9019. https://doi.org/10.1002/2017GL074691.

Swart, N. C., Gille, S. T., Fyfe, J. C., & Gillett, N. P. (2018). Recent Southern Ocean warming and freshening driven by greenhouse gas emissions and ozone depletion. *Nature Geoscience*, *11*(11), 836–841. https://doi.org/10.1038/s41561-018-0226-1.

Tapley, B. D., Bettadpur, S., Watkins, M., & Reigber, C. (2004). The gravity recovery and climate experiment: Mission overview and early results. *Geophysical Research Letters*, *31*(9). https://doi.org/10.1029/2004GL019920, L09607-4.

Taylor, P. C., Cai, M., Hu, A., Meehl, J., Washington, W., & Zhang, G. J. (2013). A decomposition of feedback contributions to polar warming amplification. *Journal of Climate*, *26*(18), 7023–7043. https://doi.org/10.1175/JCLI-D-12-00696.1.

Tedesco, M., Fettweis, X., Mote, T., Wahr, J., Alexander, P., Box, J. E., & Wouters, B. (2013). Evidence and analysis of 2012 Greenland records from spaceborne observations, a regional climate model and reanalysis data. *The Cryosphere*, *7*(2), 615–630. https://doi.org/10.5194/tc-7-615-2013.

Thompson, D. W. J., Solomon, S., Kushner, P. J., England, M. H., Grise, K. M., & Karoly, D. J. (2011). Signatures of the Antarctic ozone hole in Southern Hemisphere surface climate change. *Nature Geoscience*, *4*(11), 741–749. https://doi.org/10.1038/ngeo1296.

Truffer, M., & Motyka, R. J. (2016). Where glaciers meet water: Subaqueous melt and its relevance to glaciers in various settings. *Reviews of Geophysics*, *54*(1), 220–239. https://doi.org/10.1002/2015RG000494.

Trusel, L. D., Frey, K. E., & Das, S. B. (2012). Antarctic surface melting dynamics: Enhanced perspectives from radar scatterometer data. *Journal of Geophysical Research - Earth Surface*, *117*(2). https://doi.org/10.1029/2011JF002126.

Trusel, L. D., Frey, K. E., Das, S. B., Karnauskas, K. B., Kuipers Munneke, P., Van Meijgaard, E., & Van Den Broeke, M. R. (2015). Divergent trajectories of Antarctic surface melt under two twenty-first-century climate scenarios. *Nature Geoscience*, *8*(12), 927–932. https://doi.org/10.1038/ngeo2563.

Trusel, L. D., Frey, K. E., Das, S. B., Munneke, P. K., & Van Den Broeke, M. R. (2013). Satellite-based estimates of Antarctic surface meltwater fluxes. *Geophysical Research Letters*, *40*(23), 6148–6153. https://doi.org/10.1002/2013GL058138.

Turner, J. (2003). Antarctic climate. In J. R. Holton, J. Pyle, & J. A. Curry (Eds.), *Encyclopedia of atmospheric sciences* (pp. 135–142). Academic Press.

Turner, J., Colwell, S. R., Marshall, G. J., Lachlan-Cope, T. A., Carleton, A. M., Jones, P. D., et al. (2005). Antarctic climate change during the last 50 years. *International Journal of Climatology*, *25*(3), 279–294.

Turner, J., Hosking, J. S., Bracegirdle, T. J., Marshall, G. J., & Phillips, T. (2015). Recent changes in Antarctic sea ice. *Philosophical Transactions of the Royal Society A: Mathematical, Physical and Engineering Sciences*, *373*(2045). https://doi.org/10.1098/rsta.2014.0163.

Turner, J., Lu, H., White, I., et al. (2016). Absence of 21st century warming on Antarctic Peninsula consistent with natural variability. *Nature*, *535*, 411–415. https://doi.org/10.1038/nature18645.

Turner, J., Orr, A., Gudmundsson, G. H., Jenkins, A., Bingham, R. G., Hillenbrand, C. D., & Bracegirdle, T. J. (2017). Atmosphere-ocean-ice interactions in the Amundsen Sea Embayment, West Antarctica. *Reviews of Geophysics, 55*(1), 235–276. https://doi.org/10.1002/2016RG000532.

van den Broeke, M., Box, J., Fettweis, X., et al. (2017). Greenland Ice Sheet surface mass loss: Recent developments in observation and modeling. *Current Climate Change Reports, 3*, 345–356. https://doi.org/10.1007/s40641-017-0084-8.

van der Leun, J. C., Tang, X., & Tevini, M. (1995). Environmental effects of ozone depletion: 1994 assessment. *Ambio, 24*(3), 138. http://www.jstor.org/stable/4314318.

Van Tricht, K., Lhermitte, S., Lenaerts, J., et al. (2016). Clouds enhance Greenland ice sheet meltwater runoff. *Nature Communications, 7*, 10266. https://doi.org/10.1038/ncomms10266.

Vaughan, C., Carrasco, A. I., Kaser, K., et al. (2013). Observations: Cryosphere. In *Climate change 2013: The physical science basis. Contribution of working group I to the Fifth Assessment report of the Intergovernmental Panel on Climate Change.*

Vaughan, D. G., Marshall, G. J., Connolley, W. M., et al. (2003). Recent rapid regional climate warming on the Antarctic Peninsula. *Climatic Change, 60*, 243–274. https://doi.org/10.1023/A:1026021217991.

Veen. (2007). Fracture propagation as means of rapidly transferring surface meltwater to the base of glaciers. *Geophysical Research Letters, 34.*

Velicogna, I. (2009). Increasing rates of ice mass loss from the Greenland and Antarctic ice sheets revealed by GRACE. *Geophysical Research Letters, 36*(19). https://doi.org/10.1029/2009GL040222.

Wang, L., Davis, J. L., & Howat, I. M. (2021). Complex patterns of Antarctic ice sheet mass change resolved by time-dependent rate modeling of GRACE and GRACE follow-on observations. *Geophysical Research Letters, 48*(1). https://doi.org/10.1029/2020GL090961.

Williams, R. G., Roussenov, V., Smith, D., & Lozier, M. S. (2014). Decadal evolution of ocean thermal anomalies in the North Atlantic: The effects of ekman, overturning, and horizontal transport. *Journal of Climate, 27*(2), 698–719. https://doi.org/10.1175/JCLI-D-12-00234.1.

Winton, M. (2006). Surface albedo feedback estimates for the AR4 climate models. *Journal of Climate, 19*(3), 359–365. https://doi.org/10.1175/JCLI3624.1.

World Meteorological Organization. (2007). Scientific assessment of ozone depletion: 2006. In *Global ozone research and monitoring project report no. 52, Geneva, Switzerland.*

World Meteorological Organization. (2011). Scientific assessment of ozone depletion: 2010. In *Global ozone research and monitoring project report no. 52, Geneva, Switzerland.*

World Meteorological Organization. (2014). Scientific assessment of ozone depletion: 2014. In *Global ozone research and monitoring project-report no. 55, Geneva, Switzerland.*

World Meteorological Organization. (2018). Scientific assessment of ozone depletion: 2018. In *Global ozone research and monitoring project–report no. 58, Geneva, Switzerland.*

World Ocean Review 6. (2019). *The Arctic and Antarctic-extreme, climatically crucial and in crisis.* Hamburg, Germany: maribus gGmbH, Pickhuben. 2, 20457.

Wu, Z., & Wang, X. (2019). Variability of Arctic Sea ice (1979–2016). *Water, 11*(1), 23. https://doi.org/10.3390/w11010023.

Zahn, R. (2009). Beyond the CO2 connection. *Nature, 460*(7253), 335–336. https://doi.org/10.1038/460335a.

Zanna, L., Khatiwala, S., Gregory, J. M., Ison, J., & Heimbach, P. (2019). Global reconstruction of historical ocean heat storage and transport. *Proceedings of the National Academy of Sciences of the United States of America, 116*(4), 1126–1131. https://doi.org/10.1073/pnas.1808838115.

Zhang, L., Delworth, T. L., Cooke, W., & Yang, X. (2019). Natural variability of Southern Ocean convection as a driver of observed climate trends. *Nature Climate Change, 9*(1), 59–65. https://doi.org/10.1038/s41558-018-0350-3.

Zhang, L., Delworth, T. L., & Zeng, F. (2017). The impact of multidecadal Atlantic meridional overturning circulation variations on the Southern Ocean. *Climate Dynamics, 48*(5–6), 2065–2085. https://doi.org/10.1007/s00382-016-3190-8.

Zhang, Y., Li, J., & Zhou, L. (2017). The relationship between polar Vortex and ozone depletion in the Antarctic stratosphere during the period 1979-2016. *Advances in Meteorology, 2017.* https://doi.org/10.1155/2017/3078079.

Zhao, J., Barber, D., Zhang, S., Yang, Q., Wang, X., & Xie, H. (2018). Record low sea-ice concentration in the Central Arctic during summer 2010. *Advances in Atmospheric Sciences*, 106–115. https://doi.org/10.1007/s00376-017-7066-6.

CHAPTER

13

The degradation of the Amazon rainforest: Regional and global climate implications

Kerry W. Bowman[a], *Samuel A. Dale*[b], *Sumana Dhanani*[b], *Jevithen Nehru*[b], *and Benjamin T. Rabishaw*[b]

[a]Faculty of Medicine, School of the Environment, University of Toronto, Toronto, ON, Canada
[b]University of Toronto, Toronto, ON, Canada

HIGHLIGHTS

- A deteriorating Amazon has massive climatic and weather implications on a global scale; however, uncertainty remains in predicting the exact nature and timeline of these effects.
- Intentional fire is one of the greatest drivers of the current paradigm of degradation.
- Increasing deforestation and rising global temperatures are pushing this rainforest toward a tipping point after which the ecosystem will transform to savannah.
- Zoonotic diseases are more likely to emerge from a deteriorating Amazon.
- Respecting and restoring Indigenous stewardship over the Amazon is one of the most effective methods of maintaining an intact ecosystem.
- Action is required at the individual, national, and international level to prevent further degradation.

1 Introduction

The Amazon Rainforest is the greatest expression of life on Earth. Nourished by an equatorial cycle of constant warmth, humidity and rainfall, the Amazon region has been the genesis of some of the most complex and spectacular ecosystems on our planet. Yet here, in the early 21st century, it is under siege, and its deterioration holds significant consequences for climate and weather across the globe.

The recent fires in the Amazon region are unprecedented in their destruction and scale. Deforestation of primary forest via intentional fire is widespread, and deforestation has been on an upward trend since around 2014 (Butler, 2020; The Nature Conservancy, 2019). By the end of 2020, year-over-year deforestation in Brazil, which is the largest Amazon nation, exceeded 11,000 km^2, a 12-year high. This is an area nearly the size of Puerto Rico (US Department of Commerce, 2012). The Amazon fires received global attention (Voiland, 2020), and again they raged in 2020, but

Climate Impacts on Extreme Weather
https://doi.org/10.1016/B978-0-323-88456-3.00011-3

garnered little attention in a world gripped by a global pandemic (Global Fire Emissions Database, 2021). Despite fluctuations in public attention, the degradation of the Amazon Rainforest has profound consequences for the global climate, global health and a host of associated issues.

Intact tropical forests are foundational to maintaining planetary ecosystems and natural cycles, as they are key reservoirs of carbon and they keep many climate-influencing factors in balance. Yet the reality is that two-thirds of the world's original tropical rainforest cover has now been degraded or destroyed by people; 34% of the planet's old-growth tropical rainforests have been destroyed, and another 30% significantly degraded (Spring, 2021; Weisse & Goldman, 2020). More than half of this destruction since 2002 has been in the Amazon region of South America. This chapter will serve as a case study of the degradation of the Amazon Rainforest, and its multiple regional and global implications.

2 Methods

The authors conducted a comprehensive review of the literature using search engines and databases. Recent, peer-reviewed research papers relevant to the topics were sought, using search terms related to the Amazon Rainforest's global ecosystem services with regard to weather, climate and biodiversity; the consequences of its degradation; the importance of and particular risks faced by Indigenous groups; recommendations on how to preserve the Amazon. The papers selected were written in English between 2017 and 2021, with some exceptions. In some sections, web articles were reviewed to provide context on current events in the Amazon. All the claims in this text are informed by the principal writer's 8 years of fieldwork in the Amazon.

The discussion of the Amazon Rainforest will begin by explaining its vital ecosystem services on continental scales. While the Rainforest is significant in and of itself to protect and maintain, it is also a profoundly valuable biome for its multiple traditional key functions as a global carbon sink, a regulator of hydrological cycles, a refuge of wildlife, and a source of Indigenous cultural value. The fires and degradation of the Amazon hinder the Rainforest's ability to fulfill these vital roles to both South America and across the planet.

Next, the complex relationships between climate change and forest degradation through fires will be reviewed. This chapter will examine which mechanisms a changing climate can exacerbate fire in the Amazon biome, and which pathways anthropogenic fires in the region can contribute to climate change. The dangerous nature of the positive feedback loops will be explained, to further our understanding of how certain "tipping points" of degradation might be reached in our near future. These mechanisms are complicated, and make predicting what lies ahead for South America and the planet difficult. Yet, this chapter attempts to clarify general trends and consensus in the literature on what may happen to the Amazon Rainforest if degradation to the region continues.

The Amazon Rainforest is home to a stunning amount of biodiversity and losing this to forest fires may hold other unexpected drastic consequences for our future. The intricate pathways by which biodiversity, forest deforestation and zoonotic disease emergence are connected will be analyzed in the context of the Amazon. The various human health concerns stemming from the rising risk of zoonotic diseases cannot be ignored, especially in the Amazon.

Just as the elemental processes of a deteriorating Amazon are profoundly complex, so too are the local, regional, and geopolitical forces underpinning this assault. This complexity is owed to the fact that ecosystems are deeply influenced by political, social and cultural realities—not merely biological processes. While a discussion of human rights, geopolitics, colonialism and respect for all forms of life may seem out of place in a book otherwise devoted to climate and weather extremes, an analysis of these complex factors will help illuminate the nature of the siege on the Amazon biome as well as possible pathways to ameliorate the situation.

The story of the Indigenous Peoples and their lands is discussed to better our understanding of the long and complicated history of the Amazon biome. It is imperative to understand the history before taking a look at the current violent invasions of their lands, and the atrocities being committed against them. The people and the land are heavily intertwined, and when we lose the respect of the rights of one, we lose respect for both. Indigenous Protected Areas (IPAs) encompass huge tracts of pristine, primary rainforest and must be maintained for the protection of this biome (Blackman & Veit, 2018; Lovejoy & Nobre, 2018, 2019; Walker et al., 2009, 2020). This chapter will follow with a detailed conversation on our recommendations for building a more sustainable future—as Indigenous Peoples are foundational to the protection of the Amazon Rainforest. Solutions toward the protection of the Amazon and its importance to climate and weather are grounded in complex multifactorial initiatives which will be reviewed in our recommendations section.

This chapter will explore the degradation of the Amazon Rainforest and its implications for climate change. However, it is worth noting that the biogeochemistry of the Amazon Basin is profoundly complex. The factoring in of the geopolitical, economic, and social drivers of anthropogenic deforestation, along with the human rights concerns of the Indigenous Peoples of the Amazon, the situation becomes even more difficult to fully analyze. Regardless, much of the research done in this area points toward understanding that degrading the Rainforest will push it past certain tipping points from which it may never recover. Indeed, this has significant climate implications for both South America and the entire planet.

3 Discussion

3.1 The Amazon as a carbon sink

The massive amount of vegetation and biomass of the Amazon is excellent at storing carbon (Walker et al., 2020), as plant life exercises photosynthesis and sequesters CO_2 from the atmosphere. It is imperative to protect the Amazon Rainforest as it has acted as a vital carbon sink, but maintains the dangerous potential to turn into a carbon *faucet*. Under normal circumstances, there remains a homeostatic balance between the amount of atmospheric carbon that the Rainforest locks in and the amount of other greenhouse gases (GHGs) it emits at a baseline. For much of our contemporary history, the Rainforest has been thought to act as a net carbon sink (Rödig et al., 2018), absorbing more GHGs than it emits. This functions as a vital component of stabilizing and slowing the rapidly changing climate of the planet. Clearing the forest via fire and deforestation not only releases these locked GHGs, it also impedes on the Rainforest's capability to act as a carbon sink (Hubau et al., 2020). The reduction in forest cover has substantially reduced the Rainforest's potential for capturing and locking in CO_2 from the atmosphere.

When the Rainforest is ignited, it releases not only CO_2, but also several other potent GHGs, including methane and nitrous oxide (Covey et al., 2021). These compounds play significant roles in increasing global average surface temperatures (GAST) through the greenhouse effect. Particulate matter (PM), like black carbon from the fires, also contributes to atmospheric warming through the albedo effect, whereby darker surfaces absorb more radiant energy from the sun. Researchers have found particulates from fires in the Amazon as far as the Andean glaciers, where it deposits, darkening the glacial surface. This will lead to a decreased albedo of the snow cover and contribute to increased melting, altered hydrology, and a warmer atmosphere.

The consequences of a changing climate, including rising temperatures, altered precipitation patterns and changes in the atmosphere, can influence and reduce the efficiency of photosynthesis. This is a key factor of the ability of the Rainforest's biomass to sequester CO_2, among other potent GHGs. This shows the intricate relationship between climate change and deforestation and fire in the Amazon (Lovejoy & Nobre, 2019). Even without factoring in the burning and clearing of the Rainforest, the carbon sink capturing would be impeded upon. Further degradation to this ecosystem holds significant implications for our planet's future.

Disturbingly, recent research suggests that the Amazon has devolved into a net emitter of GHGs (Covey et al., 2021). This research points toward a general trend of decline in the Amazon's carbon storing abilities as a consequence of mass deforestation, degradation, and industrialization. The stability of the global climate is jeopardized by this transition of the Amazon Rainforest.

3.2 The Amazon's hydrology

The hundreds of billions of trees in the Amazon Rainforest release over 20 billion tons of water vapor into the atmosphere each day (Kedmey, 2015), through evapotranspiration. For most plant species, evapotranspiration is mostly controlled by atmospheric humidity and soil moisture content. The vast quantity of water vapor and moisture generated can be recycled back into the Amazon in the form of rainwater and precipitation. This cycle is a key aspect of how the Amazon is capable of watering itself and maintaining its wet biosphere (Ahlström et al., 2017). The biogenic volatile organic compounds (BVOCs) are central to this hydrological cycle and are generated and emitted by vegetation and biomass (Laothawornkitkul, Taylor, Paul, & Hewitt, 2009). While the exact roles of BVOCs are complex and difficult to clarify, they are known to influence and promote precipitation (Laothawornkitkul et al., 2009).

Deforestation will disrupt this process and hinder the Amazon's ability to emit BVOCs, resulting in shifted precipitation patterns and changes in water availability in the region. This will hold significant financial implications for the South American countries' economies which rely on agricultural exports. For example, data from 2019 shows us that soybean exports make up 11% of Brazil's total exports, and amounted to \$26 billion USD. Soybean farming is one of the key driving factors in Amazon land-use change and deforestation, yet losing forest cover may negatively impact it. The reduced forest cover and reduced precipitation levels can harm soybean and other agricultural product yields and productivity. The expected reduction in rainfall due to a 20%–25% loss of forest cover is estimated to cause hundreds of millions of dollars in losses in Brazil alone (Rogers, 2019; X-prize, 2020). The short term financial gains resulting from cutting down the forest will eventually cost the Amazonian nations far more.

The Amazon does not just produce water for itself, as the water vapor and intense quantities of moisture can be transported over incredible distances. Atmospheric anomalies, like Atmospheric Rivers (ARs) and Low-Level Jets (LLJs) have the capability to transport water vapor along with powerful winds (Gimeno, Nieto, Vázquez, & Lavers, 2014). These streams and anomalies can be found throughout the globe and are involved in contributing to precipitation levels and are even linked to flooding events (Aguilera, Gershunov, & Benmarhnia, 2019; Corringham, Ralph, Gershunov, Cayan, & Talbot, 2019; Gershunov et al., 2019). Research suggests that Amazon generated precipitation has even contributed to exacerbating floods and droughts outside of South America via these mechanisms (Blamey, Ramos, Trigo, Tomé, & Reason, 2017).

3.3 Biodiversity decline in the Amazon

Biodiversity can be thought of as the full range of life on Earth, in all forms and interactions. Biodiversity is the most complex feature of our planet and it is woven into the very fabric of climate and weather. Biodiversity is a treasure as it represents the processes learned by evolving species over millions of years about how to survive through the vastly varying environmental and climatic conditions. A comprehensive review of the true scale and implications of biodiversity loss is not fully known and also beyond the scope of this chapter. Yet it is critically important to acknowledge that cascading biodiversity loss may well represent a crisis equaling or even surpassing climate change (Legagneux et al., 2018).

The Amazon Rainforest is home to a staggering amount of biodiversity; it is home to at least 3 million different species and a new species is discovered in the Amazon, on average, every other day (Thomson, 2020; Valsecchi et al., 2017). The degradation of the Rainforest has consequently led to massive declines in biodiversity (Ellwanger et al., 2020). Moreover, ecosystem fragmentation is of grave concern in the Amazon and elsewhere. Although pockets of suitable habitat may remain, many species may not be able to access them, with their paths being blocked by settlements, dams, roads, farms and fences. This loss of biodiversity has several implications related to human health. Meta-analyses and reviews describe the complex relationships between losing biodiversity and various aspects of human health (Aerts, Honnay, & Van Nieuwenhuyse, 2018; Kilpatrick, Salkeld, Titcomb, & Hahn, 2017). Additionally, there is a concerning relationship between loss of biodiversity and increased risk of emerging zoonotic diseases like COVID-19, HIV, and Ebola. These diseases originate within animals and through species-jumping (spillover) events, may cause novel diseases in humans. The mechanisms by which zoonoses occur will be discussed in greater depth later in the chapter. Land-use changes, habitat loss and decreased biodiversity are all related in pathways that appear to promote the emergence of these novel infectious diseases (Allen et al., 2017; Sandifer, Sutton-Grier, & Ward, 2015; Wilkinson, Marshall, French, & Hayman, 2018).

A more clear observation can be made with the spread of Malaria in regions of Brazil where the Amazon has been recently deforested. As forest cover and the life within it are lost, several mosquito species expand their habitats and ranges. This is problematic, as mosquitoes are vectors for a wide variety of deadly infectious diseases, including Zika, Dengue, Chikungunya West Nile Virus, and others. Researchers have remarked that increased rates of deforestation are correlated with an increased burden of disease for Malaria (Ellwanger et al., 2020; MacDonald & Mordecai, 2019). An increasing GAST as a result of climate change may also contribute to increased transmission and burden of disease (Rossati et al., 2016). The key point to be made is that the protection of the Rainforest and its biodiversity provides a resilience to these emerging diseases and can act as a barrier to prevent the spread of pre-existing diseases.

The deforestation and degradation of the Amazon rainforest will lead to a plethora of detrimental effects to South America and to the rest of the world. As examined here, the Amazon is an invaluable carbon sink that not only traps atmospheric CO_2 but also prevents the release of locked carbon. The rainforest is home to a staggering amount of biodiversity in the continent and the globe, building a resilience against emerging zoonotic diseases. Furthermore, deforestation would severely inhibit the ability of the Amazon to produce water vapor and contribute to global water flow and hydrological cycles. Research supports the understanding that the Amazon Rainforest acts to stabilize a rapidly changing climate, which is driven by anthropogenic factors. All of this demonstrates the necessity of the Amazon rainforest, not just to Brazil or to South America, but to the entire planet. Further degradation to this vital biome must be curbed if it is to continue performing these roles.

3.4 Drought, deforestation, and fire in the Amazon

While the northern regions of the Amazon are largely seasonal, the southern Amazon experiences large seasonal variations including in temperature and precipitation regular periods of prolonged dryness (Hasler & Avissar, 2007; von Randow et al., 2004). These fluctuations are largely caused by the El Nino-Southern Oscillation (ENSO), a weather phenomenon which cycles between three phases: neutral, La Nina, and El Nino. During the El Nino period, the weakening of Walker circulation reduces the upwelling of cool water from the ocean depths resulting in warmer sea surface temperatures (SST) in the Pacific Ocean, engendering warmer weather across South America. La Nina is the cooling phase of the ENSO wherein Walker circulation strengthens and cooler water rises, increasing rainfall and allowing the dry areas of the forest to recover (Zeng et al., 2008).

While the ENSO has a range of climatic effects globally, El Nino has direct impacts on precipitation, temperatures, and other weather phenomena in South America. Most notably, El Nino was responsible for the severe droughts in South America in 1982–83 and 1997–98 (Jiménez-Muñoz et al., 2016). The 2003–05 drought in Amazonia, which was the most severe drought in over a century, however, was not caused by the ENSO but rather anomalous North Atlantic warming (Phillips et al., 2009). These drought events have been gradually increasing in intensity, which in combination with a warming climate, has exerted extreme stress on the Amazon Rainforest. Furthermore, the ENSO has been changing in association with climate change (Yeh et al., 2009).

Recent scholarship in ecology has focused on assessing the resilience of the Rainforest to drought (Phillips et al., 2009). Technically, the resilience of an ecological system refers to the maximum perturbation which it can endure without undergoing a regime shift. Three stable states (regimes) have been identified in the Amazon: forest, savanna, and treeless. A critical transition will potentially occur if extreme perturbations, such as drought and widespread fire, continue to occur in the Amazon at the present rate. The two greatest threats to the Amazon's current rainforest regime are deforestation and drought, which act synergistically to amplify risk of forest dieback by a factor of 6.6 (Staal, Dekker, Hirota, & van Nes, 2015; Yeh et al., 2009).

The Amazon Rainforest historically has had a natural resilience to seasonal dry periods and even droughts due to its self-generated humidity, classified as an understory microclimate (Staal et al., 2015). The process of evapotranspiration saturates intact closed-forest canopies with ample moisture to effectively combat bouts of drought. However, sustained droughts like the one experienced in 2005 increase tree mortality and enhance the intensity and frequency of fires. In areas of the rainforest which have already been degraded by human activity, prolonged drought could spark the mass dieback of entire regions of the Amazon (Staal et al., 2015). The degradation of closed-canopy forests through human activity can thus trigger a positive feedback loop wherein the initial vulnerability to drought and fire further erodes the effectiveness of the understory microclimate which, in turn, diminishes resilience ad infinitum (Staal et al., 2015).

Drought is a pernicious force for the Amazon, and climatic extremes increasingly exacerbate both drought and flooding. In 2021 Central Brazil experienced its worst drought in 100 years (Branford & Borges, 2021). Water volume for areas like the Paraná River Basin fell to their lowest levels in decades (Branford & Borges, 2021). The Paraná Basin area specifically is home to about a third of Brazil's population, and also includes major population centers for Paraguay and Argentina (Branford & Borges, 2021). More broadly, drought is becoming more frequent and more widespread across the Amazon (Marengo et al., 2021; The Earth Observatory, 2021). Simultaneously, La Niña is leading to increased flooding in areas of the Amazon, and the transition from El Niño and La Niña is becoming more frequent, and with more severe effects on South America, as climate change escalates (Fearnside, 2021). For example, in 2021, rivers around Manaus, the biggest city in Brazil's Amazon rainforest, swelled to unprecedented (Crispim & Jeantet, 2021). Moreover, as the rainforest's degradation continues, both drought and flooding may become more likely and more severe. This is because where rainforest is cleared, water is not collected by surrounding flora, and slides toward streams instead, where these go on to accumulate high rainy-season water flows (Fearnside, 2021). At the same time, soil in degraded areas does not retain water as effectively as the soil from healthy rainforest areas, lowering protections against drought (Fearnside, 2021). Furthermore, the mortality rates of trees was elevated for 3 years following the drought and trees located in forests that had been affected by human activity, such as selective logging and understory fires, were more likely to die following the drought and fires (Berenguer et al., 2021), indicating that human-modified forests lose resilience. This illuminates this complex and dangerous climatic feedback loop: rainforest loss caused by drought and growing dry spells could cause more trees to succumb to such stress and in turn trigger ever-increasing climate fluctuations across Latin America. Though more research is needed to distinguish in detail which parts of the Amazon are likely to see drought rather than flooding, how likely these are to persist into the future, and how far these effects reach on the global level, climate change's effects in this regard are vivid and apparent.

3.5 The Amazon and climate change

Climate change is responsible for a plethora of primary and secondary effects (Butler & Harley, 2010). It directly impacts seasonal variation, GAST, precipitation, the frequency of extreme weather events, and other biogeochemical phenomena globally. These changes to global weather and climate in turn have secondary consequences on the structure, vitality, and resilience of the interconnected network of global biomes. These consequences include habitat loss, mass migration, biodiversity loss, and increased extinctions. In addition to climate change, increased human activity in primary forests for agricultural, mining, or other economic purposes often results in ecosystem degradation through logging and intentional burning. Tropical forests are the most species-rich ecosystems globally and the Amazon is the most biodiverse region in the world, making it a high priority for conservation and greater action.

The heart of the problem of forest-fires in tropical rainforest lies not in the mere presence but rather the frequency of fire. Evidence from soil samples shows that fires have occurred naturally in rainforests in the past, but they are extremely rare, so endemic species haven't adapted sufficiently to regular burning (Cochrane, 2003). Tropical forest fires are largely dependent on disruptive activity at forest edges. While natural phenomena like lightning strikes and volcanic eruptions can cause fire in the interior region, most fires are associated with human activity and thus, occur at the edge of tropical forests.

Undamaged tropical forests are by nature fire-resistant even during periods of extreme drought. Rainforests inhibit sustained burning through the recycling of moisture within the forest canopy (Uhl, Kauffman, & Cummings, 1988). Evapotranspiration creates high levels of humidity which acts as a barrier to burning, but deforestation and fragmentation reduce the effectiveness of this barrier, in turn, increasing the likelihood and severity of sustained fires. Damaged tropical forests, on the other hand, are especially vulnerable to increases in average surface temperatures, dry periods, and weather extremes—all of which are effects of climate change. The main driver of fires in tropical forests is changes in land cover through human activity. Logging, road building, and deforestation cause fragmentation, the desiccation of forest edges, and increase the likelihood of fire in tropical forests.

3.6 The Amazon's tipping point

Biodiversity isn't merely a consequence of climate, but also has significant consequences for the climate (Salati, Dall'Olio, Matsui, & Gat, 1979). For example, the Amazon produces around half of its annual rainfall through the process of evapotranspiration. Accordingly, the hydrological cycle of the Amazon is dependent on the area of intact rainforest. Deforestation, as it erodes this area, diminishes the ability of the rainforest to produce its own rainfall. At a certain point, the rainforest will no longer generate enough evapotranspiration to support the hydrological needs of its own ecosystem and will consequently transform into a semi-arid savannah (Lovejoy & Nobre, 2018). The level of either (1) deforested area or (2) increased GAST at which this will occur is referred to as the Amazon's tipping point (Amigo, 2020; Lovejoy & Nobre, 2018, 2019).

The tipping point is twofold: either when deforestation reaches 20%–25% or at a 4°C increase in GAST, the Amazon will reach a tipping point (Sheil, 2018). These two tipping points form a positive feedback loop in that deforestation releases CO_2, methane, and other greenhouse gases and diminishes the carbon sequestering ability of forests, contributing to the greenhouse effect. Simultaneously, a warming climate exerts pressures on the rainforest, making it more vulnerable to degradation. The incidence of severe drought and flooding indicates that the Amazonia biome is oscillating and may already be showing signs of the tipping point. While scientific uncertainty may still exist regarding the exact thresholds, "there is no point in discovering the precise tipping point by tipping it" (Lovejoy & Nobre, 2018). The intact hydrological cycle of the Amazon is necessary for human wellbeing in South America and around the globe. The consequences of this ecosystem transforming into a savannah would be far-reaching, multi-faceted, and likely irreversible.

3.7 Drivers of deforestation

The primary drivers of the degradation of the Amazon are economic in nature. Agricultural and extractive industries, such as farming and mining, are key components of Brazil and the other Amazon nations' economies. The destructive methods used to expand the agricultural frontier further into primary rainforest include clear-cutting and burning, both of which are allowed by the current political leadership in Brazil. Deforestation decreased by 75% in the period between 2005 and 2014 but rose sharply following the election of the current president, Jair Bolsonaro (Nobre & Nobre, 2018). This is a clear example of the administration's priority of economic expansion and growth at the cost of environmental degradation.

The growing consumption and demand for agricultural products like beef and soybeans (which can be used for pork feed) creates an economic opportunity for Amazonian countries like Brazil. A clear example of this growing opportunity can be seen with soybean exports in Brazil, which totaled $ 2.19 billion USD and 3.81% of total exports in 2000 (OEC, 2021). The year 2010 saw Brazil export over $11 billion USD worth of soybeans, making it 5.37.% of its total exports (OEC, 2021). In 2019, $26 billion USD of soybeans were exported from Brazil, and imported mainly by China ($20.5 billion) and Europe ($2.32 billion) (OEC, 2021). The prioritization of short term economic growth over the protection of the natural environment and sustainable action is counter-intuitive. As discussed previously, the Amazon Rainforest plays a crucial role in maintaining hydrological cycles for South America and the planet. Degrading this land and ecosystem will disrupt the same natural cycles that the agricultural industries depend on. Regardless of the long term consequences, growing international demand for these products drives these industries to expand.

Another key driver of deforestation in the Amazon is the weakening of IPAs, and the degrading respect for the rights of Indigenous Peoples and their lands. Indigenous Land demarcation and violent evictions plays a role in targeting new primary Rainforest territory in the name of "economic development." In Brazil, there have been several new laws waiting to be implemented to protect and legalize unethical and violent land seizures from Indigenous Peoples (Abessa, Famá, & Buruaem, 2019; Brito, 2021; Brito, Barreto, Brandão, Baima, & Gomes, 2019; de Area Leão Pereira, de Santana Ribeiro, da Silva Freitas, & de Barros Pereira, 2020; Ferrante & Fearnside, 2019). These laws not only strip Indigenous Peoples of their lands and livelihoods, but also help speed up the process of land use change. Dramatic and rapid land use change has significant implications for the rise of zoonotic diseases as well.

The general lack of respect and regard for the protection of Indigenous Protected Areas (IPAs) and the rights of Indigenous peoples contributes to deforestation. Many laws that protect IPAs already exist, but are simply not upheld or respected. For example, the Brazilian Constitution of 1988 recognizes the rights of Indigenous Peoples and that they must be involved and heard from before any sort of development can take place. However, IBAMA, the governmental body that monitors environmental threats, was weakened by budget cuts and its representation removed from 21 of Brazil's 27 states and significantly reduced their budget (Chase, 2019; de Area Leão Pereira et al., 2020; de Area Leão Pereira, Silveira Ferreira, de Santana Ribeiro, Sabadini Carvalho, & de Barros Pereira, 2019; Fearnside, 2016). Currently, IBAMA is reputed to have the lowest performance in its history: There has been very little action taken against illegal extractivist actions, despite the soaring rates of deforestation (Chase, 2019). The culture of ignoring the cries of the Indigenous plays a significant role in continuing the practices of degradation and deforestation in the Amazon.

The intersection of global economics, human rights issues, climate change, and environmental degradation make deforestation in the Amazon Rainforest a complex and multifaceted issue. The drivers of deforestation must be resolved with a better, more sustainable framework, otherwise the steady degradation of the Amazon Rainforest can be expected to continue toward an unrecoverable tipping point. And from there on, the climatic consequences for South America and the entire planet will be dire.

3.8 Zoonosis, climate, and the Amazon

As deforestation marches forward in the Amazon, so too does the risk of disease—specifically, zoonoses. A zoonotic disease, or zoonosis, is a kind of emerging infectious disease characterized by how it spreads to humans: through contact with vertebrate animals (WHO, 2021). A zoonotic pathogen is defined as any pathogen that can be transmitted from animals to humans and in turn cause disease. When there is the rare occurrence of a pathogen being passed from an animal to a human, it is known as a spillover event. Every year, zoonoses account for millions of deaths and around one billion cases of illness (WHO, 2021). Out of all reported infectious diseases, nearly two thirds are zoonotic (WHO, 2021). Zoonoses are responsible for some of the most well-known recent disease outbreaks, with examples including Bird Flu, H1N1, SARS, and COVID-19 (UNEP, 2020).

Because of its recentness and global impact, the COVID-19 pandemic offers a compelling and high-profile example of the large-scale danger of latent diseases and of zoonoses specifically. The pandemic's costs are wide-ranging and multidimensional. By the end of 2020, 94 million people had been pushed into poverty as a direct result of the pandemic, while further increases in poverty were likely to be seen by women and girls disproportionately (UNDP, 2020). At the height of the virus' initial spread in early 2020, more than 1.5 billion students were unable to attend school, reducing the duration and quality of their learning in both the short and long-term (The World Bank, 2020). *The Economist* estimates that global GDP in 2020 was nearly six trillion dollars lower than it would have been had SARS-CoV-2, the pathogen which causes COVID-19, not spread (The Economist, 2021). These are only a few of the pandemic's consequences, and these are all on top of the direct loss of life due to COVID-19, which totaled 1.8 million deaths by the start of 2021 (WHO, 2021). The total losses as a result of the pandemic are incalculable.

In light of this, research has called for increased attention by governments and other organizations to the prevention of future pandemics (UNEP, 2020). At the same time, zoonotic diseases are becoming more common around the globe (Bowman, 2020; UNEP, 2020). Part of the reason for this increase is climate change: some of the drivers of climate change are also drivers of increased zoonotic disease spread, and climate change itself exacerbates some of these drivers (UNEP, 2020). Recent research in *Nature* surveyed approximately 6800 ecological communities across six continents, and found that land-use changes systemically contribute to harmful interfaces between humans, wildlife, and pathogens (Gibb et al., 2020). While such research highlights a dangerous trend of emerging disease, it stops short of predicting where zoonotic outbreaks may occur. However, literature on zoonosis increasingly focuses on the Amazon Rainforest as a site of potential future zoonotic spillover (Bowman, 2020; Conservation International, 2021). A number of trends in the Amazon explain this increased attention: not only is human expansion into the Rainforest through deforestation increasing, but so too is land use change and urbanization, among other drivers of zoonotic spillover. At the same time, climate change aggravates other mechanisms by which zoonoses can spread in the Amazon, including human migration and altered patterns of extreme weather. Manaus, the capital of Amazonas state in Brazil, experienced two brutal waves of COVID-19, The city's cumulative death toll, roughly 9000, is among the world's highest per capita (Grossman, 2021). Against the backdrop of climate change and the increasing destruction of crucial areas like the Amazon Rainforest, understanding the relationships between zoonoses, climate, and the Amazon can help prevent future disease spread, and even future pandemics.

One way zoonoses emerge is through the direct spread of pathogens to humans during deforestation. As is well-known, deforestation in the Amazon is widespread, and has been increasing in recent years as well (Butler, 2020; Mongabay, 2020). When an area is deforested, its flora and fauna are thoroughly removed or destroyed until the land is clear. So, the deforestation process places human beings in close proximity to endemic species, as well as to their pathogens (Ellwanger et al., 2020). This is where zoonotic "spillover" events may take place, and where zoonoses gain a foothold in humans. Though the literature on spillover during Amazon deforestation is still developing, a mechanistic model produced by Faust et al. (2018) describes which kind of deforestation provides the environment most likely for spillover events to occur. According to these simulations, spillover events are relatively uncommon at the start of the deforestation process, when a smaller amount of forest has been converted into clear land (Faust et al., 2018). As well, the likelihood of spillover events decreases nearer to the end of the deforestation process, when the amount of cleared area overtakes the amount of yet-to-be-deforested land. Instead, the likelihood for spillover effects is highest during the lengthy middle of the deforestation process, when there is a roughly equal amount of land deforested as not-yet-deforested. This is because spillover events take place at the edges of the forest, and as deforestation progresses, the length of forest edge increases (Faust et al., 2018). The model predicts that the deforestation pattern most likely to lead to spillover is a "checkerboard" pattern because this maximizes edge length (Faust et al., 2018). This checkerboard pattern is precisely how deforestation is done in the Amazon (Bowman, 2020; Dobson et al., 2020). Along the edges of this checkerboard, humans and their domesticated species remain in close proximity to the local species of the rainforest, thereby provoking spillover risk. As well, Faust et al.'s model predicts that infections are most likely to occur past 25% of area converted—as of 2021, the Amazon overall has reached around 20% deforestation (Lovejoy & Nobre, 2019).

In addition to the risks of direct interaction between human beings and wildlife along forest edges, deforestation promotes zoonotic spillover in other, more indirect ways. One of these has to do with the "dilution effect," which is the theory that areas of high biodiversity show reduced spillover to humans (Khalil, Ecke, Evander, Magnusson, & Hörnfeldt, 2016; UNEP, 2020). This is because an area of high biodiversity requires pathogens to cross between more species before they are likely to reach contact with a human, and the more species barriers there are between a pathogen and humans, the less likely it is that zoonotic spillover will occur (Ellwanger et al., 2020; Khalil et al., 2016). One corollary of the dilution effect is that areas of low biodiversity carry more risk regarding spillover, as a pathogen found in one species is then more likely to be widespread among the population, due to the absence of these barriers (Ellwanger et al., 2020; Khalil et al., 2016). At the same time, one effect of deforestation in the Amazon is a reduction in biodiversity. As flora and fauna are cleared, only the most resilient species remain, such as rodents and bats, both of which are known hosts for a range of zoonotic diseases (UNEP, 2020). Bats in particular are prominent in the Amazon: 12% of the world's 1400 bat species, known to host an incredible range of viruses, move through the Rainforest (Grossman, 2021).

One other side-effect of deforestation is the "coevolution effect" (UNEP, 2020; Zohdy, Schwartz, & Oaks, 2019), whereby the checkerboard pattern of deforestation creates isolated islands of rainforest surrounded on all sides by deforested land and disconnected from other such islands and the rainforest as a whole. This bottleneck effect, combined with biodiversity loss, affects evolutionary pressures on the species remaining in the island—the coevolution effect hypothesizes that these rainforest fragments undergo rapid diversification, and this increases the likelihood that pathogens will adapt enough to spill over (UNEP, 2020; Zohdy et al., 2019). Though empirical studies into both of these deforestation side-effects are not yet plentiful (UNEP, 2020), they reinforce the notion that deforestation functions as a driver of zoonotic spillover.

Compounding the risks of zoonotic disease emergence is how this land is often used after it is deforested, which includes agriculture and urbanization. Agribusiness industries, especially beef and soy, are widespread in the Amazon. More than two thirds of deforested lands in the Amazon go on to become cattle pasture (Cerri et al., 2018), and these industries are commonly cited as drivers for further Amazon deforestation and the overall carbon emissions of Amazonian nations (de Area Leão Pereira et al., 2020; Dias Galuchi, Rosales, & Batalha, 2019). Agriculture and land use change facilitate several specific mechanisms for zoonotic spread, including increased and sustained human activity near to the rainforest as well as interaction with peridomestic species like cattle or pigs, which historically have been notable actors in previous zoonotic outbreaks (UNEP, 2020). Some well-studied examples of this are the 1998–2005 Nipah virus outbreaks in Malaysia and then Bangladesh, where land use change allowed interaction between species which would not otherwise meet (UNEP, 2020). In the case of the Nipah virus outbreaks, peridomestic pigs encountered fluids from fruit bats originating in the forests next to pig farms. From there, farmers sold infected pigs to other commercial agriculture projects, propagating the outbreaks in humans, which ultimately required the culling of around one million pigs (UNEP, 2020). In this way, zoonotic outbreaks involve not only local scales but also international ones, and this precedent has implications for nations which share the Amazon Rainforest.

Secondly, urbanization is associated with zoonotic disease emergence. With a higher population density for both humans and urban vertebrates like mice and bats, risk increases for disease spread. Though urbanization also sometimes develops alongside improved public health infrastructure, the Amazon is an area of "de-urbanization" as well as urbanization, where de-urbanization is the abandonment of urbanized areas by sources of health infrastructure and resources, like the government (Ellwanger et al., 2020). Literature identifies a range of other interlocked drivers of zoonotic disease emergence in addition to these, including the international exotic pet trade, the expansion of extractive industries like gold mining, the construction of roads, sex work, and wildlife consumption (Ellwanger et al., 2020). Together, these mechanisms produce a wide and varied web of ways whereby zoonoses can reach humans.

Furthermore, climate change is itself a driver for some mechanisms of zoonotic disease emergence. A literature review by Nava, Shimabukuro, Chmura, and Luz (Nava, Shimabukuro, Chmura, & Luz, 2017) found that a variety of extreme weather events as well as familiar climate events like El Niño and La Niña have a relationship to infectious disease outbreaks. Specifically, increased temperature, altered precipitation, and increased likelihood of weather events like floods or drought are all associated with increased emerging infectious disease spread (Nava et al., 2017). This is particularly relevant for the Amazon area, which is simultaneously experiencing more severe drought and more severe flooding between its Eastern and Western halves, respectively, as climate change escalates (Fearnside, 2021). In addition, human migration, which is itself associated with climate change, provides increasing opportunities for pathogens to spread to and among humans (Ellwanger et al., 2020; UNEP, 2020).

3.9 Indigenous people in the Amazon

The Indigenous people of the Amazon are not a homogeneous group and represent an array of practices and interactions with the environment. There are Indigenous people living in urban settings and some are even involved in extractive industries. Our focus here, however, is on the many Indigenous people living in forest reserves or Indigenous Protected Areas (IPAs). Much of the Amazon Rainforest has been demarcated as Protected Natural Areas or Indigenous Territories in which Indigenous communities reside (RAISG, 2018; Walker et al., 2009). According to at least one study, the overlap of natural areas under Indigenous stewardship is at least 52% (RAISG, 2018). Indigenous people have lived in the Rainforest for millennia, and although the foundational purpose of IPAs was primarily to safeguard their ancestral lands and cultures, there is strong evidence to suggest that IPAs also conserve biodiversity and ecology (Blackman & Veit, 2018; Lovejoy & Nobre, 2019; Makondo & Thomas, 2018; Walker et al., 2020).

Many of the Indigenous people of the Amazon and other forest communities have been a bulwark against deforestation for decades. Many Amazon Indigenous cultures hold cosmologies grounded in the sanctity of their forests (Bowman, Dale, Dhanani, Nehru, & Rabishaw, 2021) and IPAs often feature diverse environmental interactions and the sustainable use of natural resources. There is growing recognition that these areas are critical to protect ecosystems, biodiversity, natural resources and cultural expression where formal conservation initiatives may not exist (Blackman & Veit, 2018; Londono et al., 2016; Walker et al., 2009).

As stated, the present crisis in the Amazon is grounded in illegal land clearing and (Fearnside, 2008) setting of fires to clear lands for agricultural expansion (Bowman, 2019a). Currently, unsustainable forest clearing is rapidly rising across the nine-nation Amazon region. What is clear in recent years is that Amazon IPAs are increasingly vulnerable to the less conspicuous processes of forest degradation and disturbance, which diminishes carbon storage and ecological integrity and is closely linked with land seizure and fire (Ferrante & Fearnside, 2019; Silva, da Costa, de Farias, & Wanderley, 2020; Walker et al., 2020). The trend toward weakening of environmental protections, Indigenous land rights, and the rule of law thus poses a profound existential threat to IPAs.

Important other actors in the Amazon's recent history are the *Grileiros*— "land grabbers." These are groups which operate by illegally removing trees and then igniting the rainforest, in order to forcibly clear the land. Often, this process involves the violent expulsion of Indigenous peoples from the area (Bowman, 2019a). After removing the most profitable tree species, the *Grileiros* burn and degrade the remainder enough that new settlers can move in, introducing cattle and other industrial uses for the land (Bowman, 2019a). Under Brazilian law, the transformation of the land to industrial use offers a path to obtain legitimate title for the land, thereby incentivizing *Grileiros* to be the forerunners of industry (Bowman, 2019b). Due to their confrontational and illegal practices, the *Grileiros* are in constant conflict with communities across the Amazon (Bowman et al., 2021).

The Amazon Rainforest has been occupied by Indigenous people for at least ten thousand years (Butler, 2005, 2019). There is evidence to suggest the existence of large and sustainable civilizations complete with cities, elaborate food storage, water management, and other complex technologies, well before European contact (Jarrett, Cummins, & Logan-Hines, 2017). When the European colonizers entered their lands in the 1500s, the populations which once thrived began to rapidly decline. Although the early days of contact, in which Indigenous people were given European goods to harvest the timber, were relatively peaceful, respect gradually began to erode and they were enslaved. Deforestation escalated as regions were cleared of trees and converted into plantations to grow cash crops using intensive slave-labor.

The Amazon rubber boom emerged in the late nineteenth century, resulting in a mass expansion of European colonization in the Amazon (Coomes & Barham, 1994). This wreaked havoc on Indigenous societies while generating large amounts of wealth for the Europeans. Horrendous atrocities were unleashed on the Indigenous people of the Amazon, including murder, land theft and enslavement. By the middle of the 20th century, the size of their populations had shrunk by 80%–90% and some anthropologists predicted there would be no Indigenous groups left by 1980 (Coomes & Barham, 1994). When the Amazon was opened up for development by the military in the 1960s, 70s and 80s, a new wave of hydro-electric dams, cattle ranching, mines and roads cost tens of thousands of Indigenous peoples their lands and lives. Dozens of tribes disappeared forever (Coomes & Barham, 1994).

Despite the crimes committed against them, the size of the Indigenous populations has gradually started to grow once more. Although diminished in numbers, Indigenous peoples have shown remarkable strength and resilience against more than 500 years of colonialist attack. With little local or international attention, the Indigenous people of the Amazon and local communities have impeded deforestation and associated greenhouse gas emissions for decades (Blackman & Veit, 2018; Makondo & Thomas, 2018; Walker et al., 2020). Their sustainable forest management systems create climatic and health benefits for all. This management paradigm lays a foundation for the protection of forests and ecosystems, giving everyone the benefits of stabilizing a changing climate and allowing for healthy, balanced ecosystems for human well-being. It also creates an opportunity for all to see there is another way to view the natural world beyond economics, development and industrialization.

The impacts on Indigenous territories supported by environmental and human rights focused organizations working at the local level are palpable. An example of this is the Kayapó Project (About, 2020), in which the Kayapo Indigenous people of the southeastern Amazon allied with conservation NGOs to protect almost 10 million hectares of their ratified territories. Improved surveillance and the development of resource management and income generation activities have resulted in forestland that has clearly remained intact throughout most of Kayapó (About, 2020). Yet satellite imagery shows that over 1 million hectares of Kayapó territory not aligned with environmental groups is victim to illegal invasion by gold mining and loggers (Zimmerman & Ferreira, 2020). Reversing this trend is foundational for the future of climate-buffering Amazon forests for the well-being of ecosystems and human flourishing.

3.10 Isolated indigenous groups

Although a minority, and of great surprise to many, there remain groups of uncontacted people in the world, and by far the world's most significant and controversial isolated people are located in the western Amazon (International,

2022). The existence of these groups is not just a human rights issue, but also holds environmental and climate implications. The remote Javari Valley in the Brazilian Amazon, which is approximately the size of Austria, is a protected home to the highest number of uncontacted groups on Earth (Tickle, 2013). Until recent incursions, it remained completely off limits to all economic and tourist activities. The protection of this valley is critical to the survival of these tribes as well as securing a massive swath of pristine Rainforest—which has several climatic and environmental implications.

Some governmental, religious groups and even a few anthropologists are pushing for the "planned contact" of isolated people, arguing that these people have a right to the amenities of the 21st century (Howe, 2015; Walker & Hill, 2015). Moreover, they claim that if political and economic conditions remain unchanged, tribal groups could be wiped out by unauthorized incursions into their land for mining, gas and oil and timber, which will inevitably lead to violence and degradation. Yet, national and international law guarantees these isolated peoples' the right to ownership of their lands. The most ethical way for contact to occur is if initiated by the isolated groups themselves (Ahmed-Ullah, 2017). The question of what are the responsibilities of the Amazonian nations toward these groups, who do not even know of the nations' existence, represents a larger human rights issue. It is important to remember the future of the uncontacted groups of the Amazon is a complex challenge yet holds climate significance among its considerations.

4 Recommendations

4.1 Securing indigenous stewardship

As previously stated, the amount of land in the Amazon Rainforest under the control and influence of Indigenous people is considerable. Indigenous land rights and human rights are a critical element of securing the climatic stabilizing forces of the Amazon. The Amazon Rainforest was home to agricultural and sedentary societies of great complexity long before European contact, for at least 10,000 years (Butler, 2005, 2019). Unlike current land-users, these groups were able to utilize resources from the Rainforest in a sustainable way, in which they met their food security needs while preserving ecosystem health. Many Amazonian Indigenous communities continue to priorities the sanctity of their forests (Etchart, 2017), and today, Indigenous Protected Areas (IPAs) are critical to conserve ecosystems, biodiversity, natural resources and cultural expression (Blackman & Veit, 2018; Londono et al., 2016; Walker et al., 2009). Allowing Indigenous communities to remain stewards of the Rainforest, as well as using Indigenous knowledge to formulate sustainable agricultural strategies, are necessary steps to mitigate the Amazon's degradation.

Research highlights that IPAs contribute to sustaining the Amazon biome and helping mitigate climate change (Blackman & Veit, 2018; Makondo & Thomas, 2018; Walker et al., 2020) by serving as significant carbon sinks. IPAs have been credited for storing over half of the Amazon's carbon (Walker et al., 2020), and in Brazil, IPA regions conserve around 27 times more carbon than non-protected areas (Walker et al., 2020). Indigenous land management in Brazil, Bolivia, and Colombia was found to correlate with a marked reduction in deforestation and carbon emissions (Blackman & Veit, 2018).

However, it is important to reiterate that in the 21st century, Indigenous groups are not homogenous. While some groups have remained completely uncontacted by modern societies, others do not live within IPAs at all; indeed, many such communities are shaped by, and participate in, modern market forces to different extents. Indigenous people may sometimes even engage in the burning of forest to clear land for productive use (Gray, Bilsborrow, Bremner, & Lu, 2008). Nevertheless, in general, Indigenous communities tend to value forest conservation and prefer forest management strategies that are relatively environmentally benign (Blackman & Veit, 2018). Some theories that Indigenous communities are accustomed to dealing with the challenges posed by a changing climate. Although the rapidity and magnitude posed by climate change might be unfamiliar, they nonetheless have coping and adaptation knowledge that has been passed through generations (Makondo & Thomas, 2018).

Legal systems protecting Indigenous people exist throughout the Amazon nations, however, they are often limited in scope and poorly enforced (Greaves, 2018; Merino, 2018). Participatory channels, such as prior consultation, consent processes, and environmental impact assessments (EIAs) are not given enough weight, and in turn do not give Indigenous people a substantive voice in the crafting of policies which profoundly affect them (Chase, 2019; Merino, 2018). It is very clear that when governance prefers extractive industries, Indigenous participation in decision-making declines (Eichler, 2019; Greaves, 2018).

Indigenous communities are greatly threatened by an ecological crisis which is caused by a complex set of drivers that they are often not responsible for, and have little to no control over. The human rights violations are inseparable from issues of environmental health. Securing Indigenous peoples' right to be stewards of the Amazon emerges as a solution to both crises. It is imperative that the demarcation and enforcement of Rainforest land as IPAs be a continuous process, and that Indigenous people be invited to participate in the highest levels of policy and decision making. This is the surest way to preserve the Rainforest, its global ecological benefits and the rights of the people who inhabit it.

4.2 Leveraging indigenous knowledge

Owing to Indigenous groups' history of sustainable forest management, their knowledge and perspectives can be great sources of building climate resilience strategies. They were able to achieve food security in a changing climatic and social environment through closed-canopy forest enrichment, limited clearing for crop cultivation, and low-severity fire management (Maezumi et al., 2018). Although much of what was known about their forest cultivation techniques was lost when Indigenous populations declined due to European diseases, genocidal practices, and slavery (Mann, 2005), insights can nonetheless be gained.

Cultivation within agroforestry systems offers a promising alternative and more sustainable form of development for the Amazon. Agroforestry is a method involving the deliberate growing of crops under the forest canopy. This dynamic, ecologically based, natural resource management system of land-use merges agriculture with primary forests (Weber, Günter, Aguirre, Stimm, & Mosandl, 2008). This sustains and diversifies crop production while minimizing pressure placed on the Rainforest and maintaining biodiversity, thereby increasing the social, economic and environmental outcomes (Van Noordwijk et al., 2019).

The technique was favored by Indigenous communities long before European arrival (Clement, 1999) but was replaced by more extractive methods, such as cattle ranching and monoculture, following growing industrial demands in the nineteenth century. Agroforestry is still practiced by a few in the Amazon, however, mostly by Indigenous farmers on a small, subsistence scale. For example, the Kichwa Indigenous people of the Ecuadorian Amazon cultivate staple food and medicinal plants into small plots of Rainforest land that mimic natural forest ecosystems. Over time, cash crops such as coffee and cacao came to be integrated into so-called chakra systems, as a way to sustainably increase the income of rural farmers. However, these attempts to commercialize chakra have not been successful (Jarrett et al., 2017).

Farmers are prevented from scaling up these projects by barriers involving market problems, resistance by farmers, and limited support from the government, among other factors. In addition to requiring a large initial investment, there are currently only a few economically valuable species that can be grown in an agroforestry system. It is difficult for smallholders to be competitive in the market and earn an adequate livelihood using agroforestry (Jarrett et al., 2017). There is therefore a need for equitable development policies and programs to support such initiatives. These policies could be related to land tenure, land use, rural credit, research and extension, or subsidized production in these systems (Jarrett et al., 2017; Porro et al., 2012). Agroforestry is an opportunity to link environmental opportunities with economic realities while simultaneously enhancing the livelihoods of Indigenous communities in the Amazon. A healthy, multi-national, debate needs to be ignited regarding sustainable economic development in the Amazon region, including raising awareness of agroforestry.

Around the world, Indigenous Knowledge has proven invaluable in constructing sustainable practices for forest use with positive results. For example, in Vietnam, the agricultural knowledge accumulated through centuries of observation and experimentation by the Yao people is being combined with scientific knowledge to make accurate decisions and help local communities adapt to climate change (Son, Chi, & Kingsbury, 2019). Similarly, Indigenous knowledge from the Nharira community in Zimbabwe is being implemented to conserve forest and wildlife resources (Mavhura & Mushure, 2019). Examples like these should serve as a model for future mitigation strategies against climate change.

4.3 National protection of the Amazon

Weakening environmental governance and enforcement throughout the Amazon nations exacerbates Rainforest degradation. For instance, in 2019, Bolivian farmers were permitted to use slash and burn methods to clear the rainforest for increased output (Machicao & Ramos, 2019) and in 2018, the Peruvian government permitted the construction of roads through the Amazon which was estimated to put 3000 km (Butler, 2020) of primary rainforest, including two Indigenous reserves, at risk (Finer & Novoa, 2018; Gallice, Larrea-Gallegos, & Vázquez-Rowe, 2019). These policies are accompanied by many governments' lackluster response to recent severe environmental events, such as the 2020 oil spills in Ecuador's Coca River (Recinos & Mazabanda, 2020). In Brazil, which contains the largest proportion of the Amazon, legislation was forwarded to promote less rigorous environmental impact review, a legal channel to potentially displace Indigenous communities, the potential revoking of the protection of IPAs, and the sanctioning of gold mining within Indigenous land (Abessa et al., 2019; Brito et al., 2019; de Area Leão Pereira et al., 2020; Ferrante & Fearnside, 2019). Such policies undermine national efforts to preserve the Amazon, and should be abandoned.

One important way for Amazonian countries to protect their rainforest is to maintain and strongly enforce their current environmental regulations (Arruda, Candido, & Fonseca, 2019; de Area Leão Pereira et al., 2019; Dias Galuchi et al., 2019; Gallice et al., 2019). Research indicates that the destructive impacts of industry tend to be more significant in areas where environmental regulations and their enforcement are less strong (le Polain de Waroux, Garrett, Heilmayr, & Lambin, 2016). Thus, supporting the even implementation and enforcement of these regulations through the Amazon can make a significant difference in regard to protecting the Rainforest itself, the Indigenous peoples and others living among it, and the diverse ways whereby the Rainforest stabilizes global climate.

4.4 Transnational (intra-Amazon) protection of the rainforest

There is much that can be done at the national level to address the ecological crisis in the Amazon. In systems such as the Amazon, where a natural resource spans multiple nations, it is critical that protective actions be coordinated (Wolf, 2007). Although each Amazonian nation individually formulated policies to address climate adaptation following the Paris Agreement in 2016 (UNFCCC, 2016), these strategies are often disparate and insufficient (Tigre, 2019). Furthermore, any ecologically beneficial impacts of such regulations are diminished by the passing of contradictory policies which have the opposite effect on the Rainforest.

The Amazon Cooperation Treaty Organization (ACTO) is an intergovernmental organization created to strengthen communication and coordination between the eight Amazonian nations, as well as include Indigenous participation in order to encourage sustainable development of the Rainforest (ACTO, 2010). However, from an environmental perspective, the ACTO has not been entirely successful (Tigre, 2019). Several mandates related to climate adaptation have been drafted, however only one has actually been implemented. The so-called GEF (Global Environment Facility) is supposed to support the creation of new protected areas in the Amazon, however, has been criticized for its broad mandates, lack of action-guidance, and poor efficacy (Tigre, 2017, 2019). In general, the ACTO is said to suffer from a dearth of accountability, conflict-resolution mechanisms, and, importantly, funding (Tigre, 2017).

Furthermore, any policies passed to protect the Rainforest are undermined by transnational commitments such as the South American Regional Infrastructure Integration Initiative (IIRSA) and the Pacific Alliance (Creutzfeldt, 2018). These agreements, which aim to strengthen social and economic cooperation between the Amazonian nations, have the effect of opening Rainforest land to industrial projects such as highways, waterways, railways, and ports (Walker & Simmons, 2018). Such development initiatives degrade the environment and turn IPAs into commodities (Laschefski & Zhouri, 2019).

The Amazon Rainforest is a resource shared by many nations, therefore, it is the responsibility of each to ensure its protection. The ACTO is a promising platform to mobilize a powerful, coordinated, transnational response against the Amazon destruction. It has decades of a diplomatic history and a functioning institutional structure that can be taken advantage of (Tigre, 2019). Addressing the ACTO's shortcomings to help realize its potential, as well as reassessing the value of agreements such as the IIRSA and Pacific Alliance, should be priority actions.

4.5 International protection of the Amazon

Considering the Amazon Rainforest's intrinsic value as well as its global benefits, it is vital that the international community assume responsibility for its protection. Much of the economic incentive driving agribusiness in the rainforest is fueled by the growing international demand for beef and soy, among other products. Therefore, promoting sustainable consumer practices around the world is an important aspect of protecting the rainforest. The impact of livestock could be minimized by raising awareness on the detriments of such activities. Public educational campaigns about the environmental benefits of buying meat that is sustainably produced could be very effective (Silva et al., 2020). Alternatively, countries could move toward growing their crops locally: in addition to giving nations better control of production levels this would reduce environmental impacts, and enhance economic security. It is important that these initiatives are global in scope, as China and the European Union are presently the greatest consumers of soy harvested from the Amazon basin (OEC, 2021).

The US President, Joseph Biden, has made clear that he is focused on the Amazon's protection in addressing climate change. In his first executive order on tackling climate change, he directed the U.S. Treasury and State departments, along with the U.S. Agency for International Development and the International Development Finance Corporation, to develop a plan for supporting protection of the Amazon through market-based mechanisms (The White House, 2021). President Biden appears strongly committed to climate action in his administration's foreign policy and the Amazon Rainforest is central to this strategy. Some have suggested more extreme ways through which other nations can contribute to the Amazon's protection. These include imposing pressure on Amazonian nations to comply with environmental regulations via economic sanctions or other interventions (Continente, 2019; Macedo, 2021). However, these types of measures are controversial and ethically questionable because they undermine states' sovereignty.

Fortunately, it is becoming increasingly clear that the power of consumer pressure to take action on climate is now influencing many companies, and Brazilian officials and corporations are noticing. In June 2020, a group of international institutional investors that together manage more than $3.7 trillion in assets informed the Bolsonaro administration that the group would take steps to divest from Brazil if it did not address deforestation caused by beef and soy production (Harris, 2020).

There are now mounting financial incentives to discourage deforestation. These initiatives are rapidly evolving and expanding. The first of considerable significance is REDD, an abbreviation for "reducing emissions from deforestation and forest degradation." REDD offers a new way of curbing CO2 emissions through paying for actions that prevent forest loss or degradation. These transfer mechanisms can include carbon trading, or paying for forest management. REDD has given way to REDD+, with the "plus" referring to "the role of conservation, and sustainable management of forests (UN-REDD Programme, 2021).

The largest and most recent project of this kind is the Lowering Emissions by Accelerating Forest finance (LEAF) Coalition, a public-private initiative that aims to enact beneficial climate interventions by providing finance to countries and regions that protect their tropical and subtropical forests (LEAF, n.d.). The governments of the United States, Norway, and the UK are partnering with major international corporations to support emissions reductions in an attempt to bring an end to deforestation (LEAF, n.d.; Solá & Butler, 2021). The initiative, if successful, will support sustainable development while also benefiting billions of people who depend on tropical forests, with particular safeguards and priorities to Indigenous lands. LEAFS proponents say it improves on REDD+ by working with far larger units of land, and avoiding credits being issued from forests that would have been conserved anyway.

5 Conclusion

The Amazon's inherent value, its astonishing biodiversity, and its importance to global climate and weather systems are significant reasons for humans to reevaluate our relationship with this rainforest. The Amazon serves as a global carbon sink, a regulator of hydrological cycles, a refuge of wildlife and a source of Indigenous cultural value. The incidence of recent droughts and floods in the Amazon suggests the ongoing degradation of the Rainforest by weather and climate extremes. As the biome is further compromised, the Amazon's local and global roles in climate stabilization will falter. This may lead to further exacerbated weather and climate extremes. There is an urgent need to implement sustainable solutions which will allow the Rainforest to recover from existing damage as well as prevent future degradation, in order to allow it to fulfill its important functions. Considering the complex, interconnected nature of the drivers which propel the Amazon toward its tipping point, it is evident that no single intervention will solve this ecological crisis. A multifaceted approach, involving action at the individual, national, transnational and international levels is required. This entails input from individuals of a diverse set of backgrounds and disciplines, including not just scientists, but also Indigenous people and other community members, public health practitioners, policy makers, lawyers and economists to name a few. It is important that they focus their efforts on a singular goal: to save the Amazon Rainforest, the vast amounts of life contained within, and the intricate weather systems it sustains.

References

Abessa, D., Famá, A., & Buruaem, L. (2019). The systematic dismantling of Brazilian environmental laws risks losses on all fronts. *Nature Ecology & Evolution*, *3*, 510–511.

About. (2020). *The Kayapo project*. Retrieved 24 March 2021, from https://kayapo.org/about/.

ACTO. (2010). *Amazonian Strategic Coperation Agenda*. Retrieved 15 March 2021, from http://www.otca-oficial.info/assets/documents/20170119/3d6f9b368f121d6b998800c04bb54605.pdf.

Aerts, R., Honnay, O., & Van Nieuwenhuyse, A. (2018). Biodiversity and human health: Mechanisms and evidence of the positive health effects of diversity in nature and green spaces. *British Medical Bulletin*, *127*, 5–22.

Aguilera, R., Gershunov, A., & Benmarhnia, T. (2019). Atmospheric rivers impact California's coastal water quality via extreme precipitation. *Science of the Total Environment, 671*, 488–494.

Ahlström, A., et al. (2017). Hydrologic resilience and Amazon productivity. *Nature Communications, 8*, 387.

Ahmed-Ullah, N. (2017). *From the Amazon's uncontacted people to physician-assisted deaths in Canada: U of T's Kerry Bowman probes thorny issues*. University of Toronto News. Retrieved 6 January 2021, from https://www.utoronto.ca/news/amazon-s-uncontacted-people-physician-assisted-deaths-canada-u-t-s-kerry-bowman-probes-thorny.

Allen, T., et al. (2017). Global hotspots and correlates of emerging zoonotic diseases. *Nature Communications, 8*, 1124.

Amigo, I. (2020). When will the Amazon hit a tipping point? *Nature, 578*, 505–507.

Arruda, D., Candido, H. G., & Fonseca, R. (2019). Amazon fires threaten Brazil's agribusiness. *Science, 365*, 1387.

Berenguer, E., et al. (2021). *Tracking the impacts of El Niño drought and fire in human-modified Amazonian forests*. Retrieved 23 August 2021, from https://www.pnas.org/content/118/30/e2019377118.short.

Blackman, A., & Veit, P. (2018). Titled Amazon indigenous communities cut forest carbon emissions. *Ecological Economics, 153*, 56–67.

Blamey, R. C., Ramos, A. M., Trigo, R. M., Tomé, R., & Reason, C. J. C. (2017). The influence of atmospheric Rivers over the South Atlantic on winter rainfall in South Africa. *Journal of Hydrometeorology, 19*, 127–142.

Bowman, K. (2019a). *Research notes (a)*.

Bowman, K. (2019b). *Research notes (b)*.

Bowman, K. W. (2020). *Historic Amazon rainforest fires threaten climate and raise risk of new diseases*. Retrieved 15 February 2021, from The Conversation. http://theconversation.com/historic-amazon-rainforest-fires-threaten-climate-and-raise-risk-of-new-diseases-146720.

Bowman, K., Dale, S., Dhanani, S., Nehru, J., & Rabishaw, B. (2021). Environmental degradation of indigenous protected areas of the Amazon as a slow onset event. *Current Opinion in Environment Sustainability, 50*, 260–271.

Branford, S., & Borges, T. (2021). *Amazon and Cerrado deforestation, warming spark record drought in urban Brazil*. Retrieved 28 July 2021, from https://news.mongabay.com/2021/07/amazon-and-cerrado-deforestation-warming-spark-record-drought-in-urban-brazil/.

Brito, B. (2021). *Brazil's Amazon is under threat from proposed land-use laws*. Americas Quarterly. Retrieved 11 February 2021, from https://americasquarterly.org/article/brazils-amazon-is-under-threat-from-proposed-land-use-laws/.

Brito, B., Barreto, P., Brandão, A., Baima, S., & Gomes, P. H. (2019). Stimulus for land grabbing and deforestation in the Brazilian Amazon. *Environmental Research Letters, 14*, 064018.

Butler, T. (2005). *Pre-Columbian Amazon supported millions of people*. Mongabay Environmental News. Retrieved 8 January 2021, from https://news.mongabay.com/2005/10/pre-columbian-amazon-supported-millions-of-people/.

Butler, R. (2019). *People in the Amazon rainforest*. Mongabay. Retrieved 12 February 2021, from https://rainforests.mongabay.com/amazon/amazon_people.html.

Butler, R. (2020). *What's the deforestation rate in the Amazon?*. Mongabay. Retrieved 5 January 2021, from https://rainforests.mongabay.com/amazon/deforestation-rate.html.

Butler, C. D., & Harley, D. (2010). Primary, secondary and tertiary effects of eco-climatic change: The medical response. *Postgraduate Medical Journal, 86*, 230–234.

Cerri, C. E. P., et al. (2018). Reducing Amazon deforestation through agricultural intensification in the Cerrado for advancing food security and mitigating climate change. *Sustainability, 10*, 989.

Chase, V. M. (2019). The changing face of environmental governance in the Brazilian Amazon: Indigenous and traditional peoples promoting norm diffusion. *Revista Brasileira de Política Internacional, 62*, 1–20. https://doi.org/10.1590/0034-7329201900208.

Clement, C. R. (1999). 1492 and the loss of Amazonian crop genetic resources. I. The relation between domestication and human population decline. *Economic Botany, 53*, 188.

Cochrane, M. A. (2003). Fire science for rainforests. *Nature, 421*, 913–919.

Conservation International. (2021). *Could a future pandemic come from the Amazon?*. Retrieved 9 March 2021, from https://www.conservation.org/docs/default-source/publication-pdfs/english-pandemic-prevention-in-the-amazon-2021.pdf?sfvrsn=6f7b3676_2.

Continente, P. (2019). *Economic sanctions against Brazil could save the Amazon*. Daily Forty-Niner. Retrieved 27 February 2021, from https://daily49er.com/opinions/2019/08/25/economic-sanctions-against-brazil-could-save-the-amazon/.

Coomes, O. T., & Barham, B. L. (1994). The Amazon rubber boom: Labor control, resistance, and failed plantation development revisited. *Hispanic American Historical Review, 74*, 231–257.

Corringham, T. W., Ralph, F. M., Gershunov, A., Cayan, D. R., & Talbot, C. A. (2019). Atmospheric rivers drive flood damages in the western United States. *Science Advances, 5*, eaax4631.

Covey, K., et al. (2021). Carbon and beyond: The biogeochemistry of climate in a rapidly changing Amazon. *Frontiers in Forests and Global Change, 4*, 1–20. https://doi.org/10.3389/ffgc.2021.618401.

Creutzfeldt, B. (2018). *China's engagement with regional actors: The Pacific Alliance*. Retrieved March 28, 2021, from https://pacificallianceblog.com/chinas-engagement-with-regional-actors-the-pacific-alliance/.

Crispim, F., & Jeantet, D. (2021). *Brazil's Amazon rivers rise to record levels*. Retrieved 24 August 2021, from https://malaysia.news.yahoo.com/brazils-amazon-rivers-rise-record-152024467.html.

de Area Leão Pereira, E. J., de Santana Ribeiro, L. C., da Silva Freitas, L. F., & de Barros Pereira, H. B. (2020). Brazilian policy and agribusiness damage the Amazon rainforest. *Land Use Policy, 92*, 104491.

de Area Leão Pereira, E. J., Silveira Ferreira, P. J., de Santana Ribeiro, L. C., Sabadini Carvalho, T., & de Barros Pereira, H. B. (2019). Policy in Brazil (2016–2019) threaten conservation of the Amazon rainforest. *Environmental Science & Policy, 100*, 8–12.

Dias Galuchi, T. P., Rosales, F. P., & Batalha, M. O. (2019). Management of socioenvironmental factors of reputational risk in the beef supply chain in the Brazilian Amazon region. *International Food and Agribusiness Management Review, 22*, 155–172.

Dobson, A. P., et al. (2020). Ecology and economics for pandemic prevention. *Science, 369*, 379–381.

Eichler, J. (2019). Neo-extractivist controversies in Bolivia: Indigenous perspectives on global norms. *International Journal of Law in Context, 15*, 88–102.

Ellwanger, J., et al. (2020). Beyond diversity loss and climate change: Impacts of Amazon deforestation on infectious diseases and public health. *Anais da Academia Brasileira de Ciências, 92*, 20191375.

Etchart, L. (2017). The role of indigenous peoples in combating climate change. *Palgrave Communications*, *3*, 1–4.

Faust, C. L., et al. (2018). Pathogen spillover during land conversion. *Ecology Letters*, *21*, 471–483.

Fearnside, P. M. (2008). The roles and movements of actors in the deforestation of Brazilian Amazonia. *Ecology and Society*, *13*, 23.

Fearnside, P. M. (2016). Brazilian politics threaten environmental policies. *Science*, *353*, 746–748.

Fearnside, P. (2021). *Lessons from the 2021 Amazon flood (commentary)*. Retrieved 28 July 2021, from https://news.mongabay.com/2021/07/lessons-from-the-2021-amazon-flood-commentary/.

Ferrante, L., & Fearnside, P. M. (2019). Brazil's new president and 'ruralists' threaten Amazonia's environment, traditional peoples and the global climate. *Environmental Conservation*, *46*, 261–263.

Finer, M., & Novoa, S. (2018). *Proposed road would cross primary forest along Peru-Brazil border*. MAAP. Retrieved 15 January 2021 from https://maaproject.org/2018/purus-road/.

Gallice, G. R., Larrea-Gallegos, G., & Vázquez-Rowe, I. (2019). The threat of road expansion in the Peruvian Amazon. *Oryx*, *53*, 284–292.

Gershunov, A., et al. (2019). Precipitation regime change in Western North America: The role of atmospheric Rivers. *Scientific Reports*, *9*, 9944.

Gibb, R., et al. (2020). Zoonotic host diversity increases in human-dominated ecosystems. *Nature*, *1–5*. https://doi.org/10.1038/s41586-020-2562-8.

Gimeno, L., Nieto, R., Vázquez, M., & Lavers, D. A. (2014). Atmospheric rivers: A mini-review. *Frontiers in Earth Science*, *2*, 00002.

Global Fire Emissions Database. (2021). *Amazon Dashboard*. Retrieved 22 February 2021, from https://globalfiredata.org/pages/amazon-dashboard/.

Gray, C. L., Bilsborrow, R. E., Bremner, J. L., & Lu, F. (2008). Indigenous land use in the Ecuadorian Amazon: A cross-cultural and multilevel analysis. *Human Ecology*, *36*, 97–109.

Greaves, W. (2018). Damaging environments: Land, settler colonialism, and security for indigenous peoples. *Environmental Sociology*, *9*, 107–124.

Grossman, D. (2021). *Scientists scour the Amazon for pathogens that could spark the next pandemic*. Science, AAAS. Retrieved 17 March 2021, from https://www.sciencemag.org/news/2021/04/scientists-scour-amazon-pathogens-could-spark-next-pandemic.

Harris, B. (2020). *Investors warn Brazil to stop Amazon destruction*. Retrieved 7 March 2021, from https://www.ft.com/content/ad1d7176-ce6c-4a9b-9bbc-cbdb6691084f.

Hasler, N., & Avissar, R. (2007). What controls evapotranspiration in the Amazon Basin? *Journal of Hydrometeorology*, *8*, 380–395.

Howe, R. (2015). *Controlled contact key to protecting isolated tribes*. ASU News. Retrieved 6 January 2021, from https://news.asu.edu/content/controlled-contact-key-protecting-isolated-tribes.

Hubau, W., et al. (2020). Asynchronous carbon sink saturation in African and Amazonian tropical forests. *Nature*, *579*, 80–87.

International, S. (2022). *Uncontacted tribes*. Survival International. Retrieved 6 January 2021, from https://www.survivalinternational.org/.

Jarrett, C., Cummins, I., & Logan-Hines, E. (2017). Adapting indigenous agroforestry systems for integrative landscape management and sustainable supply chain development in Napo, Ecuador. In F. Montagnini (Ed.), *Integrating landscapes: Agroforestry for biodiversity conservation and food sovereignty* (pp. 283–309). Springer International Publishing. https://doi.org/10.1007/978-3-319-69371-2_12.

Jiménez-Muñoz, J. C., et al. (2016). Record-breaking warming and extreme drought in the Amazon rainforest during the course of El Niño 2015–2016. *Scientific Reports*, *6*, 33130.

Kedmey, D. (2015). *The largest river on Earth is invisible—and airborne*. ideas.ted.com. https://ideas.ted.com/this-airborne-river-may-be-the-largest-river-on-earth/.

Khalil, H., Ecke, F., Evander, M., Magnusson, M., & Hörnfeldt, B. (2016). Declining ecosystem health and the dilution effect. *Scientific Reports*, *6*, 31314.

Kilpatrick, A. M., Salkeld, D. J., Titcomb, G., & Hahn, M. B. (2017). Conservation of biodiversity as a strategy for improving human health and well-being. *Philosophical Transactions of the Royal Society of London. Series B, Biological Sciences*, *372*, 20160131.

Laothawornkitkul, J., Taylor, J. E., Paul, N. D., & Hewitt, C. N. (2009). Biogenic volatile organic compounds in the earth system. *The New Phytologist*, *183*, 27–51.

Laschefski, K., & Zhouri, A. (2019). Indigenous peoples, traditional communities and the environment: The 'territorial question' under the new developmentalist agenda in Brazil. In V. Puzone, & L. F. Miguel (Eds.), *The Brazilian Left in the 21st Century: Conflict and conciliation in peripheral capitalism* (pp. 205–236). Springer International Publishing. https://doi.org/10.1007/978-3-030-03288-3_10.

le Polain de Waroux, Y., Garrett, R. D., Heilmayr, R., & Lambin, E. F. (2016). Land-use policies and corporate investments in agriculture in the Gran Chaco and Chiquitano. *Proceedings of the National Academy of Sciences*, *113*, 4021.

LEAF. n.d.. *The LEAF coalition*. Retrieved 30 March 2021, from https://leafcoalition.org/.

Legagneux, P., et al. (2018). Our house is burning: Discrepancy in climate change vs. biodiversity coverage in the media as compared to scientific literature. *Frontiers in Ecology and Evolution*, *5*, 00175.

Londono, J. M., et al. (2016). Protected areas as natural solutions to climate change. *Parks*, *22*, 7–12.

Lovejoy, T. E., & Nobre, C. (2018). Amazon tipping point. *Science Advances*, *4*, eaat2340.

Lovejoy, T. E., & Nobre, C. (2019). Amazon tipping point: Last chance for action. *Science Advances*, *5*, eaba2949.

MacDonald, A. J., & Mordecai, E. A. (2019). Amazon deforestation drives malaria transmission, and malaria burden reduces forest clearing. *Proceedings of the National Academy of Sciences of the United States of America*, *116*, 22212–22218.

Macedo, G. (2021). *Opinion – The responsibility to protect the Amazon*. E-International Relations. Retrieved 20 February 2021, from https://www.e-ir.info/2021/03/31/opinion-the-responsibility-to-protect-the-amazon/.

Machicao, M., & Ramos, D. (2019). *As Bolivian forests burn, Evo's bet on Big Farming comes under fire*. The Guardian. Retrieved 15 January 2021, from http://www.theguardian.pe.ca/news/world/as-bolivian-forests-burn-evos-bet-on-big-farming-comes-under-fire-350014/.

Maezumi, S. Y., et al. (2018). The legacy of 4,500 years of polyculture agroforestry in the eastern Amazon. *Nature Plants*, *4*, 540–547.

Makondo, C. C., & Thomas, D. S. G. (2018). Climate change adaptation: Linking indigenous knowledge with western science for effective adaptation. *Environmental Science & Policy*, *88*, 83–91.

Mann, C. C. (2005). *1491: New revelations of the Americas before Columbus* (1st ed.). New York: Knopf.

Marengo, J., et al. (2021). *Extreme drought in the Brazilian Pantanalin 2019–2020: Characterization, causes, and impacts*. Retrieved 24 August 2021, from https://www.frontiersin.org/articles/10.3389/frwa.2021.639204/full.

Mavhura, E., & Mushure, S. (2019). Forest and wildlife resource-conservation efforts based on indigenous knowledge: The case of Nharira community in Chikomba district, Zimbabwe. *Forest Policy and Economics*, *105*, 83–90.

Merino, R. (2018). Re-politicizing participation or reframing environmental governance? Beyond indigenous' prior consultation and citizen participation. *World Development*, *111*, 75–83.

Mongabay. (2020). *Amazon deforestation tops 11,000 sq km in Brazil, reaching 12-year high*. Mongabay Environmental News. Retrieved 27 January 2021, from https://news.mongabay.com/2020/11/amazon-deforestation-tops-11000-sq-km-in-brazil-reaching-12-year-high/.

Nava, A., Shimabukuro, J. S., Chmura, A. A., & Luz, S. L. B. (2017). The impact of global environmental changes on infectious disease emergence with a focus on risks for Brazil. *ILAR Journal*, *58*, 393–400.

Nobre, I., & Nobre, C. A. (2018). The Amazonia third way initiative: The role of technology to unveil the potential of a novel tropical biodiversity-based economy. In *Land use – Assessing the past, envisioning the future* IntechOpen. https://doi.org/10.5772/intechopen.80413.

OEC. (2021). *Brazil (BRA) exports, imports, and trade partners*. Retrieved 1 April 2021, from https://oec.world/en/profile/country/bra#trade-products.

Phillips, O. L., et al. (2009). Drought sensitivity of the Amazon rainforest. *Science*, *323*, 1344–1347.

Porro, R., et al. (2012). Agroforestry in the Amazon region: A pathway for balancing conservation and development. In P. K. R. Nair, & D. Garrity (Eds.), *Agroforestry – The future of global land use* (pp. 391–428). Netherlands: Springer. https://doi.org/10.1007/978-94-007-4676-3_20.

RAISG. (2018). *Pressões e ameaças sobre as Áreas Protegidas e Territórios Indígenas da Amazônia*. RAISG. Retrieved 9 February 2021, from https://www.amazoniasocioambiental.org/pt-br/publicacao/pressoes-e-ameacas-sobre-as-areas-protegidas-e-territorios-indigenas-da-amazonia-2/.

Recinos, A., & Mazabanda, C. (2020). *Indigenous peoples of Ecuador's Amazon file lawsuit against government and oil companies in wake of disastrous oil spill*. Amazon Watch. Retrieved 7 February 2021, from https://amazonwatch.org/news/2020/0429-indigenous-peoples-sue-the-ecuadorian-government-and-oil-companies-over-disastrous-oil-spill.

Rödig, E., et al. (2018). The importance of forest structure for carbon fluxes of the Amazon rainforest. *Environmental Research Letters*, *13*, 054013.

Rogers, K. (2019). *The Amazon is worth $8.2 billion if it's left standing, study shows*. Retrieved 1 April 2021, from https://www.vice.com/en/article/bje7wd/the-amazon-is-worth-more-money-left-standing-study-shows.

Rossati, A., et al. (2016). Climate, environment and transmission of malaria. *Le Infezioni in Medicina*, *24*, 93–104.

Salati, E., Dall'Olio, A., Matsui, E., & Gat, J. R. (1979). Recycling of water in the Amazon Basin: An isotopic study. *Water Resources Research*, *15*, 1250–1258.

Sandifer, P. A., Sutton-Grier, A. E., & Ward, B. P. (2015). Exploring connections among nature, biodiversity, ecosystem services, and human health and well-being: Opportunities to enhance health and biodiversity conservation. *Ecosystem Services*, *12*, 1–15.

Sheil, D. (2018). Forests, atmospheric water and an uncertain future: The new biology of the global water cycle. *Forest Ecosystems*, *5*, 19.

Silva, M. J. B., da Costa, M. F., de Farias, S. A., & Wanderley, L. S. O. (2020). Who is going to save the Brazilian Amazon forest? Reflections on deforestation, wildlife eviction, and stewardship behavior. *Psychology and Marketing*, *37*, 1720–1730.

Solá, A., & Butler, R. (2021). *Governments, companies pledge $1 billion for tropical forests*. Retrieved 30 March 2021, from https://news.mongabay.com/2021/04/governments-companies-pledge-1-billion-for-tropical-forests/.

Son, H. N., Chi, D. T. L., & Kingsbury, A. (2019). Indigenous knowledge and climate change adaptation of ethnic minorities in the mountainous regions of Vietnam: A case study of the Yao people in bac Kan Province. *Agricultural Systems*, *176*, 102683.

Spring, J. (2021). *Two-thirds of tropical rainforest destroyed or degraded globally, NGO says*. Reuters. Retrieved 25 March 25 2021, from https://www.reuters.com/article/idUSKBN2B00TU.

Staal, A., Dekker, S. C., Hirota, M., & van Nes, E. H. (2015). Synergistic effects of drought and deforestation on the resilience of the South-Eastern Amazon rainforest. *Ecological Complexity*, *22*, 65–75.

The Earth Observatory. (2021). *Brazil battered by drought*. Retrieved 28 August 2021, from https://earthobservatory.nasa.gov/images/148468/brazil-battered-by-drought.

The Economist. (2021). *What is the economic cost of covid-19?*. The Economist. Retrieved 16 February 2021, from https://www.economist.com/finance-and-economics/2021/01/09/what-is-the-economic-cost-of-covid-19.

The Nature Conservancy. (2019). *6 Things to know about the Amazon rainforest fires*. The Nature Conservancy. Retrieved 12 January 2021, from https://www.nature.org/en-us/what-we-do/our-priorities/provide-food-and-water-sustainably/food-and-water-stories/fires-in-the-amazon/.

The White House. (2021). *Executive order on tackling the climate crisis at home and abroad*. The White House. Retrieved 4 April 2021, from https://www.whitehouse.gov/briefing-room/presidential-actions/2021/01/27/executive-order-on-tackling-the-climate-crisis-at-home-and-abroad/.

The World Bank. (2020). *COVID-19 could lead to permanent loss in learning and trillions of dollars in lost earnings*. World Bank. Retrieved 16 February 2021, from https://www.worldbank.org/en/news/press-release/2020/06/18/covid-19-could-lead-to-permanent-loss-in-learning-and-trillions-of-dollars-in-lost-earnings.

Thomson, A. (2020). *Biodiversity and the Amazon rainforest – Greenpeace USA*. Retrieved 25 February 2021, from https://www.greenpeace.org/usa/biodiversity-and-the-amazon-rainforest/.

Tickle, A. (2013). *Uncontacted peoples of the Javari Valley, Brazil (Anthropol. Sr. Theses)*.

Tigre, M. A. (2017). Critical analysis of the act/acto. In *Regional cooperation in Amazonia* (pp. 353–392). Brill Nijhoff. https://doi.org/10.1163/9789004313507_011.

Tigre, M. A. (2019). Building a regional adaptation strategy for Amazon countries. *International Environmental Agreements: Politics, Law and Economics*, *19*, 411–427.

Uhl, C., Kauffman, J. B., & Cummings, D. L. (1988). Fire in the Venezuelan Amazon 2: Environmental conditions necessary for forest fires in the evergreen rainforest of Venezuela. *Oikos*, *53*, 176–184.

UNDP. (2020). *Impact of COVID-19 on the sustainable development goals: Pursuing the sustainable development goals (SDGs) in a World Reshaped by COVID-19*. Retrieved 14 January 2021, from https://sdgintegration.undp.org/sites/default/files/Impact_of_COVID-19_on_the_SDGs.pdf.

UNEP. (2020). *Preventing the next pandemic – Zoonotic diseases and how to break the chain of transmission*. UNEP – UN Environment Programme. Retrieved 12 January 2021, from http://www.unep.org/resources/report/preventing-future-zoonotic-disease-outbreaks-protecting-environment-animals-and.

UNFCCC. (2016). Aggregate effect of the intended nationally determined contributions: An update. In *Partnership on transparency in the Paris agreement*. Retrieved 6 January 2021, from https://www.transparency-partnership.net/documents-tools/aggregate-effect-intended-nationally-determined-contributions-update.

UN-REDD Programme. (2021). *What is REDD+?*. Retrieved 30 March 2021, from https://www.unredd.net/about/what-is-redd-plus.html.

US Department of Commerce. (2012). *United States Summary: 2010*. Retrieved 10 March 2021, from https://www2.census.gov/library/publications/decennial/2010/cph-2/cph-2-1.pdf.

Valsecchi, J., et al. (2017). *Update and compilation of the list untold treasures: New species discoveries in the Amazon 2014–15*. Retrieved 17 February 2021, from https://wwf.panda.org/?310013/381-new-species-discovered-in-the-Amazon.

Van Noordwijk, M., et al. (2019). Agroforestry into its fifth decade: Local responses to global challenges and goals in the Anthropocene. In M. Van Noordwijk (Ed.), *Sustainable development through trees on farms: Agroforestry in its fifth decade* (pp. 397–418).

Voiland, A. (2020). *Reflecting on a tumultuous Amazon fire season*. Retrieved 5 February 2021, from https://earthobservatory.nasa.gov/images/146355/reflecting-on-a-tumultuous-amazon-fire-season.

von Randow, C., et al. (2004). Comparative measurements and seasonal variations in energy and carbon exchange over forest and pasture in south West Amazonia. *Theoretical and Applied Climatology*, *78*, 5–26.

Walker, R., & Hill, K. (2015). Protecting isolated tribes. *Science*, *348*, 1061.

Walker, R., & Simmons, C. (2018). Endangered Amazon: An indigenous tribe fights Back against hydropower development in the Tapajós Valley. *Environment: Science and Policy for Sustainable Development*, *60*, 4–15.

Walker, R., et al. (2009). Protecting the Amazon with protected areas. *Proceedings of the National Academy of Sciences*, *106*, 10582–10586.

Walker, W. S., et al. (2020). The role of forest conversion, degradation, and disturbance in the carbon dynamics of Amazon indigenous territories and protected areas. *Proceedings of the National Academy of Sciences*, *117*, 3015–3025.

Weber, M., Günter, S., Aguirre, N., Stimm, B., & Mosandl, R. (2008). Reforestation of abandoned pastures: Silvicultural means to accelerate forest recovery and biodiversity. In E. Beck, J. Bendix, I. Kottke, F. Makeschin, & R. Mosandl (Eds.), *Gradients in a Tropical Mountain Ecosystem of Ecuador* (pp. 431–441). Springer. https://doi.org/10.1007/978-3-540-73526-7_41.

Weisse, M., & Goldman, E. (2020). *The latest analysis on global forests & tree cover loss*. Global Forest Review. Retrieved 29 March 2021 from https://research.wri.org/gfr/forest-pulse.

WHO. (2021). *WHO Coronavirus (COVID-19) Dashboard*. Retrieved 12 January 2021, from https://covid19.who.int.

Wilkinson, D. A., Marshall, J. C., French, N. P., & Hayman, D. T. S. (2018). Habitat fragmentation, biodiversity loss and the risk of novel infectious disease emergence. *Journal of the Royal Society Interface*, *15*, 20180403.

Wolf, A. T. (2007). Shared waters: Conflict and cooperation. *Annual Review of Environment and Resources*, *32*, 241–269.

X-prize. (2020). *Economic benefits of saving the rainforest*. XPRIZE. Retrieved 1 April 2021, from https://www.xprize.org/prizes/rainforest/articles/economic-benefits-of-saving-the-rainforests.

Yeh, S.-W., et al. (2009). El Niño in a changing climate. *Nature*, *461*, 511–514.

Zeng, N., et al. (2008). Causes and impacts of the 2005 Amazon drought. *Environmental Research Letters*, *3*, 014002.

Zimmerman, B., & Ferreira, I. (2020). *Proposed law of devastation*. ArcGIS StoryMaps. Retrieved 15 March 2021, from https://storymaps.arcgis.com/stories/f0954b0a0615405da8b40207a448bc20.

Zohdy, S., Schwartz, T. S., & Oaks, J. R. (2019). The coevolution effect as a driver of spillover. *Trends in Parasitology*, *35*, 399–408.

CHAPTER

14

Farmers' perceptions of climate hazards and coping mechanisms in Fiji

SamRoy Liligeto and Naohiro Nakamura

School of Agriculture, Geography, Environment, Ocean and Natural Sciences, The University of the South Pacific, Laucala Campus, Suva, Fiji

1 Introduction

Coping with climate hazards has always been an important factor in the survival of agriculture in rural areas not only in Fiji but globally. The agricultural sector of Fiji remains one of the important economic sectors, contributing 10.4% of the total gross domestic product (GDP) of the country (Pacific Community, 2021). It is argued that subsistence or small-scale farmers are the most vulnerable to climate change due to their limited capacity to adapt (Alpízar et al., 2020). However, there are limited empirical studies focused on the impact of climate hazards on agriculture in the Pacific Island countries (PICs) (Méheux et al., 2007; Savage et al., 2020).

The literature on the coping methods against climate hazards in small island nations has shown that traditional knowledge and methods are an important part of coping against climate hazards (Ahmad et al., 2021; Bell et al., 2016; Kumar et al., 2020; Le Dé et al., 2018). Meanwhile, some studies show that the combination of modern and traditional techniques is effective to enhance sustainability and resilience (Šūmane et al., 2018; Velten et al., 2015). Because of the scale of impact, the PICs now have rich literature on adaptation strategies to climate change, including responses to natural hazards, postmanagement strategies, and several adaptation techniques (Crumpler & Bernoux, 2020; Robinson, 2020; Savage et al., 2020). In Fiji, many empirical studies have previously been conducted in natural hazard prone areas such as Ba, Rewa, Nadi, and Ra (Brown et al., 2017; Chandra & Gaganis, 2016; Cox et al., 2018, 2020; McAneney et al., 2017; Nichols, 2019; Nolet, 2012, 2016). Nevertheless, empirical studies on farmers' perceptions of climate hazards and coping methods against the impact of climate hazards in agriculture have not been much observed, except for a few (Taylor et al., 2016). Do farmers experience the impact of climate change and in what way, what methods are they employing to cope with the impact of climate hazards, and are those methods effective? Who is particularly vulnerable under the current changing climate? This chapter looks at farmers' perceptions of climate hazards, coping methods and adaptive capacity toward climate hazards, and the effectiveness of the coping methods, using the case studies of Wainadoi and Navua in Viti Levu, Fiji. The two sites were selected because both communities are agriculture based and have experienced numerous floods, due to their location next to a river system.

2 Background

2.1 Vulnerability of Pacific Island country agriculture to climate hazards

Agriculture is an important sector in the South Pacific region. Agricultural products are a huge part of export in many PICs, with food-producing activities such as agriculture and fishing being a major portion of the labor force either commercially or for self-sufficiency (Iese et al., 2020; Sisifa et al., 2016). Many of the PICs are located in or near the cyclone belt or located near the tectonic boundaries of the Australian and Pacific plates, hence the countries are

Climate Impacts on Extreme Weather
https://doi.org/10.1016/B978-0-323-88456-3.00016-2

prone to natural hazards, including seismic activity, tsunamis and volcano eruptions (Noy, 2015). In addition, PICs heavily rely on rainfall for freshwater, for both domestic consumption and agriculture, which increases PICs' vulnerability to extreme weather changes such as heavy rainfall and droughts (Noy, 2016).

The impact of climate change on agriculture includes changes in mean temperature, rainfall, winds, and the increased intensity and frequency of extreme hazards such as cyclones, floods, droughts and heatwaves (Iese et al., 2020; Sisifa et al., 2016). According to Bell et al. (2016), the equatorial Pacific region can expect up to a 35% in increase rainfall under the current high emission scenario with smaller increases in the north, while the southern tropical Pacific with wet seasons becoming much wetter while dry seasons will become much drier.

Agriculture is highly sensitive to climate change as it depends on stable weather conditions for optimum growth (Raggi et al., 2017). However, climate change has destabilized precipitation and temperature in the Pacific region, which has increased the frequency and intensity of extreme weather events that threaten agricultural production (Iese et al., 2020; Sisifa et al., 2016). The projected temperature increase will also affect the development of fruit crops such as mango and papaya. The high temperature will increase the speed of panicle development, which results in faster periods of growth while reducing the number of days for effective pollination for mango seeds. For papaya crops, the temperature increase results in plant sterility due to flower drops among female and hermaphrodite plants and sex changes among hermaphrodite and male plants. The temperature increase will also contribute to the spread of diseases such as black leaf streak disease and dalo leaf blight, as higher temperature reduces the vitamin levels in fruit and vegetable crops (Iese et al., 2020; Sisifa et al., 2016). Factors such as recovery periods between natural hazards, high temperature and less rainfall will also increase the levels of stress affecting agriculture such as soil infertility and salinity levels. Climate change models indicate that increase in temperature of 1–2 degrees will incur a decrease in yield as heat tolerance levels of crops are exceeded in tropical and subtropical regions (Hadgu et al., 2015).

According to Lee et al. (2018), there has been a significant increase in the number of severe natural hazards in the South Pacific since 2008. For the PICs that depend heavily on rainfed agriculture and limited economic sectors, the recent extreme weather events have had a serious impact on countries' economies. For example, Tropical Cyclone Winston of 2016 in Fiji resulted in record-high prices of local agricultural products and the prices are expected to remain high (Iese et al., 2020; Sisifa et al., 2016). Human-induced climate change and associated changes in weather patterns are considered the most significant threat to global food security, including production, access, stability and utilization. The PICs are especially the most affected. Overall, the impacts of climate change on agriculture in the PICs are observed through the loss of coastal land, increased contamination of groundwater and estuaries through saltwater intrusion, and the increased severity and frequency of extreme weather events such as cyclones, drought and heat stress.

2.2 Traditional knowledge and coping methods in agriculture

Traditional Pacific societies have been religious with a heavy emphasis on "manifestations of the sacred" (Janif et al., 2016). Oral history in local communities has been a source of guidance to cope with gradual environmental change or extreme weather events (Janif et al., 2016). Over centuries, Pacific island communities have accumulated techniques and knowledge to live with the surrounding environment. Such knowledge and beliefs form people's cultural identity while having been functioning as an important source to understanding climate and meteorology, or as a means to support sustainable livelihood (Kelman et al., 2016). Scholars such as Nunn et al. (2017) and Piggott-McKellar et al. (2019) argue that past natural hazard events have enabled the Pacific region to be resilient to extreme weather events. The true strength of indigenous knowledge according to Šūmane et al. (2018) and Nunn et al. (2017) is not in simple adaptation but the ability to adopt principles from other knowledge systems and compare them with one's own to decide if these aspects are worthy to be incorporated into existing systems. It is important to note that there is a wealth of knowledge that is scientifically supported (Chambers et al., 2019; Ching et al., 2020; Johnston, 2015; Malsale et al., 2018; Nalau et al., 2018), linked to the precursors of climate hazards such as the observation of hornet's nests near the ground as a sign of an approaching cyclone or strong winds (Janif et al., 2016; Nakamura & Kanemasu, 2020). According to Fletcher et al. (2013), understanding local strategies that are used before, during, and after a climatic event would define the Pacific Islands' ability to adapt to climate-related events. To this end, the Pacific Islands Framework for Action on Climate Change (PIFACC) promotes the application of traditional knowledge into adaptive measures, calling for more research linking historical records and anecdotal evidence of local coping strategies and the integration of traditional knowledge into information management systems (Ford et al., 2016).

Likewise, PICs have rich local agricultural knowledge and techniques passed down over generations to cope with the impact of climate hazards. In Fiji, farming on hillsides and slopes to cope with flooding in low-lying areas has long been practiced, for instance. Other techniques include using bush areas surrounding a garden as a windbreak to lessen soil loss and the evapotranspiration processes. Forests function as an important part of the security system, mitigating against cyclones and drought (Taylor et al., 2016). Crops are grown in rotation with soil left to fallow without the use of chemicals. For instance, on the island of Matuku, villagers have been planting dalo (*Colocasia esculenta*) on the same piece of land for generations without affecting the yield and without the use of fertilizer (Connell & Waddell, 2006). Another technique is the use of a simple stick sharpened at one end as the sole tool to farm dalo. A corm or sucker is placed in a hole of about 20–30 cm from the surface and left unfilled to collect rainwater, which allows the dalo crop to grow (Thaman, 1999). The establishment of a garden of root crops is done under certain variables such as soil type, cover, slope and vegetation, which provides varieties of crops as needed. Yam is often planted in mounds with finely broken-up soil to prevent soil erosion (Turner, 1984). However, nowadays excessive farming on slopes of root crops, marginal sugarcane lands and traditional ginger have been common in traditional cropping systems, which has somewhat increased vulnerability because of soil erosion and the loss of soil nutrition (Wairiu, 2017).

McMaster & McGregor (1999) argue that while the traditional cropping systems in Fiji have evolved to cope with the impact of climate hazards, along with adopting new techniques such as modern technologies, the traditional systems have become less practiced, which is increasing vulnerability to climate hazards in agriculture. Mbah et al. (2021) and Janif et al. (2016) also noted that the loss of traditional knowledge and nonlearning among the younger generation are risks to the recovery from the impact of climate hazards and coping. Indeed, modern coping methods and techniques are deemed by them as threats to traditional coping methods and to the way of life in the PICs. Meanwhile, Campbell (2015) states that traditional cropping systems are not as strong against cyclones as they used to be. Prasad and Sud (2019) observed that the traditional coping method is deemed irrelevant to an ever-changing climate system nowadays. Such views are shared by Nakamura and Kanemasu (2020), who reported that as the intensity and irregularity of severe weather events increases, relying solely on traditional coping methods and knowledge for prediction and mitigation is becoming unfeasible and unreliable. For instance, after Tropical Cyclone Winston in 2016 on Koro Island, any traditional agricultural methods did not effectively function for recovery because of the intensity of the cyclone.

Some of the interesting questions that this work answers are: Can Pacific Island countries solely rely on traditional coping methods to deal with the impact of changing climate and climate hazards on agriculture? How do farmers indeed perceive climate change and hazards?

2.3 Farmers' perceptions of climate change and climate hazards

In a study on the farmers' perceptions of climate change in Pakistan, Abid et al. (2019) reported that a farmers' accurate perception of climate change significantly influences farmers' abilities to adapt. Arbuckle et al. (2015) further note that farmers' perception and level of accuracy of climate change are greatly affected by the level of one's education, beliefs, and experience. For instance, the level of education helps one understand the system of climate change and how human activities are related. According to Arbuckle et al. (2015), experience also takes a key role to form one's risk perception and to take self-protective behaviors. This is also confirmed by Burnham and Ma (2017), who conducted a study on Chinese small-scale farmers' perceptions of adaptation to climate change. According to the study, small-scale farmers with experience of changes in precipitation in the last 30 years have a strong intent to adapt their agricultural methods to cope with the changes. For those farmers, changes are potential risks of climate hazards, and they try to respond by changing their behaviors. These findings suggest that farmers need to perceive the changes in the climate to take appropriate adaptive strategies to improve the resilience of the agro ecological system. Misperceptions of climate change may result in adopting "wrong" or "inappropriate" measures, or not responding at all, which makes farmers vulnerable.

In the context of PICs, recent studies have found that people often interpret climate change and natural hazards based on religious and spiritual beliefs. That is, climate hazards are understood to be caused by the hands of God, or caused as a form of punishment for human wrongdoing (Bird et al., 2021; McKenna & Yakam, 2021). The religious background depicts gods and deities ruling over human subjects with the natural degradation of human morals, and the increased wickedness is seen as the cause of climate hazards such as floods and cyclones (Luetz & Nunn, 2020; Nunn et al., 2016). Cox et al. (2018) found that a severe flood experienced in Nadi, Fiji in 2012, was viewed by iTaukei Christians as an act of God to punish permissive sexual behaviors entertained by the tourist industry. In their study, this view was echoed by some other interviewees, who refer to the Biblical story of Sodom and Gomorrah tabooing homosexual behaviors and believe that such "wrongdoings" caused the flood. Other interviewees saw the flood as punishment by God on the current Prime Minister—Bainimarama's government, which adopted the constitution stating that Fiji is a secular state and moving away from traditional chiefly-leadership and Christianity. This political interpretation was also prominent after Tropical Cyclone Winston in 2016. Other Christian interviewees stated that areas affected by Tropical Cyclone Winston were resided by those who were committing a sin, and the cyclone path marked on the map represents such areas (Cox et al., 2018). The people of Koro Island, one of the most severely affected islands by Tropical Cyclone Winston, were labeled by Methodist interviewees as overindulgence in the consumption of grog and the nonobservance of the Sabbath (Cox et al., 2020).

The scientific interpretation of climate change is generally difficult to understand for the public and in our daily life, it is difficult to observe or predict correctly some direct impact of climate change at local levels and human time scales (Alam et al., 2017). In addition, people tend to make decision-making more based on feelings or religious beliefs than on scientific findings (Arbuckle et al., 2015). Those who have been exposed to the power of climate hazards realistically feel the limitations of the human capacity to handle such a power; thus they often interpret such events as actions of God (Cox et al., 2018). If the ontology of farmers in the PICs is more religious-based instead of science-based, farmers may not try to cope with the impact of climate change or climate hazards, because they are actions of God and there is not much one can do. This is why it is important to understand what factors shape the perception and responses of farmers to climate change (Alam et al., 2017).

Currently, there appears to be a discrepancy between scientific interpretation of climate change and local farmers' perceptions. Although there may be better-coping methods available to the PICs to battle climate hazards, farmers are not always ready to accept scientific interpretation of climate change and to move away from religious-based interpretation, or shift from traditional coping methods to science-based modern techniques (Barnett & Waters, 2016; Fischer & Connor, 2018). If farmers cannot, or hesitate to adapt, why is it, and are PIC farmers indeed struggling to adapt?

2.4 Fiji and case study sites

The Fiji Islands consist of 322 islands made up of coral reefs, uplifted ocean sediments and sand islands. Out of the 322 islands and 522 islets that make up the archipelago, 106 of these islands are permanently inhabited (Martin, 2018). Viti Levu is the largest island, which covers 57% of the land area and contains 67% of the total population with the capital city of Suva. The eastern side of Viti Levu is lush in tropical forests due to heavy rainfall, in contrast to the western low lands on Viti Levu with mountains and relatively dry.

Fiji has several resources, including mineral, forestry, fishery and water resources, making it one of the economically developed countries in the South Pacific. Tourism has been the largest industry over the past 20 years, while the sugar industry—once the largest industry in the country—is on the decline.

In Fiji, approximately 87%–91% of the land is owned by indigenous (iTaukei) Fijians (Ben & Gounder, 2019). The rest is either state land or freehold land. Customary ownership of the land is intertwined with culture and history where the inseparability of the iTaukei people to the land is based on the concept of *vanua* (Chandra & Gaganis, 2016). The land is considered to be closely tied with spiritual and family connections and is a proud aspect of iTaukei life and society (Charan et al., 2017). Subsistence farming has supported stewardship of land resources over the years when fallowing has allowed recovery of land sections preventing over-cultivation. Overexploitation and cultivation are deemed as "western" rather than traditional paradigms (Boydell, 2001).

2.4.1 *Serua/Namosi region*

The Serua/Namosi area is situated on the leeward side of Viti Levu (Fig. 14.1) and thus experiences a lot of rainfall yearly. The area is mountainous and densely forested.

As of June 2019, the combined population of the Serua/Namosi region was 20,120 (see Table 14.1). The population in the area is relatively young with the average age group in males ranging between 15 and 30 and in females 20 and 24 (Fiji Bureau of Statistics, 2017).

The Serua/Namosi region is very fertile. The region hosts primary/secondary industries, such as copper mining, logging and gravel extraction. Mahogany plantations are found in the Serua/Namosi hills (Duaibe, 2008). Agriculture is a major source of income, growing crops and then selling in major urban centers like Navua and Suva. This study's case study sites, Navua and Wainadoi are located under the path of the southeast trade winds; hence the area has more rainfall than other water catchments in Fiji. Heavy rainfall also causes large-scale floods in the area (Kostaschuk et al., 2001).

2.4.2 *Wainadoi settlement*

Wainadoi settlement in the Namosi region is situated approximately 25 km west of Suva, along the Queen's Highway. The settlement is situated 10–15 km from the sea in a valley, surrounded by mountainous terrains and lush green

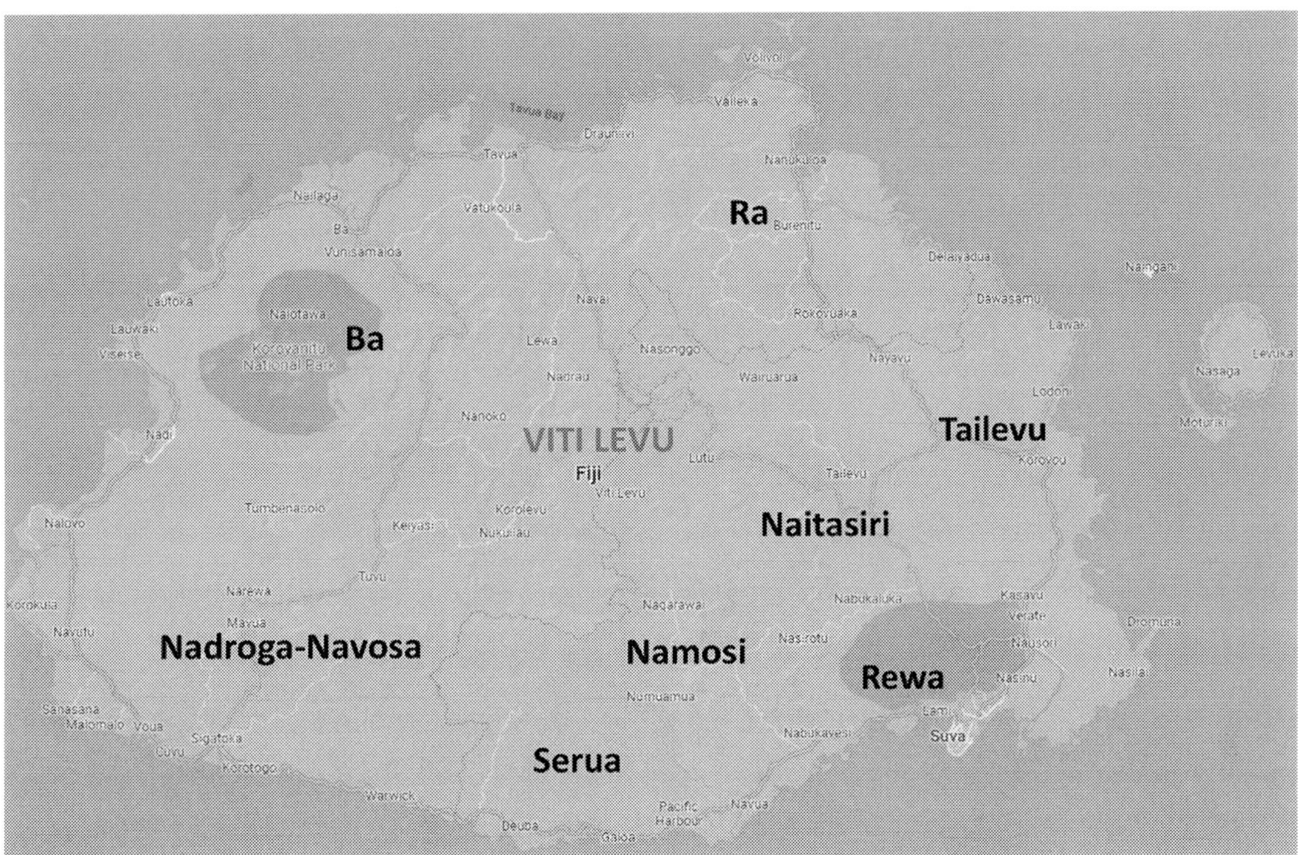

FIG. 14.1 Viti Levu Provinces and our case study sites Wainadoi and Navua. *Source: Liligeto 2022, based on Google Map, no permission required.*

TABLE 14.1 The population of Fiji provinces from the years 2007–17.

	Census Year			
Province	**2007**	**2017**	**Absolute Change**	**Percentage Change**
Ba	231,760	247,708	15,948	6.9%
Bau	14,176	15,466	1,290	9.1%
Cakaudrove	49,344	50,469	1,125	2.3%
Kadavu	10,167	10,897	730	7.1%
Lau	10,683	9,602	–1,081	–10.1%
Lomaiviti	16,253	15,657	–598	-3.7%
Macuata	72,441	65,983	–6,458	–8.9%
Nadroga/Navosa	58,387	58,931	544	0.9%
Naitasiri	160,760	177,678	16,918	10.5%
Namosi	6,898	7,871	973	14.1%
Ra	29,464	30,432	968	3.3%
Rewa	100,995	108,016	7,021	7.0%
Serua	18,249	20,031	1,782	9.8%
Tailevu	55,692	64,552	8,860	15.9%
Rotuma	2,002	1,594	–408	–20.4%
Total	**837,271**	**884,887**	**43,279**	**5.7%**

Source: Authors, based on Statsfiji. (n.d.). Retrieved 5 May 2021, from https://www.statsfiji.gov.fj/index.php/census-2017/census-2017-release-1. No permission required.

forests. Its location is important to this study as it is situated in an area prone to flood; the settlement is located next to the Wainadoi River. It is also an agriculture-driven community generally based on small-scale subsistence farming with a few commercial activities such as logging and gravel extraction.

The population of the Wainadoi settlement consists of various ethnicities with the majority of the iTaukei predominantly from the Serua/Namosi region. The population of Wainadoi is relatively small and is estimated to be around 500–600. As stated, the Wainadoi area is predominantly a farming area. Vegetables, root crops and dairy farming are predominantly grown. The sales of products at roadside stalls are a common activity in Wainadoi, while coconut selling is also increasing.

2.4.3 *Navua*

Navua is situated in a valley next to the Navua River, 35 km west of Suva. The area is predominantly flat grasslands, ideal for cattle farming and relatively near to the sea. Navua is also notorious for frequent floods that have caused significant loss and damage to residents and agriculture in the area (Duaibe, 2008; Mataki et al., 2006). The cause of frequent floods of the Navua River is due to sedimentation over time, which has enhanced the level of the river bottom and made it easier for the river to burst its banks.

There are many agricultural and commercial activities conducted in the area. According to Mataki et al. (2006), Navua town is populated by low to middle-class income earners and is rapidly urbanizing due to the growth of small-scale economic activities. The population of Navua was 5812 in June 2019 (Fiji Bureau of Statistics, 2017). According to Mataki et al. (2006), most people living in and around Navua town have on average a weekly income of $US 35–45. Before 1990, commercial rice farming was an important economic activity; however, the rice field was abandoned due to floods, pests and cheaper rice imported from Asia (Mataki et al., 2006). Agriculture remains the major industry (Fiji Bureau of Statistics, 2017) with a shift from subsistence farming to commercial farming. The loss of trees due to logging in the upper catchment area has caused many problems such as soil erosion and runoff during heavy rainfall (Duaibe, 2008). Overgrazing is another problem in Navua with many cattle farms on the outskirts of the town. Recently, there has been a large employment creation in the agricultural sector in the area, thanks to investment by the Grace Road Group, a South Korean–based Christian corporation running farming, construction, food processing, restaurant, trading, and health services, among others (http://www.grgroup.com/). The Grace Road Group reestablished commercial rice farming using heavy machinery to cultivate the land. Some local farmers still prefer to farm on hillsides as it is a common traditional practice in the area; however, because of the above-mentioned soil erosion, some farmers are facing challenges to continue hillside farming.

Of particular note in Navua is the Raiwaqa Alta district, the low-lying area particularly vulnerable to climate hazards. The majority of the residents of the district are Indo-Fijians, most of whom are former sugarcane farmers in Vanua Levu recently relocated to Raiwaqa Alta because of the expired land leases. Despite this new location being very prone to natural hazards and unfavorable conditions for agriculture, they remain there because of the lack of financial capacity to gain access to more favorable land away from the Navua River.

3 Methods

Two primary research methods were employed for this study: interview and observation. Face-to-face semistructured interviews were conducted between the 1st and 27th of May 2018, in Wainadoi and Navua. A total of 90 farmers participated in the interviews, 40 in Wainadoi and 50 in Navua. While there were some exceptions, we strategically recruited middle to old-aged farmers who have been residing in the area for 10 years or more, under the assumption that such farmers have a first-hand experience of environmental and climate change in the area. While best efforts were made to have gender-balanced participants, the majority of interviewees in this research were men. This is because men generally have the responsibility of farming in Fiji. The participants' ethnic background (iTaukei and Indo-Fijians) is comparatively balanced, except for Raiwaqa Alta in Navua, where the majority of residents are Indo-Fijians. The background of the interview participants is summarized in Tables 14.2–14.4.

Although data was not collected on participants' level of education, most participants stated that they do farming because of their low-level education, not having completed primary or secondary school. The limited level of education resulted in the participants with fewer skills and qualifications, thus farming is the only source of income and subsistence. They stated that they can only do farming but it is a lifestyle choice, aiming to be self-reliant. Such a view was shared by many participants.

The list of interview questions can be found in Appendix. Interviews were conducted either in the iTaukei language, Fiji Hindi or English, based on the preference of the participant, with the presence of a research assistant, who made a contact with potential participants and acted as a translator, both in iTaukei and Fiji Hindi. Interviews were conducted

TABLE 14.2 Participants by gender.

Settlement	Population sample		
	Male	Female	Total
Wainadoi	38	2	2
Navua area			
Batinikia	6	0	6
Vakabalea	9	1	10
Raiwaqa Alta	24	10	34
Total	77	13	90

TABLE 14.3 Participants by age groups.

Settlement	Age group				
	20s	30s	40s	50s	Total
Wainadoi	4	6	14	16	40
Navua area					
Batinikia	1	1	2	2	6
Vakabalea	1	1	3	5	10
Raiwaqa Alta	2	5	8	19	34
Total	8	13	27	42	90

TABLE 14.4 Participants by ethnicity.

Settlement	Ethnicity		
	iTaukei	Indo-Fijian	Total
Wainadoi	26	14	40
Navua area			
Batinikia	4	2	6
Vakabalea	7	3	10
Raiwaqa Alta	3	31	34
Total	40	50	90

in a *talanoa* style setting (storytime) (Halapua, 2008), where the discussion of related topics is done in a relaxed environment for the participant farmers, which enabled the free flow of discussion with the interviewer constantly keeping track of where the discussion was going based on the research questions.

Observation was also conducted in the field in the period of 1st to 17th of May 2018. We observed the locations and conditions of farms, river systems, and drainage systems. Being a resident of the Wainadoi and Navua region, Liligeto was also able to recall some changes that happened over the years in each study site.

4 Results

4.1 Farmers' perceptions of climate hazards

In both Wainadoi and Navua, interview participants were first asked about the impact of climate hazards and any other factors on agriculture. The results are summarized in Table 14.5.

TABLE 14.5 The impacts of climate hazards.

Natural hazards	Impacts of natural hazards	Wainadoi	Navua
Flooding	Damaging crops	33	50
	Fertile soil	4	7
	Longer period to restart planting	33	50
	Decreased capacity to sell products	33	50
	Harvesting mature crops	6	5
	Debri	5	18
Cyclones	Damaging crops	2	6
Heavy rainfall	Blocking drains	5	3
Change in weather	Increased rainfall	21	23
	Increased intensity of floods and cyclones	31	32
	Unpredictable weather	15	32
Some other problems and their impact		**Wainadoi**	**Navua**
Poor drainage systems	Damaging crops	12	23
	Lack of action	9	7
Height of road	Backflow of water	9	12
Lack of dredging of river	Raised river bed	11	22
	Susceptible to flooding	12	32
Prone site location	Constant flooding	33	50

It is evident that in both study sites, flooding is causing major impacts on farming. In Wainadoi, where the entire area is located along a river system, 33 interviewees identified flooding as the most common climate hazard, and the location of their farm next to the river system is indeed making them prone to a flood. They also observed the increase in flooding over the years, which is substantially affecting root crop yields such as *dalo* and cassava, the major staple food source for them. They pointed out that the change in weather and climate conditions is affecting planting times and successful harvest rates. Interviewees also stated a few positive impacts of increasing floods such as the deposit of fertile soil on the farming ground, which enhances the productivity of crops. Regarding other problems, poor maintenance of drainage systems was identified as one of the major factors of flash floods in Wainadoi (Fig. 14.2). Interview participants who live beside the Queen's Highway identified the height of the road as another major cause of a flood (Fig. 14.3). According to them, the height of the road has been elevated further because of constant road works, which in turn renders the drainage systems ineffective to handle the back flow of water.

Interviewees stated a change in weather was observed in the area over the last decade. The interview results suggest that flooding in Wainadoi has negatively affected farmers' livelihood as most households in the area rely on farming and selling products for their life. Indeed, 17 out of the 40 participants solely rely on agriculture as a means of income. They also observed an increase of pests such as rats and insects in the area.

Frequent floods also delay planting, as the soil becomes too wet to plant any crop after a flood. Nineteen out of the 40 interviewees stated that they have to wait for 1–4 months to restart planting after each flood. Although 11 participants stated that a 1–4 week waiting period should be enough to restart planting, 2 participants stated that they have to wait for more than a year. In any way, these waiting periods negatively affect farmers' livelihood.

In Navua, all interviewees are located relatively close to the Navua River ranging from 100 m to 3 km. They indicated that floods as the most common climate hazard experienced in the area. As stated, the Raiwaqa Alta district is particularly prone to floods. Heavy rainfalls and cyclones are other common climate hazards. Regarding the impact of floods, the interviewees stated that there was nothing they can do to control floods or stop their occurrence. According to them, floods are reoccurring events because of the location of the farms next to the Navua River. Besides, seven interviewees stated that poorly maintained drainage systems cause floods (Fig. 14.4). According to interviewees, constant floods of low-lying areas destroy farms, which negatively affects the livelihood of households whose only source

FIG. 14.2 A blocked drainage system in Wainadoi. Source: Picture was taken by S. Liligeto in 2018. *No permission required.*

FIG. 14.3 The elevated Queen's Highway. The elevated road causes backflow of water and blockage of natural drainage outlets. Source: Picture was taken by S. Liligeto in 2018. *No permission required.*

FIG. 14.4 A blocked drainage system in Navua. Source: Picture was taken by S. Liligeto in 2018. *No permission required.*

TABLE 14.6 Farmers' perceptions of climate hazards.

Perceptions	Wainadoi	Navua
End of the world	9	8
Natural occurrence of events	9	10
Form of punishment	13	8
Human impact on the environment	3	6
Improper use of land	6	8

Source: Authors.

of income is agriculture. Indeed, 35 out of 50 participants stated that agriculture is their only source of income. Meanwhile, similar to Wainadoi, a few interviewees identified the positive impacts of floods: seven participants stated that a flood brings fertile soil, which improves plant health and yield. Interviewees also stated the dredging of the Navua River is important to reduce the risk of flood and needs to be constantly maintained.

The participants were then asked how they understand climate hazards (Table 14.6). Firstly, climate hazards were seen as a form of punishment by a spiritual entity: "God is punishing those who are sinful"—Interviewee. This view was expressed by 13 and 8 participants in Wainadoi and Navua, respectively, all of whom are Christian. Climate hazards were also seen as the result of the improper use of land by some Christian participants, who see the human occupation of the earth as a duty of stewardship: "We don't look after the land that's why these things (natural hazards/disasters) happen"—Interviewee. Other interviewees stated that the increase in climate hazards was because of a bad omen and saw as a sign of the earth ending: "God is coming back soon, and we must prepare ourselves. The cyclones and floods we experience here in Fiji are only telling us that He is coming soon. These things must happen before He comes again"—Interviewee. This view was also prevalent among Christian participants, who believed that climate hazards would continue to increase in frequency and severity, eventually leading to the end of the world.

Others saw climate hazards as a cycle of natural events. Flood was seen as a cleansing event of the environment removing pollution and waste from the earth. An event such as a cyclone was seen as an accumulation of the increased temperature with climate returning to normal as a means of nature balancing itself. Less than a tenth of respondents (9 out of a total of 90) stated that increasing human impact on the environment in the first world countries has resulted in climate change, which is increasing the severity of cyclones and floods and changing weather patterns. Some respondents also identified deforestation and pollution as the result of human activities altering the balance of nature.

4.2 Coping methods in agriculture

The interviewee participants listed various methods to cope with the impact of climate hazards. Coping methods used in Wainadoi and Navua are summarized in Tables 14.7 and 14.8, respectively.

For this study, the methods were divided into two categories: traditional (T) and modern (M). Traditional coping methods are reliant on nature and the environment and include farming knowledge passed down from generation to

TABLE 14.7 Coping methods used in Wainadoi.

Observation of weather, insects, trees and birds to predict natural hazards (T)	9
Trimming of cassava *(Manihot esculente)* stem (T)	6
Planting on elevated beds (T)	5
Resilient crops (T)	4
Machinery (M)	4
Hybrid crops (M)	4
Rotational farming (T)	3
Crops with quicker harvesting times (M)	2
Seasonal planting (T)	2
Trees used as wind breakers (T)	1

Source: Authors.

TABLE 14.8 Coping methods used in Navua.

Hybrid crop (M)	21
Fertilizer (M)	9
Machinery (M)	7
Chemicals (M)	6
Observation of weather, insects, birds and trees to predict natural hazards (T)	2
Trimming of *cassava* and *bele* stem (T)	2
Seasonal planting (T)	1
Resilient crops (T)	1
Rotational farming (T)	1

Source: Authors.

generation. For instance, observing the weather, insects, birds and certain trees as indicators of climate hazards is a traditional method. Other traditional coping methods include farming practices such as planting on elevated beds of soil to prevent damage to crops and trimming cassava stems to reduce damage by strong winds. Traditional coping methods are considered to leave less impact on the environment and do not require a heavy capital investment, hence small-scale subsistence farmers can easily adopt them.

Modern coping methods rely heavily on technology, chemicals and pesticides. For weather prediction, government-issued warnings broadcasted by radio and television fall under modern coping methods. This heavy reliance on technology is also observed regarding the improvement of crop resilience through selective breeding of stronger and more favorable traits. The use of machinery such as diggers also greatly reduces the amount of time to clear land for the cultivation of crops compared to clearing land by hand. Heavy inputs of fertilizer, pesticides and chemicals improve the survival rate and help the growth of crops. Modern coping methods leave more impact on the environment and sometimes require a large financial investment.

Overall, both traditional and modern methods are employed in both sites. Results show the benefit to adopt each method, as the farmers from both sites state that each coping method is useful to some extent, although its effectiveness varies.

First, regarding traditional coping methods, farmers state that they are easier to implement. The methods are far less complicated and require less assistance. Knowledge and skills are formulated from the surrounding environment and experience and are easier to obtain. Also, they do not have a monetary value. Some farmers, in particular in Wainadoi, see the practice of traditional techniques and knowledge as maintaining their culture and tradition. However, traditional methods are not free from disadvantages, according to the interviewees. Such disadvantages include slower maturity times (with an irrigated system of farming for dalo), and vulnerability to severe floods and strong winds. Traditional crops are also viewed as weak and unadaptable in an ever-changing environment and their survival rates are decreasing.

Second, regarding modern coping methods, farmers see them as durable in changing weather, requiring less labor due to machinery and quicker maturing times. Those who are adopting modern coping methods state that the use of machinery, chemicals, fertilizers and hybrid crops are effective. By adopting modern techniques in farming, survival rates have increased, harvesting has become quicker, and they can till the land faster. Farmers in Navua stated a preference for modern techniques based on the growing speed of crops. In Wainadoi, hybrid plants are seen as being easier to harvest and plant due to faster maturing time, resilience to bad weather and increased survival rate. It is however deemed an expensive method because of the necessity to hire equipment or purchase chemicals and fertilizers. Interviewees also point out the disadvantages of modern methods as being easily spoiled, input reliant, expensive, nonenvironment-friendly, and not practical in small farm settings. Modern hybrid crops are seen by the interviewees to have a shorter shelf life and are stated to be smaller compared to traditional crops. The interviewees also state that modern techniques and knowledge are not custom fitted to farmers and require some special skills to employ.

Tables 14.7 and 14.8 indeed show a contrast between the two sites. In Wainadoi, traditional coping methods are more frequently employed by farmers, while modern coping methods are widely used in Navua. While 18 out of 40 interviewees prefer to adopt modern coping methods in Wainadoi, the number of those who prefer modern methods in Navua is 43 out of 50. This difference between the two sites can be explained for several reasons. While those reasons will be discussed later, here, the scale of farms is commented on. That is, in Wainadoi, the majority of farmers are small-scale subsistence farmers. Hence, they do not have the financial capacity to employ modern coping methods. In Navua, relatively large-scale commercial farmers are dominant, and they have the capacity to invest to adopt modern coping methods. Indeed, it is rather a "must" to produce a surplus to sell products in the market. Our interviews also clarify that farmers find it most difficult to restart farms after a flood or a cyclone; however, commercial farmers can restart farms quicker than subsistence farmers due to surplus monetary funds obtained through the sale of crops. Subsistence farmers do not possess this coping capacity with crops grown for subsistence use only.

A key finding from our interviews is that the participants in both sites are recognizing that traditional coping methods are becoming ineffective to cope with the increasing frequency of floods and the intensity of cyclones, even though some of them prefer to rely on them. Farmers highlighted the less reliability of traditional coping methods due to changing climate conditions. For instance, forecasting bad weather and cyclones based on the observation of the color of the sky or plant conditions is now becoming unreliable due to irregularity of the weather and the changing climate pattern. Cyclones can now occur anytime and participants highlighted an increase in the number of cyclones over the past decade. Older participants further stated that the country used to be exposed to a very intensive cyclone every two decades. In today's changing climate, the country is exposed to such intensive cyclones every year, which makes it hard for farmers to solely rely on traditional coping methods as they are basically effective only under the constant weather conditions and patterns. For instance, under the current climate conditions, the use of local breeds of dalo was unable to cope with the sudden climate changes. Local breeds of dalo planted without the aid of manure and chemicals were unable to thrive against changing climate. Very intensive cyclones such as Tropical Cyclone Harold or Ana in 2021 destroyed everything, including cassava stem, elevated beds, or resilient crops, long used as traditional coping methods.

Interviewees adopting modern coping methods were observed by us to be more vigorous to find better coping methods. Such farmers stated that there is always a need to acquire farming knowledge and improve capacity to adopt cost-effective but efficient techniques and methods. Such information can be collected from the internet, another capital. These interviewees are constantly searching for methods practiced in other parts of the world and often conducting an experiment. For example, one interviewee in Navua planted vegetables in plastic containers based on their buoyancy to float on water. Such methods although not scientifically proven to be a sustainable agricultural practice highly suggest the plight of farmers in uncovering new and better-coping methods.

Participant farmers overall recognize the need to improve current coping methods; otherwise, they would not be able to continue farming. In both sites, a general shift from traditional coping methods to modern coping methods is seen as a must, and participant farmers recognize the effectiveness and ability to adapt to climate change of modern coping methods.

4.3 Farmers' challenges in adopting effective coping methods

As stated, some farmers, especially those in Wainadoi, still prefer to rely on traditional coping methods, even though they recognize that traditional methods need some improvement. Why do then those farmers keep relying on traditional coping methods?

One of the factors which prevent farmers from adopting modern methods is that small-scale subsistence farmers—the majority in Wainadoi—do not possess enough skills, knowledge or monetary capital to acquire the equipment and materials required for modern coping methods. For instance, for small-scale farmers who mainly plant subsistence crops, the use of machines requires a huge capital input, which is considered to be an unfeasible investment. The constant use of chemicals or fertilizers also costs a lot and farmers may not receive enough returns from the investment. The use of machines, chemicals, or fertilizers and planting hybrid crops would also require some skills and knowledge, which most small-scale subsistence farmers do not possess. Small-scale subsistence farmers are wearier of new options available to them and hesitant to take unknown risks (Yegbemey, 2020). For those farmers, providing opportunities to interact and learn new technologies, techniques and financial assistance available, such as holding workshops, would be effective. Indeed, such workshops are occasionally organized by the Ministry of Agriculture nationwide to educate farmers on better coping methods against climate hazards and enhance resilience. However, Vakabalea in Navua area farmers stated that very few agricultural workshops were carried out in the area between 2015 and 2018. Even if such workshops are held, language often functions as a barrier for farmers. Most small-scale subsistence farmers find it difficult to understand workshops because they are only conducted in English. Participant farmers stated that they would better understand what is explained if workshops are delivered in their language. For example, the Raiwaqa Alta area located in Navua is dominated by Indo-Fijian farmers; therefore, workshops need to be conducted in Fiji Hindi.

The second factor is that adopting nontraditional methods is seen by some as the loss of culture. This view was observed in both study sites. These farmers tend to reject modern coping methods because they involve large-scale alteration of their way of life and destruction of the environment, as opposed to traditional coping methods with less impact on the environment. Maintaining traditional coping methods is their way and the base of their cultural identity. Seeing traditional methods as the preservation of their culture is also associated with their religious belief. Religion plays a major role to form how they perceive natural hazards (Bird et al., 2021; Cox et al., 2018; Nunn et al., 2016). In the two sites, the damage on farms by floods was often perceived as a warning for farmers to return to the life defined by their religious belief and obedience to God, in particular among those with a strong Christian background. Farmers stated that the degree of disobedience to God could be identified through the scale of damage caused by floods. According to those farmers, they were punished in the form of climate hazards for their degradation of character and the misuse of the environment. These farmers hence tend to hesitate to adopt modern coping methods, that are seen as leaving more impact on the environment, or they tend to believe that there is nothing humans can do against God's punishment, other than correcting their sins.

As stated before, Navua farmers are more commercially oriented and relatively large-scale, while Wainadoi farmers are mostly subsistent and small-scale, and do not have the financial resources to adopt modern technologies. They are also less educated while retaining strong religious beliefs and intention to maintain their culture and identity; hence they prefer to stay with traditional methods being cheaper, easy to use and deemed environmentally friendly.

5 Discussion and conclusion

The key findings of our research are threefold: (1) the traditional coping methods in agriculture are becoming less effective under the current changing climate and intensive/frequent climate hazards; (2) farmers do recognize the decreasing effectiveness of traditional coping methods; and (3) some farmers, particularly small-scale subsistent, are hesitant, or cannot adopt more effective modern coping methods due to several reasons. While those farmers often come across financial challenges to adopt new methods, farmers with less education and/or a strong religious belief see the new modern methods as the loss of their culture and identity. Overall, we conclude that under the current change of climate, small-scale subsistence farmers are becoming more vulnerable (Alpízar et al., 2020).

It is widely recognized that the poor are more prone to environmental hazards (Winsemius et al., 2018) as they tend to inhabit remote low-lying flood plains due to limited development opportunities and relatively cheaper lands. Their livelihoods are less protected (Bangalore et al., 2019) and thus have a lower chance of coping with property loss when exposed to hazards such as floods (Thanh Thi Pham et al., 2020). If the poor are defined as those who lack access to financial capital, our findings are parallel with those previous findings. Also, our findings may be relevant to other climate hazard prone areas in Fiji.

What are then practical and feasible measures to enhance the resilience of small-scale subsistence farmers, in other words, to support small-scale farmers to convert their coping methods to more resilient ones under the current climate change? Given that many communities in Fiji are prone to climate hazards (Neef et al., 2018; Nichols, 2019), coping methods in agriculture with climate hazards need to be redeveloped, reassessed, and reorganized for the resilience of future generations facing similar circumstances (Bryant-Tokalau, 2018; Pan et al., 2018).

One possible measure is the implementation of an alternative farming system similar to a traditional farming system, which may maintain farmers' cultural identities while adopting a better and improved method of farming (Clissold & McNamara, 2020; Šūmane et al., 2018). The broad bed and furrow system is one such method. It consists of raised farming beds constructed in the shape of an inverted trapezium by heaping soil dug from drains onto raised beds to cater to seasonal flooding (Velmurugan et al., 2015). It is quite similar to a traditional method of planting dalo on raised beds of soil surrounded by a drainage system. Compared to traditional methods, the broad bed and furrow system enables high-value vegetables in the monsoon season to grow faster and is stated to increase cropping intensity from 100% to 125%. This technique capitalizes on excess water to farming systems while harvesting water in the dry season. The combination of raised beds and the stable supply of water maintains the mean temperature of the soil until the harvest season. Excavated zones can be used for rice farming while in raised beds farmers can cultivate seasonal vegetables (Sharma et al., 2020; Velmurugan et al., 2015). If this system is employed in Wainadoi and Navua, farmers would practice a "new" method that maintains the characteristics of a traditional method of planting on elevated beds while more efficiently managing the impact of a flood.

Planting resilient crops and seasonal planting are already employed as traditional coping methods in both sites. There may be a possibility to improve these two methods. Participants identified yam (*Dioscorea*), giant swamp *dalo*

(*Cyrtosperma chamissonis*), breadfruit (*Artocarpus altilis*), or sweet potato (*Ipomoea batatas*) as "hurricane foods." Although a strong cyclone may destroy the foliage and stem system, these crops are highly durable to wind and remain intact, hence they can be consumed even after some time. Planting red grams, mung beans, and peanuts is also effective to recover lost nitrogen from soil erosion or excess flooding (Isbell et al., 2017). These short-term maturing crops are easy to replant compared to root crops. Other crops such as watercress (*Nasturtium officinal*) and pumpkin (*Cucurbita*) also grow well in wet weather and ensure a higher return on investment over a shorter period. Farms with a high crop diversity have no adverse impacts but may increase the likelihood of at least one crop surviving from the impact of a climate hazard.

Seasonal planting is also stated to be an efficient way to maintain a constant supply of produce year-round. Changing crop patterns, replacing existing crops and introducing new resistant crops will improve resilience against changing climate and intensifying climate hazards. The interaction of many species results in the mixture and rotation of grasses and legumes over yield to produce more biomasses compared to monoculture yields (Isbell et al., 2017). If hurricane foods are cultivated together with seasonal planting methods, it will ensure a ready supply of food available when gardens are destroyed by cyclones or floods (Handmer & Iveson, 2017).

In any case, such "new" methods should not be implemented by a top-down approach. It is ideal if someone from the area becomes a specialist in agricultural coping methods and works with farmers to identify their needs and concerns (Yegbemey, 2020). Such a community-engaged and participatory approach would be more effective as the community is heavily involved in decision-making and implementation processes (Bourne et al., 2021; Thaler & Seebauer, 2019; Walshe et al., 2017).

Appendix

In-depth interview questions

1. Agricultural activities

What type of agriculture are you involved in?
Is your farm attended solely by your family members or together with other community members?
How many people from your household participated in the activity?
What is your reason for farming?
What are things you have to think about before you start planting?
When you went into farming what was your main goal? For example: to feed your family.
Apart from farming, do you have other sources of income? Example fishing.

2. Disaster preparation

What hazards/disasters are typically observed around the area?
What were the most severe hazards or disasters that you have experienced in your life?
How do you prepare your farm when there is:
a. cyclone approaching.
b. Heavy rain.
c. Flood.
When there is a natural hazard, who helps you prepare your farm for it?
Do you visit the agricultural site to identify the risk? If yes, how often? and if no, why not? And what are the risks?
How do you address the likely risk that may face your agricultural activity?

3. Tendencies of flooding and perception

Is your farm close to where you reside?
What is the general weather like daily?
How does flooding affect the type of crops you cultivate?
How often do you experience flooding?
What months in a year do you often experience flooding?
Do you feel that heavy rainfall is the only cause of the flooding or are there other causes?
Do you feel that the intensity of floods has increased or decreased?
Do structures like the main road affect the intensity of floods?
Do you feel that the timing of floods within a year has changed recently?
When was the last time you experienced a natural hazard that has damage your crops?

4. Pre and post disaster management

How were you informed or warned about an approaching natural hazard?

What was your reaction?
What was the impact of the hazard on your farm?
How long did it take for you to restart farming in the place where you had engaged in farming before the hazard?
What was the challenge?
During cyclone Winston, what were the measures that you took to minimize damage on your farm?

5. Local and traditional agricultural coping techniques for natural hazards

What do the following natural hazards indicate, according to legend/traditional knowledge?

a. Floods
b. Cyclones
c. Landslides

Please name at least 5 (or 10) traditional skills you know that will help you in a natural hazard.
Do you use your indigenous and local skills and knowledge to help you protect your farm when there is a natural hazard? If so name these techniques.
Who taught you those skills and knowledge?
How do you obtain or pass on traditional skills in your community? For example, is it passed on from father to son or is it a community effort to teach new farmers or is it acquired through observing others?
Is there a reason why you use traditional coping methods?
Have you applied those traditional skills to better protect your farm against a natural hazard and has it been practical?
Are you currently using any modern techniques and knowledge on your farm? If so name these techniques.
Which knowledge and skills are more practical and useful to protect your farm, traditional or modern, can you explain why?

6. Family and community involvement in local agriculture

Do people assist you individually or is it a communal effort?
How do the family and the community assist when they hear an approaching hazard?
What assistance do they give during and after the hazard?

7. Natural hazard aid from the government

What does the government support during a natural hazard?
What type of assistance did they give?
Did you receive any support from Fiji National Provident Fund?

Acknowledgment

The authors thank the farmers of Wainadoi and Navua, Viti Levu, Fiji, who participated in our project. This work was funded by The University of the South Pacific.

References

Abid, M., Scheffran, J., Schneider, U. A., & Elahi, E. (2019). Farmer perceptions of climate change, observed trends and adaptation of agriculture in Pakistan. *Environmental Management*, *63*(1), 110–123. https://doi.org/10.1007/s00267-018-1113-7.

Ahmad, D., Afzal, M., & Rauf, A. (2021). Flood hazards adaptation strategies: A gender-based disaggregated analysis of farm-dependent bait community in Punjab, Pakistan. *Environment, Development and Sustainability*, *23*(1), 865–886. https://doi.org/10.1007/s10668-020-00612-5.

Alam, G. M. M., Alam, K., & Mushtaq, S. (2017). Climate change perceptions and local adaptation strategies of hazard-prone rural households in Bangladesh. *Climate Risk Management*, *17*, 52–63. https://doi.org/10.1016/j.crm.2017.06.006.

Alpízar, F., Saborío-Rodríguez, M., Martínez-Rodríguez, M. R., Viguera, B., Vignola, R., Capitán, T., et al. (2020). Determinants of food insecurity among smallholder farmer households in Central America: Recurrent versus extreme weather-driven events. *Regional Environmental Change*, *20*, 22.

Arbuckle, J. G., Morton, L. W., & Hobbs, J. (2015). Understanding farmer perspectives on climate change adaptation and mitigation: The roles of Trust in Sources of climate information, climate change beliefs, and perceived risk. *Environment and Behavior*, *47*(2), 205–234. https://doi.org/10.1177/0013916513503832.

Bangalore, M., Smith, A., & Veldkamp, T. (2019). Exposure to floods, climate change, and poverty in Vietnam. *Economics of Disasters and Climate Change*, 79–99. https://doi.org/10.1007/s41885-018-0035-4.

Barnett, J., & Waters, E. (2016). Rethinking the vulnerability of small island states: Climate change and development in the Pacific Islands. In *The Palgrave handbook of international development* (pp. 731–748). Palgrave Macmillan. https://doi.org/10.1057/978-1-137-42724-3_40.

Bell, J., Taylor, M., Amos, M., & Andrew, N. (2016). Climate change and Pacific Island food systems. In *508. CCAFS reports*.

Ben, C., & Gounder, N. (2019). Property rights: Principles of customary land and urban development in Fiji. *Land Use Policy*, *87*. https://doi.org/10.1016/j.landusepol.2019.104089, 104089.

Bird, Z., Wairiu, M., Combes, H. J. D., & Iese, V. (2021). Religious and cultural-spiritual attributions of climate-driven changes on food production: A case study from North Malaita, Solomon Islands. In *Climate change management* (pp. 39–56). Springer Science and Business Media Deutschland GmbH. https://doi.org/10.1007/978-3-030-67602-5_3.

Bourne, M., de Bruyn, L. L., & Prior, J. (2021). Participatory versus traditional agricultural advisory models for training farmers in conservation agriculture: A comparative analysis from Kenya. *The Journal of Agricultural Education and Extension*, *27*(2), 153–174. https://doi.org/10.1080/1389224X.2020.1828113.

Boydell, S. (2001). *Philosophical perceptions of Pacific property-land as a communal asset in Fiji* (pp. 21–24). Pacific Rim Real Estate Society.

Brown, P., Daigneault, A., & Gawith, D. (2017). Climate change and the economic impacts of flooding on Fiji. *Climate and Development*, *9*(6), 493–504. https://doi.org/10.1080/17565529.2016.1174656.

Bryant-Tokalau, J. (2018). *Indigenous Pacific approaches to climate change: Pacific Island countries.*

Burnham, M., & Ma, Z. (2017). Climate change adaptation: Factors influencing Chinese smallholder farmers' perceived self-efficacy and adaptation intent. *Regional Environmental Change*, *17*(1), 171–186. https://doi.org/10.1007/s10113-016-0975-6.

Campbell, J. R. (2015). Development, global change and traditional food security in Pacific Island countries. *Regional Environmental Change*, *15*(7), 1313–1324. https://doi.org/10.1007/s10113-014-0697-6.

Chambers, L., Lui, S., Plotz, R., Hiriasia, D., Malsale, P., Pulehetoa-Mitiepo, R., et al. (2019). Traditional or contemporary weather and climate forecasts: Reaching Pacific communities. *Regional Environmental Change*, 1521–1528. https://doi.org/10.1007/s10113-019-01487-7.

Chandra, A., & Gaganis, P. (2016). Deconstructing vulnerability and adaptation in a coastal river basin ecosystem: A participatory analysis of flood risk in Nadi, Fiji Islands. *Climate and Development*, *8*(3), 256–269. https://doi.org/10.1080/17565529.2015.1016884.

Charan, D., Kaur, M., & Singh, P. (2017). Customary land and climate change induced relocation—A case study of Vunidogoloa Village, Vanua Levu, Fiji. In *Climate change management* (pp. 19–33). Springer. https://doi.org/10.1007/978-3-319-50094-2_2.

Ching, A., Morrison, L., & Kelley, M. (2020). Living with natural hazards: Tropical storms, lava flows and the resilience of island residents. *International Journal of Disaster Risk Reduction*, *47*. https://doi.org/10.1016/j.ijdrr.2020.101546.

Clissold, R., & McNamara, K. E. (2020). Exploring local perspectives on the performance of a community-based adaptation project on Aniwa, Vanuatu. *Climate and Development*, *12*(5), 457–468. https://doi.org/10.1080/17565529.2019.1640656.

Connell, J., & Waddell, E. (2006). *Between local and global: Environment, change and development in the Asia Pacific region.*

Cox, J., Finau, G., Kant, R., Tarai, J., & Titifanue, J. (2018). Disaster, divine judgment, and original sin: Christian interpretations of tropical cyclone Winston and climate change in Fiji. *The Contemporary Pacific*, *30*(2), 380–410. https://doi.org/10.1353/cp.2018.0032.

Cox, J., Varea, R., Finau, G., Tarai, J., Kant, R., Titifanue, J., et al. (2020). Disaster preparedness and the abeyance of agency: Christian responses to tropical cyclone Winston in Fiji. *Anthropological Forum*, *30*(1–2), 125–140. https://doi.org/10.1080/00664677.2019.1647833.

Crumpler, K., & Bernoux, M. (2020). Climate change adaptation in the agriculture and land use sectors: A review of nationally determined contributions (NDCs) in Pacific Small Island developing states (SIDS). In *Climate change management* (pp. 1–25). Springer. https://doi.org/10.1007/978-3-030-40552-6_1.

Duaibe, K. (2008). *Human activities and flood hazards and risks in the south West Pacific: A case study of the Navua catchment area, Fiji Islands (*Unpublished thesis submitted to). Victoria University of Wellington.

Fiji Bureau of Statistics. (2017). *Rural population.* https://www.statsfiji.gov.fj/index.php/census-2017/census-2017-release-1. Accessed.

Fischer, R. A., & Connor, D. J. (2018). Issues for cropping and agricultural science in the next 20 years. *Field Crops Research*, *222*, 121–142. https://doi.org/10.1016/j.fcr.2018.03.008.

Fletcher, S. M., Thiessen, J., Gero, A., Rumsey, M., Kuruppu, N., & Willetts, J. (2013). Traditional coping strategies and disaster response: Examples from the south pacific region. *Journal of Environmental and Public Health*, *2013*. https://doi.org/10.1155/2013/264503.

Ford, J. D., Cameron, L., Rubis, J., Maillet, M., Nakashima, D., Willox, A. C., et al. (2016). Including indigenous knowledge and experience in IPCC assessment reports. *Nature Climate Change*, *6*(4), 349–353. https://doi.org/10.1038/nclimate2954.

Hadgu, G., Tesfaye, K., & Mamo, G. (2015). Analysis of climate change in northern Ethiopia: Implications for agricultural production. *Theoretical and Applied Climatology*, *121*(3–4), 733–747. https://doi.org/10.1007/s00704-014-1261-5.

Halapua, S. (2008). *Talanoa process: The case of Fiji.* http://unpan1.un.org/intradoc/groups/public/documents/un/unpan022610.pdf.

Handmer, J., & Iveson, H. (2017). Cyclone pam in Vanuatu: Learning from the low death toll. *Australian Journal of Emergency Management*, *32*(2), 60–65. https://ajem.infoservices.com.au/downloads/AJEM-32-02-22.

Iese, V., Halavatau, S., N'Yeurt, A. D. R., Wairiu, M., Holland, E., Dean, A., et al. (2020). Agriculture under a changing climate. In *Springer climate* (pp. 323–357). Springer. https://doi.org/10.1007/978-3-030-32878-8_9.

Isbell, F., Adler, P. R., Eisenhauer, N., Fornara, D., Kimmel, K., Kremen, C., et al. (2017). Benefits of increasing plant diversity in sustainable agroecosystems. *Journal of Ecology*, *105*(4), 871–879. https://doi.org/10.1111/1365-2745.12789.

Janif, S. Z., Nunn, P. D., Geraghty, P., Aalbersberg, W., Thomas, F. R., & Camailakeba, M. (2016). Value of traditional oral narratives in building climate-change resilience: Insights from rural communities in Fiji. *Ecology and Society*, *21*(2). https://doi.org/10.5751/ES-08100-210207.

Johnston, I. (2015). Traditional warning signs of cyclones on remote islands in Fiji and Tonga. *Environmental Hazards*, *14*(3), 210–223. https://doi.org/10.1080/17477891.2015.1046156.

Kelman, I., Gaillard, J. C., Lewis, J., & Mercer, J. (2016). Learning from the history of disaster vulnerability and resilience research and practice for climate change. *Natural Hazards*, *82*, 129–143. https://doi.org/10.1007/s11069-016-2294-0.

Kostaschuk, R., Terry, J., & Raj, R. (2001). Cyclones tropicaux et crues aux Fidji. *Hydrological Sciences Journal*, *46*(3), 435–450. https://doi.org/10.1080/02626660109492837.

Kumar, P., Debele, S. E., Sahani, J., Aragão, L., Barisani, F., Basu, B., et al. (2020). Towards an operationalisation of nature-based solutions for natural hazards. *Science of the Total Environment*, *731*. https://doi.org/10.1016/j.scitotenv.2020.138855, 138855.

Le Dé, L., Rey, T., Leone, F., & Gilbert, D. (2018). Sustainable livelihoods and effectiveness of disaster responses: A case study of tropical cyclone pam in Vanuatu. *Natural Hazards*, *91*(3), 1203–1221. https://doi.org/10.1007/s11069-018-3174-6.

Lee, D., Zhang, H., & Nguyen, C. (2018). The economic impact of natural disasters in Pacific Island countries: Adaptation and preparedness. *IMF Working Papers*, *18*(108), 1. https://doi.org/10.5089/9781484353288.001.

Luetz, J. M., & Nunn, P. D. (2020). Climate change adaptation in the Pacific Islands: A review of faith-engaged approaches and opportunities. In *Climate change management* (pp. 293–311). Springer. https://doi.org/10.1007/978-3-030-40552-6_15.

Malsale, P., Sanau, N., Tofaeono, T. I., Kavisi, Z., Willy, A., Mitiepo, R., et al. (2018). Protocols and partnerships for engaging pacific island communities in the collection and use of traditional climate knowledge. *Bulletin of the American Meteorological Society, 99*(12), 2471–2489. https://doi.org/10.1175/BAMS-D-17-0163.1.

Martin, P. C. (2018). *How sea-level and climate change affect island people: Examples from Yadua Island, Fiji.* (Doctoral dissertation).

Mataki, M., Koshy, K., & Nair, V. (2006). *Implementing climate change adaptation in the Pacific Islands: Adapting to present climate variability and extreme weather events in Navua (Fiji).*

Mbah, M., Ajaps, S., & Molthan-Hill, P. (2021). A systematic review of the deployment of indigenous knowledge systems towards climate change adaptation in developing world contexts: Implications for climate change education. *Sustainability, 13*(9), 4811. https://doi.org/10.3390/su13094811.

McAneney, J., van den Honert, R., & Yeo, S. (2017). Stationarity of major flood frequencies and heights on the Ba River, Fiji, over a 122-year record. *International Journal of Climatology, 37*, 171–178. https://doi.org/10.1002/joc.4989.

McKenna, K., & Yakam, L. T. (2021). Signs of "the end times": Perspectives on climate change among market sellers in Madang, Papua New Guinea. In *Climate change management* (pp. 139–155). Springer Science and Business Media Deutschland GmbH. https://doi.org/10.1007/978-3-030-67602-5_8.

McMaster, J., & McGregor, A. (1999). *The Fiji service sector: Opportunities for growth.* A consulting report prepared for the Ministry of Commerce, Fiji Government.

Méheux, K., Dominey-Howes, D., & Lloyd, K. (2007). Natural hazard impacts in small island developing states: A review of current knowledge and future research needs. *Natural Hazards, 40*(2), 429–446. https://doi.org/10.1007/s11069-006-9001-5.

Nakamura, N., & Kanemasu, Y. (2020). Traditional knowledge, social capital, and community response to a disaster: Resilience of remote communities in Fiji after a severe climatic event. *Regional Environmental Change, 20*(1). https://doi.org/10.1007/s10113-020-01613-w.

Nalau, J., Becken, S., Schliephack, J., Parsons, M., Brown, C., & Mackey, B. (2018). The role of indigenous and traditional knowledge in ecosystem-based adaptation: A review of the literature and case studies from the Pacific Islands. *Weather, Climate, and Society, 10*(4), 851–865. https://doi.org/10.1175/WCAS-D-18-0032.1.

Neef, A., Benge, L., Boruff, B., Pauli, N., Weber, E., & Varea, R. (2018). Climate adaptation strategies in Fiji: The role of social norms and cultural values. *World Development, 107*, 125–137. https://doi.org/10.1016/j.worlddev.2018.02.029.

Nichols, A. (2019). Climate change, natural hazards, and relocation: Insights from Nabukadra and Navuniivi villages in Fiji. *Climatic Change, 156*(1–2), 255–271. https://doi.org/10.1007/s10584-019-02531-5.

Nolet, E. (2012). A tsunami from the mountains. The floods of. In *Pacific climate cultures: Living climate change in Oceania* (pp. 60–72). University of St Andrews.

Nolet, E. (2016). "Are you prepared?" representations and management of floods in Lomanikoro, Rewa (Fiji). *Disasters, 40*(4), 720–739. https://doi.org/10.1111/disa.12175.

Noy, I. (2015). *Natural disasters and climate change in the Pacific island countries: New non-monetary measurements of impacts. School of Economics and Finance Working paper: 08.*

Noy, I. (2016). Natural disasters in the Pacific Island countries: New measurements of impacts. *Natural Hazards, 84*, 7–18. https://doi.org/10.1007/s11069-015-1957-6.

Nunn, P. D., Mulgrew, K., Scott-Parker, B., Hine, D. W., Marks, A. D. G., Mahar, D., et al. (2016). Spirituality and attitudes towards nature in the Pacific Islands: Insights for enabling climate-change adaptation. *Climatic Change, 136*(3–4), 477–493. https://doi.org/10.1007/s10584-016-1646-9.

Nunn, P. D., Runman, J., Falanruw, M., & Kumar, R. (2017). Culturally grounded responses to coastal change on islands in the Federated States of Micronesia, Northwest Pacific Ocean. *Regional Environmental Change, 17*(4), 959–971. https://doi.org/10.1007/s10113-016-0950-2.

Pacific Community. (2021). *The Fiji Agriculture Policy Bank.* https://pafpnet.spc.int/policy-bank/countries/fiji#:~:text=Agriculture%20is%20a%20mainstay%20of,estimated%2010.4%25%20of%20the%20GDP.

Pan, Y., Smith, S. C., & Sulaiman, M. (2018). Agricultural extension and technology adoption for food security: Evidence from Uganda. *American Journal of Agricultural Economics, 100*(4), 1012–1031. https://doi.org/10.1093/ajae/aay012.

Piggott-McKellar, A. E., McNamara, K. E., Nunn, P. D., & Watson, J. E. M. (2019). What are the barriers to successful community-based climate change adaptation? A review of grey literature. *Local Environment, 24*(4), 374–390. https://doi.org/10.1080/13549839.2019.1580688.

Prasad, R. S., & Sud, R. (2019). Implementing climate change adaptation: Lessons from India's national adaptation fund on climate change (NAFCC). *Climate Policy, 19*(3), 354–366. https://doi.org/10.1080/14693062.2018.1515061.

Raggi, L., Ciancaleoni, S., Torricelli, R., Terzi, V., Ceccarelli, S., & Negri, V. (2017). Evolutionary breeding for sustainable agriculture: Selection and multi-environmental evaluation of barley populations and lines. *Field Crops Research, 204*, 76–88. https://doi.org/10.1016/j.fcr.2017.01.011.

Robinson, S., & a. (2020). Climate change adaptation in SIDS: A systematic review of the literature pre and post the IPCC fifth assessment report. *Wiley Interdisciplinary Reviews: Climate Change, 11*(4). https://doi.org/10.1002/wcc.653.

Savage, A., Schubert, L., Huber, C., Bambrick, H., Hall, N., & Bellotti, B. (2020). Adaptation to the climate crisis: Opportunities for food and nutrition security and health in a pacific small island state. *Weather, Climate, and Society, 12*(4), 745–758. https://doi.org/10.1175/WCAS-D-19-0090.1.

Sharma, P., Dupare, B. U., & Khandekar, N. (2020). Economic impact assessment of broad-bed furrow seed drill for soybean. *Agricultural Research, 9*(3), 392–399. https://doi.org/10.1007/s40003-019-00444-4.

Sisifa, A., McGregor, A., Fink, A., & Dawson, B. (2016). Pacific communities, agriculture and climate change. In *Vulnerability of Pacific Island agriculture and forestry to climate change* (pp. 5–45). Pacific Community.

Šūmane, S., Kunda, I., Knickel, K., Strauss, A., Tisenkopfs, T., Rios, I. D. I., et al. (2018). Local and farmers' knowledge matters! How integrating informal and formal knowledge enhances sustainable and resilient agriculture. *Journal of Rural Studies, 59*, 232–241. https://doi.org/10.1016/j.jrurstud.2017.01.020.

Taylor, M., McGregor, A., & Dawson, B. (2016). Vulnerability of staple food crops to climate change. In A. McGregor, M. Tayor, R. Bourke, & V. Lebot (Eds.), *Vulnerability of Pacic Island agriculture and forestry to climate change* (pp. 161–238). Pacific Community.

Thaler, T., & Seebauer, S. (2019). Bottom-up citizen initiatives in natural hazard management: Why they appear and what they can do? *Environmental Science and Policy*, *94*, 101–111. https://doi.org/10.1016/j.envsci.2018.12.012.

Thaman, R. R. (1999). Pacific Island biodiversity on the eve of the 21st century: Current status and challenges for its conservation and sustainable use. *Pacific Science Association Information Bulletin*, *51*, 1–37.

Thanh Thi Pham, N., Nong, D., Raghavan Sathyan, A., & Garschagen, M. (2020). Vulnerability assessment of households to flash floods and landslides in the poor upland regions of Vietnam. *Climate Risk Management*. https://doi.org/10.1016/j.crm.2020.100215, 100215.

Turner, J. (1984). "True food" and first fruits: Rituals of increase in Fiji. *Ethnology*, *23*(2), 133–142. https://doi.org/10.2307/3773698.

Velmurugan, A., Swarnam, T. P., & Lal, R. (2015). Effect of land shaping on soil properties and crop yield in tsunami inundated coastal soils of southern Andaman Island. *Agriculture, Ecosystems and Environment*, *206*, 1–9. https://doi.org/10.1016/j.agee.2015.03.012.

Velten, S., Leventon, J., Jager, N., & Newig, J. (2015). What is sustainable agriculture? A systematic review. *Sustainability*, *7*(6), 7833–7865. https://doi.org/10.3390/su7067833.

Wairiu, M. (2017). Land degradation and sustainable land management practices in Pacific Island countries. *Regional Environmental Change*, *17*(4), 1053–1064. https://doi.org/10.1007/s10113-016-1041-0.

Walshe, R. A., Chang Seng, D., Bumpus, A., & Auffray, J. (2017). Perceptions of adaptation, resilience and climate knowledge in the Pacific. *International Journal of Climate Change Strategies and Management*, *10*(2), 303–322. https://doi.org/10.1108/ijccsm-03-2017-0060.

Winsemius, H. C., Jongman, B., Veldkamp, T. I. E., Hallegatte, S., Bangalore, M., & Ward, P. J. (2018). Disaster risk, climate change, and poverty: Assessing the global exposure of poor people to floods and droughts. *Environment and Development Economics*, *23*(3), 328–348. https://doi.org/10.1017/S1355770X17000444.

Yegbemey, R. N. (2020). Farm-level land use responses to climate change among smallholder farmers in northern Benin, West Africa. *Climate and Development*. https://doi.org/10.1080/17565529.2020.1844129.

CHAPTER

15

People's management of risks from extreme weather events in the Pacific Island region

Eberhard Weber

School of Agriculture, Geography, Environment, Oceans and Natural Sciences, The University of the South Pacific, Suva, Fiji

1 Introduction

Climate change and adaptation to its impacts has become a new paradigm. With it the focus of climate change–related research and policy advice shifts from developed countries, where a mitigation paradigm was rooted, to countries of the Global South, which have great adaptation needs, but limited capacities to adapt.

The International Panel on Climate Change (IPCC) highlights that "Small Island States (SIS) in the Pacific, Indian, and Atlantic Oceans have been identified as being among the most vulnerable to climate change and climate extremes" IPCC (2012). The analysis of being exposed to natural hazards and disasters in the Pacific Island region stands at the beginning of the chapter. It compares disaster in this part of the world with other regions and asks, if the Pacific Islands are hotspots of disasters, or if disasters do not play much of a role in this part of the world.

Risk, vulnerability, and exposure are closely related. While risk refers to the likeliness that adverse external events (i.e., events where people have no power to avoid) happen, vulnerability looks at these events and people's ability to cope with or even adapt to them. Resilience, the ability to bounce back to the state before a disaster stroke or even better ("building back better") is closely related to reflections on risk, exposure, and vulnerabilities.

The IPCC (2012) reflects on how these components of disaster risk are connected when it highlights that disaster risk relates both to external events (hazards) and people's exposure to these events and their ability to adequately respond to hazards without experiencing damaging losses. The concept is very similar to Robert Chambers definition of (social) vulnerability, which has two sides: an external side of adverse events that can have negative impacts on people's lives and livelihoods, and an internal sides, which reflects on people's abilities to adequately respond to the external side (Chambers, 1989).

The chapter looks first at the external side of vulnerability, the side that is characterized by events that pose severe adverse stress on people's lives. The chapter looks exclusively at natural hazards, although Chambers and others include economic and political hazards as well (Birkmann & McMillan, 2020). A vital element of vulnerability is exposure. One can interpret exposure as a social-spatial category which brings location and social group together. An urban area (location) might experience a Tropical Cyclone. People living in informal settlements (social group) with improvised houses are more severely exposed compared to those living in stable houses. Spatial exposure does not always lead to vulnerability.

The chapter then reflects how people adapt to hazards, especially those closely related to climate change. Two aspects are crucial: (1) what can be done to avoid that hazards turn into disasters. Here, people's capacities and capabilities to better respond to hazards stand in the foreground. (2) How is mobility part of these reflections on adaptation? Is it part of adaptation processes, or is mobility, moving away from possible threats, an indication that adaptation failed or is not possible?

Climate Impacts on Extreme Weather
https://doi.org/10.1016/B978-0-323-88456-3.00013-7

2 Climate change, natural hazards and disasters in Pacific Island countries and territories

Situated in the center of the Pacific Ring of Fire, the Pacific Island region is prone to and has been exposed to geological hazards, such as earthquakes, volcanic eruptions and tsunamis for thousands of years. More recently, impacts of climate change have put other natural hazards (flooding, tropical storms, drought, and heat waves) in the focus of attention. This includes sea-level rise that causes coastal erosion, salinization of freshwater resources in coastal areas and atolls as well as increase of intensity of extreme weather events. While hazards with geological or meteorological backgrounds occur in a sudden and unforeseeable manner and often release huge amounts of energy in a short period of time, sea-level rise is a slow-onset, low-intensity hazard that takes decades to build up. Still it leaves locations uninhabitable by eroding coastal areas, destroying freshwater resources and finally submerging larger land areas. Atolls and low-lying limestone islands are particularly exposed to such impacts (Duvat et al., 2021).

2.1 The Pacific Islands region

There are 14 independent states in the Pacific Island region and seven territories of developed countries (Fig. 15.1). The ratio between the Exclusive Economic Zone (EEZ) and land is around 44:1 in favor of the ocean. Countries and territories have 550,344 km^2 in land. Papua New Guinea dominates land size with 462,840 km^2 leaving 88,655 km^2 to the remaining 13 Small Island Developing States and seven territories. Far bigger is the EEZ, the parts of the ocean legally under the control of countries. It is 24,412,561 km^2 (Weber, 2017). The open ocean (some 141 million km^2), ocean waters that are not assigned to any country, is not included in this figure. The PICT are small in population. By mid-2020 a total of 11.5 million people live in independent Pacific Island countries and another 853,000 in the territories. The only country with more than 1 million inhabitants is Papua New Guinea with a population of 8.9 million by mid-2020. Second is Fiji with 895,000 inhabitants. The smallest states are Niue with 1600 and Tokelau with 1500 inhabitants respectively (SPC, 2020).

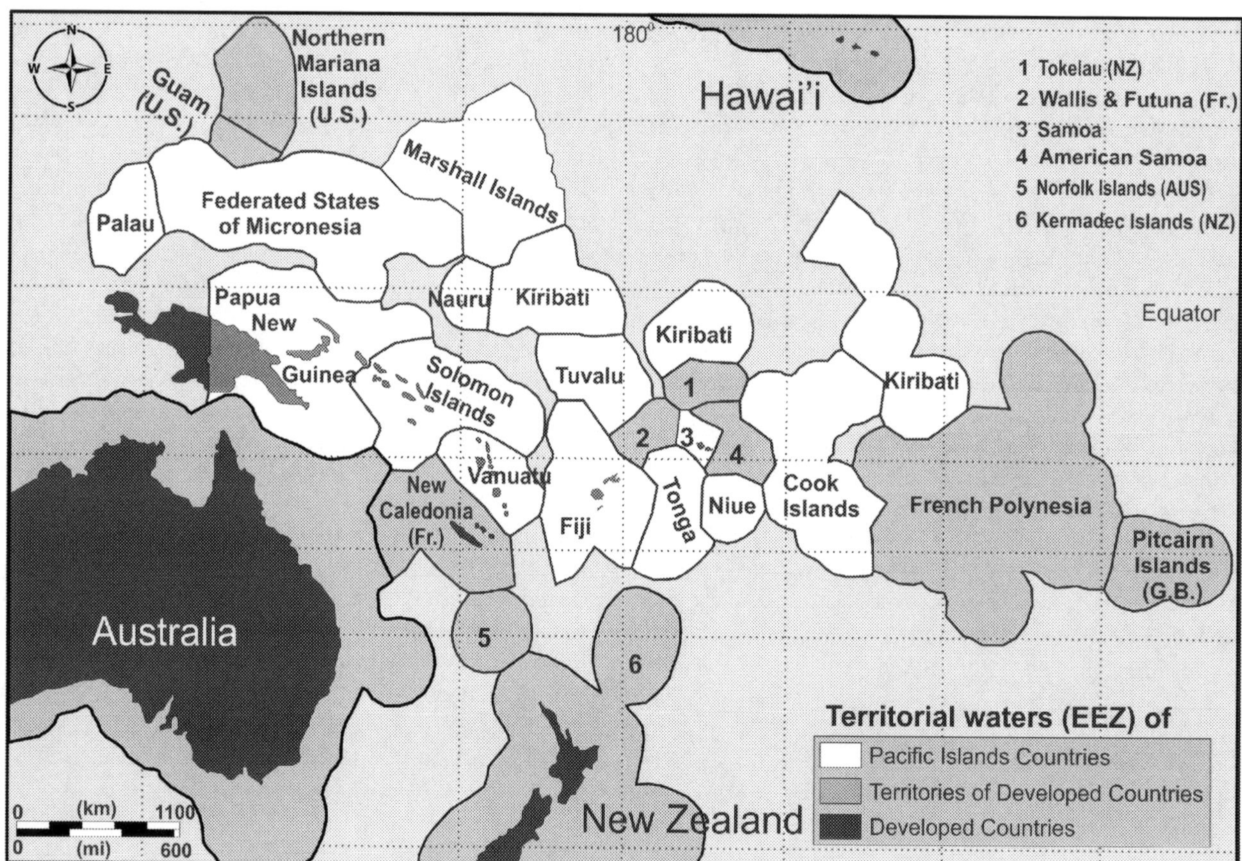

FIG. 15.1 The Pacific Island countries and territories (PICT). *No permission required.*

Some islands have extremely high population densities. The most western islet of South Tarawa atoll in Kiribati, Betio, had a population of 15,646 in 2010 living on 1.54 km^2. This has been a population density of just over 10,000 people/km^2. Until 2025 Betio's population is expected to increase to 27,419 people. Then the population density will be almost 18,000 people/km^2 (Government of Kiribati, 2012). The population density of Ebeye islet of Kwajalein Atoll in the Marshall Islands is even higher: 15,000 inhabitants living on 0.36 km^2; make a density of around 45,000 people/km^2 (Weber & Koto, 2021).

Compared to developed and larger developing countries, economies of Pacific Island Countries and Territories (PICT) are little diversified. Even economically, more advanced countries such as Fiji and Papua New Guinea have rather small formal sectors (Table 15.1). Labor markets are not dynamic enough to accommodate all people seeking work into the formal sector. In Papua New Guinea alone, more than 160,000 people enter the labor market annually. In Fiji, the estimated number is around 15,000. In formal employment the public sector is dominant. In addition, there are wholesale businesses and other service providers. Modern industries are restricted to bigger PICT. They cover a tiny range of economic activities such as food processing (incl. Alcoholic and nonalcoholic beverages), furniture, textiles and construction materials. The production of investment goods, consumer electronics, cars and other motor vehicles does not exist in any PICT (Weber, 2017).

TABLE 15.1 Selected demographic and physical characteristics of PICTs.

Country	Subregion	Population (mid-2020)	Increase since 1960s (%)	Land area (km^2)	Population density (per/km^2)	Atolls/ Coral Islands	Raised Islands	Volcanic Islands	Total
Cook Islands	Polynesia	15,300	(17)	237	65	7	0	8	15
Niue	Polynesia	1600	(68)	259	6	0	1	0	1
Samoa	Polynesia	1,98,700	73	2934	68	0	0	2+8 islets	2
Tonga	Polynesia	99,800	77	749	133	a few	>100	a few	~170
Tuvalu	Polynesia	10,600	95	26	408	9	0	0	9
Federated States of Micronesia	Micronesia	1,05,500	169	701	150	~600	0	>10	607
Kiribati	Micronesia	1,18,700	174	811	146	32	1	0	33
Marshall Islands	Micronesia	54,600	292	181	302	34	0	0	34
Nauru	Micronesia	11,700	154	21	557	0	1	0	1
Palau	Micronesia	17,900	92	444	40	<300	>10	>5	340
Fiji Islands	Melanesia	8,84,887	156	18,333	56	2	a few	<100	~320
Papua New Guinea	Melanesia	89,35,000	309	4,62,840	13	a few	a few	>600	>600
Solomon Islands	Melanesia	7,12,100	474	28,230	20	a few	0	>900	>990
Vanuatu	Melanesia	2,94,700	277	12,281	25	0		82	82
Territory									
American Samoa	Polynesia	56,800	183	199	285	2	0	5	7
French Polynesia	Polynesia	2,78,900	230	3521	73	~80	a few	~40	130
Tokelau	Polynesia	1500	(20)	12	97	3	0	0	3
Wallis & Futuna	Polynesia	11,400	33	142	93	0	0	2	2
Guam	Micronesia	1,76,700	164	541	327	0	0	1	1
Northern Mariana Islands	Micronesia	56,600	583	457	124	0	0	15	15
New Caledonia	Melanesia	2,73,000	216	18,576	67	0	0	7	7

No permission required.

PICTs are exposed to various natural hazards. Their small size make them more vulnerable to such hazards, also to such intensified by climate change. They cause great damage to countries and territories, put pressure on people's lives and livelihoods, and strain governments' resources.

2.2 Natural hazards in the Pacific Islands region

Natural hazards can cause severe disasters in PICTs. Assessments differ, if Pacific Islands are hotspots of disasters, or if disasters are less prevalent than in other parts of the world. In a recent brochure the European Union made even wrong and misleading statements "The Pacific island countries (including Papua New Guinea, Fiji Islands, Solomon Islands, Cook Islands, and Vanuatu) rank among the world's worst affected in terms of casualties and people impacted by disasters" (European Union, 2020). Nothing could be further from reality.

PICT do not have huge disasters by international standards. The numbers of people killed/affected are "tiny" compared to the world's biggest disasters. Damages in absolute money terms do not reach dimensions like elsewhere. Between 1900 and mid-March 2021 there have been 409 disasters were recorded in the Pacific Island region (Table 15.2). They caused a death toll of 13,860 people. Two disasters caused more than 600 death: The largest was the eruption of Mt. Lamington, PNG, in 1951 with 3000 deaths. At second place is an earthquake/tsunami in Sandaun Province, PNG, causing 2182 deaths in July 19. Over the past 120 years there have been 24 disasters recorded alone in Asia that caused more fatalities than the combined death toll in PICT in this period (Centre for Research on the Epidemiology of Disasters (CRED), 2021).

In this period (1900–2021) Asia had by far the biggest number of disasters (10,367) and biggest number of disaster deaths (26.8 million). Respective numbers for the Americans are 5407 disaster events with 892,000 deaths and for Africa 5154 disaster events with 1.5 million deaths. Europe is second lowest with 3146 disasters events that caused the death of 9.2 million people. Three individual events contributed to almost 95% of the overall death toll in Europe: (1) Famine in the Soviet Union (1932/3.5 million deaths), (2) typhoid epidemic in the Soviet Union (1917–22, 2.5 million deaths), and (3) drought in the Soviet Union (1921, 1.2 million deaths).

Tables 15.3 and 15.4 show that the numbers of disasters in PICT between 1900 and mid-March 2021 are not very high, nor casualties and impacts in money terms, even if one considers that—especially in years far in the past- the database has by far not recorded all disasters. Many important disasters are missing, e.g., the Spanish flu between 1918 and 1920, which affected Samoa like no other country causing around 8500 deaths (more than 20% of the population) (Weber, Kopf, & Vaha, 2021).

For Fiji, the disaster of February 1931, so far the most serious disaster in Fiji, is classified as tropical cyclone killing 200 people. Reports on the event state that although "the 1931 flood was generated by a hurricane, the disaster is regarded primarily as a flood disaster because drowning caused the vast majority of deaths" (Yeo & Blong, 2010). These few examples should suffice that disasters in the Pacific Island region are small compared to disasters elsewhere, but are they insignificant?

In relative terms, disasters in PICT are looming threats to societies and economies with severe economic and social consequences. Capacities to deal with such events are frequently overstretched. The small size of islands, their populations, economies, and capacities of governments and people to cope with disasters constitute severe vulnerabilities making disasters major, disturbing events, which have the potential to undo development efforts of decades in the blink of an eye.

2.3 Disasters in PICT compared to the world's biggest recent disaster events

The most expensive disaster in modern history has been the 2011 Tōhoku earthquake and tsunami. The costs are estimate around USD 200 billion (Kim, 2011). The damage has been about 3.2% of Japan's GDP in 2011. An analysis of 204 disaster events in 12 PICs between 1980 and 2016 shows that the average damage was 14.4% of the GDP. Samoa (47.7% of GDP) and Vanuatu (42.8% of GDP) are on top of the list. The most severe of these 204 disaster events brought economic damages of 161.8% of the GDP in the case of Samoa and 131.2% of the GDP in Vanuatu. Statistically, PICT are hit by a disaster almost every second year. For Fiji the figure is more than 70%/year. Only PNG has a higher disaster risk per year with 81%. Also above 50% are Vanuatu (57%) and Solomon Islands (51%) (Lee, Zhang, & Nguyen, 2018).

In Samoa a tsunami killed 143 people in September 2009. The death toll per 1000 people of Samoa's population was higher than for all countries affected by the 2004 Indian Ocean Tsunami except Sri Lanka. The argument is not that the Samoan tsunami was a bigger disaster in absolute terms, but it affected a bigger share of the Samoan population. The

TABLE 15.2 Reported disasters in the Pacific Islands (1900–mid-March 2021).

Hazard	Pacific Island countries														Territories						
	Polynesia						Melanesia				Micronesia				P	P	P	Me	Mi	Mi	
	Cook Islands	**Niue**	**Samoa**	**Tokelau**	**Tonga**	**Tuvalu**	**Fiji**	**Papua New Guinea**	**Solomon Islands**	**Vanuatu**	**Kiribati**	**Palau**	**Marshall Islands**	**Micronesia (FED. Rep.)**	**American Samoa**	**French Polynesia**	**Wallis & Futuna**	**New Caledonia**	**Guam**	**Northern Mariana**	total
Windstorm	11	5	11	6	18	7	44	7	19	29	2	2	2	7	4	4	4	16	9	4	211
Drought			1		1	1	3	3	3		1		2	2							17
Flood			1				12	19	5	2	2		2	1	1	1					46
Earthquake			1		2		2	18	9	8					1		1		1		43
Volcanic					1			17	1	7										1	27
Technological	1				1	1	1	10			1	1				1	1		1		19
Others	3	1	2		2		1	24	2	3	1	1	1	1		3		1			46
	15	**6**	**16**	**6**	**25**	**9**	**63**	**98**	**39**	**49**	**7**	**4**	**7**	**11**	**6**	**9**	**6**	**17**	**11**	**5**	**409**
Melanesia	249														17						266
Polynesia	77														21						98
Micronesia	29														16						45
	355														54						409

Data from Centre for Research on the Epidemiology of Disasters (CRED). (2021). The international disaster database. https://public.emdat.be/.

TABLE 15.3 Disaster, fatalities, people affected, and damages recorded in the Pacific Island region (1900–2021).

	Number	Reported fatalities	Population affected	Reported losses (in 000'US$)
Windstorm	211	1945	3,814,650	4,038,880
Drought	17	84	3,662,019	94,900
Flood	46	233	1,003,064	350,375
Earthquake	43	6024	642,312	328,615
Volcanic	27	3515	296,032	110,000
Technological	19	865	12,153	12,000
Others	46	1194	62,858	31,650
Melanesia	267	12,591	8,374,650	2,448,406
Polynesia	97	834	768,730	1,374,469
Micronesia	45	435	349,708	1,143,545
TOTAL	409	13,860	9,493,088	4,966,420

Data from Centre for Research on the Epidemiology of Disasters (CRED). (2021). The international disaster database. https://public.emdat.be/.

TABLE 15.4 Reported disasters in the PICTs according to decades (1900–mid March 2021).

Hazard	1900–1909	1910–1919	1920–1929	1930–1939	1940–1949	1950–1959	1960–1969	1970–1979	1980–1989	1990–1999	2000–2009	2010–2019	since 2020	Total
Windstorm		1	1	2	6	10	18	22	37	30	36	40	8	211
Drought									2	6		9		17
Flood							1		4	3	21	14	3	46
Earthquake				1		1		5	5	13	11	7		43
Volcanic				1	1	2	1	1	1	4	11	5		27
Technological										11	4	4		19
Others								3	5	9	17	10	2	46
	0	**1**	**1**	**4**	**7**	**13**	**20**	**31**	**54**	**76**	**100**	**89**	**13**	**409**
Melanesia				4	3	11	8	24	37	49	65	55	11	267
Polynesia		1	1		4	2	11	4	15	19	19	19	2	97
Micronesia							1	3	2	8	16	15		45

Data from Centre for Research on the Epidemiology of Disasters (CRED). (2021). The international disaster database. .

damage was between 15% and 20% of the Samoa's GDP in 2009 (World Bank, 2012). Already earlier, in 1991, Samoa's economy suffered damages from Tropical Cyclone (TC) Val amounting to 231% of the country's GDP (CRED, 2021).

In Vanuatu TC Nigel caused a damage of USD173 million in 1985, about 131% of Vanuatu's GDP (Lee et al., 2018). Thirty years later, in March 2015, the country was struck by TC Pam, the most severe tropical cyclone ever recorded in the Southern Hemisphere. The damage was some USD 449 million, around 60% of Vanuatu's GDP (CRED, 2021). A year later Fiji experienced a similarly strong tropical cyclone. TC Winston devastated major parts of the country in February 2016 causing damage of some USD 900 million, about 20% of the GDP. If one includes the value of destroyed environmental assets and losses in environmental services the total damage has been USD 1.3 billion (Government of Fiji, 2016).

Niue is a tiny island-nation of 260 km^2 and around 1600 inhabitants. The country has been suffering from cyclones more than once in the past few decades (Barker, 2000). At times damages were so overwhelming that many suggested evacuating the entire population to New Zealand (Wade, 2005).

Damages relative to a country's GDP exceeds dimensions of even the world's biggest disasters. This makes PICTs hotspots of disasters. Even when damages are relatively small in absolute terms they are huge measured in countries' and people's capacities to cope with such events. In many cases a single disaster put serious stress on governments and people likewise. It is certainly realistic to assume that exposure to natural hazards will increase with climate change. Societies have to better prepare and find ways to reduce disaster risks.

2.4 People's vulnerabilities to disasters and the search for security

Reflections about vulnerability have a long tradition, which started outside climate change discourses. In his analysis of famines in India, Bangladesh, and Africa, Amartya Sen observes declines of food entitlements and shows that not everybody is affected by hunger and famine in the same way and to the same extend (Sen, 1981). He distinguishes between groups in society, their exposure to famine, and the implications/impacts on their lives. This "discovery" may sound simple or even simplistic, but for long time it was not really an issue of social science research to distinguish between those who are adversely affected by famine, those who are well protected and those who might even benefit. When introducing differentiations the principle of vulnerable and less-vulnerable groups in societies have been articulated; famines became events of societal rather than natural causation.

(Watts & Bohle, 1993a, 1993b) reflect on vulnerability to food insecurity from a structural perspective. What constitutes people's vulnerabilities to such events? Around the same time, Blaikie, Cannon, Davis, and Wisner (1994) used vulnerability in concepts around natural hazards and disasters. Since the beginning of the 21st century, more and more authors have used climate change discourses and people's vulnerabilities to show how impacts of climate change are felt on grassroots levels (Adger, 2006; Kasperson et al., 2006; Leichenko & O'Brien, 2002).

"Social" vulnerability is often used as synonym to "poverty" (Sivadas & Ismail, 2020). This ignores that "poverty" relates to deprivation while vulnerability reflects on exposure to external risks, shocks, stresses and how well a person or a group of people can cope/adapt to such external pressures. According to Chambers vulnerability has two sides: "an external side of risks, shocks and stress to which an individual or household is subject; and an internal side which is defenselessness, meaning a lack of means to cope without damaging loss" (Chambers, 1989). Vulnerability is the combination of adverse, external events and the capacity of a person/group of persons to cope with such events.Concepts of coping played important roles in anthropological urban studies in Third World contexts (McFarlane & Silver, 2017; Moser, 1998; Thieme, 2018). The concepts experienced a revival in food security research and livelihood analyses (Chambers, 1989; Chen, 1991; Davies, 1996; Krishnamurthy et al., 2020; Oskorouchi & Sousa-Poza, 2021). Research on famines in the Sahel and the Horn of Africa in the 1980s was often centered around "coping" to better understand why some people survived periods of extreme food stress while others did not (Watts & Bohle, 1993a, 1993b); see (Koch-Laier et al., 1996; Mekuyie & Mulu, 2021; Ncube et al., 2018). Later, when perspectives shifted from food to livelihood security and finally to sustainable livelihoods, the term "coping" was used to deal with different sorts of hazards, natural as well as human-made.

Vulnerabilities to disaster relate closely to social and economic circumstances of people's everyday lives and the position they hold in the structural make-up of their societies. Certain categories of people, such as the poor, the elderly, women-headed households and others are at great(er) risk to be adversely affected by hazards. There is much evidence that simultaneous exposure to multiple hazards enhance social vulnerability. Multiple hazards of differing categories (natural, social, economic, political) and differing time dimensions (sudden, long-term, etc.) need to be factored in when assessing overall vulnerabilities. Research on exposure to multiple hazards is scarce; far more is needed to more comprehensively map vulnerable people's realities (Anderson et al., 2021; Leichenko & O'Brien, 2002; O'Brien & Leichenko, 2000; Pagliacci & Russo, 2020). Worthwhile research could be conducted where severe natural hazards stroke at the time when communities were also exposed to the covid-19 pandemic.

2.4.1 Small islands and their vulnerabilities to natural hazards

Small islands, their isolation and remoteness can become a matter of life and death. In 2002 it remained long unclear, if the people of Tikopia and Anuta, two isolated islands in the eastern parts of the Solomon Islands, had escaped TC Zoe. Only surveillance flights by the Australian Air Force confirmed that all people were fine although 70% of their houses had been completely destroyed (Courtney, 2005).

Societies in the Pacific Island region have dealt with hazards since centuries. Cyclones, floods, droughts and other hazards have become part of people's lives. People developed mechanism to cope with such hazards (Bertana, 2020; Beyerl et al., 2018; Campbell, 2006; Latai-Niusulu et al., 2020; Neef et al., 2018; Nunn & Kumar, 2019; Nunn, 2007). Modernization processes might have weakened traditional knowledge systems, but knowledge to respond to hazards is still frequently applied.

Social vulnerability to hazards has distinct structural and time dimensions. (Watts & Bohle, 1993a, 1993b) hint at consequences when people are not able to recover from adverse events quickly and sufficiently. When a disaster strikes, it is often the result of insufficient preparation/reduction of disaster risks. The accumulation of negative events can enhance vulnerability when different adverse events come together, when people have not yet fully recovered from a disaster when the next strikes. People's vulnerabilities increase as their capacities to cope/adapt decline. A number of Tropical Cyclones in PICT that happened simultaneously with COVID-19 pandemic are recent examples of such multiple exposure (Farrell et al., 2020; Frausto-Martínez et al., 2020; Holland, 2020; Mahul & Signer, 2020; Steenbergen et al., 2020).

A hazard type, its intensity and frequency is not enough to make predictions about the likeliness, scope and dimension of a disaster. It is essential to analyze, who are the most vulnerable sections of society, how are they exposed to a hazard and what constitutes their vulnerability. People's capacities to successfully respond to a hazard/disaster can become weaker through coping (Watts & Bohle, 1993a, 1993b). After a disaster they may sell assets to cope. They become impoverished. Disasters render them sick or injured. This weakens their capabilities to cope in future. People's social networks are destroyed when they flee and/or when members of their family die in the event of a disaster. The end of a hazard does not mean the end of a disaster. People are often much weaker than before. People are more susceptible to future hazards. The more often people have to go through disasters the less is needed to trigger another disaster.

Various Sustainable Livelihood Approaches (SLA) borrow much from Sen's and Chambers' insights. They add distinct applied elements making them tools to analyze vulnerabilities and to enhance livelihood security. The approaches provide guidelines of how to analyze vulnerability contexts and people's coping/adaptation capacities. Specific livelihood capitals play a major role to support risk reduction: natural capital (access to land, marine and other natural resources and the quality of these resources); physical capital (access to infrastructure, communication and information); human capital (skills, formal qualification, health); financial capital (incomes of all sorts, savings and assets one easily can convert into money). A fifth capital, which received particular importance is social capital, the membership in formal and informal social networks that provide support and security to overcome adverse situations (Yila et al., 2013). Some authors have recommended to add legal capital as separate category. Legal capital understood as supportive structures and processes that have legal foundations (e.g., Human Rights, social security legislation entitlements versus voluntary social welfare measures) and that enable people to enforce claims (Weber, 2017).

2.4.2 Adaptive capacities in Pacific Island societies

Small island societies have low adaptive capacities (de Águeda Corneloup & Mol, 2014; Lohmann et al., 2019). Tropical and subtropical islands are particularly exposed to the impacts of climate change (Basel et al., 2020; Leal Filho et al., 2020; Nand & Bardsley, 2020). Governments took notice of these challenges only slowly. While they have slowly "mainstreamed" climate change and disaster risk reduction, the general public lack behind, not necessarily because challenges are unknown, but because other priorities receive more attention (Agrawala & Van Aalst, 2008). It is not only people's willingness to pay for adaptation efforts, but their ability to pay that make a difference (de-Graft Acquah & Onumah, 2011; Akter, 2020; O'Garra & Mourato, 2016; Tulone et al., 2020).

In recent decades, there has been rapid change in PICT. Urbanization, individualization, and commercialization have weakened community solidarity, subsistence production and exchange in small rural communities (Kronen & Bender, 2007; Weber, 2014). Earlier people lived in big family networks. Obligations to support each other were not individual decisions, but made by groups (Fleming, 1997; McLeod, 2010). They were ritualistic, enforced network outcomes. At all times, however, people were agents of their destinies. Even today they are neither victims of natural hazards, nor of social structures. People try to have impacts on both. Social structures are the outcome of agency and social action (Hempenstall, 2011). Such anthropological insights are crucial for the activation of traditional, indigenous disaster risk reduction mechanisms (Fletcher et al., 2013; Hosen et al., 2020; Kelman et al., 2012; Lesa, 2020).

In his report on traditional disaster risk reduction in Pacific Island communities (Campbell, 2006) provides valuable insights in people's knowledge to deal with natural hazards in systematic ways. The report gives details of traditional coping mechanism and how societies adapt to hazards. Societies based on subsistence production generate surpluses to bridge critical times, either through cash that has been generated selling surpluses or food preservation and storage. What Campbell calls "Agricultural Resilience" uses crop and location diversity to minimize or spread risk. This is a dominating principle of many subsistence or peasant economies: to put emphasis on livelihood security rather than on profit maximization (Feeny, 1983; Sulewski et al., 2020).

In such societies, mutual help are strong features, which includes the membership to social as well as social-spatial networks. Membership in social groups are constantly reinforced through rituals (birth, weddings, funeral, rites of passage and initiation) that usually have reciprocal components. Thaman stresses the need to preserve "time-tested, socially and environmentally suited Pacific Island food systems" that strengthen group relationships and enhance cohesion (Thaman, 1982). Some societies, especially on small isolated islands have once been almost entirely autarkic.

Authors describe them as subsistence affluence societies (Bainton & McDougall, 2021; Maiava, 2020; Sahlins, 1972). Such communities have been regarded highly resilient to natural hazards and external intervention (e.g., in form of inappropriate disaster relief). Authors stress that increasing external influence weakened resilience, undermining community solidarity, people's social capital. External interventions in disaster risk reduction and postdisaster recovery need to pay more attention to local contexts, capacities and constraints. Efforts should be designed in complementary ways that *strengthen* local populations' agency and resilience and not *weaken* or *replace* them (Campbell, 2006; Ford et al., 2020; Piggott-McKellar et al., 2020).

Well established institutions can reduce damaging impacts of natural hazards by enhancing people's capabilities and capacities. Social networks and social capital derived from them enable communities to confront hazards and help reduce disaster risks. Functioning social networks also facilitate information flows. Such networks can be used in planning disaster risk reduction measures as well as recovery and rehabilitation processes (Yila et al., 2013).

Specific socio-cultural context deserve particular attention. Ethnic differentiation like in the Western Province of the Solomon Islands after the 2007 tsunami (indigenous Melanesian vs immigrant Micronesian) or in Fiji after virtually every disaster (indigenous Fijians vs Indo-Fijians) can disturb cooperation when sensitive ethnic issues are ignored. Such differences can be helpful as different ethnic groups have different knowledge and skills in disaster risk management and postdisaster recovery. They also have access to different networks (Barnes-Mauthe et al., 2015; MacGillivray, 2018; Ngin et al., 2020). Remittances from relatives in the Pacific diaspora can prove crucial for sustaining the recovery from natural hazards (Bennett et al., 2020; Brent Vickers, 2018; Petzold, 2017).

2.4.3 *Remoteness, isolation and displacement as challenges to disaster management*

The example of Tikopia and Anuta after TC Zoe in 2002 shows that locations, and connected to it, accessibility, are crucial for disaster management and postdisaster recovery (Korovulavula et al., 2020; Petzold & Magnan, 2019). Another example of severe isolation in the Pacific Island region is Kiribati, stretching almost 5000 km from East to West. The distance from the most western island (Banaba) to the country's most eastern island (Caroline Island in the Southern Line Group) is more than 4500 km. In December 1999, it took a group of dancers three weeks to travel by boat from Kiribati's capital on Tarawa Atoll in the very west to Caroline Island to celebrate the new Millennium (Associate Press Archive, December 31, 1999).

Temporary displacements happen connected to disasters. Evacuation centers provide safe places to people exposed to floods and cyclones. Usually they return to their own houses once the hazard is over. Permanent resettlement might happen more often in future, even within the same country. After the tsunamis in the Solomon Islands (2007) and Samoa (2009) challenges around disaster rehabilitation differed greatly as a result of land tenure. In the Solomon Islands, people who had moved to higher grounds after the tsunami became squatters on government land. The provincial government tried to relocate them back to the coastal zone, where schools destroyed by the tsunami were rebuilt. In Samoa, government provided various support (heavy machinery, building material/tools, financial support) to enable people from the coastal zone to build new villages at higher grounds. New houses, schools, churches, shops, etc. emerged within a short time. The Polynesian land tenure system in Samoa meant that people relocated within their own land.

A challenge to informal elements of social capital is that the creation of formal safety nets is difficult (Adams & Neef, 2019). Micro-finance initiatives can provide such security, but they usually have elements of formalization (Feeny, 1983). Where protection remains informal and depends on community cohesion and solidarity safety nets can become less efficient through modernization processes and increasing social and economic differentiation of communities (Diedrich & Aswani, 2016). When some no longer depend on social capital as they have become well-off over time, traditional support structures erode or even collapse. Most challenging are transitions from traditional to modern systems of social and economic security. While the traditional systems have weakened, the new systems are not (yet) fully established to provide sufficient support (Weber, 1995). Such challenges happen when social security shifts from informal arrangements to formalized systems (Hu, 2020; Solimano, 2021).

2.4.4 *Migration as successful adaptation to climate change*

A particular discourse has emerged about the role migration in the adaptation to the impacts of climate change. Is migration an element of climate change adaptation (Hochachka, 2021; Maharjan et al., 2021; Mohammed et al., 2021; Nalau & Verrall, 2021), or does the need to migrate show that adaptation has failed (Nand & Bardsley, 2020).

There are instances environmental change including climate change impacts make people move. Not always are such migrants welcome in the places they try to reach. The notion of migration with dignity refers to the asymmetries between migrants that are welcome in the destination and such that are seen as unwanted immigrants, which are even criminalized (former Kiribati President Tong 2006, per. communication).

The social construction of good and bad migrants is the result of socio-economic and cultural dynamics that create real or perceived antagonism between migrants and people, media and political ideologies in the countries the migrants go to (Browne & Reilly, 2020; Hirsch, 2019). For the past two decades, terminologies such as "environmental refugee"/"climate (change) refugee" have dominated the debate. Strictly speaking, there are only a few legal systems that recognize "refugees" caused by environmental deterioration. The major legal framework, the Convention Relating to the Status of Refugees, does not provide such protection (Derham et al., 2020; Scott, 2020).

Pacific Islands have a long history of migration (Weber, 2014). Initially migration was driven by people's own motivations and efforts to settle the vast expanse of the Pacific Ocean and to travel to neighboring and far away islands (Finney et al., 2007; Nunn & Kumar, 2019; Nunn, 2007). Then people were taken around by outside powers to distribute labor according to colonial requirements (Scarr, 1967; Singe, 2020). Islands were emptied for nuclear testing (Weber & Koto, 2021). Today migration has become a means for many Pacific Islanders to seek greener pastures outside their home countries. The notion of being "environmental" or "climate change refugees" is rather strange to the people of the Pacific. Most of them even reject not only the terminology, but the entire concept (Smith & McNamara, 2015).

2.4.4.1 Migrants as actors and translocality

Migrants are actors. They use their agency when moving to other places. We rarely can identify a single cause for migration, but migrants' expectations are as complex and multilayered as their reasons to move (Marino & Lazrus, 2015; Sakdapolrak et al., 2016). It is difficult to disentangle environmental/climate reasons from economic improvement, livelihood security, or poverty, ethnic discrimination/persecution, and other reasons. There are many examples where two, three or even more of such reasons mix.

Looking at people's strategies to diversify livelihoods (Adzawla & Baumüller, 2021; Kimengsi et al., 2020; Mohammed et al., 2021; Salam & Bauer, 2020; Sodokin & Nyatefe, 2021) migration, livelihood diversification and climate change adaptation connect in a positive way. When people experience that climatic changes make their efforts to gain secure livelihood difficult they decide to move elsewhere, at least part of their household. Climate change leading to higher intensity of extreme events, negative impacts of food production, and/or sea-level rise make prevailing economic conditions harder and aggravate marginalization and poverty (O'Brien & Leichenko, 2000). Often vulnerable people live in marginal lands, informal settlements and other locations that are particularly exposed to the impacts of climate change (Weber, 2017). In such situations livelihood diversification coincides with migration, the diversification of locations. Translocal households (McMichael et al., 2021; Peth & Sakdapolrak, 2020; Steinbrink & Niedenführ, 2020; Vibert, 2020).

Different members of the same household live in various places to optimize work. This enhances livelihood security. Those who leave send remittances to their loved ones back home in the village. They enhance structures of reciprocity and mutual support. When this increases security and well-being then migration has been a positive way of adaptation (Tacoli, 2011).

2.4.4.2 Labor mobility in PICT

Many Pacific Islanders experienced positive impacts on their livelihoods as well those of family members back home when they became *Gastarbeiter* in Australia, New Zealand, and the USA right after World War II. Migration opportunities varied according to economic booms and depressions in receiving countries. Some PICs have formalized access to other states, either through citizenship or particular arrangements their governments have entered. People living in Pacific Island Territories have citizenship of their respective metropolitan country. Formalized labor migration schemes are summarized in Table 15.5.

In addition to Pacific Islanders, who can reside and work in countries at the rim of the Pacific Ocean there are other arrangement that make permanent immigration easier for people from PIC. New Zealand and Samoa have an annually quota that allows 1100 Samoans and their core families to permanently emigrate to New Zealand to take up work there (Samoan Quota). With Fiji (250 per year), Tonga (250), Kiribati (75) and Tuvalu (75) similar agreements exist, although with lower numbers (Pacific Access Category). Eligible applicants are drawn by lot (Weber, 2017).

Australia and New Zealand have also temporary unskilled labor schemes that allow every year thousands of Pacific Islanders work in horticulture and the tourism industry of both countries. In 2017/18 the Seasonal Worker Program (SWP) issued some 8500 visa to Pacific Islanders. Some 80% of seasonal workers came from Tonga and Vanuatu. Samoa and Timor Lesté contributed almost 20%. A few other PIC had smaller percentages (Gibson & Bailey, 2021).

In 2017/18 New Zealand's Recognized Seasonal Employer (RSE) program had a cap of 12,500 intakes. More than 11,000 workers actually came for the 2017/18 season. Between 2007 and 2018 around 38% of seasonal workers to New Zealand came from Vanuatu, 19% from Tonga, 16% from Samoa, 5% from Kiribati, 2% from Solomon Islands, and 1% each from Fiji, PNG, and Tuvalu. The remaining 18% were seasonal workers from Asia. In July 2018 Australia started a Pacific Labour Scheme (PLS). Similar to the Seasonal Worker Program it issues visa to citizens of PIC (incl. Timor Lesté). Under this scheme low- and semiskilled workers can come to Australia for up to 3 years (Gibson & Bailey, 2021).

TABLE 15.5 Migration data and formalized schemes for Pacific Islander labor migration.

Country (bold) territory (governed by)	Population (mid-2020)	Stock of emigrants (in '000)	Stock of emigrants as % of population	Pacific access category (PAC)/Samoa quota (SQ) (annual quota)	Skilled labor schemes	Unskilled labor schemes	Migration opportunities
Polynesia							
American Samoa (USA)	56,800	–					Full access to USA (incl. USA Pacific Territories)
Cook Islands	15,300	28.4	61.4%		PICTA-TIS		Full access to NZ
French Polynesia (France)	278,900	–					Full access to NZ
Niue	1600	9.2	75.6%		PICTA-TIS		Full access to NZ
Samoa	198,700	120.4	67.3%	1100 (SQ)	PICTA-TIS, PLS	SWP, RSE, PSWPS2	
Tokelau (New Zealand)	1500	–					Full access to NZ
Tonga	99,800	47.4	45.4%	250	PICTA-TIS, PLS	SWP, RSE, PSWPS1	
Tuvalu	10,600	n.a.	n.a.	75	PICTA-TIS, PLS	SWP, RSE, PSWPS2	Special Deal with German merchant marine
Wallis and Futuna (France)	10,600	n.a.	n.a.				Full access to France (incl. French Pacific territories)
Micronesia							
Fed. States of Micronesia	105,500	21.9	19.7%		PICTA-TIS		Full access to USA
Guam (USA)	176,700	–					Full access to USA (incl. USA Pacific Territories)
Kiribati	118,700	6.4	6.5%	75	PICTA-TIS, PLS	SWP, RSE, PSWPS1	Special Deal with German/Japanese merchant marine
Marshall Islands	54,600	10.5	16.6%		PICTA-TIS		Full access to USA
Nauru	11,700	n.a.	n.a.		PICTA-TIS, PLS	SWP, PSWPS2	
Northern Mariana (USA)	56,600	–					
Palau	17,900	8.0	38.8%		PICTA-TIS		Full access to USA (incl. USA Pacific Territories)

Continued

TABLE 15.5 Migration data and formalized schemes for Pacific Islander labor migration—cont'd

Country (bold) territory (governed by)	Population (mid-2020)	Stock of emigrants (in '000)	Stock of emigrants as % of population	Pacific access category (PAC)/Samoa quota (SQ) (annual quota)	Skilled labor schemes	Unskilled labor schemes	Migration opportunities
Melanesia							
Fiji Islands	884,887	182.2	21.3%	[a]	PICTA-TIS, SMS, PLS	RSE[a]	
New Caledonia (France)	273,600	–					Full access to France (incl. French Pacific territories)
Papua New Guinea	8,935,000	61.2	0.9%		PICTA-TIS,, PLS	SWP, PSWPS1	
Solomon Islands	712,100	5.4	1.0%		PICTA-TIS, SMS, PLS	SWP, PSWPS2	
Vanuatu	294,700	3.9	1.6%		PICTA-TIS, SMS, PLS	SWP, RSE, PSWPS1	
Timor Leste (2010)	1,066,409	23.0	2.2%		PLS	SWP, PSWPS2	Applied for ASEAN membership (2011)
Polynesia	673,800	205.4					
Micronesia	541,700	37.8		Full access to developed countries		1,049,600	
Melanesia	12,166,696	252.7					
total (incl. Timor Leste)	14,448,605	518.9					

SWP, Seasonal Worker Program from 1 July 2012 to 30 June 2016.
PSWPS1, Pacific Seasonal Worker Pilot Scheme which started in 2008 with four countries (Kiribati, Papua New Guinea, Tonga and Vanuatu; PSWPS1).
PSWPS2, 2011 four more Pacific Island countries were invited to join the program (Nauru, Samoa, Solomon Islands and Tuvalu; PSWPS2). East Timor participated in a small trial of seasonal mobility in the accommodation sector.
RSE, Recognized Seasonal Employer (RSE) Scheme started in 2007. The NZ Department of Labour has Inter-Agency Understandings with the following Pacific Island states: Kiribati, Samoa, Tuvalu, Tonga, Solomon Islands and Vanuatu, although the scheme is open world-wide.
PLS, Pacific Labor Scheme for low and semiskilled workers, up to three years in one contract.
SMS, Skilled Labor Mobility Scheme of the Melanesian Spearhead Group started in 2012
PICTA-TIS, Pacific Islands Countries Trade Agreement-Trade in Services, negotiations concluded in February 2012; not yet in force
[a] *Excluded for political reasons between 2007 and 2014.*
No permission required.

Beyond these migration opportunities Pacific Islanders can migrate to developed countries at the rim of the Pacific Ocean through obtaining regular work permits and visa for the purpose family reunion. From Fiji, e.g., a notable number of citizens work in the British Army, making it the biggest ethnic section of the British Army (King, 2021). Some 7000 Fijians live in the U.K., most of them as active or retired soldiers of the British Army (Hulkenberg, 2015). Finally thousands of rugby players from predominately Fiji, Papua New Guinea, Samoa, and Tonga work and live in Europe (France, U.K.), New Zealand and Australia and the USA (Guinness, 2018; Hawkes, 2018).

Of the 12.3 million people who lived in PICs (11.4 million) and territories (854,000) by mid-2020 some 1.05 million have a right to reside and work in a metropolitan country without the need of a residential or work permit (SPC, 2020).

2.4.5 *Disaster management and its integration with climate change adaptation*

Impacts of climate change often relate to hazards that easily can trigger disasters. Tropical storms and floods stand at the top of hazard events that turned into disasters, but also droughts do play some role. All these hazards can increase in intensity through climate change. Over the past two decades major efforts have been taken up from Pacific Island countries' governments, regional organizations and nongovernmental organizations to integrate climate change and disaster risk reduction/management efforts (Gero et al., 2010). Effort to integrate climate change and disaster management includes attempts to mainstream both.

In 2005 Pacific Island leaders made commitment to address disaster issues systematically when they endorsed the Pacific Disaster Risk Reduction and Disaster Management Framework for Action, 2005–2015: The FfA intended to create resilient countries and communities. It had been a process driven globally by the Hyogo Framework for Action 2005–15 and the International Strategy for Disaster Reduction (ISDR).

In the same process Pacific Leaders endorsed the Pacific Islands Framework of Action on Climate Change (PIFAC), 2006–15, which also calls for specific actions against the impacts of climate change, including to the effects of extreme climatic events.

Between 2006 and 2016 the Pacific Disaster Risk Management Partnership Network (PDRMPN), an "open-ended, voluntary" membership of international, regional and national government and nongovernment organizations, supported Pacific countries toward mainstreaming DRM through addressing their disaster risk reduction and disaster management priorities.

Mainstreaming is transforming an idea into strategic intervention. When referring to mainstreaming in the context of disaster risk management the effort is to integrate key principles of disaster risk management into national development goals and their implementation such as policies and strategies. It also means integration into development planning and budgetary processes. In reality this requires that all new legislation is tested, if it is cohesive with the ideas and priorities of disaster risk reduction. It means in addition that budgets of relevant government sections have resources allocated that help to address disaster risk reduction. Some governments combined disaster and climate change perspectives through Joint National Action Plans (JNAPs).

The annual meetings of the PDRMPN were essential in the integration of disaster related fields and activities with those concerning climate change. Information sharing provided opportunity to relevant NGOs, government ministries, donors and transnational bodies at regional as well as international levels such as various UN agencies and the World Bank. Although the idea of exchanging ideas stood in the forefront of these meetings they were also a platform to facilitate concrete actions among participants to reduce vulnerability and enhance resilience to disasters and the effects of climate change. In particular the open meetings provided a wide scope to civil society organizations to have a voice in important decision making.

PDRMPN facilitated cooperation and the formation of partnerships and coordinated activities. It worked excellent on inter-country levels within the Pacific islands and to link them to international efforts, mainly within the United Nations system. Weaknesses were in linking to subnational levels, particularly grassroots organizations and activities. Applied activity at the grassroots are often separated from the policy development efforts of efforts organized under PDRMPN.

3 The 2020/2021 cyclone seasons and covid-19 in the Pacific Islands region

In 2020 and 2021 Solomon Islands, Vanuatu, Fiji and Tonga experienced a number of severe Tropical Cyclones (TC) that coincided with the covid-19 pandemics. Multiple hazards affected people at the same time. These caused severe hardship to people, who already suffered much from restrictions and job losses from covid-19. The pandemics put also severe challenges to disaster response. Countries had to deal with severe tropical cyclones just at the time when governments tried to implement measures against COVID-19 (Weber et al., 2021). Relevant authorities had to reduce infection risks as much as possible, which affected virus containment measures.

The strongest of the cyclones was TC Harold, which caused severe damages in early April 2020. TC Harold was a powerful cyclone reaching highest category 5. It caused widespread destruction in Solomon Islands, Vanuatu, Fiji, and Tonga. The cyclone caused 30 deaths, 2 in Fiji, 1 in Vanuatu, and 27 in the Solomon Islands. Three countries had not yet recovered from damaging cyclone disasters just a few years earlier (Vanuatu, Cyclone Pam in 2015, Fiji, Cyclone Winston in 2016, and Tonga, Cyclone Gita in 2018).

When TC Harold was approaching the Solomon Islands the country had no covid-19 cases. The government advised many, who were temporarily staying in the country's capital Honiara to return to their home islands to reduce over-crowding especially in informal settlements to support social distancing measures. On April 4 more than 600 persons from Malaita Island wanted to return home on MV Taemarehu. Strong winds caused by TC Harold washed 37 people off the ferry. All are still reported missing and are assumed to be indirect casualties of covid-19 (Holland, 2020).

In Vanuatu TC Harold destroyed around 80% of houses on the northern islands of Espiritu Santo and Pentecost. Around 80,000 people got displaced, crops destroyed and livestock perished. As a result of covid-19 restrictions foreign aid workers had left the country just a few weeks prior to TC Harold and were to quarantine before they could return. As a result Vanuatu was handling disaster relief efforts entirely through own staff in order to avoid that the coronavirus could enter the covid-free country.

In Fiji TC Harold caused damage of some USD 44 million, destroyed at least 635 houses and damaged an additional 2100 homes, many in locations that were under lockdown because of covid-19 (e.g., Kadavu Island, Vatulele, Southern Lau islands). Special arrangements for relief efforts were needed to reach these places. Relief supplies, e.g., were not shipped from Suva, which was under lockdown then, but from Natovi jetty some 50 km north of Fiji's capital. All officers involved in relief efforts traveling to outer islands have been screened by the Ministry of Health. Nobody from the Suva containment area could be included in relief operations on outer islands.

Of more than 10,000 temporarily displaced persons some 6200 sought refuge in 250 evacuation centers despite social distancing restrictions in place. The assessment of damages and providing relief to affected people right after TC Harold were difficult as measures to avoid the spread of the coronavirus had to be observed (Mangubhai et al., 2021). Social distancing rules were in place, people were told to stay at home, and advice given was to regularly wash one's hands. It is close to impossible to follow such crucial advice to protect oneself and others from a deadly virus, when house are blown away by unimaginable strong winds, when people have to find refuge in overcrowded evacuation centers.

TC Harold reach Solomon Islands and Vanuatu at a time when both countries had no recorded covid-19 cases. Effort of the governments of these two countries was to avoid that international aid personnel brings the coronavirus to their countries. In Fiji 17 covid-19 cases had been reported by April 7, 2020. It was uncertain, if unreported cases in the community existed. To avoid the spread of covid-19 to outer islands, travel to these islands had been discontinued.

Less than 8 months after TC Harold Fiji was struck by TV Yasa, one of the strongest cyclones ever observed in the South Pacific and the most powerful since Cyclone Winston in February 2016. TC Yasa hit Fiji on December 17, 2020. The cyclone reached wind gusts of up to 345 mph. It was the fourth cyclone of the highest storm intensity (Category 5) in 4 years that reached Fiji. Another 6 weeks later Tropical Cyclone Ana made landfall in Fiji on January 31, 2021 causing much damage to informal settlements in Suva before proceeding to Kadavu, where also severe damages were reported. Both TC reached Fiji when it was covid-free. Relief operations were easier than after TC Harold.

4 Conclusion

Natural hazards have happened in PICT since long. Disasters arising from these hazards have been small by international standards, but huge considering the smallness of PICT, their limited capacities and capabilities to deal with such disasters. PICT have contributed little to global greenhouse gas (GHG) emissions, but climate change is a reality in this part of the world with considerable impacts. There is little uncontested evidence that extreme events have become more frequent in PICT. With a number of severe Tropical Cyclones between 2015 and 2020 it appears that cyclone hazards have intensified. It is predictable that this trend continues into the future. Hazards will continue to become stronger as higher ocean water temperatures provide more energy when cyclones generate and then intensify.

It is essential to reduce disaster risks, and optimize rehabilitation and recovery after disasters. This includes better preparedness to extreme events. Traditional experience, knowledge and skills are no longer enough to reduce disaster risks. Still people's knowledge play important roles to provide sound foundations on which external interventions and ways to deal with disasters can be based on. To devaluate people's knowledge will not only reduce effectiveness, but also take away from them responsibility and agency: people's ability to make decisions concerning their own lives and to implement them. Already today, such implementation has become concerted action between insiders and outsiders; between traditional and modern ways. Such combination of efforts can strengthen resilience and people's ability to face hazards well enough that they do not turn into disasters.

The greatest deficit is that adaptation is greatly understood as an instrument of Public Work Departments, building codes and engineers. Adaptation widely ignores vulnerable people's aspiration to reduce risks by diversifying livelihoods. This long standing strategy is the most important mechanism people have to deal with natural hazards and disasters. While the analysis of livelihood diversification has become a major paradigm in research of climate change and disaster risk reduction, policies and government action in these areas are too often ignoring the messages of such research. There is certainly scope to improve.

Still situations will become more challenging when in situ adaptation is not enough. When pressures become so severe that there is a need to evacuate people for good, to bring them to safety. Here it is essential to safeguard that they can live in dignity elsewhere. To integrate requirements arising from climate change and disaster risk reduction with development efforts is urgent and timely. In this context, it is essential to include climate change concerns into negotiations about unskilled and skilled labor, whether permanent or temporary. Such measures can strengthen adaptive capacities.

Policies that aim at disaster risk reduction exist in PICT on regional, national, and subnational levels. Sections of governments that deal with climate change and disaster risk reduction are well connected. The exchange of information has improved a lot. Much weaker are links between governments and local communities. An urgent requirement for the near future is a better connection between decision making on national levels and the grassroots. This includes logistical links, the further integration between climate change and disaster risk reduction, and the mainstreaming of policies, funds and efforts into development planning.

Local-level institutions and networks strengthening solidarity and mutual support, both formal and informal, are central to people's agency and interaction. Such efforts cannot neglect or even disguise that natural hazards happen to people, who often are exposed to other challenges, such as poverty, unemployment, livelihood insecurity and tenure insecurity. It is particular damaging when natural hazards affect people, who face other disadvantages at the same time. Such exposure to multiple hazards makes disaster risk reduction particularly challenging. Here more sensitivity, but also creativity is asked from state actors.

References

Adams, C., & Neef, A. (2019). Patrons of disaster: The role of political patronage in flood response in the Solomon Islands. *World Development Perspectives, 15*. https://doi.org/10.1016/j.wdp.2019.100128.

Adger, W. N. (2006). Vulnerability. *Global Environmental Change, 16*(3), 268–281. https://doi.org/10.1016/j.gloenvcha.2006.02.006.

Adzawla, W., & Baumüller, H. (2021). Effects of livelihood diversification on gendered climate vulnerability in Northern Ghana. *Environment, Development and Sustainability, 23*(1), 923–946. https://doi.org/10.1007/s10668-020-00614-3.

Agrawala, S., & Van Aalst, M. (2008). Adapting development cooperation to adapt to climate change. *Climate Policy, 8*(2), 183–193. https://doi.org/10.3763/cpol.2007.0435.

Akter, S. (2020). Social cohesion and willingness to pay for cyclone risk reduction: The case for the coastal embankment improvement project in Bangladesh. *International Journal of Disaster Risk Reduction, 48*. https://doi.org/10.1016/j.ijdrr.2020.101579, 101579.

Anderson, C. C., Renaud, F. G., Hagenlocher, M., & Day, J. W. (2021). Assessing multi-hazard vulnerability and dynamic coastal flood risk in the Mississippi Delta: The global delta risk index as a social-ecological systems approach. *Water (Switzerland), 13*(4). https://doi.org/10.3390/w13040577.

Bainton, N., & McDougall, D. (2021). *Unequal Lives in the Western Pacific* (pp. 1–46). ANU Press. https://doi.org/10.22459/ue.2020.01.

Barker, J. C. (2000). Hurricanes and socio-economic development on Niue Island. *Asia Pacific Viewpoint, 41*(2), 191–205. https://doi.org/10.1111/1467-8373.00115.

Barnes-Mauthe, M., Gray, S. A., Arita, S., Lynham, J., & Leung, P. S. (2015). What determines social capital in a social–ecological system? Insights from a network perspective. *Environmental Management, 55*(2), 392–410. https://doi.org/10.1007/s00267-014-0395-7.

Basel, B., Goby, G., & Johnson, J. (2020). Community-based adaptation to climate change in villages of Western Province, Solomon Islands. *Marine Pollution Bulletin, 156*. https://doi.org/10.1016/j.marpolbul.2020.111266, 111266.

Bennett, K., Neef, A., & Varea, R. (2020). Embodying resilience: Narrating gendered experiences of disasters in Fiji. In *Vol. 22. Community, environment and disaster risk management* (pp. 87–112). Emerald Group Holdings Ltd. https://doi.org/10.1108/S2040-726220200000022004.

Bertana, A. (2020). The impact of faith-based narratives on climate change adaptation in Narikoso, Fiji. *Anthropological Forum, 30*(3), 254–273. https://doi.org/10.1080/00664677.2020.1812050.

Beyerl, K., Mieg, H. A., & Weber, E. (2018). Comparing perceived effects of climate-related environmental change and adaptation strategies for the pacific small island states of Tuvalu, Samoa, and Tonga. *Island Studies Journal, 13*(1), 25–44. https://doi.org/10.24043/isj.53.

Birkmann, J., & McMillan, J. M. (2020). *Linking hazard vulnerability, risk reduction, and adaptation*. Oxford University Press (OUP). https://doi.org/10.1093/acrefore/9780199389407.013.145.

Blaikie, P., Cannon, T., Davis, I., & Wisner, B. (1994). *At risk—Natural hazards, people's vulnerability and disasters*. Routledge. https://doi.org/10.4324/9780203974575.

Brent Vickers, J. (2018). More money, more family: The relationship between higher levels of market participation and social capital in the context of adaptive capacity in Samoa. *Climate and Development, 10*(2), 167–178. https://doi.org/10.1080/17565529.2017.1291404.

Browne, R., & Reilly, K. (2020). *Disrupting imagined geographies: Media, power and representation in contemporary migration* (pp. 176–199). Edward Elgar Publishing. https://doi.org/10.4337/9781788115483.00016.

Campbell, J. R. (2006). Traditional disaster reduction in pacific island communities. In *GNS science report*.

Centre for Research on the Epidemiology of Disasters (CRED). (2021). *The international disaster database*. https://public.emdat.be/.

Chambers, R. (1989). Editorial introduction: Vulnerability, coping and policy. *IDS Bulletin*, 1–7. https://doi.org/10.1111/j.1759-5436.1989.mp20002001.x.

Chen, M. A. (1991). *Coping with seasonality and drought*. Sage Publications.

Courtney, J. B. (2005). The South Pacific and southeast Indian Ocean tropical cyclone season 2002-03. *Australian Meteorological Magazine, 54*(2), 137–150.

Davies, S. (1996). *Adaptable livelihoods. Coping with food insecurity in the Malian Sahel*. Springer.

de Águeda Corneloup, I., & Mol, A. P. J. (2014). Small island developing states and international climate change negotiations: The power of moral "leadership". *International Environmental Agreements: Politics, Law and Economics, 14*(3), 281–297. https://doi.org/10.1007/s10784-013-9227-0.

de-Graft Acquah, H., & Onumah, E. E. (2011). Farmers perception and adaptation to climate change: An estimation of willingness to pay. *Agris On-Line Papers in Economics and Informatics*, *3*(4), 31–39. http://online.agris.cz/files/2011/agris_on-line_2011_4_acquah_onumah.pdf.

Derham, T., Mathews, F., & Palmer, C. (2020). Elephants as refugees. *People and Nature*, 103–110. https://doi.org/10.1002/pan3.10070.

Diedrich, A., & Aswani, S. (2016). Exploring the potential impacts of tourism development on social and ecological change in the Solomon Islands. *Ambio*, *45*(7), 808–818. https://doi.org/10.1007/s13280-016-0781-x.

Duvat, V. K. E., Magnan, A. K., Perry, C. T., Spencer, T., Bell, J. D., Wabnitz, C. C. C., et al. (2021). Risks to future atoll habitability from climate-driven environmental changes. *Wiley Interdisciplinary Reviews: Climate Change*, *12*(3). https://doi.org/10.1002/wcc.700.

European Union. (2020). *European civil protection and humanitarian aid operations pacific region*. https://ec.europa.eu/echo/printpdf/where/asia-and-pacific/pacific_en.

Farrell, P., Thow, A. M., Wate, J. T., Nonga, N., Vatucawaqa, P., Brewer, T., et al. (2020). COVID-19 and Pacific food system resilience: Opportunities to build a robust response. *Food Security*, *12*(4), 783–791. https://doi.org/10.1007/s12571-020-01087-y.

Feeny, D. (1983). The moral or the rational peasant? Competing hypotheses of collective action. *The Journal of Asian Studies*, *42*(4), 769–789. https://doi.org/10.2307/2054764.

Finney, B., Howe, K. R., Irwin, G., Low, S., Neich, R., Salmond, A., et al. (2007). *Vaka Moana. Voyages of the ancestors: The discovery and the settlement of the Pacific*. Honolulu: University of Hawaii Press.

Fleming, R. (1997). *The common purse: Income sharing in New Zealand families*. Auckland University Press.

Fletcher, S. M., Thiessen, J., Gero, A., Rumsey, M., Kuruppu, N., & Willetts, J. (2013). Traditional coping strategies and disaster response: Examples from the south pacific region. *Journal of Environmental and Public Health, 2013*. https://doi.org/10.1155/2013/264503.

Ford, J. D., King, N., Galappaththi, E. K., Pearce, T., McDowell, G., & Harper, S. L. (2020). The resilience of indigenous peoples to environmental change. *One Earth*, *2*(6), 532–543. https://doi.org/10.1016/j.oneear.2020.05.014.

Frausto-Martínez, O., Aguilar-Becerra, C. D., Colín-Olivares, O., Sánchez-Rivera, G., Hafsi, A., Contreras-Tax, A. F., et al. (2020). COVID-19, storms, and floods: Impacts of tropical storm cristobal in the western sector of the Yucatan Peninsula, Mexico. *Sustainability (Switzerland)*, *12*(23), 1–17. https://doi.org/10.3390/su12239925.

Gero, A., Méheux, K., & Dominey-Howes, D. (2010). Disaster risk reduction and climate change adaptation in the Pacific: The challenge of integration. In *ATRC-NHRL miscellaneous report 4* Australian Tsunami Research Centre – Natural Hazards Research Laboratory.

Gibson, J., & Bailey, R.-L. (2021). Seasonal labor mobility in the Pacific: Past impacts, future prospects. *Asian Development Review*, *38*(1), 1–31. https://doi.org/10.1162/adev_a_00156.

Government of Fiji. (2016). *Post-disaster needs assessment. Tropical cyclone Winston, February 20, 2016*. Government of Fiji. https://www.gfdrr.org/sites/default/files/publication/Post%20Disaster%20Needs%20Assessments%20CYCLONE%20WINSTON%20Fiji%202016%20(Online%20Version).pdf.

Government of Kiribati. (2012). South Tarawa island report. In *Island reports no. 6* Government of Kiribati.

Guinness, D. (2018). Corporal destinies: Faith, ethno-nationalism, and raw talent in Fijian professional rugby aspirations. *HAU: Journal of Ethnographic Theory*, *8*(1–2), 314–328. https://doi.org/10.1086/698267.

Hawkes, G. L. (2018). Indigenous masculinity in sport: The power and pitfalls of rugby league for Australia's Pacific Island diaspora. *Leisure Studies*, *37*(3), 318–330. https://doi.org/10.1080/02614367.2018.1435711.

Hempenstall, P. (2011). Cultural change in Oceania: Remembering the historical questions. In *Changing contexts, shifting meanings: Transformations of cultural traditions in oceania* (pp. 313–322). University of Hawai'i Press. http://muse.jhu.edu/books/9780824860141.

Hirsch, S. (2019). Racism, 'second generation' refugees and the asylum system. *Identities*, *26*(1), 88–106. https://doi.org/10.1080/1070289X.2017.1361263.

Hochachka, G. (2021). Integrating the four faces of climate change adaptation: Towards transformative change in Guatemalan coffee communities. *World Development*, *140*. https://doi.org/10.1016/j.worlddev.2020.105361, 105361.

Holland, E. (2020). Tropical cyclone Harold meets the novel coronavirus. *Pacific Journalism Review: Te Koakoa*, *26*(1), 243–251. https://doi.org/10.24135/pjr.v26i1.1099.

Hosen, N., Nakamura, H., & Hamzah, A. (2020). Adaptation to climate change: Does traditional ecological knowledge hold the key? *Sustainability*, *12*(2), 676. https://doi.org/10.3390/su12020676.

Hu, A. (2020). *The early rise of social security in china: Ideas and reforms, 1911–1949* (pp. 55–90). Springer Science and Business Media LLC. https://doi.org/10.1007/978-3-030-54959-6_2.

Hulkenberg, J. (2015). The cost of Being Fijian in the United Kingdom. *Anthropological Forum*, *25*(2), 148–166. https://doi.org/10.1080/00664677.2014.995152.

IPCC. (2012). *Managing the risks of extreme events and disasters to advance climate change adaptation*. Cambridge University Press.

Kasperson, R. E., Dow, K., Archer, E., Caceres, D., Dow, K., Downing, T., et al. (2006). *Vulnerable peoples and places* (pp. 143–164). Island Press.

Kelman, I., Mercer, J., & Gaillard, J. (2012). Indigenous knowledge and disaster risk reduction. *Geography*, *97*(1), 12–21. http://www.geography.org.uk/download/GEOGRAPHY_vol97_part1_KELMAN.pdf.

Kim, V. (2011). *Japan damage could reach $235 billion*. Los Angeles Times: World Bank estimates.

Kimengsi, J. N., Mukong, A. K., & Balgah, R. A. (2020). Livelihood diversification and household well-being: insights and policy implications for forest-based communities in Cameroon. *Society & Natural Resources*, *33*(7), 876–895. https://doi.org/10.1080/08941920.2020.1769243.

King, A. (2021). Decolonizing the British Army: A preliminary response. *International Affairs*, *97*(2), 443–461. https://doi.org/10.1093/ia/iiab001.

Koch-Laier, J., Davies, S., Milward, K., & Kennan, J. (1996). *Gender, household food security and coping strategies*. IDS (Institute of Development Studies), Development Bibliography 14.

Korovulavula, I., Nunn, P. D., Kumar, R., & Fong, T. (2020). Peripherality as key to understanding opportunities and needs for effective and sustainable climate-change adaptation: A case study from Viti Levu Island, Fiji. *Climate and Development*, *12*(10), 888–898. https://doi.org/10.1080/17565529.2019.1701972.

Krishnamurthy, P. K., Choularton, R. J., & Kareiva, P. (2020). Dealing with uncertainty in famine predictions: How complex events affect food security early warning skill in the Greater Horn of Africa. *Global Food Security*, *26*. https://doi.org/10.1016/j.gfs.2020.100374.

Kronen, M., & Bender, A. (2007). Assessing marine resource exploitation in Lofanga, Tonga: One case study—Two approaches. *Human Ecology*, *35*(2), 195–207. https://doi.org/10.1007/s10745-006-9084-3.

Latai-Niusulu, A., Binns, T., & Nel, E. (2020). Climate change and community resilience in Samoa. *Singapore Journal of Tropical Geography, 41*(1), 40–60. https://doi.org/10.1111/sjtg.12299.

Leal Filho, W., Otoara Ha'apio, M., Lütz, J. M., & Li, C. (2020). Climate change adaptation as a development challenge to small Island states: A case study from the Solomon Islands. *Environmental Science and Policy, 107*, 179–187. https://doi.org/10.1016/j.envsci.2020.03.008.

Lee, D., Zhang, H., & Nguyen, C. (2018). *The economic impact of natural disasters in Pacific Island countries: Adaptation and preparedness.* International Monetary Fund.

Leichenko, R. M., & O'Brien, K. L. (2002). The dynamics of rural vulnerability to global change: The case of southern Africa. *Mitigation and Adaptation Strategies for Global Change, 7*(1), 1–18. https://doi.org/10.1023/A:1015860421954.

Lesa, F. A. (2020). Trouble in pacific paradise: A call for merging traditional and modern tools of climate protection. *Georgetown Journal of International Affairs*, 21–27. https://doi.org/10.1353/gia.2020.0021.

Lohmann, P., Pondorfer, A., & Rehdanz, K. (2019). Natural hazards and well-being in a small-scale island society. *Ecological Economics, 159*, 344–353. https://doi.org/10.1016/j.ecolecon.2018.12.023.

MacGillivray, B. H. (2018). Beyond social capital: The norms, belief systems, and agency embedded in social networks shape resilience to climatic and geophysical hazards. *Environmental Science and Policy, 89*, 116–125. https://doi.org/10.1016/j.envsci.2018.07.014.

Maharjan, A., Tuladhar, S., Hussain, A., Mishra, A., Bhadwal, S., Ishaq, S., et al. (2021). Can labour migration help households adapt to climate change? Evidence from four river basins in South Asia. *Climate and Development*. https://doi.org/10.1080/17565529.2020.1867044.

Mahul, O., & Signer, B. (2020). The perfect storm: How to prepare against climate risk and disaster shocks in the time of COVID-19. *One Earth*, 500–502. https://doi.org/10.1016/j.oneear.2020.05.023.

Maiava, S. (2020). *A clash of paradigms: Response and development in the south pacific*. Routledge.

Mangubhai, S., Nand, Y., Reddy, C., & Jagadish, A. (2021). Politics of vulnerability: Impacts of COVID-19 and Cyclone Harold on Indo-Fijians engaged in small-scale fisheries. *Environmental Science & Policy, 129*, 195–203.

Marino, E., & Lazrus, H. (2015). Migration or forced displacement?: The complex choices of climate change and disaster migrants in Shishmaref, Alaska and Nanumea, Tuvalu. *Human Organization, 74*(4), 341–350.

McFarlane, C., & Silver, J. (2017). Navigating the city: dialectics of everyday urbanism. *Transactions of the Institute of British Geographers, 42*(3), 458–471. https://doi.org/10.1111/tran.12175.

McLeod, D. (2010). Potential impacts of climate change migration on Pacific families living in New Zealand. *Climate Change and Migration.*

McMichael, C., Farbotko, C., Piggott-McKellar, A., Powell, T., & Kitara, M. (2021). Rising seas, immobilities, and translocality in small island states: Case studies from Fiji and Tuvalu. *Population and Environment*. https://doi.org/10.1007/s11111-021-00378-6.

Mekuyie, M., & Mulu, D. (2021). Perception of impacts of climate variability on pastoralists and their adaptation/coping strategies in Fentale district of Oromia region, Ethiopia. *Environmental Systems Research, 10*(1). https://doi.org/10.1186/s40068-020-00212-2.

Mohammed, K., Batung, E., Kansanga, M., Nyantakyi-Frimpong, H., & Luginaah, I. (2021). Livelihood diversification strategies and resilience to climate change in semi-arid northern Ghana. *Climatic Change, 164*(3–4). https://doi.org/10.1007/s10584-021-03034-y.

Moser, C. O. N. (1998). The asset vulnerability framework: Reassessing urban poverty reduction strategies. *World Development, 26*(1), 1–19. https://doi.org/10.1016/S0305-750X(97)10015-8.

Nalau, J., & Verrall, B. (2021). Mapping the evolution and current trends in climate change adaptation science. *Climate Risk Management, 32*. https://doi.org/10.1016/j.crm.2021.100290, 100290.

Nand, M. M., & Bardsley, D. K. (2020). Climate change loss and damage policy implications for Pacific Island Countries. *Local Environment, 25*(9), 725–740. https://doi.org/10.1080/13549839.2020.1825357.

Ncube, A., Mangwaya, P. T., & Ogundeji, A. A. (2018). Assessing vulnerability and coping capacities of rural women to drought: A case study of Zvishavane district, Zimbabwe. *International Journal of Disaster Risk Reduction, 28*, 69–79. https://doi.org/10.1016/j.ijdrr.2018.02.023.

Neef, A., Benge, L., Boruff, B., Pauli, N., Weber, E., & Varea, R. (2018). Climate adaptation strategies in Fiji: The role of social norms and cultural values. *World Development, 107*, 125–137. https://doi.org/10.1016/j.worlddev.2018.02.029.

Ngin, C., Grayman, J. H., Neef, A., & Sanunsilp, N. (2020). The role of faith-based institutions in urban disaster risk reduction for immigrant communities. *Natural Hazards, 103*(1), 299–316. https://doi.org/10.1007/s11069-020-03988-9.

Nunn, P., & Kumar, R. (2019). Measuring peripherality as a proxy for autonomous community coping capacity: A case study from Bua Province, Fiji Islands, for improving climate change adaptation. *Social Sciences, 8*(8). https://doi.org/10.3390/socsci8080225.

Nunn, P. D. (2007). *Climate, environment, and society in the Pacific during the last Millennium*. Elsevier.

O'Brien, K. L., & Leichenko, R. M. (2000). Double exposure: Assessing the impacts of climate change within the context of economic globalization. *Global Environmental Change, 10*(3), 221–232. https://doi.org/10.1016/S0959-3780(00)00021-2.

O'Garra, T., & Mourato, S. (2016). Are we willing to give what it takes? Willingness to pay for climate change adaptation in developing countries. *Journal of Environmental Economics and Policy*, 249–264. https://doi.org/10.1080/21606544.2015.1100560.

Oskorouchi, H. R., & Sousa-Poza, A. (2021). Floods, food security, and coping strategies: Evidence from Afghanistan. *Agricultural Economics, 52*(1), 123–140. https://doi.org/10.1111/agec.12610.

Pagliacci, F., & Russo, M. (2020). Be (and have) good neighbours! Factors of vulnerability in the case of multiple hazards. *Ecological Indicators, 111*. https://doi.org/10.1016/j.ecolind.2019.105969.

Peth, S. A., & Sakdapolrak, P. (2020). Resilient family meshwork. Thai–German migrations, translocal ties, and their impact on social resilience. *Geoforum, 114*, 19–29. https://doi.org/10.1016/j.geoforum.2020.05.019.

Petzold, J. (2017). Social capital and small-island resilience. In *Climate change management* (pp. 17–61). Springer. https://doi.org/10.1007/978-3-319-52225-8_3.

Petzold, J., & Magnan, A. K. (2019). Climate change: Thinking small islands beyond Small Island Developing States (SIDS). *Climatic Change, 152*(1), 145–165. https://doi.org/10.1007/s10584-018-2363-3.

Piggott-McKellar, A. E., McNamara, K. E., & Nunn, P. D. (2020). Who defines "good" climate change adaptation and why it matters: A case study from Abaiang Island, Kiribati. *Regional Environmental Change, 20*(2). https://doi.org/10.1007/s10113-020-01614-9.

Sahlins, M. (1972). *Stone age economics*. Aldine-Atherton.

Sakdapolrak, P., Naruchaikusol, S., Ober, K., Peth, S., Porst, L., Rockenbauch, T., et al. (2016). Migration in a changing climate. Towards a translocal social resilience approach. *Die Erde, 147*(2), 81–94. https://doi.org/10.12854/erde-147-6.

Salam, S., & Bauer, S. (2020). Rural non-farm economy and livelihood diversification strategies: Evidence from Bangladesh. *GeoJournal*. https://doi.org/10.1007/s10708-020-10269-2.

Scarr, D. (1967). Recruits and recruiters: A portrait of the pacific Islands Labour trade. *The Journal of Pacific History*, *2*(1), 5–24. https://doi.org/10.1080/00223346708572099.

Scott, M. (2020). *Climate change, disasters and the refugee convention*. Cambridge University Press.

Sen, A. (1981). *Poverty and famines: An essay on entitlement and deprivation*. Oxford University Press.

Singe, J. (2020). *Blackbird story: A gripping tale of intrigue and adventure in the South Seas and Torres Strait*. Australia: Xlibris.

Sivadas, S., & Ismail, N. W. (2020). Increasing urban vulnerability in Malaysia: A myth or reality? *Journal of Islamic, Social Economics and Development (JISED)*, *5*(28), 130–140.

Smith, R., & McNamara, K. E. (2015). Future migrations from Tuvalu and Kiribati: Exploring government, civil society and donor perceptions. *Climate and Development*, *7*(1), 47–59. https://doi.org/10.1080/17565529.2014.900603.

Sodokin, K., & Nyatefe, V. (2021). Cash transfers, climate shocks vulnerability and households' resilience in Togo. *Discover Sustainability*, *2*(1). https://doi.org/10.1007/s43621-021-00010-5.

Solimano, A. (2021). *A history of the privatized pension system in Chile*.

SPC. (2020). *Pacific Island 2020 populations poster*. Pacific Community (SPC). https://sdd.spc.int/digital_library/pacific-islands-2020-populations-poster.

Steenbergen, D. J., Neihapi, P. T., Koran, D., Sami, A., Malverus, V., Ephraim, R., et al. (2020). COVID-19 restrictions amidst cyclones and volcanoes: A rapid assessment of early impacts on livelihoods and food security in coastal communities in Vanuatu. *Marine Policy*, *121*. https://doi.org/10.1016/j.marpol.2020.104199, 104199.

Steinbrink, M., & Niedenführ, H. (2020). Translocal livelihoods: New perspectives in livelihood research. In *Springer geography* (pp. 35–52). Springer. https://doi.org/10.1007/978-3-030-22841-5_3.

Sulewski, P., W⊠s, A., Kobus, P., Pogodzińska, K., Szymańska, M., & Sosulski, T. (2020). Farmers' attitudes towards risk—An empirical study from Poland. *Agronomy*, 1555. https://doi.org/10.3390/agronomy10101555.

Tacoli, C. (2011). *Not only climate change: mobility, vulnerability and socio-economic transformations in environmentally fragile areas of Bolivia, Senegal and Tanzania*. International Institute for Environment and Development (IIED). https://pubs.iied.org/sites/default/files/pdfs/migrate/10590IIED.pdf.

Thaman, R. R. (1982). Deterioration of traditional food systems, increasing malnutrition and food dependency in the Pacific Islands. *Journal of Food and Nutrition*, *39*, 109–121.

Thieme, T. A. (2018). The hustle economy: Informality, uncertainty and the geographies of getting by. *Progress in Human Geography*, *42*(4), 529–548. https://doi.org/10.1177/0309132517690039.

Tulone, A., Crescimanno, M., Vrontis, D., & Galati, A. (2020). Are coastal communities able to pay for the protection of fish resources impacted by climate change? *Fisheries Research*, *221*. https://doi.org/10.1016/j.fishres.2019.105374, 105374.

Vibert, E. (2020). Translocal lives: Gender and rural mobilities in South Africa, 1970–2020. *Politikon*, *47*(4), 460–478. https://doi.org/10.1080/02589346.2020.1840027.

Wade, H. (2005). *Pacific regional energy assessment: An assessment of the key energy issues, barriers to the development of renewable energy to mitigate climate change, and capacity development needs to removing the barriers*. Apia: Pacific Islands Renewable Energy Project.

Watts, M. J., & Bohle, H. G. (1993a). Hunger, famine and the space of vulnerability. *GeoJournal*, *30*(2), 117–125. https://doi.org/10.1007/BF00808128.

Watts, M. J., & Bohle, H. G. (1993b). The space of vulnerability: The causal structure of hunger and famine. *Progress in Human Geography*, *17*(1), 43–67. https://doi.org/10.1177/030913259301700103.

Weber, E. (1995). *Globalisierung und Politische Ökonomie der Armut in Indien: Die Auswirkungen wirtschaftlichen und politischen Wandels auf die Ernährungssicherheit von Armutsgruppen am Beispiel einer Kleinfischersiedlung in der südindischen Metropole Madras*.

Weber, E. (2014). *Environmental change and (im)mobility in the South* (pp. 119–148). Springer Science and Business Media LLC. https://doi.org/10.1007/978-94-017-9023-9_6.

Weber, E. (2017). Trade agreements, labour mobility and climate change in the Pacific Islands. *Regional Environmental Change*, *17*(4), 1089–1101. https://doi.org/10.1007/s10113-016-1047-7.

Weber, E., Kopf, A., & Vaha, M. (2021). *COVID-19 pandemics in the pacific island countries and territories* (pp. 25–47). Springer Science and Business Media LLC. https://doi.org/10.1007/978-3-030-68120-3_3.

Weber, E., & Koto, C. (2021). Disturbing the creation of a spatial system—Outside intervention and urbanization in the Republic of the Marshall Islands. In *The Routledge handbook of small towns* Routledge.

World Bank. (2012). *Samoa post tsunami reconstruction project*. World Bank. https://www.worldbank.org/en/results/2012/04/16/samoa-post-tsunami-reconstruction-project.

Yeo, S. W., & Blong, R. J. (2010). Fiji's worst natural disaster: The 1931 hurricane and flood. *Disasters*, *34*(3), 657–683. https://doi.org/10.1111/j.1467-7717.2010.01163.x.

Yila, O., Weber, E., & Neef, A. (2013). The role of social capital in post-flood response and recovery among downstream communities of the BA river, Western Viti Levu, Fiji Islands. *Community, Environment and Disaster Risk Management*, *14*, 79–107. https://doi.org/10.1108/S2040-7262(2013)0000014010.

CHAPTER

16

Management of extreme hydrological events

Ruth Katui Nguma[a] *and Veronica Mwikali Kiluva*[b]
[a]Kenya Meteorological Department, Nairobi, Kenya
[b]Masinde Muliro University of Science and Technology, Kakamega, Kenya

1 Introduction

The impacts of climate change have been felt across the world. However, it is still a grand challenge for scientists and scholars to understand the nature of these impacts on the hydrological cycle. This is the case considering the spatial–temporal dimensions of water and how its movement, distribution and quality is and will respond to climate change. Research evidence (*Global Change and Extreme Hydrology: Testing Conventional Wisdom,* 2011) observed that climate change has exacerbated the occurrence of extreme hydrological hazards with the most common hazards being extreme floods and droughts. Whereas floods and droughts occur as a result of natural climate processes, human interventions equally impact the nature of these hydrological hazards. In fact, the complexities of the interactions between the atmosphere and the systems on the land surface have impeded a clear understanding of hydrological extremes. In essence, the occurrence of extreme floods and droughts is defined by different processes that occur at different spatial–temporal scales in the atmosphere, catchments, socio-economic processes and river systems that feed back to each other. Giuntoli, Vidal, Prudhomme, and Hannah (2015) assert that the understanding of future the occurrence of extreme floods and droughts is enabled by use of global hydrological and climate models. However, as the author further notes, there are major uncertainties that limit this understanding. Tabari (2019) ascends to the same opinion, further noting that the rarity of these events along with their small sample size limits the ability to carry out their analysis. Furthermore, the author posits that misrepresentation of the associated processes at the mesoscale levels, the course spatial–temporal resolution of the global models and the interconnectedness existing between different extremes, further complicates the analysis. Whereas this is the case, the available evidence shows that hydrological extremes will continue to increase in the future, as Giuntoli et al. (2015) further argue. This will further be worsened by the increasing world population. This increase is deemed to increase both risk and vulnerability as there would be higher demand for a scarce commodity and increased conversion of forests to urban centers that will likely experience extreme urban flooding incidents.

Risks of weather and extreme climate events affect virtually all natural and nonnatural systems (GoK, 2018; Sivakumar et al., 2014). In reference to Murray and Ebi (2012), responding to climate extremes does not only depend on the hazard, but also on other dimensions of risk which include the factors of vulnerability and exposure. Following this point, the Sendai Framework 2015–30 was brought up requiring that the member states enact concrete actions for reducing disaster risk while at the same time safeguarding development gains. Previously in 2013, the Warsaw International Mechanism (WIM) of losses and Damage came into force as a result of the small Island developing states (SIDs), and Least Developed Countries (LDCs) demanding that the residual and unavoidable losses and damages that are not addressed by disaster risk reduction and adaptation strategies are addressed (Stabinsky & Hoffmaister, 2015). This was particularly the case following the observation that these countries were the most vulnerable to climate change impacts including extreme events, and low onset disasters. In essence, climate change extremes have resulted in damage and losses that cannot be addressed through adaptation and mitigation strategies. Hence, WIM sought to address this issue by providing financial assistance, enabling data collection, its analysis and creation of its awareness, and provision of coordination and oversight role in dealing with the impacts of climate change, and risk management. It is important to note that managing the risk of extreme events requires the understanding of the risk, and this is often

Climate Impacts on Extreme Weather
https://doi.org/10.1016/B978-0-323-88456-3.00009-5

realized through assessment. However, according to Liu, Li, and Attarod (2021), higher uncertainties have been highlighted in view of the occurrence of climate extremes. The understanding and assessment of climate extremes is still in its infancy stage and, as a result, this may compromise risk management approaches. This point has been echoed by Schewe et al. (2019), indicating that at both regional and global levels, the understanding, assessments, and attribution of hydrological extremes along with the associated impacts remains a great challenge. This challenge is largely attributed to the fact that human interventions have, to a greater extent, contributed to the occurrence and the nature of hydrological extremes being observed in recent times.

2 Management measures

2.1 Mitigation

Disaster mitigation has been defined by Field, Barros, and Stocker (2012) as the actions that attempt to limit further adverse conditions once a certain disaster has materialized. Extreme weather and climate events have greater impacts on those sectors that have closer links to climate, such as water, agriculture and food security, forestry, health, and tourism. This situation calls for more stringent and comprehensive mitigation measures.

In the context of climate change, mitigation initiatives have been largely employed in order to limit their undesirable impacts (Barnett & Webber, 2009). In particular, mitigation techniques and mechanisms have been instituted to effectively respond to, and reduce disaster risks. From the local to the global scale, weather and climate risks have been characterized by more frequent and catastrophic extreme events such as droughts, floods, landslides, among others. In particular, there has been an increased worldwide concern in regard to hydro-climatic extremes (Liu et al., 2021). This is the case considering hydrological extremes, with a main focus on droughts and floods.

Mård and Baldassarre. (2016) observed that management and response to hydrological events have been made more complex by the fact that hydrological systems were largely interfered with by human interventions. To this end, the authors further assert that feedback mechanisms between society and the hydrological extremes warrant more research and scientific inquiry. This is because society has been shaping hydrological extremes even as these extremes shape society. As a result of this interaction, the evolution and mechanism of hydrological extremes has become even more complex, and without a clear understanding of this complexity, the relevant risk management approaches may be limited.

One of the most significant hydrological events viewed from a local to global scale is floods. In the past decade, floods have been responsible for approximately 64% of all disaster events in Africa. This status is followed by severe storms that are experienced at the level of 15% *(Disasters in Africa: 20 Year Disasters in Africa: 20 year review,* 2000*)*. On the other hand, widespread occurrence of droughts has also been reported across the globe. In particular, the Greater Horn of Africa has experienced lengthier and severe droughts with the example of the 2010–11 drought event that resulted to losses of billions of dollar in economic output and even loss of lives and livelihoods. The most recent drought in the Greater Horn of Africa was also reported in the year 2015–16, which left a large percentage of the population in starvation.

Elsewhere in Australia, the Millennium drought that occurred since 2003–12, resulted in death and drying of the main Australian rivers. In addition to this, the famous California 2012–15 droughts resulted to severe water shortages with adverse effects not only to the agricultural sector, but also in other important sectors of the economy. In view of the above cases of hydrological extremes, Kenya has made attempts to develop policy guidelines (GoK, 2018; Sivakumar et al., 2014) to help mitigate the impacts of weather and climate extreme events. This has been the case in deploying disaster mitigation interventions elsewhere in places like the Netherlands, California and Australia among others regions.

According to Cabrera, Daniel, Guillermo, and David (2009), the inherent risks associated with these hydrological extreme events must be effectively managed. This calls for concerted and dedicated efforts in harnessing mitigation options and integrating climate resilience measures to reduce risks. Many nations of the world, particularly in Africa and Asia, have already welcomed the process of mainstreaming climate change management initiatives into the national planning and development agenda, guided by accurate and reliable weather and climate data, analysis, and policy frameworks.

From the local to global scale, there exists documented evidence of the increasing trend in the rate at which the global climate is changing (Devkota & Gyawali, 2015). The available evidence shows that among other weather and climate extremes, hydrological events are bound to increase in severity and frequency in future (Diaz & Corzo, 2018). With the observed increase in frequency and severity of hydrological events being confirmed by more research evidence (Royan, Hannah, Reynolds, Noble, & Sadler, 2014), the necessity for a spatiotemporal analysis of these extremes has been highlighted and emphasized. This case in point has been further informed by the remarkable local and regional variability in floods and drought hazards that has been pointed out through climate model projections across the globe and further pointed out by Kundzewicz and Matczak (2015).

Durack, Wijffels, and Matear (2012) concluded that the water cycle was intensifying in relation to hydrological processes of key importance such as precipitation, runoff and evaporation. The authors further assert that hydrological extremes were characterized by increasing trend, frequency, magnitude and duration. This situation calls for more robust disaster risk management approaches given that the ever changing weather and climate situation is likely to cause more intensified droughts and floods events. This is to suggest that drier and wet conditions that correspond to drought and flooding events respectively, are likely to be experienced across the globe.

In fact, Qing, Wang, Zhang, and Wang (2020) posit that the occurrence of extreme hydrological events poses a great threat to both water and food security situations across the globe. Notwithstanding, the predicted future increase in hydrological extremes has been attributed farther to not only climate change, but also to the complex interaction of climate change, population growth, increasing demand for urban development coupled with increasing water demand, and other socio-economic factors of interest.

Mitigation and adaptation measures to risks of hydrological extremes have been widely utilized in the context of risk management. The management strategies that are associated with the hydrological events such as flooding events have been highlighted by Apel, Aronica, Kreibich, and Thieken (2009), Thieken et al. (2002), and Thieken and Menzel. (2004). On an account of research by Dovers and Hezri (2010), mitigation remains a difficult challenge although if the world would adhere to the agreed targets to reduce carbon emissions, more realistic disaster risk reduction would be realized.

With reference to flood risk management, Genovese and Thaler (2020) observed that flood mitigation is often more structural than drought measures of mitigation. In fact, the mitigation initiatives of both drought and floods risk require an integrated approach since they are both extreme events that feed to each other.

As has been interrogated and described by Kiluva (2012), structural and no-structural mitigation measures are deployed together for effective flood risk management. Largely, structural measures are commonly aimed at reducing the harm of floods through landscape reconstruction. In addition, structural flood risk reduction measures make use of flood control structures such as levees, dykes, culverts, dams, and diversion canals, among other structures. Whereas this is the case for the structural mitigation measures, nonstructural measures are often deployed with the aim of ensuring that damage to the elements at risk is reduced. In many instances, nonstructural mitigation is done by evacuating people and their property from those areas that are deemed risky.

Further nonstructural mitigation according to Kiluva (2012) can be achieved by instituting measures such as flood forecasting and early warning systems, regulations on land use practices, flood proofing measures, and in the larger perspective, mitigation policies that encompass preparedness, emergency response protocol development and recovery mechanisms.

Generally, a combination of both structural and nonstructural mitigation measures has been cited to be more effective as opposed to just relying on either structural or nonstructural strategies. Following this point, a recent study by Yun et al. (2021) argued that reservoir regulation can be used to mitigate risks of future hydrological extreme events in the Lancang-Mekong River Basin. However, as the authors further indicate, realizing total mitigation of hydrological extreme events through reservoir regulation may be limited by the fact that there exists limited knowledge in relation to the changes of hydrological extremes under climate change. This situation is justified by the fact that hydrological extreme events that are triggered by the ever changing weather and climate occur in different spatial–temporal scales whose extent has not yet been well captured and mimicked by the global climate and hydrological models (Tabari, 2019). Adaptation to the footprints that have been created by the ever changing weather and climate remains key in reducing the inherent risks.

2.2 Adaptation

Adaptation to extreme hydrological events contributes to disaster risk management. The relationships between coping, coping capacity, adaptive capacity, and the coping range have been very well discussed in the Field et al. (2012) report. More research output on adaptation to climate change impacts has been documented by Abegg, Agrawala, Crick, and De Montfalcon (2007) and Adger et al. (2007). In addition to this, there is need to assess the various available adaptation practices, options, constraints and capacity to adapt for effective and sustainable adaptation (Beg et al., 2002; Birkmann & von Teichman, 2010).

An extreme event in the present climate may become more common, or more rare, under future climate conditions. When the overall distribution of the climate variables changes, whatever happens to the mean that the climate may be different from what happens to the extreme events at either end of the distribution. Understanding the indicators of vulnerability and adaptive capacity are hence important in instituting effective adaptation measures (Adger et al., 2007).

According to a report documented by Field et al. (2012), adaptation in human systems refers to the process of adjustment to actual or expected climate and its effects, in order to moderate harm or exploit beneficial opportunities. The authors continue to report that in natural systems, the process of adjustment to the actual climate and its effects; human intervention may facilitate adjustment to the expected climate.

At the local to global scale, the existing inequalities influence local coping mechanisms and adaptive capacity, and this farther poses disaster risk management and adaptation challenges. Research findings by Simonovic (2010) highlight that climate change adaptation strategies are hence significant and critical processes for both reduction and management of risks posed by extreme hydrological events. In an attempt to contribute to what adaptation entails, Murray and Ebi (2012) pointed out that the processes of adjusting to the actual effects, as well as the expected impacts of climatic extremes, constitute adaptation. This is the case considering that such processes are aimed at reducing, moderating and even ensuring that beneficial opportunities are exploited. Murray and Ebi (2012) further observed that understanding of the dimensions of exposure and vulnerability was a key aspect for effective strategies of adapting to climatic extremes. In fact, the main aim of adaptation as a risk management approach is to reduce vulnerability. Moreover, it is essential to note that adaptive actions are known to offer both short and long term benefits.

Whereas a variety of climate change adaptation measures have been put in place across the globe, a particular interest in adapting to the risks of hydrological extreme events are highlighted herein. This is the situation despite research evidence pointing out that adaptation is a more messier prospect as much as it may seem easier (Dovers & Hezri, 2010). The reason and explanation for the mess is attributed to the factors of how much adaptation is justifiably needed, given that the occurrence of droughts and floods is highly marred by complexity and uncertainty that arise from this complexity.

Another important reason for this mess in adaptation has been attributed to the role that is played by the contextual matters that are often marred by large extents of variations in view of local vulnerabilities, differences in policies and economic factors in the larger perspective. More to this point, the third reason for the adaptation being more messier than often thought relates to the how to adapt. This is the case considering the key role that is played by both policy and institutional mechanisms that are deployed to enable adaptation measures to take place. In actual sense, whereas mitigation initiatives target a global phenomenon, adaptation initiatives are more of local response to a problem that is occurring and experienced at a global scale.

The need for climate change adaptation varies from place to place, a situation that is highly dependent on the sensitivity and vulnerability to environmental impacts. To this end, Kundzewicz and Matczak (2015) noted that unless the world adapts to the increasing occurrence and extremity of floods and droughts, higher socio-economic losses are likely to be realized. According to Jongman (2018), flood risk management through adaptation will be more effective if an integrated approach is applied. Further to this, the author continues to highlight that a combination of adaptive strategies such as flood control infrastructure, risk financing schemes such as insurance, and nature-based solutions would be an ideal form of integration. The author further, however, notes that integration of different adaptation strategies varies from place to place. In addition, this variation is further affected by the prevailing risk levels, political goodwill and funding capacity. A case in point involves the use of nature-based flood risk management approaches, such as making room for the river system in the Netherlands, and also ensuring that physical protection structures such as dyke are put in place. Such a scenario is often limited by the fact that large capital investments are required, and without political goodwill, such good plans may not be realized. This is especially the case considering the developing countries whose competition for the available resources allocation limits engagement in such approaches that are aimed at adapting to floods.

Contrary to the argument by Jongman (2018) that flood control infrastructure were still great strategies of adapting to the extreme flood events, Kundzewicz and Matczak (2015) note that the traditional structural controls against floods have proved to be ineffective, especially when extreme and rare events prove more disastrous even on the flood control structures themselves.

Cases of residual risks that are characterized by flood control dam failure have been reported across the globe, a factor that exposes the limitations of deploying structural risk reduction measures. Of great concern to note is the argument by Kundzewicz and Matczak (2015) that highlights that when a dyke fails during an extreme event, it no longer acts as a risk reducing dyke, but rather it amplifies the problem. On the other hand, in the event a dam fails, when it was designed for storage of water to use as a buffer during droughts, this may instead enhance drought impacts as the reservoir storage capacity is depleted.

The process of adapting to climate change involves taking accurate actions that are geared toward preparing for, and adjusting to the present and future extreme events and their subsequent effects on the environment. In reference to a research account by Mubaya and Mafongoya (2017), adapting to droughts and floods as the main hydrological extreme events requires that both formal and informal institutional arrangements are put in place to facilitate the adaptation measures. In essence, the informal arrangements have been highlighted for facilitating more collective action by communities as opposed to the formal institutions. In addition to the role that is played by institutions in enabling adaptation measures and the associated implementation, policy frameworks are essential tools that are used by governments to create an enabling environment for institutional arrangements to work. Equally important, adaptation practices in view of droughts, include and are not limited to the supply and demand sides strategies. The former seeks to increase and augment water supplies while the latter seeks to improve water use efficiency.

Incremental adaptation initiatives aim at maintaining the essence and integrity of a system. Those adaptation strategies that are transformational aim at changing the fundamental attributes of a system in response to climate change and its adverse impacts to cope with the resultant consequences. Kundzewicz and Matczak (2015) further point out that drought and flood early warning systems act as the soft adaptation measures and are particularly used for triggering the mobilization of adaptation measures to the extremes.

2.3 Early warning systems

The success of any early warning system relies on the free flow of information. According to Rogers and Tsirkunov (2011), an effective early warning system must empower individuals, communities and businesses to respond timely and appropriately to hazards in order to reduce the risk of death, injury, property loss and damage. These early warnings must get the message across and stimulate those at risk to take action in form of invoking response measures. The costs and benefits associated with the individual early warning systems must also be accurately evaluated (Rogers & Tsirkunov, 2010).

The development of effective early warning systems could contribute to fostering livelihood resilience by improving coping mechanisms and even enhancing adaptive capacity. However, Baudoin, Henly-Shepard, Fernando, Sitati, and Zommers (2014) caution that the current shortcomings in early warning systems' conception and applications undermine risk reduction at the grassroots level, which contributes to loss of lives and shocks to livelihoods. The authors further recommend the need to significantly improve the way in which early warning systems are designed and applied. The same research findings proceed on to suggest deployment of an integrated cross-scale approach that ensures the involvement of the at-risk population in all phases; from the risk detection phase, through to all emergency management processes.

The obvious intensification of the hydrological cycle as a result of global warming raises serious concerns about obvious future floods and their potential impacts on socio-systems and eco-systems, where the degree of exposure to these extreme events has also increased. Research findings by de la Fuente, Meruane, and Meruane (2019) confirm to the world that the development of adequate adaptation solutions such as early warning systems is crucial for effective forecasting aimed at providing timely and sufficient information and support for real-time decision-making purposes.

A report by the United Nations entitled (*Establishing drought early warning systems in West Asia and North Africa (UN-DESA/DSD) Intern*, 2013) emphasizes that joint responsibility is encouraged to involve more power in face of the battle of combating extreme hydrological events. This objective can be achieved through cooperation and coordination at local, national, regional and global levels and through partnerships with a multitude of stakeholders, ranging from the citizens to policymakers to the private sector. The same approach is supported by the Warsaw International Mechanism (WIM) of losses and Damage, a framework that encourages leadership, cooperation and coordination in addressing loss and damage associated with climate change impacts. This is through financial assistance, improving data collection, analysis, and creation of awareness of weather and climate data and information, and further to this, providing leadership and oversight role in order to eliminate overlap and duplication of efforts (Stabinsky & Hoffmaister, 2015). In the current sense, WIM further seeks to enable pooling of resources and create synergies in order to deal with climate change impacts that cannot be addressed by adaptation and mitigation measures. This includes extreme events such as hydrological extremes like drought and floods.

Adaptation to the changing weather and climate in many countries on the planet is achieved by using early warning systems (EWS) (Willis, Wilgen, Everson, Abreton, & Pero, 2001). In view of the aforementioned, it is, however, noted that early warning information does not always reach the people who need it most, despite a warning being issued, and the reliability of the early warning information also needs improvement. The foregoing authors further found that the packaging of the early warning information needs to be improved, including carrying out translation of the message into local languages to support local level response strategies.

According to Field et al. (2012), early warning systems refer to the set of capacities that are needed to generate and disseminate timely and meaningful warning information to enable individuals, communities, and organizations that are threatened by a hazard to prepare and to act appropriately and in sufficient time to reduce the possibility of harm or loss. More research to support the role of early warning systems in disaster risk reduction in the context of extreme weather and climate has been conducted by Abegg et al. (2007), Barnett and Webber (2009), Birkmann and von Teichman (2010), Simonovic (2010), and Sivakumar et al. (2014).

Integrated communication systems are deployed to facilitate vulnerable communities with early warning messages to reduce risks from extreme weather and climate events. The enhancements of both short-and long-term coping strategies, climate monitoring and early warning systems, flood control infrastructure, and other disaster preparedness measures at all levels, including subregional, national, and local levels could help in reducing vulnerability to climate change and climate variability.

As has been reported by Huho, and Kosonei (2014), understanding and managing extreme weather and climate events is achieved by use of early warning systems. To support this argument, for instance, Kenya has a sophisticated early warning system that utilizes modern technology that alerts vulnerable communities of potential floods and drought related risks. The risk information is effectively communicated to risk prone regions through the Short Messaging Service (SMS) and electronic mail. The recipients of the early warning messages are hence able to make informed decisions for better preservation of their investments and livelihoods (GoK., 2004). In the case of flooding, timely evacuation to safer grounds is instituted and lives and property are saved.

As documented by Alfieri, Salamon, Pappenberger, Wetterhall, and Thielen (2012), early warning systems use scientific models that integrate the components of risk knowledge, monitoring and predicting, dissemination of information and response to the subsequent generated early warnings. Communities that make use of early warning systems end up reducing disaster risks and hence ensuring preparedness and rapid response to extreme weather events. It is critical to note that many countries in Africa have outdated and inadequate hydro-meteorological stations that are in most cases not updated, leading to ineffective means of generating and disseminating disaster early warning information. To this effect, Hallegatte (2012) recommends early warning systems for the developing world, whereby most countries still lag in installing the systems. This situation calls for governments to install and reactivate all existing automatic weather and agro-meteorological stations and automatic hydrological stations in their respective countries. Those countries that have invested in robust hydro-meteorological stations and hydrological systems are able to provide the needed weather and climate risk information to the end users such as farmers, academic institutions, research institutions, among others.

Several traditional and modern early warning systems in relation to managing the risk of droughts and floods have been largely deployed by both the governments and communities to increase their adaptive capacity. Early warning systems entail capacities that are developed by different authorities, be it in a modernized or traditional way for the purposes of both generation and dissemination of meaningful information for warning of potential extreme events in a timely manner. According to Alfieri et al. (2012), early warning systems play a critical role in disaster risk management in the sense that they enhance preparedness levels. This is a key factor that contributes in reducing the risk of the impact of droughts and floods.

Largely, numerical weather prediction has been emphasized as a basis for the development of flood early warning systems across the globe. In this sense, early warning systems are known to enable the detection of extreme events early enough, giving those at risk enough lead-time to prepare and respond effectively. Important to note, is the fact that hydrological hazards do not actually have political boundaries, as highlighted by a research by Niebla et al. (2011). The author makes reference to a concept adopted in Europe known as the Alert4All. It is a scalable and innovative approach whose aim is to improve the communication systems on alerts and disaster warnings. The system creates synergies between the aspects of technology, human and management in order to align alerting procedures and structures with the contemporary types of disasters across boundaries. This approach enables preparedness across borders. The level of preparedness that is realized may be in relation to plans on response and emergency.

The Alert4All concept in Europe was made in line Common Alerting Protocols (CAP) by WMO (Yu-fai & Cheng, 2018). The CAP approach recommends a standardized format of a message that is designed for all hazards, all media, and communications. Simply stated, it is a standard way of ensuring that a message on an alert is understood across borders in the way for all hazards, and through all media and that the communication is standardized all across. This is the case considering that hydrological hazards among other weather and climate extremes know no boundaries and as such, they need to be managed in an integrated manner. Further making reference to a research by Chaves and De Cola, (2017), the effective and efficient management of hydrological extremes, requires timely communication with the population at risk. More to this, the relevant information and recommendations for the appropriate actions to be taken should be highlighted. The authors noted the Alert4All concept as an example that showcases efficiency and effectiveness in communicating alerts on hydrological hazards. Following research evidence by Hallegatte (2012), hundreds of lives are saved annually through early warning systems and the generated warnings and information. Furthermore, more than 2.5 billion of Euros are saved through these early warning systems as author further notes. With the assured early availability of accurate climate risk information to users, accurate and reliable future planning is enabled for disaster risk reduction purposes and effective response.

2.4 Response

Weather and climate-induced hydrological extremes, such as floods and droughts are witnessed locally and globally. Understanding and managing these connected extreme events is critical for the effective risk reduction. Effective response to hazardous hydrological extreme events such as floods and droughts is critical for the global Sustainable Development Goals (SDGs) to be realized. According to a study by Devkota and Gyawali (2015), extreme hydrological events can be addressed through adhering to prudent water resources management practices. Response to extreme hydrological extremes such as droughts and floods occurs in various ways. First, it entails the responses that human populations deploy with the aim of reducing the risk of these events, and on the other hand, the responses by the environment to these events. With reference to the former, both formal and informal responses to droughts and floods have been documented. Following the formal responses to hydrological extreme events, governments across the world have adopted adequate policy initiatives to deal with the associated risks of droughts and floods.

Floods and droughts with extreme consequences result from the superposition of various processes at different space- and time scales. A report by Dovers and Hezri (2010) states that through institutions, humans are able to respond to droughts and floods. To facilitate decision-making and deliberations in view of these events, policy processes are employed as a deliberate mechanism that enables the workings of the weather and climate risks responding institutions. Formal institutional responses to droughts and floods are often backed by the state and in other cases, the private sector, not forgetting the role that is played by international bodies such as the United Nations and the World Bank among other institutions. Informal institutional arrangements are commonly found among communities in the local context, and even when formal institutions are present, they still operate. This is the reason why researchers such as Agrawal (2010) and Mubaya and Mafongoya (2017) recommend the integration and recognition of informal institutions and individual local efforts in response to droughts and floods extreme events. A close collaboration and inclusion of all stakeholders has been supported by research evidence as documented by Somanathan et al. (2014) and Wilhite (2019).

Whereas the human responses to droughts and floods are defined through the lens of policies and institutional arrangements, the environmental responses are more complex. This is the case considering that, in places where flooding occurs, the immediate environment may respond by becoming uninhabitable due to the ecosystem's changes resulting to growth of unfriendly environmental features. In view of drought hazards, response by the environment may be the extinction of some species that are important for the environment, which may further serve to amplify the potential impact of droughts and floods risk in future. The feedback mechanisms of droughts and floods may result to either ameliorating or worsening impacts of drought and flood risk.

Initiatives, whether at local or global scale, that involve mitigation efforts and adaptation strategies through deployment of early warning systems are geared toward ensuring effective response to the extreme hydrological events. The overall goal is to reduce the potential risks that are associated with the extreme events.

2.5 Policy

Policy alludes to the set principles, and guidelines that are adopted and followed by a government to guide decision-making for better outcomes for a society. Dovers and Hezri (2010) define policy as the position that is taken and communicated by a government in recognition of a problem that affects the society, and how the said problem should be dealt with. To this end, a government specifies, and implements policy interventions with the sole aim of influencing the behavior of the society to realize the goals of the policy established.

According to Wilhite (2019) on drought risk management, the main aim of policy response to drought is to encourage the pollution and the economic sectors perceived to be vulnerable to deploy measures of risk management that are reliable. It is also within the drought policy framework that sustainable use of natural resources and the agricultural resources is largely promoted and pursued. In addition to this, through policy frameworks, recovery from drought and floods as the most prevalent hydrological extremes are propagated. In essence, proactive planning and proactive measures are highlighted in view of risk and resource management and public outreach programs.

Dovers and Hezri (2010) further assert that policy implementation may take the form of policy programs such as the dissemination of weather information, and the support to conserve water through local funded projects in view of a drought policy. Considering management of risks of weather and extreme climate events, a policy framework is required to guide the mitigation and adaptation initiatives. This is in line with the opinion of Pielke (1998) that climate change policies focus mainly on mitigation and adaptation risk management options.

Various nations across the globe have enacted policy frameworks to guide them when responding to effects of climate change and extreme events. According to Sivakumar et al. (2014), risks of climate change and extreme events have been on the increase in frequency and severity across the globe. Efforts to either mitigate and/or adapt to the effects have been realized through policy frameworks. It is within these frameworks that risks are defined, and modes of reducing, mitigating and even adapting to the effects on the pipeline are addressed. Cochrane, Young, Soto, and Bahri (2009) highlight that public policy and institutions play a major role in promoting and guiding both the adaptation and mitigation of the risks of weather and climate extreme events.

In reference to management of risks that are associated with hydrological extremes, more challenges especially in policy response have been highlighted. This is the case considering that hydrological extremes in view of droughts and floods pose a great challenge to scientists. According to recent research evidence by Liu et al. (2021), understanding the hydrological extreme events such as droughts and floods is marred by higher uncertainties. These uncertainties are largely associated with climate change projections, and the human interventions that either ameliorate or exacerbate hydrological extremes risk. Elsewhere in Yarlung Zangbo river basin in China, management of hydrological extreme events had been challenged by limited scientific understanding (Lutz, Immerzeel, Shrestha, & Bierkens, 2014). This is the case due to uncertainties arising from climate change projections, and on the other hand, human interventions which are difficult to observe and integrate in the hydrological models.

Policy responses to hydrological extreme events such as drought and floods have been implemented across the world. Drawing lessons from the case of the 2014–15 drought that occurred in Brazil affecting water supplies to the Sao Paulo city, it is evident that political context affects policy effectiveness. In response to this drought, Grover and Lucinda (2020) noted that the major water utility serving the city responded by implementing two economic instruments, that is, a financial reward and penalty for reducing and increasing water consumption, respectively.

Interestingly, the penalty-based policy instrument, which was implemented after the national elections in Brazil, realized better water conservation levels as opposed to the reward-based one. Commenting on this situation, Grover and Lucinda (2020) argue that the choice of the reward policy instruments was influenced by the political landscape of the day. This is a case of how policy response to hydrological extreme events is not only a function of scientific research principles but is largely influenced by political context.

Policy response to floods and droughts should be based on the sound understanding of the associated risks. Particularly, droughts are slow creeping disasters whose impacts take a longer time before being manifested in various socio-economic sectors. Following this point, policy responses to droughts have been rather more reactive as opposed to being proactive responses. Echoing this statement, Kamara, Sahle, Agho, and Renzaho (2020) noted that in the Kingdoms of Eswatini and Lesotho, policy response to drought risk was largely reactive. This was the case as the national policies on drought did not incorporate the resilience to recurring droughts in their goals. This is against the call for enhancing resilience against disasters as provided for in the Sendai Framework for disaster risk reduction.

Kamara et al. (2020) criticize countries' policy response to drought by citing that these countries are more concerned with disaster management and recovery. To this end, the authors recommend an adoption of a policy response that seeks to adhere to the principles of the Sendai Framework on disaster risk reduction as opposed to emergency management and recovery. Wilhite (2019) further supports the opinion that policy response to drought has been a reactive one. The author notes that policies that promote crisis management approach to droughts risk increasing vulnerability to future drought events. Furthermore, the author points that policy response to drought should engender a sense of cooperation and coordination across all the levels of administration. This should be the case in view of asserting national government leadership in ensuring that proactive measures are put in place and measures to increase adaptive capacities of communities to survive lengthened periods of drought episodes.

Wilhite (2019) continues to point out that in policy response to drought, reliance on crisis management approach in view of drought risk, may not be viable in case of extended periods of droughts. This is the case considering that cases of more frequent, severe, and prolonged drought events have been on the increase with concern for the management of hydrological drought being highlighted. This sentiment was further echoed by a research by Van Loon (2015) that although hydrological drought affected more sectors than other droughts, research attention was not directed to it. This was equally the case considering that policy response only recognized the meteorological and hydrological categories of drought. This is to suggest that mitigation and adaptation policies often negated the importance of considering hydrological droughts, especially when considering lengthened periods of drought that affect the hydrological system, and even the ground water resources.

Focusing on flood risk management policies, a tendency for many governments to adopt proactive response initiatives is largely highlighted. Presumably, the far-reaching tangible impacts of floods may be attributed to this tendency. However, as Shah, Rahman, and Chowdhury (2018) observed in their research, flood risk management remains a daunting task to countries situated in both the developed and developing world. This is the case considering that flood risk management policies have largely promoted structural flood risk reduction measures such as diversion channels, dams, dykes, embankments, and other related control measures.

Together with the nonstructural response measures such as early warning, land-use changes, and reduction of exposure and vulnerability among others, as such have been highlighted in flood risk management policies. In most cases, control measures of flood risk are often deployed as mitigation measures, while the nonstructural ones have been used mostly for adaptation purposes. Considering this point, van Buuren, Ellen, and Warner (2016) observed that in the Netherlands, flood risk management had gone through a vigorous policy learning process. The authors argue that, although the Dutch flood risk management policy had promoted protection-oriented measures against floods, through a process of policy learning, they deployed a risk approach as witnessed in the last decade.

In fact, the 2009 National Water Plan in the Netherlands was introduced on the concept of multilayered safety. This was done in recognition of the shortcomings of over-dependency on protection measures against floods, and the need to develop resilience. This was in view of adapting to the inevitable occurrence of future flood events, especially when considering that as much as climate change contributed to trigger hydrological extremes, human contribution could not be ignored. In this connection, socio-economic development, coupled with an increased population and impervious surfaces especially in urban centers, have brought about new challenges for flood management policy makers. Following this situation, Hurlbert and Gupta (2016) recommend adaptive governance as the ultimate solution for the increasingly evolving nature of flood risk across the globe. Echoing this opinion, Haasnoot, Kwakkel, Walker, and ter Maat (2013) further recommend the deployment of adaptive and dynamic policies that have the capability to incorporate uncertainties and unforeseen future occurrences.

2.6 Institutions

In any society, there are laws, processes, arrangements, and customs structures underpinning the social, cultural, economic, and political relationships and the associated transactions. These structures are termed as institutions, and they may exist in multiple levels of organizations within a society. Agrawal (2010) noted that these organizations exist in categories of public, private, and civic, and as such, they are generally categorized as formal or informal. The formal institutions are characterized by organizational structures with tangible governance frameworks.

On the other hand, the informal institutions relate to uncodified rules of the game that are based on social and cultural norms. Altogether, both the formal and informal institutions are responsible factors that shape the behavior of societal interactions that further influence the adaptive capacity of individuals. Arguably, institutional arrangements are known as mediators of livelihood outcomes in communities, especially when dealing with climate change risks and extreme weather events. This position can further be explained by higher risks and poor livelihood outcomes that are reported among the poor dwelling in slums in the event of extreme hydrological events such as floods. In such a situation, the impact on communities living in slums will be magnified against a backdrop of those that live in other areas in the urban setting. Such situations are an outcome of institutional arrangements as argued by Agrawal (2010). In this context, the author asserts that social exclusion of vulnerable groups such as women and children may be due to institutional arrangements that favor men over women. Arguing for the same opinion, Mubaya and Mafongoya (2017) observed that institutional arrangements may favor some groups over others, and as such, when dealing with climate change risks and extreme weather events, some groups may access institutional support than others. Such a situation brings about differentiated livelihood outcomes, vulnerabilities, and adaptive capacities. Whereas this may be the negative outcome of institutional arrangements, such a situation can be mediated by making integrated institutional arrangements that recognize the vulnerable and secluded groups whose adaptive capacity may be limited.

Essentially, institutional arrangements exist at international, regional, and national levels. At the international level, the United Nations has been on the forefront and so to speak, it gives support to the National Adaptations Plans (NAP). Such support has been realized through recommendations of policy, and the political agenda development. More so, international institutions provide scientific information and guidance to guide policy on mitigation and adaption measures. Needless to state, such institutions as well provide the technical and financial support along with the capacity building initiatives for national adaptation plans.

Besides the international institutional arrangements, regional institutional arrangements have been used for the support of cross-border issues highlighted by Alfieri et al. (2012). A case in point that supports regional institutional arrangements relates to the scenarios in the Nile River Basin, whose management during both cases of flood and drought scenarios requires institutional support both in the upstream and downstream countries. Largely, the national institutions depend on the regional institutional arrangement to meet their objectives in view of weather and climate change risks and extreme events management. These regional institutional arrangements mostly steer the action of National Adaptation Plans (NAP) by providing technical as well as financing support of the same. This position is further echoed by cases in the European Union (EU) strategy in view of adapting to climate change risk. The EU framework provides the much-needed information and financial resources that are needed to boost the capacities of national adaptation plans for actions in managing risks of climate and extreme weather events.

In reference to the national institutional arrangements, nations across the globe often designate a national institution or so to speak, an agency. At times, it may be a designation of several governmental institutions to lead the national adaptation plans up to the local level by engaging different actors across board. This is to suggest that institutions play a critical role in ensuring that there is a successful implementation of mitigation and adaptation strategies. According to Dovers and Hezri (2010), institutions create an enabling environment for organized and collective efforts that are geared toward realizing shared societal goals. Such goals take in the efforts to meet and address common societal challenges such as climate change risks and extreme weather events. It is also through these institutions that differences in collective efforts are reconciled. Although this is the case, it is important to note that institutions are generally persistent in their role but marred with constant evolution.

By and large, institutional arrangements influence vulnerability by structuring risks and the eventual impacts (Agrawal, 2010). Accordingly, these institutions further play a mediating role in view of individual and community response to climate change and extreme weather events risks and their anticipated impacts. Interestingly, in most cases these institutions act as the means through which external facilitative resources are channeled. In actual sense, these institutions govern the accessibility of resources for mitigation, thus facilitating effective adaptation to climate change risks and extreme weather events.

In connection to the foregoing, Somanathan et al. (2014) highlight that institutional arrangements have a greater impact on both the feasibility and choice of policy options to be implemented in view of mitigation measures against climate change risks and extreme weather events. The authors further argue that the economic decision-making processes are mediated by the institutional structures. Equally important, they further shape the political decision-making context upon which some interests are suppressed while others are empowered. Moreover, dominant policies are enabled by institutions which play a major role in shaping thoughts and understanding of policy choices.

3 Challenges

Considerably, the management of risks of extreme hydrological events which have been accelerated by climate change, and further compounded by anthropogenic activities, has proved to be a challenging task. This has been the case considering that research evidence points out that hydrological extreme events are bound to increase in both frequency and severity (Giuntoli et al., 2015). This is in line with the projected changes as accounted for by global climate models (GCMs), as well as by global hydrological models.

The output of global climate models is used as a forcing to simulate future hydrological extreme values. However, as Giuntoli et al. (2015) observed, these models are characterized by uncertainties that blur the understanding of hydrological extremes risk. Risk is defined through the elements of the hazard, exposure and vulnerability. However, it is important to note that the hazard data for hydrological extremes is limited (Bayazit, 2015). According to Bayazit (2015), the nonstationarity characteristic of hydrological extreme records presents a challenge for the climate models, which are often inclined toward stationarity of data. Therefore, as the author further observes, estimation of changes in the extreme tails of data, that is, low and high ends, is much more difficult as compared to estimating changes realized in the mean conditions. This is not just the case when considering the data on the extremes, but as well, the ability of the climate models to reproduce these extremes is deficient. In this case, therefore, the available climate models remain limited in helping to estimate the extremes. In view of both the exposure and vulnerability data that is required to perform a comprehensive risk assessment, research evidence indicates that such data is largely unavailable (Trajkovic et al., 2016). In addition to such data being unavailable, the world is currently grappling with unprecedented changes to land use, urbanization and increased population, which is constantly altering the exposure and vulnerability components. Populations move from one area to another, voluntarily or being assisted, others are displaced, and others even migrate, and they may settle either permanently or temporarily in areas that have different sets of risk. Such risks are different in the sense that these people have to disengage and engage with different sets of institutions, risk management strategies, knowledge, and resources, among others, as such. As well, hazards are dynamic in nature and therefore, when people move from one place to another, they may experience differences of hazards in time and space. This further undermines the ability of scientists to assess the risk of hydrological extremes for better management thereof. In fact, the uncertainties in the hazard, exposure and vulnerability components are defined by Hurlbert and Gupta (2016) as the risk, alluding to the fact that risk is the effect of uncertainty. By so defining risk, the authors allude to the fact that uncertainty should be accounted for when managing risks. Therefore, consideration of the dynamics of the risk components (or their proxy indicators) and their interactions is paramount. Making reference to the dynamic nature of risk components, extreme hydrological events force the population to move from one area to another, a factor that shapes both their exposure and vulnerability in a complex manner. Hurlbert and Gupta (2016) further recommend adaptive governance as a response measure to climate change risk, and hydrological extremes.

According to a research account by Webster et al. (2003), policy response to climate change impacts is marred by challenges. The authors attribute these challenges to the uncertainties that characterize climate change science in relation to climate projections. This position is echoed by research evidence by Brandt (2014) who observed that policies that are directed toward reducing perceived risks of extreme events receive inconsistent response from both the public and the government. Simply stated, governments and the general-public response to extreme events related to climate change may be low since these events do not occur frequently, and that the risk-reducing policies directed to these extreme events are only based on perceptions and climate projections which are, in the actual sense, marred with uncertainties.

The foregoing situation results in the case of having policy implementation programs that are not in tandem with the changes that occur in view of climate risk and the occurrence of extreme events. To address this challenge, Haasnoot et al. (2013) suggest that a dynamic adaptive policy is critical in agreement with the previously mentioned suggestion by Hurlbert and Gupta (2016). Such a policy framework gives room for accounting and incorporating uncertainties.

Whereas uncertainty has been largely blamed for the largely ineffective policy response to the risk caused by extreme hydrological events, Helm (2008) attributes the gap between climate science and policy knowledge to problems associated with the political and economic environment. The author argues that the complexity of the political economic environment tends to favor some policy instruments aligned with some interests, and at the same time may disregard some other policies that are not in line with the political and economic interests of the time. This situation may be further illustrated by exploring the 2014–15 drought that hit Sao Paulo city in Brazil. In this case, the autonomy of water management utilities in the city was compromised by political influence and as a result, an ineffective policy instrument of financially rewarding the residents for conserving water was implemented (Grover & Lucinda, 2020).

Important to note also is an observation that was made by Wilhite (2019) that policy makers often fall short of understanding the technical and scientific issues that arise when addressing hydrological extremes. This is also the case with scientists, who as well have a poor understanding of the constraints that exist in relation to policy processing when dealing with extreme hydrological events. Wilhite (2019) notes that limited integration of science and policy in the planning process, inhibits effective policy response to droughts and floods. Therefore, the author recommends overcoming this limitation by enhanced communication between scientists and policy makers with the view to ironing out the limiting factors. It will also enhance understanding of the related risks and, as such, be able to identify the best way of managing the risk.

Another important point to note is that adaptation and mitigation policies interact with other aspects of policy, a situation that can bring both negative and/or positive impacts on the policy goals. This argument is in line with an account by Haasnoot et al. (2013) whose study of flood risk policy response in the Netherlands, found out that management of hydrological extreme events was largely affected by the European Union (EU) policy framework. To incorporate the objectives of the EU Water Directive Framework, the Delta program had to introduce a clause noting that water resources management in the Delta would meet the criteria for ecological and water quality objectives as set out by the directive. To meet the arising policy requirements from other policies, Haasnoot et al. (2013) observed that the management of hydrological extreme events in the Rhine Delta in the Netherlands under the Delta program pursued a dynamic adaptive policy framework.

Notably, governance and institutional arrangements are complex domains affecting the coordination of adaptation and mitigation policies that are meant to deliver adaptive capacities to actors. This complexity undermines the efforts directed toward developing policies in view of either adapting or mitigating risks of droughts and floods. Further noting the effect of this complexity, Grobicki, MacLeod, and Pischke (2015) posit that an integrated risk management policy directed to both floods and droughts is urgently needed. This is because a situation of overlapping policy responses results in wasted resources. In this case, the authors point out that separately managing flood and drought risks was bound to bring about conflicting objectives in the environmental policy.

In line with Kane and Shogren (2000), limited understanding of the cross-link between adaptation and mitigation policies can undermine the general effectiveness of public policies and programs. This position was further highlighted by an interesting observation by Di Baldassarre, Martinez, Kalantari, and Viglione (2017) in relation to droughts and floods feedback mechanisms. The authors further agree that treating droughts and floods as separate events was bound to affect risk management effectiveness. In consideration of this point, it is therefore imperative for both policy makers and institutions working to manage the risk of hydrological extremes to consider an integrated approach to the management of these hydrological extreme events. The challenge here is to ensure that the separate policies adopted for drought and flood risks do not result in unproductive counter-effects. This situation was later illustrated by Di Baldassarre et al. (2018) pointing out that construction of reservoirs for flood risk control worsened water shortages elsewhere within a river basin. To overcome this challenge, higher levels of institutional coordination and collaboration have been highlighted and argued out by Wilhite (2019).

Birkholz, Muro, Jeffrey, and Smith (2014) point out that risk perception largely affects the effectiveness of flood risk management in various places across the globe. In considering the case of California residents, Birkholz et al. (2014) observed that their perception of flood risk was compromised by the presence of levees giving them a false hope that they were secure from flooding. This case highlights the need for effective communication of risk, and the creation of awareness with the aim of shaping behavior of society to support the available policy objectives.

In agreement with Birkholz et al. (2014) and Hurlbert and Gupta (2016), the argument that risk is both real and a product of social construction has been highlighted. This is to suggest that the realities of risk, coupled with how society talks about the risk, have a mutually influencing effect on each other. As a result of the foregoing, it therefore occurs that risks are brought about through social processes that are debated, selected, and eventually transformed to conform with the views of society.

4 Recommendations and a case study

Having extensively explored the management of risks of weather and extreme climate events with a particular focus on hydrological extreme events, various lessons were drawn in view of making recommendations. The need for effective management of hydrological extreme events in view of floods and droughts has been largely explored. Among the most applied strategies of risk management are mitigation and adaptation options.

Interestingly, as Hurlbert and Gupta (2016) research findings indicate, a fragmented approach to risk management of extreme hydrological events has been adopted in the years across the globe. The authors further state that mitigation and adaptation policies have been missing in view of their effectiveness. This is the case considering that such policies are formulated in response to drought and flood risk definitions that are highly flawed. In most cases, these definitions are more technically coined and, as such, they fail to incorporate the risk perspective from the point of view of policymakers.

To this end, it is highly recommended that the gap between the scientific definitions of both drought and flood risks, and policymakers' understanding of risks is closed. More to this point, Hurlbert and Gupta (2016) suggest that the dimensions of framing policy, defining risk, and in the larger perspective, the consideration of the cross-links that exist between the policy problems of drought and floods should be considered.

Considerably, an issue of uncertainty that is inherent in climate change projections has been subjected to scholarly criticism. This is because making decisions under uncertain claims of climate change impacts may be a taunting task for a policymaker. To this end, it is hereby recommended that development of a holistic risk analysis framework considering the dynamics of the risk components (or their proxy indicators), and their interactions which are missing in the literature are pursued. This is in line with recent research by Viner et al. (2020) highlighting the need to account for the evolving nature of risk components, and their interaction in the management of climate extremes. This is in view of considering planning for an uncertain future in the face of increased competition for the available resources. At least, efforts to make uncertainty clear to policymakers and the vulnerable communities at large should be stepped up in order to shape the right perception toward extreme hydrological events among stakeholders. Notwithstanding, accounting for the effect of social processes and how they shape and are shaped by risk perception, would be an added advantage to realize effective management of the risk of hydrological extreme events.

Arguably, cases of drought and flood risk management have been present both in the past and in current times. In view of this, the drought risk management approach following the United States of America's (USA) California 2012–15 multiyear drought will be explored. The state of USA July 2015 drought impact and intensity have been highlighted and discussed by Cody, Folger, and Brown (2015) as shown in Fig. 16.1 below.

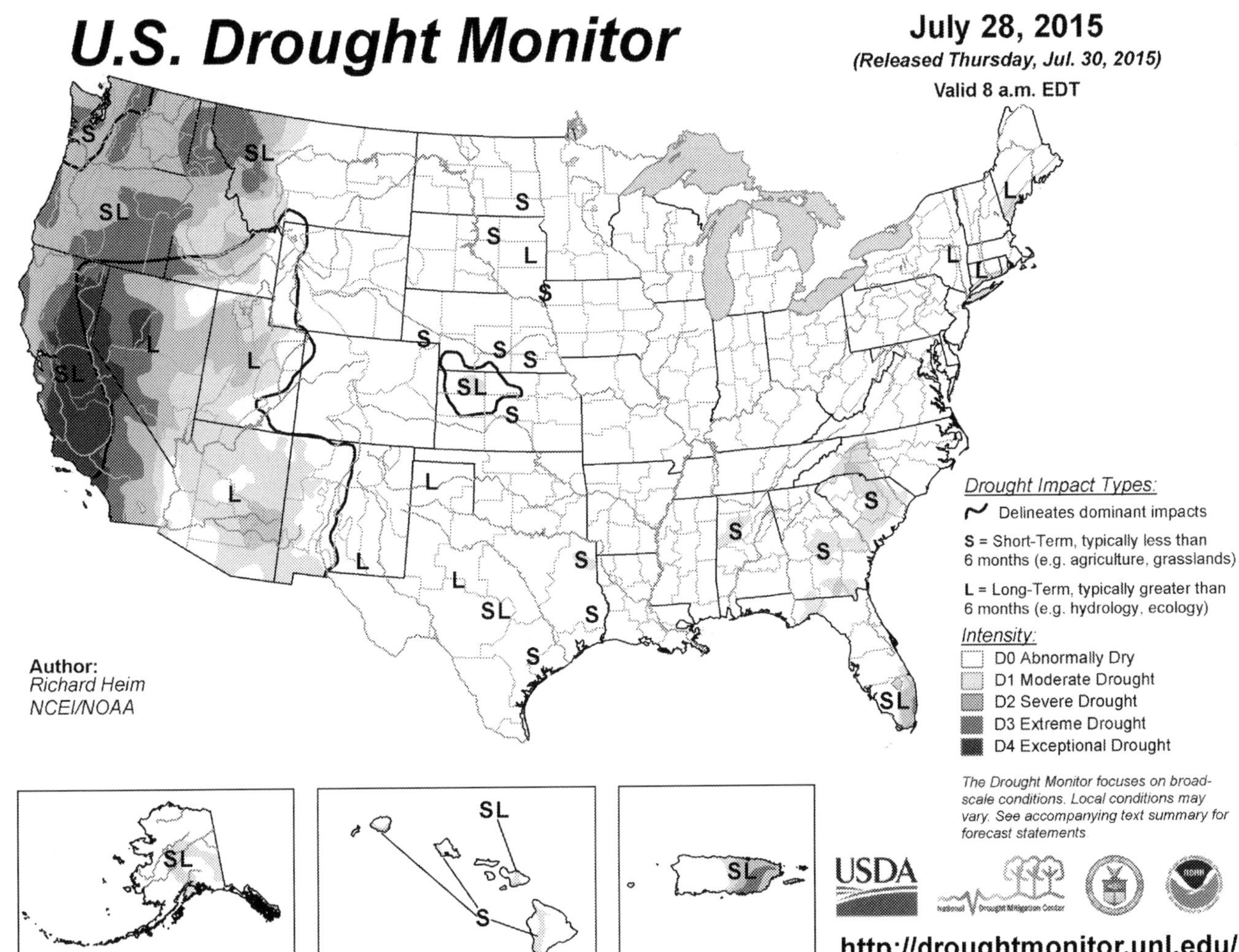

FIG. 16.1 Drought conditions in the United States of America on the July 28th, 2015. *No permission required.*

According to Lund, Medellin-Azuara, Durand, and Stone (2018), the 5-year drought in California, brought about innovations, and lessons for drought risk management in the present time and future. The author observed that exposure to such severe droughts, reveals the weaknesses that exist in management of risk in view of the existing measures, be it in line with mitigation or in adaptation.

Wilhite (2019) posits that there are four areas of interest to consider when managing the risk of drought as a hydrological extreme event. Primarily, there is a great need to enhance risk and early warning information systems, analysis of the vulnerabilities of different sectors and the population, assessment of impact and general communication of the risk. Secondly, careful application of well selected practices in relation to preparedness and mitigation are highly supported. Moreover, creation of awareness and education of stakeholders on risk management through a participatory process is hereby suggested. Notwithstanding, governance of the policy framework coupled with political commitment and goodwill along with responsibility is further highlighted.

An observation by Lund et al. (2018) states that California drought risk management was largely influenced by the lessons learned in the previous droughts. The authors indicate that both policy and institutional preparedness are largely informed by the performances in the previous droughts. This was the case even to the infrastructure and water shortages that were experienced in a previous drought. As a mitigation measure, California boasts of water projects that enable water deliveries during droughts. Such deliveries are enabled by a well-planned and managed network of water suppliers who store water for use and deliveries during droughts. Unfortunately, due to the severity and length of the 5-year California drought, less water in stores was realized along with less deliveries being reported, with even some suppliers recording no deliveries as of the year 2014. This case illustrates a point known as "policy tipping point."

Although Wilhite (2019) suggested that early warning systems in the preparedness for drought were critical in risk management, Mount et al. (2015) point out that the information on early warning was not well integrated in this case.

Mount et al. (2015) assert that the water monitoring systems of California were primitive and characterized with significant gaps in relation to monitoring of critical information such as water use, flows, storages, and diversions outside the urban areas. Lack of this critical piece of information compromised the ability of California agencies to reduce the drought risk of the time.

Whereas this was the case for information and preparedness, Mount et al. (2015) further state that there lacked clarity in view of the drought management policy objectives, expectations, and priorities. This point can further be illustrated by the 2014 actions that curtailed water management as proposed by the water board. The amount of water that is required for meeting all needs of the public health and safety issues did not comply with Wilhite (2019) suggestion that science and policy ought to be integrated coupled with clear communication, which in essence should be understood by all stakeholders. In this case, Mount et al. (2015) highlight that this curtailment as argued by critics and many water users in California was not well articulated and effectively communicated to the stakeholders.

Whereas California is known to have experienced other drought events prior to the 2012–15 droughts, few lessons were learned. This is according to an observation made by Mount et al. (2015) that despite prior experience in dealing with droughts, California was yet to learn and apply the lessons on developing a robust water supply network, and the need to effectively manage demand. Beside this point, the authors further found that California still lagged in its effort to modernize and develop an effective environmental drought management policy. This was the case considering that as the author puts it, little scientific research, and integration of the same was lacking. Allocation of environmental water was not considered, a factor that put the biodiversity and the environment at a great risk of the drought.

Generally, the California case study clearly illustrates the need to understand the risks derived from hydrological extreme events from the scientific point of view, and then integrate it with policy. Furthermore, the case highlights the urgent need for information on risk, vulnerability, and exposure in the management of floods and droughts risks. Besides, the case further highlights the dangers of not incorporating the uncertainties generated from science and communicating them to stakeholders. In fact, the case further spells out the need to consider a robust risk management approach, which has the capability to incorporate not only uncertainties, but also the complexity of occurrence of hydrological extreme events in view of climate change as noted by Hurlbert and Gupta (2016).

Hurlbert and Gupta (2016) recommend consideration of adaptive risk management approach in relation to policy framing, risk definition and response to climatic extremes such as floods and drought hazards. This position is as well held by Wilhite (2019) who argued for an integrated approach in managing drought risk in this human-dominate era. Further agreeing with the foregoing statement and referring to the Dutch Delta flood risk management approach, Haasnoot et al. (2013) promote the adoption of a dynamic, and adaptive approach to the risk management of hydrological extremes events. These extreme events are further accelerated by a growing world population, climate change and the complexity occasioned by a growing socio-economic environment.

References

Abegg, B., Agrawala, S., Crick, F., & De Montfalcon, A. (2007). Climate change impacts and adaptation in winter tourism. In *Vol. 9789264031692. Climate change in the European Alps: Adapting winter tourism and natural hazards management* (pp. 25–60). Organisation for Economic Cooperation and Development (OECD). https://doi.org/10.1787/9789264031692-en.

Adger, W. N., Agrawala, Mirza, M. M. Q., Conde, O'Brien, K., Pulhin, et al. (2007). Assessment of adaptation practices, options, constraints and capacity. In *Climate change 2007: Impacts, adaptation and vulnerability. Contribution of Working Group II to the fourth assessment report of the intergovernmental panel on climate change* (pp. 717–743).

Agrawal, A. (2010). *Local institutions and adaptation to climate change. Social dimensions of climate change: Equity and vulnerability in a warming world. Vol. 2* (pp. 173–178). World Bank Group.

Alfieri, L., Salamon, P., Pappenberger, F., Wetterhall, F., & Thielen, J. (2012). Operational early warning systems for water-related hazards in Europe. *Environmental Science and Policy, 21*, 35–49. https://doi.org/10.1016/j.envsci.2012.01.008.

Apel, H., Aronica, G. T., Kreibich, H., & Thieken, A. H. (2009). Flood risk analyses – How detailed do we need to be? *Natural Hazards, 49*(1), 79–98. https://doi.org/10.1007/s11069-008-9277-8.

Barnett, J., & Webber, M. (2009). *Accommodating migration to promote adaptation into climate change. A policy brief prepared for the Secretariat of the Swedish Commission on climate change and development and the World Bank World Development report 2010 Team.*.

Baudoin, M., Henly-Shepard, S., Fernando, N., Sitati, A., & Zommers, Z. (2014). Early warning systems and livelihood resilience: Exploring opportunities for community participation. In *UNU-EHS working paper series.*

Bayazit, M. (2015). Nonstationarity of hydrological records and recent trends in trend analysis: A state-of-the-art review. *Environmental Processes, 2*(3), 527–542. https://doi.org/10.1007/s40710-015-0081-7.

Beg, N., Morlot, J. C., Davidson, O., Afrane-Okesse, Y., Tyani, L., Denton, F., et al. (2002). Linkages between climate change and sustainable development. *Climate Policy, 2*(2–3), 129–144. https://doi.org/10.1016/S1469-3062(02)00028-1.

Birkholz, S., Muro, M., Jeffrey, P., & Smith, H. M. (2014). Rethinking the relationship between flood risk perception and flood management. *Science of the Total Environment, 478*, 12–20. https://doi.org/10.1016/j.scitotenv.2014.01.061.

Birkmann, J., & von Teichman, K. (2010). Integrating disaster risk reduction and climate change adaptation: Key challenges-scales, knowledge, and norms. *Sustainability Science*, *5*(2), 171–184. https://doi.org/10.1007/s11625-010-0108-y.

Brandt, U. S. (2014). The implication of extreme events on policy responses. *Journal of Risk Research*, *17*(2), 221–240. https://doi.org/10.1080/13669877.2013.794151.

Cabrera, V. E., Daniel, S., Guillermo, A. B., & David, L. (2009). *Managing climate risks to agriculture: Evidence from El Nino.*

Chaves, J., & De Cola, T. (2017). Public warning applications: Requirements and examples. *Wireless Public Safety Networks.*

Cochrane, K., Young, Soto, D., & Bahri, T. (2009). *Climate change implications for fisheries and aquaculture. FAO Fisheries and aquaculture technical paper. Vol. 530.*

Cody, B. A., Folger, P., & Brown, C. (2015). *California drought: Hydrological and regulatory water supply issues.* Congressional Research Service.

de la Fuente, A., Meruane, V., & Meruane, C. (2019). Hydrological early warning system based on a deep learning runoff model coupled with a meteorological forecast. *Water*, *11*(9), 1808. https://doi.org/10.3390/w11091808.

Devkota, L. P., & Gyawali, D. R. (2015). Impacts of climate change on hydrological regime and water resources management of the Koshi River basin, Nepal. *Journal of Hydrology: Regional Studies*, *4*, 502–515. https://doi.org/10.1016/j.ejrh.2015.06.023.

Di Baldassarre, G., Martinez, F., Kalantari, Z., & Viglione, A. (2017). Drought and flood in the Anthropocene: Feedback mechanisms in reservoir operation. *Earth System Dynamics*, *8*(1), 225–233. https://doi.org/10.5194/esd-8-225-2017.

Di Baldassarre, G., Wanders, N., AghaKouchak, A., Kuil, L., Rangecroft, S., Veldkamp, T. I. E., et al. (2018). Water shortages worsened by reservoir effects. *Nature Sustainability*, 617–622. https://doi.org/10.1038/s41893-018-0159-0.

Diaz, V., & Corzo, G. (2018). Large-scale exploratory analysis of the spatiotemporal distribution of climate projections: Applying the STRIVIng toolbox. In *Spatiotemporal analysis of extreme hydrological events* (pp. 59–76). Elsevier. https://doi.org/10.1016/B978-0-12-811689-0.00003-3.

Disasters in Africa: 20 year review. (2000).

Dovers, S. R., & Hezri, A. A. (2010). Institutions and policy processes: The means to the ends of adaptation. *Wiley Interdisciplinary Reviews: Climate Change*, *1*(2), 212–231. https://doi.org/10.1002/wcc.29.

Durack, P. J., Wijffels, S. E., & Matear, R. J. (2012). Ocean salinities reveal strong global water cycle intensification during 1950 to 2000. *Science*, *336*(6080), 455–458. https://doi.org/10.1126/science.1212222.

Establishing drought early warning systems in West Asia and North Africa (UN-DESA/DSD) Intern. (2013). United Nations.

Field, C. B., Barros, V., & Stocker, T. F. (2012). *Managing the risks of extreme events and disasters to advance climate change adaptation: Special report of the intergovernmental panel on climate change.*

Genovese, E., & Thaler, T. (2020). The benefits of flood mitigation strategies: Effectiveness of integrated protection measures. *AIMS Geosciences*, 459–472. https://doi.org/10.3934/geosci.2020025.

Giuntoli, I., Vidal, J. P., Prudhomme, C., & Hannah, D. M. (2015). Future hydrological extremes: The uncertainty from multiple global climate and global hydrological models. *Earth System Dynamics*, *6*(1), 267–285. https://doi.org/10.5194/esd-6-267-2015.

Global change and extreme hydrology: Testing conventional wisdom. (2011).

GoK. (2018). *National Climate Change Action Plan (Kenya) 2018–2022. Ministry of Environment and Forestry.*

GoK. (2004). *Strategy for flood management for Lake Victoria Basin, Kenya. A government of Kenya (GoK) publication. Prepared under Associated Programme on Flood Management (APFM).*

Grobicki, A., MacLeod, F., & Pischke, F. (2015). Integrated policies and practices for flood and drought risk management. *Water Policy*, *17*, 180–194. https://doi.org/10.2166/wp.2015.009.

Grover, D., & Lucinda, C. R. (2020). An evaluation of the policy response to drought in the City of São Paulo, Brazil: An election cycle interpretation of effectiveness. *Journal of Development Studies*, 1–18. https://doi.org/10.1080/00220388.2020.1786061.

Haasnoot, M., Kwakkel, J. H., Walker, W. E., & ter Maat, J. (2013). Dynamic adaptive policy pathways: A method for crafting robust decisions for a deeply uncertain world. *Global Environmental Change*, *23*(2), 485–498. https://doi.org/10.1016/j.gloenvcha.2012.12.006.

Hallegatte, S. (2012). *A cost-effective solution to reduce disaster losses in developing countries: Hydro-meteorological services, early warning, and evacuation.*

Helm, D. (2008). Climate-change policy: Why has so little been achieved? *Oxford Review of Economic Policy*, *24*(2), 211–238. https://doi.org/10.1093/oxrep/grn014.

Huho, J. M., & Kosonei, R. C. (2014). Understanding extreme climatic events for economic development in Kenya. *IOSR Journal of Environmental Science, Toxicology and Food Technology*, 14–24. https://doi.org/10.9790/2402-08211424.

Hurlbert, M., & Gupta, J. (2016). Adaptive governance, uncertainty, and risk: Policy framing and responses to climate change, drought, and flood. *Risk Analysis*, *36*(2), 339–356. https://doi.org/10.1111/risa.12510.

Jongman, B. (2018). Effective adaptation to rising flood risk. *Nature Communications*, *9*(1). https://doi.org/10.1038/s41467-018-04396-1.

Kamara, J. K., Sahle, B. W., Agho, K. E., & Renzaho, A. M. N. (2020). Governments' policy response to drought in Eswatini and Lesotho: A systematic review of the characteristics, comprehensiveness, and quality of existing policies to improve community resilience to drought hazards. *Discrete Dynamics in Nature and Society*, 1–17. https://doi.org/10.1155/2020/3294614.

Kane, S., & Shogren, J. F. (2000). Linking adaptation and mitigation in climate change policy. *Climatic Change*, *45*(1), 75–102. https://doi.org/10.1007/978-94-017-3010-5_6.

Kiluva, V. M. (2012). *Fundamentals of disaster management and humanitarian assistance. A textbook for disaster management and humanitarian assistance practitioners and trainees* (p. 9966).

Kundzewicz, Z. W., & Matczak, P. (2015). Hydrological extremes and security. *Proceedings of the International Association of Hydrological Sciences*, *366*, 44–53. https://doi.org/10.5194/piahs-366-44-2015.

Liu, X., Li, Z., & Attarod, P. (2021). Understanding hydrological extremes and their impact in a changing climate: Observations, modeling and attribution. *Frontiers in Earth Science*, *8*, 657.

Lund, J., Medellin-Azuara, J., Durand, J., & Stone, K. (2018). Lessons from California's 2012–2016 drought. *Journal of Water Resources Planning and Management*, *144*(10), 04018067-1–04018067-13.

Lutz, A. F., Immerzeel, W. W., Shrestha, A. B., & Bierkens, M. F. P. (2014). Consistent increase in high Asia's runoff due to increasing glacier melt and precipitation. *Nature Climate Change*, *4*(7), 587–592. https://doi.org/10.1038/nclimate2237.

Mård, J., & Baldassarre. (2016). Understanding hydrological extremes in the Anthropocene. In *EGU general assembly conference abstracts.*

Mount, J., Hanak, E., Chappelle, Gray, B., Lund, J., Moyle, et al. (2015). *Policy priorities for managing drought.*

Mubaya, C. P., & Mafongoya, P. (2017). The role of institutions in managing local level climate change adaptation in semi-arid Zimbabwe. *Climate Risk Management, 16*, 93–105. https://doi.org/10.1016/j.crm.2017.03.003.

Murray, V., & Ebi, K. L. (2012). IPCC special report on managing the risks of extreme events and disasters to advance climate change adaptation (SREX). *Journal of Epidemiology and Community Health, 66*(9), 759–760. https://doi.org/10.1136/jech-2012-201045.

Niebla, Weber, T., Skoutaridis, P., Hirst, P., Ramírez, J., Rego, D., et al. (2011). *Alert4All: An integrated concept for effective population alerting in crisis situations.*

Pielke, R. A. (1998). Rethinking the role of adaptation in climate policy. *Global Environmental Change, 8*(2), 159–170. https://doi.org/10.1016/S0959-3780(98)00011-9.

Qing, Y., Wang, S., Zhang, B., & Wang, Y. (2020). Ultra-high resolution regional climate projections for assessing changes in hydrological extremes and underlying uncertainties. *Climate Dynamics, 55*(7–8), 2031–2051. https://doi.org/10.1007/s00382-020-05372-6.

Rogers, D. P., & Tsirkunov, V. (2010). *Costs and benefits of early warning systems. Global assessment report on disaster risk reduction. ISDR and World, Bank..*

Rogers, D. P., & Tsirkunov, V. (2011). *Implementing hazard early warning systems GFDRR WCIDS report 11-03.* Available on the PreventionWeb.

Royan, A., Hannah, D. M., Reynolds, S. J., Noble, D. G., & Sadler, J. P. (2014). River birds' response to hydrological extremes: New vulnerability index and conservation implications. *Biological Conservation, 177*, 64–73. https://doi.org/10.1016/j.biocon.2014.06.017.

Schewe, J., Gosling, S. N., Reyer, C., Zhao, F., Ciais, P., Elliott, J., et al. (2019). State-of-the-art global models underestimate impacts from climate extremes. *Nature Communications, 10*(1). https://doi.org/10.1038/s41467-019-08745-6.

Shah, M. A. R., Rahman, A., & Chowdhury, S. H. (2018). Challenges for achieving sustainable flood risk management. *Journal of Flood Risk Management, 11*, S352–S358. https://doi.org/10.1111/jfr3.12211.

Simonovic, S. P. (2010). *Systems approach to management of disasters: Methods and applications. Vol. 348.* Wiley.

Sivakumar, M. V. K., Stefanski, R., Bazza, M., Zelaya, S., Wilhite, D., & Magalhaes, A. R. (2014). High level meeting on national drought policy: Summary and major outcomes. *Weather and Climate Extremes, 3*, 126–132. https://doi.org/10.1016/j.wace.2014.03.007.

Somanathan, E., Sterner, T., Sugiyama, T., Chimanikire, D., Dubash, N. K., Essandoh-Yeddu, J. K., et al. (2014). *National and sub-national policies and institutions.*

Stabinsky, D., & Hoffmaister, J. P. (2015). Establishing institutional arrangements on loss and damage under the UNFCCC: The Warsaw international mechanism for loss and damage. *International Journal of Global Warming, 8*(2), 295–318. https://doi.org/10.1504/IJGW.2015.071967.

Tabari, H. (2019). Statistical analysis and stochastic modelling of hydrological extremes. *Water, 11*(9). https://doi.org/10.3390/w11091861.

Thieken, A., Grunewald, U., Merz, B., Pethrow, T., Schumberg, S., Kreibich, H., et al. (2002). Flood risk reduction in Germany after the Elbe. In *Aspects of hazard mapping and early warning systems In Proceedings of the international symposium on cartographic cutting-edge technology for natural hazard management* (pp. 145–156).

Thieken, A. H., & Menzel. (2004). Scenario-based modelling studies to assess the impact of climate change on floods in the German Rhine catchment. In *Climate change and Yangtze floods* (pp. 168–181). Science Press.

Trajkovic, S., Kisi, O., Markus, M., Tabari, H., Gocic, M., & Shamshirband, S. (2016). Hydrological hazards in a changing environment: Early warning, forecasting, and impact assessment. *Advances in Meteorology, 2016*. https://doi.org/10.1155/2016/2752091.

van Buuren, A., Ellen, G. J., & Warner, J. F. (2016). Path-dependency and policy learning in the dutch delta: Toward more resilient flood risk management in the Netherlands? *Ecology and Society, 21*(4). https://doi.org/10.5751/ES-08765-210443.

Van Loon, A. F. (2015). Hydrological drought explained. *WIREs Water, 2*(4), 359–392. https://doi.org/10.1002/wat2.1085.

Viner, D., Ekstrom, M., Hulbert, M., Warner, N. K., Wreford, A., & Zommers, Z. (2020). Understanding the dynamic nature of risk in climate change assessments—A new starting point for discussion. *Atmospheric Science Letters, 21*(4). https://doi.org/10.1002/asl.958.

Webster, M., Forest, C., Reilly, J., Babiker, M., Kicklighter, D., Mayer, M., et al. (2003). Uncertainty analysis of climate change and policy response. *Climatic Change, 61*(3), 295–320. https://doi.org/10.1023/B:CLIM.0000004564.09961.9f.

Wilhite, D. A. (2019). Integrated drought management: Moving from managing disasters to managing risk in the Mediterranean region. *Euro-Mediterranean Journal for Environmental Integration*. https://doi.org/10.1007/s41207-019-0131-z.

Willis, C., Wilgen, T., Everson, C., Abreton, D., Pero, P., & Fleming. (2001). *Client: Department of Water Affairs and Forestry, and Department for International Development.*

Yu-fai, T., & Cheng, Y. C. A. (2018). Warnings on tropical cyclone for WMO global multi-hazard alert system. *Tropical Cyclone Research and Review, 7*(4), 230–236.

Yun, X., Tang, Q., Li, J., Lu, H., Zhang, L., & Chen, D. (2021). Can reservoir regulation mitigate future climate change induced hydrological extremes in the Lancang-Mekong River basin? *Science of the Total Environment, 785*. https://doi.org/10.1016/j.scitotenv.2021.147322, 147322.

Index

Note: Page numbers followed by *f* indicate figures and *t* indicate tables.

A

ACTO. *See* Amazon Cooperation Treaty Organization (ACTO)
Advances in weather and climate extremes
  evolution of definitions, 50–51
  frequency change, 53–55
  intensity change, 55–56
  large-scale features, 56–57
  spatiotemporal scales of measurement, 52
  subseasonal to seasonal prediction, 57
  timing change, 56
African Rainfall Climatology (ARC), 65–66
Agribusiness industries, 225
Agricultural drought, 9
Agricultural resilience, 260
Albedo effect, 219
Alert4All, 276–277
Amazon
  biodiversity decline in, 220
  as carbon sink, 219
  and climate change, 222
  deforestation, drivers of, 222–223
  drought, deforestation, and fire in, 221
  hydrology, 219–220
  Indigenous knowledge, leveraging, 228
  Indigenous people in, 225–226
  Indigenous stewardship, securing, 227–228
  international protection, 229–230
  isolated Indigenous groups, 226–227
  methods, 218–219
  national protection of, 228–229
  tipping point, 222
  transnational (intra-Amazon) protection of rainforest, 229
  zoonosis, climate, and, 223–225
Amazon Cooperation Treaty Organization (ACTO), 229
AMO. *See* Atlantic Multidecadal Oscillation (AMO)
Annual mean precipitation, projected changes in, 121
Antarctic climate, ozone hole in, 196–198
Antarctic sea ice, paradox in, 200–201
ARC. *See* African Rainfall Climatology (ARC)
Arctic and Antarctic, climate extremes over
  as cooling champers of earth, 193–194
  different glaciers and ice sheets, changes in, 201–206
  heat tracking in polar seas, 198–200
  meltwater pathways, 206–207
  ozone hole in Antarctic climate, 196–198
  paradox in Antarctic sea ice, 200–201
  polar climate warming, 194–196
  polar regions, 191–192
Arid and semiarid land (ASAL), 146, 159
ARs. *See* Atmospheric Rivers (ARs)
ASAL. *See* Arid and semiarid land (ASAL)
Atlantic Multidecadal Oscillation (AMO), 161–162
Atmosphere, 2
Atmospheric Rivers (ARs), 220

B

Bhalme and Mooley Drought Index (BMDI), 9–10
Biogenic volatile organic compounds (BVOCs), 219
Biosphere, 2
"Black swan" event, 139

C

CAP. *See* Common Alerting Protocols (CAP)
Carbon sink, Amazon as, 219
CC equation. *See* Clausius–Clapeyron (CC) equation
CDO. *See* Climate Data Operators (CDO)
CEIs. *See* Climate Extremes Indices (CEIs)
Central India extreme rainfall event, 89
Chennai flood in 2015, 91–92
CHIRPS. *See* Climate Hazards Group InfraRed Precipitation with Station data (CHIRPS)
Clausius–Clapeyron (CC) equation, 55, 133, 135
Climate, defined, 1–2
Climate Data Operators (CDO), 22–27
Climate Datasets, 20
Climate Extremes Indices (CEIs), 20–39
  existing datasets of, 39–41
  existing global datasets suitable for computation of, 36–39
    global datasets, 39
    regional datasets, 39
  Expert Team on Climate Change Detection and Indices (ETCCDI)/Expert Team on Sector-specific Climate Indices (ET-SCI), 20–28
    available software routines, 27–28
    input meteorological variables and quality checks for assembling, 22–27
  future research, scope and recommendation for, 41–42
  gridded climate data, 33–35
  historical background of ETCCDI core and noncore indices, 21–22
  reanalysis data, 35–36
  recent advancements in understanding climate extremes, 41
  station-derived meteorological observations, 33
Climate Hazards Group InfraRed Precipitation with Station data (CHIRPS), 65–66, 154–155
  CHIRPS v2, 67–68
Climate indices, 162
Climate phenomenon influencing the occurrence of extremes, 7
  drought, 9–10
  extra-tropical cyclones, 8
  extreme sea level, 10
  flood, 9
  hydrological extremes, 9
  storms, 8
  tornadoes, 8–9
  tropical cyclones, 8
  variability, modes of, 7
Climate Prediction Center (CPC), 65–66
Climate Research Unit (CRU), 154–155
Climate variability, drought across East Africa under
  area of study, 160–161
  climate indices, 162
  self-calibrating Palmer Drought Severity Index (scPDSI), 161–162
Climate variability in the past millennium, 131–133
ClimPACT2, 67
CMI. *See* Crop Moisture Index (CMI)
CMIP. *See* Coupled Model Intercomparison Project (CMIP)
Coefficient of variation (CV), 164
Coevolution effect, 224–225
Colocasia esculenta, 236–237
Common Alerting Protocols (CAP), 276–277
Compound and simultaneous extremes, 10
Compound Extreme Events, 42
Concurrent Extremes, 42
Connected Extreme Events, 42
Coordinated Regional Climate Downscaling experiment (COR-DEX), 52

Coupled Model Intercomparison Project (CMIP), 52
CMIP5, 94
CMIP6, 41
COVID-19 pandemic, 223–224
in the Pacific Islands region, 265–266
CPC. *See* Climate Prediction Center (CPC)
CPC MORPHing technique (CMORPH), 65–66
Crop Moisture Index (CMI), 9–10
CRU. *See* Climate Research Unit (CRU)
Current Warm Period (CWP), 132
CV. *See* Coefficient of variation (CV)
CWP. *See* Current Warm Period (CWP)

D

Data mining, 104–105, 112
Deciles, 9–10
Deforestation, 222–224
Delta program, 281
De-urbanization, 225
Disaster management, 264–265
Diurnal cycle indices, 51*t*
Drought, 9–10, 139–140, 145, 221
Drought across East Africa under climate variability
area of study, 160–161
climate indices, 162
self-calibrating Palmer Drought Severity Index (scPDSI), 161–162
Duration seasonality of weather, and climate extremes, 56
Dutch Delta flood risk management approach, 284

E

Early warning systems (EWS), 186, 275
Earth energy balance, 2–3
Earth Energy Budget, 2–3
Earth system, 2
atmosphere, 2
biosphere, 2
hydrosphere, 2
lithosphere, 2
ECVs. *See* Essential Climate Variables (ECVs)
EIAs. *See* Environmental impact assessments (EIAs)
El Nino Southern Oscillation (ENSO), 4, 7, 56, 83, 103, 131, 146, 159–160, 221
Empirical orthogonal function (EOF), 162, 171
Environmental effects of weather extremes, 11–12
Environmental impact assessments (EIAs), 227
EOF. *See* Empirical orthogonal function (EOF)
EREs. *See* Extreme rainfall events (EREs)
Essential Climate Variables (ECVs), 35
ETCCDI. *See* Expert Team on Climate Change Detection and Indices (ETCCDI)
ETCRSCI. *See* Expert Team on Climate Risk and Sector-specific Indices (ETCRSCI)
ET-SCI. *See* Expert Team on Sector-Specific Climate Indices (ET-SCI)
Evapotranspiration, 222
Evolution of definitions of weather and climate extremes, 50–51
EWS. *See* Early warning systems (EWS)
Expert Team on Climate Change Detection and Indices (ETCCDI), 50, 118
Expert Team on Climate Change Detection and Indices (ETCCDI)/Expert Team on Sector-specific Climate Indices (ET-SCI), 20–21
available software routines, 27–28
quality control and homogenization of input data, 27
subdaily or daily meteorological data for computing CEIs, 22–27
Expert Team on Climate Risk and Sector-specific Indices (ETCRSCI), 50–51
Expert Team on Sector-Specific Climate Indices (ET-SCI), 21–22, 66, 68*t*
Extra-tropical cyclones, 8
Extreme events, estimation of, 87
Extreme precipitation indices, 73–75
Extreme rainfall events (EREs), 84–85, 90
Extreme sea level, 10
Extreme weather climate variable, 5–7
precipitation extremes, 5–7
temperature extremes, 5
wind extreme, 7

F

Farmers' perceptions of climate hazards and coping mechanisms in Fiji
coping methods in agriculture, 244–246
farmers' challenges in adopting effective coping methods, 246–247
farmers' perceptions of climate change and climate hazards, 237–238
farmers' perceptions of climate hazards, 241–244
traditional knowledge and coping methods in agriculture, 236–237
vulnerability of Pacific Island country agriculture to climate hazards, 235–236
Fiji and case study sites, 238–240
Navua, 240
Serua/Namosi region, 238
Wainadoi settlement, 238–240
Flood-rich periods in Europe, 138*t*
Floods, 9, 137–139
Food security, 185
Frequency change of weather and climate extremes, 53–55
Frequency/threshold indices, 51*t*

G

GAST. *See* Global average surface temperatures (GAST)
Gauge rainfall, 77–81
GCDs. *See* Gridded Climate Datasets (GCDs)
GCMs. *See* Global climate models (GCMs)
GEF. *See* Global Environment Facility (GEF)
General indices, 51*t*
GEWEX activities. *See* Global Energy and Water Exchanges (GEWEX) activities
GFI. *See* Global Flood Inventory (GFI)
GHCN. *See* Global Historical Climate Network (GHCN)
GHCNDEX. *See* Global Historical Climatology Network daily climate extremes (GHCNDEX)
GHGs. *See* Greenhouse gases (GHGs)
Glacier mass balances, 135–136
Global average surface temperatures (GAST), 219
Global climate models (GCMs), 94, 280
Global Energy and Water Exchanges (GEWEX) activities, 67–68
Global Environment Facility (GEF), 229
Global Flood Inventory (GFI), 84–85
Global Historical Climate Network (GHCN), 35
Global Historical Climatology Network daily climate extremes (GHCNDEX), 52
Global Precipitation Climatology Center (GPCC), 65–66
Global Precipitation Climatology Project (GPCP), 65–66
Global Precipitation Measurement (GPM), 181
Global Satellite Mapping of Precipitation (GSMAP), 65–66
Global Subdaily Rainfall dataset (GSDR), 52
Global warming, 56
on extreme rainfall events over India, 84
precipitation extremes in the United States under
annual mean precipitation, projected changes in, 121
maximum 5-day precipitation, projected changes in, 123
maximum consecutive dry days, projected changes in, 122
present-day annual dry and wet extremes, evaluation of, 119–121
GPCC. *See* Global Precipitation Climatology Center (GPCC)
GPCP. *See* Global Precipitation Climatology Project (GPCP)
GPM. *See* Global Precipitation Measurement (GPM)
Grace Road Group, 240
Gravity Recovery and Climate Experiment (GRACE), 202
Greenhouse gases (GHGs), 20, 101, 196, 219
Greenland Ice Sheet, 204
Gridded Climate Datasets (GCDs), 34
Gridded Meteorological Datasets, 34
Gridded products, 67–68
Ground Observations, 33
GSDR. *See* Global Subdaily Rainfall dataset (GSDR)
GSMAP. *See* Global Satellite Mapping of Precipitation (GSMAP)

H

HadEX, 52
Heat tracking in polar seas, 198–200
Himalayan extreme rainfall event, 89–90
Hurricane foods, 247–248
Hydroclimatic extreme events, observed historical changes in
climate variability in the past millennium, 131–133

droughts, 139–140
floods, 137–139
glacier mass balances, 135–136
past changes, 140–141
precipitation extremes, 133–135
Hydroclimatic extreme events over Iran
data, 107–108
methodology, 108–109
study area, 105–107
Hydrological cycle, 3
Hydrological drought, 9, 145
Hydrological events, extreme
challenges, 280–282
management measures, 272–280
adaptation, 273–275
early warning systems, 275–277
institutions, 279–280
mitigation, 272–273
policy, 277–279
response, 277
recommendations and case study, 282–284
Hydrological extremes, 9
Hydrosphere, 2

I
IBAMA, 223
Ice shelves, 205
IDW approach. *See* Inverse distance weighted (IDW) approach
IMD. *See* India Meteorological Department (IMD)
IMERG. *See* Integrated Multi-Satellite Retrievals (IMERG)
Impacts of climate change and extreme events, 12–13*t*
India Meteorological Department (IMD), 86
Indian monsoon rainfall extremes, features of
case studies, 90–94
Chennai flood in 2015, 91–92
Kerala flood 2018, 90–91
Mumbai flood 2005, 92
Uttarakhand flood 2013, 93–94
data, 86
extreme events, estimation of, 87
extreme events, history of, 87–88
global warming on extreme rainfall events, 84
Indian monsoon variability in extreme rainfall events, 83
moisture transport, 86
percentile departure of rainfall (PDR), 86
probable maximum precipitation (PMP), 87
projections of extreme rainfall events, 94
regional rainfall extremes, 88–90
Central India extreme rainfall event, 89
extreme rainfall events in the north east, 90
Himalayan extreme rainfall event, 89–90
peninsular India extreme rainfall event, 88–89
socio-economic impacts of extreme rainfall events, 84–86
Indian Ocean Dipole (IOD), 7, 83, 89, 161–162
Indian summer monsoon rainfall (ISMR) variability, 83
Indigenous people in the Amazon, 225–226
Indigenous Protected Areas (IPAs), 218, 223, 225, 227
Infrared Radiation, 194*f*
Innovations in weather and climate extremes prediction, 57
In situ measurements, 33
Integrated Multi-Satellite Retrievals (IMERG), 181
INTENSE project, 52
Intensity change of weather and climate extremes, 55–56
Intergovernmental Panel on Climate Change (IPCC), 19, 50, 201, 253
International Strategy for Disaster Reduction (ISDR), 265
Inter-Tropical Convergence Zone (ITCZ), 69, 146, 161
Inter-Tropical Discontinuity (ITD), 69
Inverse distance weighted (IDW) approach, 162
IOD. *See* Indian Ocean Dipole (IOD)
IPAs. *See* Indigenous Protected Areas (IPAs)
IPCC. *See* Intergovernmental Panel on Climate Change (IPCC)
ISDR. *See* International Strategy for Disaster Reduction (ISDR)
ISMR variability. *See* Indian summer monsoon rainfall (ISMR) variability
Isolated Indigenous groups, 226–227
ITCZ. *See* Inter-Tropical Convergence Zone (ITCZ)
ITD. *See* Inter-Tropical Discontinuity (ITD)

J
Joint National Action Plans (JNAPs), 265

K
Kayapó Project, 226
Kenya Drought Management System, 146
Kenya Meteorological Department (KMD), 147–148
Kerala flood 2018, 90–91

L
Land grabbers, 226
Large-scale features associated with weather and climate extremes, 56–57
LEAF Coalition. *See* Lowering Emissions by Accelerating Forest finance (LEAF) Coalition
Least Developed Countries (LDCs), 271–272
Lithosphere, 2
Little Ice Age (LIA), 131
Lowering Emissions by Accelerating Forest finance (LEAF) Coalition, 230
Low-Level Jets (LLJs), 220
Low-pressure system (LPS), 89

M
Madden-Julian Oscillation (MJO), 56, 161
Magdalena flood, 133
Malawi, impacts of tropical cyclone Idai on, 181
Maximum 5-day precipitation, projected changes in, 123
Maximum consecutive dry days, projected changes in, 122
Maximum monthly precipitation (MMP) events, 103–105
MCA. *See* Medieval Climate Anomaly (MCA)
Mean daily monsoon, evolution of, 69–73
Mean precipitation, 69
Measurement, detection and attribution of extremes, 11
Measurement of weather and climate extremes, spatiotemporal scales of, 52
Measuring climate extremes, 5
Medieval Climate Anomaly (MCA), 131
Meltwater lakes, 207
Meltwater pathways, 206–207
Meteorological drought, 9, 145
in semiarid eastern Kenya
data analysis, 148–149
data collection, 147–148
gridded verses in situ data, 152–155
limitations of the study, 155
spatiotemporal occurrence, 149–152
study area, 146–147
Migration as successful adaptation to climate change, 261–264
labor mobility in PICT, 262–264
migrants as actors and translocality, 262
MJO. *See* Madden-Julian Oscillation (MJO)
MMP events. *See* Maximum monthly precipitation (MMP) events
Moisture transport, 86
Monsoon jump, 69–73
Monthly maximum indices, 51*t*
Mozambique, impacts of tropical cyclone Idai on, 182
Multi-Source Weighted-Ensemble Precipitation (MSWEP), 65–66
Mumbai flood 2005, 92

N
NAO. *See* North Atlantic Oscillation (NAO)
National Adaptations Plans (NAP), 279–280
National Drought Emergency Fund (NDEF), 146
National Meteorological and Hydrological Services (NHMSs), 51
National Oceanic and Atmospheric Administration (NOAA), 35
National Rainfall Index (NRI), 9–10
National Water Plan 2009, 279
Natural disasters, 175, 186
Natural hazards in Pacific Islands region, 256
Natural water, 145
Navua, 240
NDEF. *See* National Drought Emergency Fund (NDEF)
Network Common Data Form (NetCDF) file format, 22–27
NHMSs. *See* National Meteorological and Hydrological Services (NHMSs)
Nigerian Meteorological Agency (NiMet), 67
Nile River Basin, 280
Niño 3 index, 162
NOAA. *See* National Oceanic and Atmospheric Administration (NOAA)

"Noncore" indices, 22
North Atlantic Oscillation (NAO), 7
NRI. *See* National Rainfall Index (NRI)

O

Old World Drought Atlas (OWDA), 140
Open Global Glacier Model (OGGM), 136
Overgrazing, 240
Ozone, 196
Ozone hole in Antarctic climate, 196–198

P

Pacific Disaster Risk Management Partnership Network (PDRMPN), 265
Pacific Island countries (PICs), 235–237
Pacific Island Countries and Territories (PICT), 254*f*, 255–256
  selected demographic and physical characteristics of, 255*t*
Pacific Island country agriculture, 235–236
Pacific Island region, 254–256
  disasters in PICT compared to the world's biggest recent disaster events, 256–259
  natural hazards in, 256
  people's vulnerabilities to disasters and the search for security, 259–265
    adaptive capacities in Pacific Island societies, 260–261
    disaster management, 264–265
    migration, 261–264
    remoteness, isolation and displacement, 261
    small islands and their vulnerabilities to natural hazards, 259–260
  2020/2021 cyclone seasons and COVID-19 in, 265–266
Pacific Islands Framework for Action on Climate Change (PIFACC), 236
Pakistan flood in 2010, 89–90
Palmer Drought Severity Index (PDSI), 9–10, 139, 166–168
Paraná River Basin, 221
PDR. *See* Percentile departure of rainfall (PDR)
PDRMPN. *See* Pacific Disaster Risk Management Partnership Network (PDRMPN)
PDSI. *See* Palmer Drought Severity Index (PDSI)
Peak water, 136, 137*f*, 138
Peninsular India extreme rainfall event, 88–89
Percentile departure of rainfall (PDR), 86
PICs. *See* Pacific Island countries (PICs)
PICT. *See* Pacific Island Countries and Territories (PICT)
PIFACC. *See* Pacific Islands Framework for Action on Climate Change (PIFACC)
PMP. *See* Probable maximum precipitation (PMP)
Polar regions, 191–192
Polar seas, heat tracking in, 198–200
Precipitation
  extreme precipitation indices, 73–75
  mean precipitation, 69
Precipitation extremes, 5–7, 133–135
  in the United States under global warming
    annual mean precipitation, projected changes in, 121
    maximum 5-day precipitation, projected changes in, 123
    maximum consecutive dry days, projected changes in, 122
    present-day annual dry and wet extremes, evaluation of, 119–121
Probable maximum precipitation (PMP), 87

Q

Quasi-biennial oscillation (QBO), 161–162

R

Rainfall Anomaly Index (RAI), 9–10
Rainfall extremes, 133–134
RCM. *See* Regional Climate Models (RCM)
RDI. *See* Reconnaissance Drought Index (RDI)
Recognized Seasonal Employer (RSE) program, 262
Reconnaissance Drought Index (RDI), 151
Reference climatology, 42
Regional Climate Models (RCM), 65–66, 135
Retrospective analysis, 35
RSE program. *See* Recognized Seasonal Employer (RSE) program

S

SADC. *See* Southern African Development Community (SADC)
SAM. *See* Southern Annular Mode (SAM)
scPDSI. *See* Self-calibrating Palmer Drought Severity Index (scPDSI)
SDGs. *See* Sustainable Development Goals (SDGs)
Seasonal planting, 248
Sea surface temperature (SST), 7, 86, 102–103, 177–178, 198–199, 221
Self-calibrating Palmer Drought Severity Index (scPDSI), 139–140, 161–162, 164, 166–168, 171
Serua/Namosi region, 238
SGLI. *See* Standardized Groundwater Level Index (SGLI)
Shared socioeconomic pathways (SSP), 54–55
Short Messaging Service (SMS), 276
SIDs. *See* Small Island developing states (SIDs)
SIS. *See* Small Island States (SIS)
SLA. *See* Sustainable Livelihood Approaches (SLA)
Small Island developing states (SIDs), 271–272
Small Island States (SIS), 253
SMB. *See* Surface mass balance (SMB)
SMS. *See* Short Messaging Service (SMS)
SNIPE. *See* Standardized Nonparametric Indices of Precipitation and Evapotranspiration (SNIPE)
"Social" vulnerability, 259
Socio-economic impacts of extreme rainfall events over India, 84–86
Solar radiation, 193, 193–194*f*
Southern African Development Community (SADC), 176
Southern Annular Mode (SAM), 7
South West Indian Ocean (SWIO) basin, 176
  historical and future characteristics of tropical cyclones in, 179–180
  tropical cyclones in, 178–179
Spatiotemporal scales of measurement of weather and climate extremes, 52
SPEI. *See* Standardized Precipitation Evapotranspiration Index (SPEI)
SPI. *See* Standardized Precipitation Index (SPI)
SRI. *See* Standardized Runoff Index (SRI)
SSP. *See* Shared socioeconomic pathways (SSP)
SST. *See* Sea surface temperature (SST)
Standardized Groundwater Level Index (SGLI), 166–168
Standardized Nonparametric Indices of Precipitation and Evapotranspiration (SNIPE), 166–168
Standardized Precipitation Evapotranspiration Index (SPEI), 9–10, 139–140, 166–168
Standardized Precipitation Index (SPI), 9–10, 139–140, 148–149, 166–168
Standardized Runoff Index (SRI), 139–140
Station gauges, 66–67
Station Observations, 33
Station Records, 33
Storms, 8
Subdaily precipitation indices, 51*t*
Subseasonal to seasonal prediction, 57
Sun's radiation, 204, 208*f*
Surface mass balance (SMB), 135–136
Surface Water Supply Index (SWSI), 9–10
Sustainable Development Goals (SDGs), 277
Sustainable Livelihood Approaches (SLA), 260
SWIO basin. *See* South West Indian Ocean (SWIO) basin
SWSI. *See* Surface Water Supply Index (SWSI)

T

TCs. *See* Tropical cyclones (TCs)
Teleconnection patterns, 104–105, 110, 112
Temperature extremes, 5
Timing change of weather and climate extremes, 56
TMPA. *See* Tropical Multi-satellite Precipitation Analysis (TMPA)
Tornadoes, 8–9
TRMM. *See* Tropical Rainfall Measuring Mission (TRMM)
Tropical Cyclone Idai in Southern Africa
  formation and evolution of, 180
  impacts of, 181–184
    on Malawi, 181
    on Mozambique, 182
    on Zimbabwe, 183–184
  methodology, 177
  South West Indian Ocean (SWIO) basin
    historical and future characteristics of tropical cyclones in, 179–180
    tropical cyclones in, 178–179
  study area, 176–177
  understanding tropical cyclones, 177–178
Tropical cyclones (TCs), 8, 176
Tropical Cyclone Winston of 2016, 236–238

Tropical Multi-satellite Precipitation Analysis (TMPA), 86
Tropical Rainfall Measuring Mission (TRMM), 65–66, 86
Troposphere, 2

U

Ultraviolet radiation, 196
Uncertainties in daily rainfall over West Africa
  data and methods, 66–68
    gridded products, 67–68
    methodology, 68
    station gauges, 66–67
  extreme precipitation indices, 73–75
  mean daily monsoon, evolution of, 69–73
  mean precipitation, 69
United Nations Environment Programme (UNEP), 19
Urbanization, 225
Uttarakhand flood 2013, 93–94

V

Variability, modes of, 7
Varimax method, 163

W

Wainadoi settlement, 238–240
Warsaw International Mechanism (WIM), 271–272, 275
WCRP. *See* World Climate Research Program (WCRP)
Weather, defined, 1–2
WGCM. *See* Working Group of Coupled Modeling (WGCM)
WIM. *See* Warsaw International Mechanism (WIM)
Wind extreme, 7
WMO. *See* World Meteorological Organization (WMO)
Working Group of Coupled Modeling (WGCM), 41
World Climate Research Program (WCRP), 4, 49–50, 67–68
World Meteorological Organization (WMO), 19

Z

Zimbabwe, impacts of tropical cyclone Idai on, 183–184
Zoonotic pathogen, 223

Printed in the United States
by Baker & Taylor Publisher Services